Global Sustainable COMMUNITIES HANDBOOK

Global Sustainable COMMUNITIES HANDBOOK

GREEN DESIGN TECHNOLOGIES AND ECONOMICS

Edited By

WOODROW W. CLARK II

ELSEVIER

AMSTERDAM • BOSTON • HEIDELBERG • LONDON
NEW YORK • OXFORD • PARIS • SAN DIEGO
SAN FRANCISCO • SINGAPORE • SYDNEY • TOKYO

Butterworth-Heinemann is an imprint of Elsevier

Butterworth-Heinemann is an imprint of Elsevier
225 Wyman Street, Waltham, MA 02451, USA
The Boulevard, Langford Lane, Kidlington, Oxford OX5 1GB, UK

© 2014 Elsevier Inc. All rights reserved.

No part of this publication may be reproduced or transmitted in any form or by any means electronic, mechanical, photocopying, recording or otherwise without the prior written permission of the publisher.

Permissions may be sought directly from Elsevier's Science & Technology Rights Department in Oxford, UK: phone (+44) (0) 1865 843830; fax (+44) (0) 1865 853333; email: permissions@elsevier.com. Alternatively you can submit your request online by visiting the Elsevier web site at http://elsevier.com/locate/permissions, and selecting Obtaining permission to use Elsevier material.

Notice
No responsibility is assumed by the publisher for any injury and/or damage to persons or property as a matter of products liability, negligence or otherwise, or from any use or operation of any methods, products, instructions or ideas contained in the material herein. Because of rapid advances in the medical sciences, in particular, independent verification of diagnoses and drug dosages should be made.

British Library Cataloguing-in-Publication Data
A catalogue record for this book is available from the British Library

Library of Congress Cataloging-in-Publication Data
Clark, Woodrow W.
 Global sustainable communities handbook: green design technologies / Woodrow W. Clark II.
    pages cm
 Includes bibliographical references and index.
 ISBN 978-0-12-397914-8
1. Sustainable development--Handbooks, manuals, etc. 2. Community development--Handbooks, manuals, etc. 3. Sustainable design--Handbooks, manuals, etc. 4. Green technology--Handbooks, manuals, etc. 5. Renewable energy sources--Handbooks, manuals, etc. 6. Environmental policy--Handbooks, manuals, etc. 7. Economic policy--Handbooks, manuals, etc. I. Title.
 HC79.E5C565 2014
 338.9'27--dc23
                                                             2013050939

For information on all Butterworth-Heinemann publications
visit our web site at store.elsevier.com

ISBN: 978-0-12-397914-8

Printed and bound in United States of America
14 15 16 17   10 9 8 7 6 5 4 3 2 1

 Working together to grow libraries in developing countries

www.elsevier.com • www.bookaid.org

CONTENTS

Preface · xiii
List of Contributors · xv
Overview · xix

1. Introduction · 1
Woodrow W. Clark II

References · 12

2. The Green Industrial Revolution · 13
Woodrow W. Clark II, Grant Cooke

Introduction · 13
The First Industrial Revolution · 18
The Second Industrial Revolution · 20
Not Enough Supply, too Much Demand · 24
The Green Industrial Revolution · 26
Sustainability is the Key · 28
Renewable Energy: From Central Grid to On-Site and Distributed Power · 29
Storage and Intermittent Technologies · 32
Additive Manufacturing · 34
The Next Economics: Social Capitalism · 35
The Green Industrial Revolution has Started · 38
References · 39

3. Public Policy and Leadership: Infrastructures and Integration · 41
Michael "Hakeem" D. Johnson

Introduction · 41
Background · 42
The Shift · 43
Almere Zoneiland: The Sun Island, Kingdom of the Netherlands · 45
Rizhao, China · 46
Gaviotas, Colombia · 47
The Birth of a Nation: The Republic of South Sudan · 49
References · 51

4. **China: Resource Assessment of Offshore Wind Potential**   53
   Lixuan Hong

   Introduction   53
   Methodology: Geographic Information System-Based Spatial Modeling   55
   Offshore Wind Potential under Technical, Spatial, and Economic Constraints   57
   Economic Costs of Tropical Cyclone Risks   63
   National and Provincial Cost–Supply Curves   67
   Conclusions   75
   References   76

5. **Sustainable Planning of Open Urban Areas in Developing Countries**   79
   A Lesson from a Case Study in Tel Aviv, Israel
   Tali Hatuka, Hadas Saaroni

   What is the Challenge? Quick Urbanization and Climate Change in Developing Countries   80
   A Lesson from the Design of Jaffa Slope Park: Contradictions and Gaps   84
   Toward Establishing Design Codes for Outdoor Urban Spaces in Developing Countries   89
   The Future of Publicness and its Meaning   91
   References   91

6. **India: Issues for Sustainable Growth/Innovation for Sustainability**   93
   Namrita S. Heyden

   Introduction   93
   Society and Governance   94
   Business/Corporate Role   96
   Trends in Energy Use   99
   Conclusion   103
   References   103

7. **Germany's Energiewende: Community-Driven Since the 1970s**   105
   Craig Morris

   A Fledgling Movement from the 1970s   106
   The Energiewende Today   109
   Community Ownership Today   110
   Conclusions   112

8. **Multidisciplinary Integrated Development** — 115
   M.K. Elahee

   Sustainability — 116
   Energy — 116
   Limitations — 119
   Engineering — 120
   Toward Interdisciplinarity — 121
   Conclusion — 121
   Bibliography — 122

9. **Think Globally, Act Locally, and Plan Nationally An Evaluation of Sustainable Development in Indonesia at National, Regional, and Local Levels** — 125
   Lacey M. Raak

   Executive Summary — 126
   Methodology — 129
   Background — 132
   National Sustainable Development Strategies — 138
   Indonesia's National Sustainable Development Strategy — 140
   Analysis of Sustainable Development Indicators — 156
   Results from Evaluation of Planning Priorities and Budget Allocations — 157
   Analysis of Planning Priorities and Budget Allocations — 158
   Recommendations — 170
   Conclusion — 179
   Acknowledgments — 180
   Translation Acknowledgment — 180
   Photo Acknowledgment — 180
   References — 180
   Appendix 1: NSDS Process Evaluation — 184
   Appendix 2 — 190
   Appendix 3: Indonesia Planning and Budget Analysis Charts — 197
   Appendix 4: East Kalimantan Planning and Budget Analysis Charts — 203
   Appendix 5: Balikpapan Planning and Budget Analysis Charts — 210

10. **Financial Investments for Zero Energy Houses: The Case of Near-Zero Energy Buildings** — 217
    Natalija Lepkova, Domantas Zubka, Rasmus Lund Jensen

    Introduction — 217
    Concept of a Zero Energy Building — 219

Importance of Renovating Residential Buildings — 223
Evaluating Investments for Renovation of a Detached House into a Zero Energy
   Building in Denmark — 225
A Survey on Possibilities to Implement Zero Energy Buildings — 236
Conclusions — 245
References — 247
Appendix — 248

11. The Canadian Context: Energy — 255
Kartik Sameer Madiraju

Introduction — 256
Energy Conservation and Efficiency — 258
Renewable Energy Research and Development — 270
Agile, Sustainable Communities — 279
Conclusion — 287
References — 288

12. Energy Management in a Small-Island Developing Economy — 293
The Case of Mauritius
Khalil Elahee

Introduction — 293
Energy Management and Climate Change — 294
Buildings: Caught in a Vicious Circle — 296
Assessing Energy Efficiency — 299
The Way Forward — 302
References — 303

13. Energy System of the Baltic States and its Development — 305
Jurate Sliogeriene

Infrastructure of the Energy Industry and Its Economic Significance — 306
Objectives of Sustainable Development of the Energy Industry — 309
Energy System of the Baltic Countries — 316
Renewable Energy Production Technologies in the Baltic States:
   The Case of Lithuania — 333
References — 344

14. Renewable Energy Generation: Incentives Matter — 347
A Comparison Between Italy and Other European Countries
D. Chiaroni, V. Chiesa, F. Frattini

Introduction: Forms of Incentives for Renewable Energy Generation — 347

Renewable Energy Generation in Italy: An Overall Picture	351
The Case of Photovoltaics in Italy	354
The Case of Biomasses in Italy	362
Renewable Energy Generation: Incentives Matter	366

15. Germany's Energiewende — 369
Eric Borden, Joel Stonington

Primary Goals and History of the Energiewende	369
Background and History of the Energiewende	370
Renewable Energy Law and Feed-in Tariff Mechanism	373
Cap and Trade: Putting a Price on Carbon in Europe	374
Achievements of the Energiewende	376
Challenges and Impediments to the Energiewende	379
Conclusions and Going Forward	385

16. Educational Programs for Sustainable Societies Using Cross-Cultural Management Method — 387
A Case Study from Serbia

Jane Paunković

Introduction	387
Educational Program "Strategic Leadership for Sustainable Societies" in Serbia	390
Research Objective and Study Design	397
Results	398
Discussion	398
References	402
Web Sites	404

17. Business Ventures and Financial Sector in the United Arab Emirates — 405
Robert Ruminski

Introduction	405
Small and Medium Enterprises and Government Financial Support	409
Ease of Doing Business in the United Arab Emirates	411
Investment Bodies: Key Players and Contributors	413
Legal Aspects of Conducting Business Activity	415
Banking Sector: Structure and Recent Developments	421
Other Key Players of the Financial Market	434
Business Cooperation between the UAE and Poland	439
Market Strengths, Weaknesses, Opportunities, and Threats (SWOT) Analysis of the UAE Market: Business Perspective	446
Conclusions and Summary	448

18. Development of a Sustainable Disabled Population in Countries of the Cooperation Council for the Arab States of the Gulf   451
Simon Hayhoe

Introduction	451
The Context of the Study	453
Findings from Available Literature	456
Conclusion	464
References	466

19. Political–Economic Governance of Renewable Energy Systems: The Key to Creating Sustainable Communities   469
Woodrow W. Clark II, Xing Li

Corporate and Business Influences and Power	469
International Cases	470
China Leapfrogs Ahead	470
China has "Leapfrogged" into the Green Industrial Revolution	471
The Western Economic Paradigm Must Change	472
Introduction and Background	473
European Union Policies	478
Japan and South Korea are Leaders in the Green Industrial Revolution	482
Distributed Renewable Energy Generation for Sustainable Communities	486
Developing World Leaders in Energy Development and Sustainable Technologies	487
Costs, Finances, and Return on Investment	489
Conclusions and Future Research Recommendations	492
Acknowledgments	493
References	493

20. Sustainable Agriculture: The Food Chain   495
Attilio Coletta

Introduction	495
Social Implications	496
Economic Implications	497
Environmental Implications	500
Developing New Solutions	504
References	505

21. Development Partnership of Renewable Energies Technology and Smart Grid in China — 507
A.J. Jin, Wenbo Peng

Introduction	507
Solar Electricity Systems and Their Relationship with the Grid	509
Wind Power	513
Data Response and Power Transmission Lines: Examples in the United States	516
Smart Grid and Market Solution	518
China Rebuilds a Power System and Smart Grid	521
References	526

22. The Regenerative Community Régénérer: A Haitian Model and Process Toward a Sustainable, Self-Renewing Economy — 527
Carl Welty

Background	528
Old and New Paradigms	529
Régénérer's Business Model	529
Claremont Environmental Design Group's Team and Methods	530
Ecosystematic Analysis	531
Regenerative Design	531
Geographic Information Systems and Data Management	531
Evaluation Metrics	532
Régénérer's 25-Acre Pilot Project	532
Partners in Research to Build a Regenerative Haiti	536
Haitians and Californians Benefit and Learn from Each Other	537

23. Microcities — 539
Naved Jafry, Garson Silvers

Microcities: Helping Mitigate the Rise of the Underground Society	540
How the City Will Be Sustained Economically	545
The Importance of Implementing Good Laws	548
Infrastructure and Environmental Planning	550
The Health Aspect of a Microcity	553
Discussion	556
References	557

24. Conclusion 559
Woodrow W. Clark II
Lessons Learned from Developed and Developing Nations 560
Economic Themes, Strategies as Opportunities for Renewable Energy 561
Conclusions as Opportunities 568
References 569

Index *573*

PREFACE

A very special thanks to my wife (Andrea) and 6-year-old son (Paxton) for working with me while I undertook this book which focuses upon sustainable communities in a very real yet innovative manner that describes how such communities can be created now and updated constantly in the future is a lesson for everyone. The world changes and with it what we need to take action now and in the future. My family understands this need for change and was the inspiration for my spending over 2 years preparing the book: it is dedicated to them and our global family future.

Every time that I speak, give lectures and keynote talks, I end with a picture of myself holding Paxton when he was 3 years old. My point is that the reason I am there speaking, and doing this book, is that his life is at stake. Either we parents and grandparents need to do something about climate change today or we will destroy our childrens' future health, environment, and their climate. We cannot wait.

There is no time to delay or tell our family and children. We have to lower emissions and pollution now. The clock is ticking so we can not wait to waste, dream, or even imagine that the issues of climate change will disappear. We must act now. As my son puts it, "Daddy, be sure to turn off the light when you leave." And he recycles bottles at his school and we go to recycle locations nearby our home to recycle. I tell him that we need to do this for the Earth and he gets some money for his college education.

Hence the book is organized into regions around the world and specific areas of expertise needed to create sustainable communities. The book starts with an overview of the chapters and then an introduction to the regions and areas of specific needs for every sustainable community using examples that have worked and are being implemented now.

Woodrow W. Clark II

LIST OF CONTRIBUTORS

Eric Borden
German Chancellor Fellow, Alexander von Humboldt Foundation, Berlin, Germany; Email: borden.eric@gmail.com

D. Chiaroni
Dipartimento di Ingegneria Gestionale, Politecnico di Milano, Piazza Leonardo da Vinci, Milan, Italy; Email: davide.chiaroni@polimi.it

V. Chiesa
Dipartimento di Ingegneria Gestionale, Politecnico di Milano, Piazza Leonardo da Vinci, Milan, Italy; Email: vittorio.chiesa@polimi.it

Woodrow W. Clark II, Ph.D.
Qualitative Economist, Managing Director, Clark Strategic Partners, Beverly Hills, CA, USA; Email: wwclark13@gmail.com; Website: www.clarkstrategicpartners.net

Attilio Coletta, Ph.D.
Department. DAFNE – Università degli Studi della Tuscia – Viterbo, Italy; Email: Coletta@unitus.it

Grant Cooke
Founder and CEO, Sustainable Energy Associates, LLC., Benicia, CA, USA; Email: gcooke@sustainableenergyassc.com; grantcooke11@gmail.com; Website: http://www.sustainableenergyassc.com

M.K. Elahee, Ph.D.
Faculty of Engineering, The University of Mauritius, Reduit, Mauritius; Email: elahee@uom.ac.mu

Khalil Elahee
Chairman, Energy Efficiency Committee Member, National Energy Commission Associate Professor Mechanical and Production Engineering Department, Faculty of Engineering, The University of Mauritius, Mauritius; Email: elahee@uom.ac.mu

F. Frattini
Dipartimento di Ingegneria Gestionale, Politecnico di Milano, Piazza Leonardo da Vinci, Milan, Italy; Email: federico.frattini@polimi.it

Tali Hatuka, Ph.D.
Geography and the Human Environment Department, Tel Aviv University, Ramat Aviv Tel Aviv, Israel; Email: hatuka@post.tau.ac.il

Simon Hayhoe, Ph.D.
Faculty (Educational Technology), Sharjah Women's College, UAE; Research Associate, Centre for the Philosophy of Natural & Social Sciences, London School of Economics, UK; Email: simon.hayhoe@hct.ac.ae

Namrita S. Heyden, Ph.D.
University of Chapel Hill, NC, USA; Email: namrita.singh@gmail.com

Lixuan Hong, Ph.D.
China Energy Group, Environmental, Energy Technologies Division, Lawrence Berkeley National Laboratory, Berkeley, CA, USA; Email: lixuanhong@lbl.gov

Naved Jafry
Chairman, ZEONS Group, La Jolla, CA, USA; Email: NJ@zeons.org

Rasmus Lund Jensen
Department of Civil Engineering, Head of Indoor Environmental Engineering Laboratory, Aalborg University, Aalborg, Denmark; Email: rlj@aau.com

A.J. Jin, Ph.D.
Chief Scientist, China Huaneng Clean Energy Research Institute China Huaneng Group, Beijing, P.R. China; Managing Director Haetl Ltd, a Clean Solar Thermal Electricity Company, Lafayette, CA USA; Email: ajjin@ieee.org; ajjin@hnceri.com

Michael "Hakeem" D. Johnson, Ph.D.
Acting Country Director, Adeso-African Development Solutions Juba, South Sudan; Email: mdj97@hotmail.com; Website: www.adesoafrica.org

Natalija Lepkova, Ph.D.
Department of Construction Economics and Property Management, Vilnius Gediminas Technical University, Vilnius, Lithuania; Email: Natalija.Lepkova@vgtu.lt; natalijal2000@yahoo.com

Xing Li
Professor and Director, Research Center on Development and International Relations, Department of Culture and Global Studies, Aalborg University Aalborg, Denmark; Email: xing@cgs.aau.dk

Kartik Sameer Madiraju
Department of Bioresource Engineering, McGill University, Montreal, Quebec, Canada; Email: kartik.madiraju@gmail.com

Craig Morris, Ph.D.
Director of Petite Planète, Krozingerstr., Freiburg, Germany; Email: cm@petiteplanete.org

Jane Paunković, Ph.D.
Professor, Faculty of Management Zajecar, Megatrend University, Serbia; Email: jane.paunkovic@fmz.edu.rs; janep@fulbrightmail.org

Wenbo Peng, Ph.D.
China Huaneng Group, Huaneng Clean Energy Research Institute, Haidian District, Beijing, P.R. China; Email: pengwenbo2012@hotmail.com

Lacey M. Raak
Sustainability Director, University of California, Santa Cruz, CA, USA; Email: laceymae29@gmail.com

Robert Ruminski, Ph.D.
Chair of Banking and Comparative Finance, Faculty of Management and Economics of Services, University of Szczecin, Szczecin, Poland; Email: robertruminski@wp.pl

Hadas Saaroni, Ph.D.
Geography and the Human Environment Department, Tel Aviv University, Ramat Aviv Tel Aviv, Israel; Email: saaroni@post.tau.ac.il

Garson Silvers
CEO, ZEONS Group, Sustainable Developer, Beverly Hills, CA, USA;
Email: garson@zeons.org; garsonsilvers@sbcglobal.net;
Website: http://zeons.org/; www.garsonsilvers.com

Jurate Sliogeriene, Ph.D.
Associate Professor at Vilnius Gediminas Technical University, Department of Construction Economics and Property Management, Lietuva, Vilnius, Lithuania;
Email: jurate.sliogeriene@vgtu.lt; sliogeriene.j@gmail.com;

Joel Stonington
Visiting journalist, Spiegel Online International and German Chancellor Fellow, Germany;
Email: jstonington@gmail.com

Carl Welty
Principle Architect, Claremont Environmental Design Group, Claremont, CA, USA;
Email: c.welty@cedg-design.com; c.weltycedg@gmail.com

Domantas Zubka
Master of Management and Business Administration, Vilnius Gediminas Technical University, Vilnius, Lithuania; Email: Domantas.Zubka@gmail.com

OVERVIEW

The book took over 2 years to do because of many reasons. Some of it was due to my wanting to have former and current Fulbright Fellow (like myself) write chapters for the book and then continue after that with more contributions, edits, and additions. As a Fulbright Fellow myself in 1994, I spent 6 months at Aalborg University (AAU) in Aalborg, Denmark. My role was to teach "entrepreneurship" to the business and economic students there, as well as work with students on their graduation projects at the end of the spring semester. I did that.

But I learned far more from the students and faculty, as well as local community members. Denmark in the early 1990s had emerged as one of the leading countries in the newly formed European Union (EU), which was moving away from natural gas and into biomass and wind energy for the entire country and every community. I was astonished when I arrived in Aalborg in January 1994 to see all the wind turbines that surrounded not only the large areas of the northern Jutland region, but also in towns, on farms, and in residential communities surrounding the towns. At the time, there were none at AAU, a question that I asked constantly, but never got a good answer about. Nonetheless, the AAU Faculty of Economics and Business invited me back in the fall and then every year after that for a least 1 week of lectures. By 1998, I moved to AAU to be a full-time visiting professor.

As a footnote, while working with the AAU faculty and students, we even shared the development of programs in the United States and at AAU. After the summer of 1994 when I returned to California, I was the manager of technology transfer for the Lawrence Livermore National Laboratory (LLNL) in northern California (then part of the University of California system, but funded primarily by the U.S. Department of Energy). In that role, we began several international collaborations with Denmark, along with other Nordic countries, which also led to collaborative programs being developed in China.

Two significant incidents occurred during that period in the 1990s of collaboration between AAU and LLNL: (1) the exchange of information and work on renewable energy systems, including biomass, wind, ocean, and even hybrid renewable energy systems, and (2) the need for on-site or distributed power but *not* deregulation of the energy systems needed for every person in all communities. In 1998, I had a delegation from AAU come to Sacramento (sponsored by some legislators and the LLNL) for a conference on deregulation (known as "liberalization" or "privatization" in the EU).

Both issues became extremely important then, especially in early 2000. But the significance of deregulation hit California especially strong then, as I was one of the few people in California who predicted an "energy crisis" that turned into rolling brownouts, which impacted businesses, homes, students, and hospitals. The state was in turmoil. I was asked to come back from AAU and be one of Governor Davis' energy advisors. I did so in December 2000. The rest is history and is described in detail in my book with Ted Bradshaw, *Agile Energy Systems: Global solutions for the California Energy Crisis* (Elsevier Press 2004).

After that, for the next decade, I did several books on the solutions to climate change, which are sustainable communities. Each book would take cases from around the world as examples of how sustainable communities look today and plan for the future. Then at the end of the first decade of the 21st century, I got in touch with Elsevier Press again and met Ken McCombs, my publisher, at his office in New York City. We decided there and then to do a handbook about the designs and needs for creating sustainable communities.

Then Ken led an effort within Elsevier in this area as well as making this book and others far more interactive via the Internet, updates on case chapters, and new cases on a regular basis. Hence the idea was to make the *Global Sustainable Communities Design Handbook* into a useful tool for people to use that ranged in information, actual cases, measurements, and criteria to economics, law, planning, design, engineering, and green technologies. This book is a tribute to Ken and his dedicated colleagues at Elsevier.

The Fulbright Fellow experience and the consequences after my first 6 months at AAU taught me many lessons—learning from my students and faculty, doing research together, and also publishing articles and books together. It was a strong belief that other Fulbright Fellows, researchers, and students had had the same experiences. Hence they too would be great authors of chapters for this book. I was asked to join the board of the then Fulbright Academy of Science and Technology (FAST) and worked with the then executive director, Eric Howard, to get chapters for the book, based on the FAST annual conference in 2011, which focused on sustainable communities.

While the idea was great, the reality was difficult. The idea of a Fulbright conference focused on sustainable communities with papers and panels could mean a strong global contribution by authors who were giving papers at the conference. A few chapters in the book were a result of this opportunity. However, we had hopes for many more. The problem was that the conference had both underfunded itself and then had few paper presenters. It is an honor to have those papers included in this book. It is hoped that there will be more Fulbright Fellows who will add chapters to future volumes.

This book is the first in a series of books about sustainable communities from around the world. Significantly, there are cases in almost every region and country. Sustainable communities are not just in developed nations. In fact, there are more in developing and emerging nations. Even in so-called third world or underdevelopment nations there are sustainable communities that do not pollute the environment and are very conscious about their use of nature, water, and land resources. In a recent book that I did on *The Next Economics* (Springer Press 2012), I had a recommendation from an American Indian tribal community to emphasis what that community and other tribes are doing to sustain, protect, and eliminate pollution within the confines and boundaries of the United States.

The United States and other developed nations need to learn from themselves, others, and successful cases around the world. And in the end, it all starts at home and on the local level if we are to provide hope for future generations. We must stop pollution today in our homes, communities, and places of work and play.

The book is organized to reflect its subtitle "Green Design, Engineering, Health, Technologies, Education, Economics, Contracts, Policy, Law, and Entrepreneurship." However, we wanted the book to also use cases from around the world in which readers whose specialty is in these areas could see and learn from what others have done. Hence the book has global sections from regions, communities, nations, and continents. In order to provide the book with some structure for updates on chapters, new chapters, and combination of the specialty areas, the book is divided into part 1 containing chapters that are an overview and introduction into more specific content. For the overview and introduction, Clark did a chapter as an "Introduction" and then another one (Chapter 2) with Grant Cooke on "The Green Industrial Revolution."

Then to set the stage for the in-depth cases, a section on the mechanisms for sustainable development within part 1 was set up to provide a framework for Chapter 3 on "Public Policy and Leadership" as they impact infrastructures and integrate them. Chapter 4, "China: Resource Assessment of

Offshore Wind Potential," looks closely at offshore wind potential in China as an important resource for the nation's rising energy demands.

Next is part 2 on continent areas with Chapter 5, "Sustainable Planning of Open Urban Areas in Developing Countries," and a case study of Tel Aviv in Israel. "India: Issues for Sustainable Growth/Innovation for Sustainability" is explored then in Chapter 6 with Chapter 7 focused on Germany and its "Energiewende" plans that have been community driven over the last four decades.

Critical for any region or nation around the world, public policy and leadership must be one of the uppermost focus areas. Chapter 8, "Multidisciplinary Integrated Development," discusses the Mauritius Islands. Chapter 9 provides another broader approach to "Think Globally, Act Locally, and Plan Nationally" from an evaluation of sustainable development in Indonesia.

Moving then to another case from the Baltic States in central Europe for "Financial Investments in Zero Energy Houses" is Chapter 10. While the chapter covers ways to make the economics of zero energy buildings work, the next four chapters provide more details. Thus in the next section on energy conservation and efficiency, Chapter 11 looks at the "Canadian Context: Energy" followed by Chapter 12 on "Energy Management in a Small-Island Developing Economy: The Case of Mauritius."

Chapters 13 covers the Baltic States again but with a focus on "Energy System of the Baltic States and Its Development," providing a measurable engineering model. Chapter 14, "Renewable Energy Generation: Incentives Matter," then looks at Italy and other European countries to see how renewable energy generation can be compared for in-depth analyses. The analytical results are not only informative but also a model for others to use.

Part 3 on the finance and economics of sustainable communities gets into how all of this funded. It answers the basic question always asked about "how do we pay for all of this?" Chapter 15 focuses on "Germany's Energiewende" from economics and financing perspectives. Chapter 16 then covers a case from Serbia on "Educational Programs for Sustainable Societies Using Cross-Cultural Management Methods."

This section then moves into Chapter 17 on "Business Ventures and Financial Sector in the United Arab Emirates" in the Middle East. However, a large part of the concern for energy and its impact on the environment relates to the health costs from carbon, greenhouse gases, and pollution itself. Chapter 18 details these concerns with specifics from "Development of a Sustainable Disabled Population in Countries of the Cooperation Council for the Arab States of the Gulf."

Part 4 of the book covers culture, history, and public policy, which covers critical issues and areas and then provides some cases of sustainable communities. Chapter 19 sets the key element to create sustainable communities as "Political–Economic Governance of Renewable Energy Systems." Then agriculture and farming are covered in Chapter 20 as "Sustainable Agriculture: The Food Chain." Chapter 21 discusses the "Development Partnership of Renewable Energies Technology and Smart Grid in China."

The two communities that are cited continue to be developed but are clearly along the road to becoming sustainable. Chapter 22 covers "The Regenerative Community Régénérer: A Haitian Model and Process Toward a Sustainable, Self-Renewing Economy." Chapter 23 provides a case from India for "Microcities." The book then concludes with Chapter 24 on how the future is *now*. Indeed it is now. Our families are depending on us to make dramatic changes today—not 15–20 years from now.

Woodrow W. Clark II

CHAPTER 1

Introduction

Woodrow W. Clark II
Qualitative Economist, Managing Director, Clark Strategic Partners, Beverly Hills, CA, USA

Where do we start on such a vast area to cover in one book? For one thing, the book has 23 chapters by different authors. Only the author and one other author have 2 chapters. In the author's case, he has 3 chapters, including this introduction and the conclusion. The book intends to be done annually and then twice and even four times a year in hard cover, but the online chapters of the book will be updated and improved with the latest information and data very regularly over a long period of time. This is critical, as most books (including this one) take time and by the publication of the book, the content is dated. Most reports, published papers, and written chapters of books are dated in terms of months, if not years, before readers get the materials. By that time, the information in the chapter is old.

Now with this book, we are setting a far different standard for reporting on communities around the world. In short, the book is in "real time." Also, its significance is the impact that a reader in need of up-to-date information can get from that availability of data, measurements, evaluations, and results. Consider the second chapter on "The Green Industrial Revolution" that Grant Cooke and I coauthored.

At first we did a book on The Third Industrial Revolution due to the concerns expressed by a colleague of mine, Jeremy Rifkin, who mentioned this concept in his book on the *European Dream* (2006). When I urged Jeremy to do a book with this title, he said that he was too busy and did not want to do so. So Grant and I did a book with that title. Then, to our amazement in the spring of 2011, I get an urgent phone call from Jeremy. He is upset because my book with the title of the *Third Industrial Revolution* was out on Amazon along with his book by the same title. One of us had to back off.

Ironically, Grant and I had moved beyond that title to *The Green Industrial Revolution* for our next book. So, for the current one with the *Third Industrial Revolution* title, we decided to change it to *Global Energy Innovation* (Preager Press, 2011). Meanwhile we pursued the *Green Industrial Revolution* (GIR), which we finished and had, with two Chinese coauthors, translated into Mandarin. The book will be published in China before it is in the

United States or other developed countries. There is a very good strong reason for that strategy: we want to send a message to the United States in particular that it must stop being caught (trapped really) in the fossil fuel eras of the First and Second Industrial Revolutions while China, Asia, and other nations move aggressively into the GIR.

The United States is beholding to the fossil fuel companies and the wealthy people who control those industries, as well as now American politics. Climate change will continue—and get worse. While China now leads the world in greenhouse gas (GHG) pollution, having replaced the United States, GHG pollution is still lower if done on a per capital basis. Nonetheless, China is taking far more aggressive steps to stop and curtail GHG along with other pollution, waste, and environmental problems than the United States. They will soon lead the world in both the GIR and its related green economics through 5-year plans that factor in measurable goals along with finance and economic resources.

Then, in Chapter 3, under the mechanisms for sustainable development, Michael (Hakeem) Johnson, Ph.D., gives the basics for "Public Policy and Leadership for the Integration of Infrastructures." Basically, all communities have systems of governance that have cultural and historical roots. The issue that will face the leadership in these communities for their future will be how they interact or be involved with other communities close by or separated by land, mountains, or seas. Since the earliest times of recorded history, the way in which governments make decisions is different—be they villages, Indian tribes, large cities, and communities to college campuses, shopping malls, residential complexes, and even resort areas.

What the leadership must do is provide a vision and set up a plan with goals and measurements upon which its public can agree. This chapter does just that with cases from around the world—Columbia, China, The Netherlands, and the Middle East. However, in today's world, there was a "shift" to another key player global player in local, national, and regional leadership: the United Nations. The chapter provides an overview and then groups within the United Nations, which are concerned with sustainable development and its leadership.

Offshore wind has been a common part of the northern European implementation of renewable energy over the last three or four decades. The Netherlands and Germany started the trend but then the Nordic countries got very involved, especially with wind turbines, their technology, and advanced software systems. Vestas in south Jutland, Denmark, was and is now the world largest wind turbine company. Offshore winds in the Baltic Sea and Atlantic

Ocean were key to the expansion of wind power and collections of wind turbines into "farms" offshore. Enough wind, along with other renewable energy sources, is being generated now to help Denmark achieve its 100% green energy plans by 2025. Today in 2013, the nation is over 50% there.

What does this offshore wind turbine model present as a model for other countries? In the United States, there are no offshore wind farms. While areas exist along the extensive U.S. coastlines, there are none offshore as of 2013. This may change. Meanwhile in China, also with an extensive coastline, a higher energy demand, and the need for renewable energy sources, offshore wind is becoming a viable cost-effective option. Chapter 4, "China: Resource Assessment of Offshore Wind Potential" by Lixuan Hong, Ph.D., provides a detailed analysis of the potential for wind offshore in all of China, but with a detailed focus on one coastal region in particular, the Jiangsu Provence north of Shanghai. China has expanded dramatically its construction of wind turbines into massive farms in the western region of the nation. However, the problem is the lack of transmission for all that power to the newly heavily populated eastern coastal areas of China. Hence the idea that Dr. Hong explored in this chapter, which is part of her dissertation at AAlborg University in Denmark, was to see the potential of wind power offshore and thus close to the increasing energy demand for sustainable communities along the Chinese coastline.

The next section of the book looks at regions of the world that are rapidly being developed with green technologies for creating sustainable communities. These communities serve as cases studies for other sustainable communities around the world. The key is look at them and see the multidisciplinary approach that they used in order to create, expand, and interconnect the communities so that they reduce carbon, greenhouse gases, and climate change.

The first case takes a case from Israel, by Professor T. Hatuka and H. Saaroni with lessons learned from Tel Aviv, Israel, that provide the basic experiences discussed in Chapter 5 for "Sustainable Planning of Open Urban Areas in Developing Countries: A Lesson from a Case Study in Tel Aviv, Israel." The authors take those experiences that focus on the need for regional designs in order to provide codes that can be charted in an analysis, evaluation, and regulation for outdoor urban areas. The point is to have a "general framework" to address climate and environmental issues in an "era of global warming design codes for outdoor public spaces."

"India: Issues for Sustainable Growth/Innovation for Sustainability" by N. S. Heyden is Chapter 6, which looks at India's growth since the 1990s to becoming of the Brazil Russia, India, and China (BRIC) members of counties that have emerged from the "third-world" label for developing countries into dynamic progress nations. India has the need for expanding energy resources as its awareness at the local level for sustainable communities grows. The issue is what energy sources are there now and will there be any in the future that can be generated without doing further damage to the environment? A recent national student indicated, for example, that the increased need for energy from "hydro-based" sources is important, along with efficiency. However, in remote areas, more renewable sources for energy will be needed to meet demand. Such is not the case in urban areas, where the need is dominated by grid dependency on energy from traditional sources. Yet new technologies exist for generation from geothermal and tidal energy, along with fuel cell storage and alternative fuels for surface transportation.

Then, Chapter 7 by Craig Morris, looking at the European Union(EU) with "Energiewende: Germany's Community-Driven Since the 1970s," shows how national leadership and legislation are needed for creating sustainable communities. Unfortunately, many times these national leadership initiatives are based on tragic events. For Germany, it was the nuclear power plant disaster in Fukushima that led the chancellor to close out nuclear power in Germany and also question the need for central power plants. Distributed energy became a national perspective and goal. Yet the impact and need for feed-in tariffs (since 1991) continue. Then the Renewable Energy Act in 2000 and then an amendment in 2004 put both wind and solar, respectively, into the market by unlinking them to "retail rates" so that small profits could be made. The renewable energy power systems grew rapidly and became a global trend.

The need to look at public policy and leadership at the national level immediately gets to the local level or communities for implementation. The key factor, however, is to have a multidisciplinary and integrated approach to doing so. In the book *Agile Energy Systems*, Clark and Bradshaw (2004) documented that argument as they reviewed the problems and mistakes of deregulation and then noted that both central plant and local on-site or distributed power systems were needed. It was not one or the other, but both. Agile in that energy infrastructure must be "flexible" in using both means of generating power. The use of renewable energy, therefore, which needs to replace fossil fuels, would require the use of central power generation on solar and wind "farms," as well as power on roof tops of buildings, parking lots, and vacant space, along with use in windows and shade and protection over public walks and roadways.

In Chapter 8, Professor M.K. Elahee makes that argument with a case study of "Multidisciplinary Integrated Development" in the Mauritius Island nation off the east coast of Africa. While Mauritius is a model for all nations, it is particularly important for other island nations that have enormous amounts of renewable energy available from the sun and wind but also ocean, tides, and geothermal. Mauritius, for example, has 38% of its energy from coal imported from Australia. That cost, dependency, and security for energy resources need not happen. The key is to look at energy resources for making communities sustainable not only from an engineering perspective, but also as economic, design, and health disciplines as well.

That perspective of multidisciplinary focus connects directly to Chapter 9 by Raak on the need to "Think Globally, Act Locally, and Plan Nationally." Her perspective focuses on how Indonesia needs must be based on an evaluation of sustainable development from national, regional, and local levels. All are needed. In that context, this chapter is the longest and most detailed, as the city of Balikpapan and its province of East Kalimatan were used as the case study to show relationships on all three levels within a nation.

The study documents a decade of analyses along with multiple disciplinary levels of design, planning, engineering, technologies, health, environment, and economics needed for creating and improving sustainable communities. However, all levels of government are needed, as "sustainable development is continually evolving."

With these basic chapters providing the overview of what needs to be done as a multidisciplinary approach to creating sustainable communities, there are many cases. Chapter 10 concerns the "Financial Investments for Zero Energy Houses: The Case of Near-Zero Energy Buildings in the Baltic States" by Professor N. Lepkova, D. Zubka, and R.L. Jensen. Lithuania is the case in point, as this region of the EU is moving rapidly into the green industrial revolution rather than just following its historical pathways from certain areas of the western EU.

The concept and even implementation of "zero energy buildings" are beyond the traditional construction and building real estate development in the United Kingdom and United States, for example. Even Germany and Denmark do not have such standards. Such zero energy buildings, pointed out by Professor Lepkova and her colleagues, include but are not limited to specific, climate-adapted design, passive solar design, active renewable energy systems, thermal mass superinsulation, tightly sealed envelope, low-emission windows, and energy-saving appliances. The results are "very high energy performance" buildings, which are key factors for sustainable communities.

Then, Chapter 11, "The Canadian Context: Energy" by K.S. Madiraju from McGill University in Quebec, Canada, shows how Canada—with its vast renewable energy resources—can make the entire nation a sustainable community. In particular, the current growth to exploit natural gas, along with coal and other fossil fuels in Canada, is not needed. Hence the negative impact on the climate will be reduced, along with the high costs for health and environmental problems that are being created. Meanwhile throughout Canada, a "passive decentralization approach to energy" has taken hold in the various provinces, but within the "guidance and direction provided by a national strategy."

Professor M.K. Elahee, who wrote Chapter 8, also provided the economics on how Mauritius as a small island nation could provide as a model case for other nations. His Chapter 12 on "Energy Management in Small-Island Developing Economy: The Case of Mauritius" provides the answers to people and politicians, who often state, "This costs too much." They are wrong and it is just the opposite that is true. With 82% of energy in Mauritius coming from imported fossil fuels, the nation could save billions of U.S. dollars by just using renewable energy sources there. That shifting of funds from fossil fuels to renewables would mean a vast difference in the national financial resources. In 2011, this process started with the Energy Efficiency Act, which created a national energy efficiency management office. Results have started throughout the island–nation and gained global attention, including the World Bank.

Professor Sliogeriene then goes back to the "Energy System of the Baltic States and Its Development" in Chapter 13 to look at Estonia in particular. The need for an energy infrastructure using renewable sources is critical so that the nation can become independent and not continue to rely on other nations, such as Russia. Historically, that had been the case there but should not continue as—aside from being energy dependent on the former Soviet Union—the technologies and environmental impact were dependent on fossil fuels and behind the times as Estonia "inherited a technologically inefficient and resource-guzzling centralized energy sector." Moreover, Russia could then exert economic pressure and political concessions upon Estonia and other Baltic nations. So the case is made for Estonia to develop an energy infrastructure that is based on sustainable development.

Chapter 14 on "Renewable Energy Generation: Incentives Matter—A Comparison between Italy and Other European countries" from D. Chiaroni, V. Chiesa, and F. Frattini provides insight on how Italy compares to other EU nations in the use and expansion of renewable energy for creating sustainable communities. Starting with the Kyoto Protocol, which was established in 1997 and then a decade later when it began to be enforced, the EU established an action plan. They focus on renewable energy for Italy and compare them with other countries in the EU. Incentives from the government are key factors. Photovoltaics in Italy are particularly important, especially with the use of feed-in tariff incentives, created in Germany. Biomass was another area that is focused upon in the chapter as the authors provide evidence about their success but with cautionary notes and the need for changes.

Chapter 15 takes a different look at "Germany's Energiewende" by E. Borden and J. Stonington. The translation of "energiewende" into English means "energy transition" or "transformation," which is the context for Germany moving off of its dependency on nuclear power. However, the background and history of the German energy sector go far more into the nuclear power plant problems in the United States with Three-Mile Island in 1979 to the Chernobyl power plant disaster in the Ukraine when it was part of the Soviet Union in 1986. Today the German solution to nuclear power is not just the feed-in tariff, but other science and economic programs. The authors review, for example, the U.S. "cap and trade" program, which puts a price on carbon in the EU. Yet since 2008 when this market "solution" to climate change started, it has failed on many levels. Instead, Germany has seen success with lowering carbon and greenhouse gases through its sustainable communities, which function in the "energiewende" model.

Chapter 16 on "Educational Programs for Sustainable Societies Using a Cross-Cultural Management Method: A Case Study from Serbia" by J. Paunkovic takes the issue and need for sustainable communities even further by arguing that people need to be aware, educated, and trained in what sustainable communities are and could be. The case of Serbia is given in considerable detail with the needs but also the programs for the entire nation to go beyond the UN and other organizations in setting its standards and goals. Paunkovic outlines and discusses many of the programs and activities that are now in place and moving ahead since 2008 that provide an educated public, leadership, and new research avenues and areas around the world.

The next section of the book goes into more detail about how communities pay for becoming sustainable. Throughout the book and in various chapters, authors make references to "pillars" ranging from needs for economics, technologies, and innovation to infrastructures and plans for communities.

The concern about new business and hence economic models can be seen emerging rapidly around the world. Professor R. Ruminski provides a case in Chapter 17 of the "Business Ventures and Financial Sector in the United Arab Emirates." Starting with the definition of sustainable communities established in the Brundtland Commission (1987) as "development that meets the needs of the present without compromising the ability of future generations to meet their own needs," he moves into the combination of macro- and microeconomic processes, as the case of the United Arab Emirates (UAE) provides ranging from its strategic location in the Middle East oil and gas supplies to investment and capital that its leaders envision for future generations. And it all starts with the UAE being "committed to maintaining the policy of economic openness, seeking actively to develop economic/business projects that are in harmony with the changes taking place in the world" despite the global economic crisis in 2008–2009, which it is now recovering from with more innovation and economic development for sustainable communities.

Finally in this section is the concern for health care and medical costs. Again looking at the Middle East, the challenge has been difficult, yet achievable, providing a model for all nations that have similar concerns. In Chapter 18, Hayhoe provides a case in "Development of a Sustainable Disabled Population in Countries of the Cooperation Council for the Arab States of the Gulf." A growing number of disabled people in the Middle East have become a focus for all Arabic nations. The concern involves every community and nation. Today the Cooperative Council for Arab States of the Gulf founded in 1981 has targeted sustainable communities as a necessary core solution to the needs of disabled people. Four decades later this concern has continued and spread through examples of issues ranging from infrastructures to safety in cars by using seat belts to reduce harm in case of automobile accidents. All of these practical concerns are tied into the history and culture of nations throughout the Middle East and have helped significantly to reduce the health costs.

The last part of the book concerns the culture, history, and traditions of nations with how they create and implement public policy today. The UAE touched on that, but then very specific examples provide more cases. For example, Chapter 19 by Clark and Li places the "Political–Economic Governance of Renewable Energy Systems: The Key to Creating Sustainable Communities." Both authors have taken the issue of renewable energy, particularly from different perspectives in their past publications, together (Clark & Li 2004; Li & Clark 2009, 2013). This time they look at what

China has done to "leapfrog" into what Clark and colleagues (2013) call the green industrial revolution.

A cautionary note, written well before in draft form for this chapter, is that the new Chinese Central Government administration does not use Western economics as its model for sustainable development. That warning is significant in that China today has experienced vast real estate development but not in as an aggressive sustainable manner that it should have done over the last decade. Instead central government political leaders have followed the Western real estate development model that gives landowners and developers large amounts of valuable land and financial support but with little care for the environment, health, safety, and people to make the communities sustainable. The new Chinese Central Government has an important and unique mandate before itself that is what Clark and Li call "social capitalism."

Now how does social capitalism impact sustainable communities? Professor Coletta starts with agriculture and farming in Chapter 20, "Sustainable Agriculture: The Food Chain," stating the fact that farm animals are the second biggest contributor to climate change after the use of fossil fuels for vehicles and buildings "due to the use of their bodies for food and the results of their waste." Most people are not aware of this fact and the impact on the climate. Additionally, there are "social implications" to this fact in how and what people eat depending on their national gross domestic product and living standards and impacting "food security," as the cost of food is a critical factor in every country. Historically, the food sector has a "yield gap" or the difference between actual production and achievable food consumption using current technologies. Today, "recent dynamics (such as climate change) show scarcities that stem from the development path of modern society." Urbanization and economic growth in developed nations and now BRIC, as well as developing nations, cause serious problems with agriculture, irrigation, food supply, and its security, let alone the issues of fresh water, land, waste, and pollution.

There are solutions to some of these problems in agriculture, urbanization, and real estate development, as Jin and Peng argue in Chapter 21 on the "Development Partnership of Renewable Energies Technology and Smart Grid in China." With comparisons between the United States and China, the authors provide examples of smart girds, not for large utilities and their infrastructures, but with a focus on solar power for buildings in communities. The generation of power from the sun allows its use for buildings and infrastructures that comprise sustainable communities, as well as being "smart" by feeding power back into the utility grid for its distribution of that power to other groups. Then the power utility becomes the smart supplier of power to the sustainable communities at night or when there is no sun during the day.

Other renewable technologies and those that also store power (such as batteries, fuel cells, inverters, and capacitors) allow the communities to use little or no fossil fuel and then become sustainable in terms of no impact on the environment. Jin and Peng provide other cases, such as wind power and its impact on communities in need of power. Major companies see this link with the smart grid to be larger than the Internet in the near future. In closing, the authors point out how China is rebuilding its power system in the context of the smart green grid through its current 5-year plan, which includes massive amounts of government financing and support.

Yet China, some nations in the EU, and even to some extent the United States are struggling with cases of sustainable communities as defined by the Brundtland Report in the 1980s and other reports since then from the UN Intergovernmental Council on Climate Change, Framework Convention on Climate Change, and many others, including C40 (which is now C63) cities around the world seeking to be sustainable in order to reduce the climate change. The problem is always the same. How do these communities themselves know that they are sustainable? Based on what criteria and measurements? And then how are they compared to other "sustainable communities" around the world?

There are organizations, such as the U.S. Green Building Council, that have set measurable standards for buildings, such as Leadership in Environment and Energy Design (LEED), from the lowest level of just being "certified" to the highest level, which is "platinum." Each level has measurable requirements that buildings must meet. And there are trained LEED people who do, verify, and certify the LEED levels. Nonetheless, there need to be actual communities that can serve as cases of sustainability. Some experts argue that the communities in developing countries come the closest to meeting the criteria and even LEED standards. Hence, let's consider two such cases to end the book.

Chapter 22 is about a region in Haiti that is being developed in a sustainable manner through U.S. architect Carl Welty entitled "Regenerative Community Régénérer in Haiti: Model and Process Toward a Sustainable, Self-Renewing Economy." Welty is associated with a southern California company (Regenerative Development Group) that advises, designs, and then constructs sustainable communities. He reviews the historical and cultural background to Régénérer while then revising and criticizing the old and new paradigms about building communities. The owner of the 3000 acres that comprise Régénérer is a wealthy Haitian whose land is in a valley far from the urban area of Port-au-Prince destroyed in 2011 by the hurricane that struck Haiti and killed thousands of people.

Even today (2013) that region is struggling to rebuild itself despite the millions of U.S. dollars given to the nation and that region in particular. Haiti is the poorest country in the Western Hemisphere. So what is needed is a new "business model" to rebuild the Haiti that is now being implemented in Régénérer, which the land owner and U.S. development advisors argue consists of a "quadruple bottom line" consisting of "economic prosperity, environmental regeneration, social progress and cultural vitality." The chapter sets up measurable criteria, standards, and factors that make Régénérer a sustainable community. With that is also the need for international partnerships, support, collaborations, and finance.

Chapter 23 is about "Microcities" with a "case of India" by N. Jafry and G. Silvers where the authors set a perspective as in other BRIC countries that looks at the rapidly increasing demand for goods and services that places enormous social and economic problems on nations, particularly local communities. India has difficult socioeconomic structures and can only obtain change at the local level. That is, however, not unusual. The authors see this as a "bottoms-up approach" to creating sustainable communities through the use of being energy efficient, as well as using biogas from animal waste and other biomass sources for energy use.

Nonetheless, the regional and national perspective is "aware of its stake in promoting sustainability initiatives in India" so that some examples are given. In that context, the authors review India's use of conventional energy resources of which coal and oil comprise over 62%. Nonetheless, India has almost 30% of its energy from renewable sources. India is now into its 11th 5-year plan where it now targets 60 cities that are growing rapidly as in need for energy efficiency and renewable energy "by motivating and supporting local governments to drive local innovation and investment though cities." The basic problem will be is there are finance and economic resources to achieve these end goals.

Clark concludes the book in Chapter 24 by challenging the readers to make changes for creating sustainable communities now—today. The key areas are economic changes through looking for new ways to finance sustainable communities. In his book, *The Next Economics* (2012), there were chapters on how to finance the green industrial revolution. The fact is that climate change has gotten more and more destructive and results in a great intensity of weather changes, which destroy land, kill people, and paralyze entire regions and nations. The costs today are astonishing and growing more each day. Because no value can be placed on an individual's life, the issue is *not* to adjust or even tolerate climate change. People at the local level need to take action *now*. They cannot afford to wait for national governments and even international organizations to act. It is too late.

REFERENCES

Clark, W.W. (Ed.), 2012. The next economics: global cases in energy, environment, and climate change. Springer Press. December.

Clark, W.W., Xing, L., 2004. Social capitalism: transfer of technology for developing nations. International Journal of Technology Transfer and Commercialization, Inderscience 3 (1).

Xing, L., Clark, W.W., 2009. Crises, opportunities and alternatives globalization and the next economy: a theoretical and critical review. Chapter 4, In: Xing & Gorm Winther, Li (Ed.), Globalization and transnational capitalism. Aalborg University Press, Denmark.

Xing, L., Clark, W.W., 2013. Energy concerns in China's policy-making calculations: from self-reliance, market-dependence to green energy. Special Issue, Contemporary Economic Policy Journal Western Economic Association.

CHAPTER 2

The Green Industrial Revolution

Woodrow W. Clark II[1], Grant Cooke[2]
[1]Qualitative Economist, Managing Director, Clark Strategic Partners, Beverly Hills, CA, USA
[2]Founder and CEO, Sustainable Energy Associates, LLC., Benicia, CA, USA

Contents

Introduction	13
The First Industrial Revolution	18
The Second Industrial Revolution	20
Not Enough Supply, too Much Demand	24
The Green Industrial Revolution	26
Sustainability is the Key	28
Renewable Energy: From Central Grid to On-Site and Distributed Power	29
Storage and Intermittent Technologies	32
Additive Manufacturing	34
The Next Economics: Social Capitalism	35
The Green Industrial Revolution has Started	38

INTRODUCTION

In San Diego, California, the sun is powering over 4500 buildings. The sunny Southern California coastal region has nearly 37 MW of rooftop solar power installed on homes and commercial and government buildings. With 1 MW able to power 750–1000 homes, San Diego is powering the equivalent of 8000 homes, making this city of three million America's ground zero when it comes to per capital use of solar energy (Environment California study 1/24/2012).

Children in Ghana's mud huts study their school lessons after dark by the light of a shiny light-emitting diode light, powered by a solar collector about the size of pack of cards. The inexpensive reading light, which was invented by a California university professor, comes from a group called Unite to Light that is bringing these small solar-powered lights to rural Africa. The group is intent on furthering children's education and health by replacing the noxious kerosene lamps used by Africa's rural poor. In addition to Africa, Unite to Light is supplying solar-based lighting to 33 developing nations on four continents (www.unite-to-light.org).

In Norway, a hydrogen highway stretches 375 miles or 600 km from Oslo to Stavanger. It opened in 2009 with the first hydrogen station in

Stavanger followed by one in Porsgrunn. A modified Toyota Prius can travel 200 km using just 2 kg of hydrogen. Statoil Hydro retails hydrogen for $6.28 per kilo for a fuel cost of around 7.4 cents/km (11.8 cents a mile). In Norway, regular gasoline costs around $1.40 a liter and a normal Prius gets around 22 km to the liter (6.4 cents/km, 10.3 cents a mile). While slightly more expensive than regular gas, the vehicles are odorless, noiseless, and free of CO_2 emissions. Japan, Sweden, Denmark, California, and especially Germany have built hydrogen-refueling stations in major cities with plans for more. To provide vehicles for the hydrogen highways, car companies are quietly producing hydrogen fuel cell cars that electrolyze renewable energy sources into hydrogen for refueling in the home, office, or public buildings (http://www.greenmuze.com/climate/cars/1149-norways-hydrogen-highway.html).

The Chinese have taken cutting-edge German technology that was originally licensed from the United State and created the first commercial high-speed magnetic-levitation train. Called Maglev, this carbonless technology uses controlled tension from magnets to form an electromagnetic suspension system for lift and propulsion. The Shanghai train connects downtown and the Pudong International Airport, making the 30-km trip in just 7 minutes. Other trains that can travel up to 1000 km, or about 620 miles per hour, are in development. These Maglev trains will travel in underground airless tubes.

A Canadian cement company is running their fleet of trucks on goopy oil made from CO_2-consuming algae. Ontario's St. Mary's Cement is using the photosynthesis of algae to convert the CO_2 produced by the cement manufacturing process into a biofuel. The CO_2-consuming algae are being harvested and dried using waste heat from the plant. Then it is burned as fuel inside the plant's cement kilns as well as being used as fuel to power the cement trucks.

Throughout the world, the Green Industrial Revolution (GIR) is budding, even though the planet is becoming hotter, smokier, more crowded, and seemingly more dependent on fossil fuels. As oil supplies peak and global warming threatens—and the planet groans as the seventh billion person is born—the world is turning to environmentally friendly, carbonless energy sources. While frustratingly slow for those concerned about an environmental catastrophe, there is an emerging green economy. It is destined to be the most significant social and economic transition in world history.

This book's authors are calling this extraordinary new era the Green Industrial Revolution. As it takes hold globally, it will result in a complete restructuring of the way energy is generated, supplied, and used. The GIR will become the largest social and economic megatrend of the postmodern era. It will be an era of extraordinary potential and opportunity, with remarkable innovation in science and energy that will lead to sustainable and carbonless economies powered by advanced technologies such as hydrogen fuel cells and nonpolluting technologies such as wind and solar. Small community-based and onsite renewable energy generation will replace massive coal and nuclear-powered central plant utilities, and smart green grids will deliver energy effortlessly and efficiently to intelligent appliances.

This new era encompasses changes in technology, economics, businesses, manufacturing, jobs, and consumer lifestyles. The transition will be as complete as when the steam-driven first industrial revolution gave way to the second industrial revolution, which was based on fossil fuels, internal combustion engines, mechanical engineering, and neoclassical economics. This monumental shift is already underway in parts of the world, where renewable energy, sustainable communities, smart green grids, and environmentally sound technologies are becoming common.

In *The European Dream* (2004), Jeremy Rifkin suggests the concept of the third industrial revolution. He noted that the modern world was marked by extraordinary changes brought about by the nexus of digital communications and renewable energy. Rifkin's book is Europe-centric and lavish in its praise of European leadership in business, economics, and environmental development at the end of the 20th century.

Unfortunately for Rifkin's view of European superiority, the European Union (EU) and the 17 nations using the euro as currency ran into a political and economic "brick wall" with the 2011 euro crisis. The crisis was triggered by the enormous debt that the southern EU countries—Greece, Italy, Portugal, and Spain—had generated as they instituted excessive social programs that were unsustainable. Until the 2008 subprime mortgage conflagration, the EU's convoluted taxing and public funds processes essentially hid the actual cost of these debt-ridden programs.

In 2012, as the European Union struggles to remain whole, Asia, especially China, is emerging as the economic and political region challenging the United States. China is now the world's second largest economy and is on track to overtake America by 2015. Nobel economist Robert Fogel estimates that by 2040 China's gross domestic product (GDP) will reach \$123 trillion. This will be about 40% of global GDP compared to America's 14% (Foreign Policy in MarketWatch 1/31/2011 Paul Farrell column).

While Rifkin coined the term the Third Industrial Revolution, we prefer the Green Industrial Revolution as more accurate for what is emerging. While Europe was and is environmentally conscious and aware of being "green," the GIR actually began in Japan and South Korea many years before Rifkin noticed it in Europe. Our 2011 book, *Global Energy Innovation: Why America Must Lead*, documented this remarkable economic and technological paradigm change.

As a small and densely populated island nation of 130 million people, Japan has a tradition of "no waste" that dates back to the Middle Ages. For centuries, Japan relied on its own natural resources for energy and development. Natural resources were exploited, but because 70–80% of Japan is mountainous or forested, the development of land for commercial, farm, and residential use was limited. Even today, human waste is recycled for fertilizer. It is no wonder that for three decades Japan led in the creation of photovoltaic (PV) and other renewable energy systems. Its concerns for water conservation led to the success of Toto, one of the greatest and most efficient water use companies in the world.

By the 1980s, Japan and South Korea were concerned with the need to become energy secure and, as a result, developed national policies and programs to reduce their growing dependency on foreign fuels. These countries realized after World War II and the Cold War that their futures were not rooted in the same carbon-intensive economies that had built the United States and western Europe. These Asian nations were too highly dependent on imported fuels and energy to grow rapidly.

Decades later, Japan is once again struggling with an energy crisis, created by a devastating earthquake and tsunami in the northeast coastal region that destroyed one of the key Fukushima nuclear power plants. From this tragedy, Japan may leap even further ahead in developing a carbonless economy as it expands renewable energy generation to compensate for the loss of nuclear power. Other Asian nations are rapidly developing large-scale renewable energy generation as well.

Despite all the activity in Europe and Asia, few Americans (outside of a small circle of scholars and a handful of prescient venture capitalists and investment bankers) saw this new global megatrend looming. Even many people within the green industry have remained oblivious. For the most part, America's dependence on fossil fuels has clouded its ability to see that the carbon-based Second Industrial Revolution has ended. Today, the corporations and people vested in fossil fuels and related products from the

Second Industrial Revolution are holding America back, preventing it from competing and advancing into the GIR.

The Green Industrial Revolution, with its extraordinary new technologies and the promise of thousands of new green jobs, is trying to come to America. It is hampered by the lack of a national energy policy and a political process that is beholden to the fossil fuel industry. Big oil has been America's "elephant in the room" for over a hundred years, exploiting the nation's resources, pushing the country into a dependence on foreign oil producers that are politically destabilizing and not aligned with our national interests.

The United States remains in the Second Industrial Revolution, when in January 2010, the U.S. Supreme Court institutionalized the problem. In *Citizens United v. Federal Election Commission*, the high court ruled that large corporations, with unlimited financial resources, are to be considered as "individuals." This means that they have no limits of freedom of speech in terms of financial and political influences. The powerful interests that buttressed America's lavish carbon-intensive lifestyle are using their enormous resources to influence public opinion and politics, trying to keep America clinging desperately to an era that the rest of the world is leaving.

The planet is threatened by an environmental and climate catastrophe of unimaginable proportions. Population is the ticking time bomb. The United Nations predicts that we will increase from today's 7 billion people to 10 billion by 2050. In other words, we will add 3 billion people in less than 40 years. China will add 320 million for 1.4 billion, India will add 600 million to about 1.5 billion, while the United States will add 120 million for about 400 million total by midcentury.

All natural resources, particularly fossil fuels, are finite. Experts are warning that there are not enough resources to feed 10 billion people and that a global population over 8 billion is inviting environmental collapse.

The Earth needs all it inhabitants to come to grip with these future realities. We need leadership and action, particularly from the United States. America was the democratic and technological leader of the 20th century, and its political leaders need to end their dysfunctional behavior. America needs to rapidly commit to a carbon-less economy and embrace and lead the Green Industrial Revolution.

Human history is defined by extraordinary leaps in technology. When a technological window opens, it triggers remarkable social, political, scientific, commercial, and artistic advances. Archaeologists point to the ancients rise as they discovered fire and entered the Stone Age. Suddenly stones provided the basis for new technology applications that increased survival—prey was easier to kill and the odds were better against predators. Similar strides in human development accompanied the technological advances as humans learned to make copper, bronze, and iron tools.

So it went. Hunters domesticated dogs to help catch game. Technology responded to human needs, and the wheel was discovered, making transportation and the distribution of goods easier. Oxen were taught to pull a plow; horses were tamed to be ridden. Stunning advances in technology, science, and engineering allowed the great civilizations of the Egyptians, Mayans, Chinese, and other ancients to flourish. Superior battle technology aided Hannibal as he conquered Italy and it made the Venetians the rulers of the Mediterranean.

Breakthroughs in technology opened the door to the glories of the Renaissance, an era when all social elements—architecture, art, science, politics, and economics—raced ahead after the despair of the Dark Ages.

THE FIRST INDUSTRIAL REVOLUTION

In the middle of the 15th century, a German named Johannes Gutenberg and two friends, Andreas Deritzehn and Andreas Heilmann, became interested in a crude invention. Called a printing press, it pressed images on paper from a flat wood cut. A goldsmith by training, Gutenberg began to work with a new form of type made from an alloy of lead, tin, and antimony. To create these lead types, Gutenberg used a special matrix that allowed for quick and precise type blocks from a uniform template. To cover the type, he developed an oil-based ink that was durable and lasting, and he used paper as well as vellum, a high-quality parchment.

Gutenberg printed a beautifully executed Bible in 1455. Each page of the Gutenberg Bible, as it became known, had 42 lines. Copies sold for 30 florins, or about 3 years worth of wages. Through various iterations of his Bible, Gutenberg refined his technique for mechanical movable type printing. His inventions led to an explosion of printing activities in Europe, dubbed the Printing Revolution.

From Gutenberg's small shop in Mainz, Germany, printing spread to hundreds of cities throughout Europe, helping to stimulate the Renaissance and later the scientific revolution of the 16th and 17th centuries. Historians estimate that by the end of the 15th century, European printing presses had produced 20 million books, and by the end of the 16th century, the output rose above 150 million books (Febvre and Martin, 1976).

Gutenberg and other early printers unlocked the human mind on an unprecedented scale. New ideas could travel beyond the rock walls of monasteries and the laboriously hand copying of scripts by monks. Major Greek and Latin texts that had survived the Inquisition and the book burnings of the Middle Ages were printed and poured over by scholars. Literacy was spreading as economies grew, and readership soon extended beyond the clergy and royal clerks. Ideas crossed borders, capturing the masses in the Reformation and threatening the power of the political and religious authorities. Literacy broke the monopoly of the elite and bolstered a middle class. Mass communication had arrived, and ideas were suddenly accessible. Another incredible advance in technology had pushed civilization forward.

Because the printing press was a European phenomenon, the rest of the world fell behind in science and social development. The rapid exchange of new and startling ideas eventually led to the Age of Enlightenment, that extraordinary era of cultural, social, and scientific advancement that was nudged along by Isaac Newton.

In the early 1700s, as the Age of Enlightenment emerged in Great Britain, a young Scottish inventor and mechanical engineer named James Watt started to tinker with the design of a crude steam engine that was used for pumping water from mines. Called the Newcomen engine, it had been in use for about 50 years when Watt came across it. Immediately, Watt realized that the original engine wasted most of its steam. As a remedy, he caused the steam to condense in a separate chamber apart from the piston. Watt's separate condenser radically improved the power, efficiency, and cost-effectiveness of steam engines. Then he added a rotary shaft and gears to his machine and a new world was created.

In 1775, Watt entered a partnership with Matthew Boulton, a businessman who quickly recognized the commercial viability of this new energy-generating source. Boulton and Watt were enormously successful with their steam engine, providing the power for a change in the way people lived, worked, and played. Watt's steam engine also triggered the need for experts who understood machines, opening the door for a whole new field of science. Modern engineering was born, and mechanical engineering dominated the field of practical and applied technologies to this day.

By the end of the Napoleonic Wars in 1815, wood was scarce in Britain. Most of the nation's magnificent oak trees had been cut into planks and masts for the naval war ships that held Napoleon at bay until his eventual defeat at Waterloo. However, the country was blessed with an abundance of coal, which contained twice as much energy as wood. Coal quickly displaced wood as the fuel for the steam engine. Soon it was used to produce heat for industrial processes, to drive engines, and to create propulsion, as well as to warm buildings.

The combination of Gutenberg's printing press, which made the quick dissemination of innovation and ideas possible, and Watt's coal-driven steam engine, which created a whole new energy source, triggered an explosion in industrial activity. Chemical energy from coal was transferred to thermal energy and then to mechanical energy. The steam engine powered industrial machinery and steam locomotives. Suddenly, a world was created that needed people who understood engineering and combustion engines. New education, business, and employment sectors were created to support these new technologies.

The First Industrial Revolution (1IR) was surging. It was a turning point in human history that started around 1760 and ran through the last part of the 19th century. Great Britain led the world as it transitioned from a manual labor, draft animal-based economy to machine-based manufacturing. Coal was the prime energy source and the United Kingdom had plenty of it. Society was altered forever, as major changes in agriculture, manufacturing, mining, warfare, transportation, philosophy, and communications took place.

Watt's engine helped usher in Britain's great textile and manufacturing industries. It helped the British navy cement its authority and opened new commercial trade routes. Soon, all of Western civilization was entrenched in the 1IR, driven by steam, and surrounded by innovative ideas, which were distributed with the help of the printing press.

The First Industrial Revolution hastened the drive toward large-scale manufacturing. At first, machines started to replace manual labor, horsepower, and wind and waterpower. This transition eventually reached North America, hastened by the westward expansion and the discovery of gold in California in 1849. American communities, which were based on trade and agriculture and dependent on tools and animals, began to rely more and more on machines and engines.

THE SECOND INDUSTRIAL REVOLUTION

America had its own version of the Enlightenment driven by the brilliant Ben Franklin. One of the original Founding Fathers of the United States, Franklin was a prolific scientist and inventor. Around 1752, he conducted the famous kite experiment in Philadelphia, and the era of electricity was ushered in by this pudgy physicist standing in a muddy field waving a kite with a key at lightening clouds. Fortunately, Franklin avoided electrocution, which was the fate of others trying to duplicate his experiment.

This scientific experiment led to a series of ideas, inventions, and breakthroughs. Eventually, people began to understand, harness, and commercialize electricity, which had been studied since 1600. Along with the discovery and nascent efforts to harness electricity, power and energy generation started to shift from steam and coal to oil and the internal combustion engine. Eventually, the internal combustion engine powered by fossil fuel would take its place at the heart of the Second Industrial Revolution (2IR).

In 1851, Samuel Kiers, an American from Pennsylvania, began selling kerosene to local coal miners, calling it carbon oil. Petroleum was seeping into some salt mines that Kier owned, fouling the salt, and he wanted to find a use for the oil. He invented a process to distill the crude and a lamp that burned the oil efficiently. Kier established America's first oil refinery in Pittsburgh in 1853. Using the crude oil from his salt mines, his five-barrel still produced the first commercial illuminating oil from petroleum.

Called the grandfather of the American oil industry, Kiers' efforts to rid his salt mines of petroleum led to commercial oil drilling and production. With the commercial availability of oil, the internal combustion engine became the force for modern industrialization.

In its simplest form, an internal combustion engine harnesses a small intense explosion to drive a secondary object—usually a set of wheels, such as on a car, or another machine, such as a turbine or propeller. A small amount of high-energy fuel such as gasoline is placed in a small, enclosed space. Then air is mixed in, the mixture is ignited, and energy is released in the form of expanding gas. This miniexplosion applies force to a part of the engine, usually to pistons, turbine blades, or a nozzle. This force pushes the engine part, converting chemical energy into useful mechanical energy.

A Belgian engineer, Jean J. Lenoir, developed the first commercialized internal combustion engine. He developed a single-cylinder, two-stroke engine in 1860. His engine burned a mix of coal gas and air, which was ignited by a "jumping spark" ignition system. In 1863 the Hippomobile with a hydrogen gas-fueled, one-cylinder internal combustion engine made a test drive from Paris to Joinville-le-Pont in about 3 hours. The auto had a top speed of about 6 miles per hour.

Lenoir refined the engine and added a three-wheeled carriage. While noisy and given to seizing up when overheated, 143 of the vehicles were sold by 1865. Leaping to the conclusion that oil and gas would soon be the new energy source, the Parisian newspaper *Cosmos* pronounced the steam age over (*Scientific American* 1860, p. 193).

Engineers, like the German Nikolaus Otto, began making improvements in internal combustion technology, which soon rendered the Lenoir design obsolete. In 1872, an American named George Brayton invented Brayton's Ready Motor and went into commercial production. His engine used constant pressure combustion and was the first commercial liquid-fueled internal combustion engine (Improvement in gas-engines, Patent no. 125166).

By the end of the 19th century, the first of the world's major auto manufacturers was founded in Germany. Wilhelm Maybach designed an engine built at Daimler Motoren Gesellschaft—following the specifications of Emil Jellinek—who required the engine to be named *Daimler-Mercedes* after his daughter. In 1902, automobiles with that engine were put into production by DMG (Georgano 1990).

As the 2IR took hold, the internal combustion engine, fueled by various types of carbon and oil-based gases, became the driving force behind this new age of extraordinary machines—cars, trains, boats, space ships, etc.—that soon defined modern life. Educational institutions trained mechanical engineers to operate and maintain the technologies, which created jobs and new businesses worldwide.

Manufacturing grew and the assembly line made it possible to mass produce goods and products. Except for some brutal wars in Europe and the relentless genocide of the world's aboriginal natives, the planet's overall population expanded. With this expansion came a greater distribution of products and an increased dependence on fossil fuels.

Along with machines, electricity, and transportation came the telephone—a technology that revolutionized the daily lives of ordinary people. There is some controversy surrounding its invention, including a claim that an Italian, Antonio Meucci, was the actual inventor rather than Alexander Graham Bell. However, it was Bell who spoke the first complete sentence from one telephone to another, which was transmitted on March 10, 1876. To Thomas Watson, his assistant, Bell said across the line, "Watson, come here; I want you." The commercialization of the telephone was an iconic example of American entrepreneurship from a new invention and technology.

This revolutionary technology provided the 2IR with the same explosion in communications as Gutenberg's printing press did for the 1IR. It was a new world of analog communications. Driven by electricity, ideas, concepts, and images could travel faster than ever before. Science and knowledge exploded exponentially.

Fossil fuels of the 2IR opened the world to the wonders of the personal transportation device. At first, fossil fuels allowed for the transition from an agrarian society to an urban one and provided a way to make electricity. But when used to power a car, fossil fuels allowed urban workers to leave their city apartments and settle in the suburbs. This led to the need to construct highways and build housing developments, and America's car culture took root. Not only did the car become the means to get to work and how we measured success, but also, ironically, it became a symbol of rebellious freedom, such as the spontaneous romantic exhilaration of Jack Kerouac's *On the Road*.

Unfortunately, since the birth of the 1IR, the Western world's improving lifestyle—and the human passion for autos—has been dependent on fossil fuels. At one time, the coal, oil, and natural gas that powered this new economic model and the prosperity it entails seemed relatively inexpensive, inexhaustible, and presumably harmless. More and bigger homes and buildings were built, more concrete poured, and more fossil fuels extracted and burned. Frankly, there wasn't much to stop this social and economic juggernaut.

As industrialization led to urbanization, and urbanization to suburbanization, America built national highways and thousands of miles of freeways that circled and interlaced our cities. The more concrete that was poured, the faster the suburbs grew and the American lifestyle was forever changed and thoroughly dependent on fossil fuels.

The world's undeveloped countries followed and soon were building highways and suburbs that sprawled along concrete ribbons, creating congestion, generating pollutants, and producing an atmospheric overhang of smog. These developing nations bought second-hand and unsold American and European cars. Most of these nations then became dependent on Western-controlled supplies of fossil fuels, although many had their own fossil fuel resources, which were developed decades later under the financing and control of Western nations.

Societies around the world evolved to require greater and greater amounts of energy for light, heat, locomotion, mechanical work, communications, and then for smart phones, computers, televisions, microwaves, washing machines, coffee makers, and all the other technology and gadgets of modern living.

In fact, world energy consumption is predicted to grow by 53% from 2008 to 2035, according to the U.S. Energy Information Administration. Total world energy use rises from 505 quadrillion British thermal units (BTU) in 2008 to 619 quadrillion BTU in 2020 and 770 quadrillion BTU in 2035 (USEIA *International Energy Outlook 2011* 20554.135.7/forecasts/ieo/).

Much of the increase in energy consumption comes from developing nations such as China, India, and Brazil, where demand is driven by strong long-term economic growth. The Energy Information Administration(EIA) is predicting that energy use in non- Organization for Economic Co-operation and Development nations will increase 85% by 2030.

In the United States, power demand is going up 15–20% in the next decade because of the electrification of our society (*Fortune*, April 30, 2012, p. 96). Throughout the 2IR, these energy requirements have come primarily from fossil fuel—coal, natural gas, and oil.

To keep its energy-intensive lifestyle going, the United States uses about 19 million barrels of oil a day, or about 7 billion barrels of oil a year. In 2011, this worked out to about 22% of the world's total consumption of oil, which was about 32 billion barrels in 2011, according to the U.S. EIA (http://www.eia.gov/tools/faqs/faq.cfm?id=33&t=6).

This high-level of consumption is not sustainable.

NOT ENOUGH SUPPLY, TOO MUCH DEMAND

For over a hundred years, oil and gas discoveries have made fortunes for those optimistic and smart enough to understand the earth's geology and exploit it or, in the case of the Middle East, lucky enough to live on top of enormous hydrocarbon deposits.

Now, scientists believe that the world's oil and natural gas supplies have peaked and are declining rapidly because of the demands from the technologies of the 2IR. As M. King Hubbert, a Shell Oil geophysicist, observed in his startling prediction first made in 1949, the fossil fuel era will be of very short duration. In 1956 he predicted that U.S. oil production would peak about 1970 and then decline. At the time he was scoffed at; now, in the early 21st century, Hubbert looks extraordinarily prescient (Hubbert 1956).

Richard Sears, a geophysicist and former vice president for exploration and deep water technical evaluation at Shell, appeared at the 2010 Technology Entertainment and Design conference in Long Beach, California. In his talk on the future of energy, Sears said that there is only 30 to 50 years left before a broad gap opens between worldwide oil supply and demand. Sears then held up a pincushion of the globe with red thumbtacks. "This is it," he said. "This is all the oil in world. Geologists have a pretty good idea of where it is."

Sears makes the point that throughout the world, oil supplies are in decline. For example:
- Kuwait's oil supplies have been in decline since 1970.
- U.S. oil supplies have been in decline since 1971, hence the need to protect those resources around the world.
- Iran's oil production has been in decline since 2008, while demand is soaring.
- Indonesia's oil supplies have been in a free fall since 1991, and this former OPEC exporter is now an importer.
- The European North Sea oil reserves have been declining since 1999, and the decline is accelerating. The United Kingdom is no longer an oil exporter.
- Norway's oil production has been in decline since 2001, and for environmental reasons it has limited and cut back on offshore drilling.
- Mexico's oil production has been dropping since 2005.

While oil supply decreases, demand increases. In early 2011, China released customs data that showed that oil imports rose 18% in 2010. Platts, an oil industry research company, reported that China's oil consumption in March 2011 averaged 9.2 million barrels per day. Platts calculated that China's apparent oil demand is up 10.5% year to year (Energy Sector Investing 2011).

Driving this consumption is China's adoption of the automobile. Once a nation where everyone commuted by bicycle, China now has 60 million cars on the road, with 12 to 18 million more new cars predicted for 2011 (British Petroleum report 2011).

China's immediate solution to the demand problem, while it implements its next 5-year plan, has been for its state-controlled oil and gas company to buy massive amounts of oil and gas from around the world on a long-term basis and import them. However, in China's 13th 5-year plan there were discussions about only allowing electric and hydrogen fuel vehicles to be sold in the country. That would be earth shaking for China's own manufacturing of vehicles, and it would force other nations to convert rapidly to nonfossil fuel modes of transportation.

India is not far behind China in oil consumption. India consumes nearly 3 million barrels per day as car sales have jumped. India is expected to be the fourth largest car market in the next 3 years (Asian Age 2010).

The rest of the world's undeveloped societies are modernizing and using more oil for cars, trucks, airplanes, and boats. For example, Saudi Arabia is the world's largest oil producer. However, it is also the sixth largest oil consumer and internal consumption is growing rapidly. Increasing population and fuel subsidies are pushing internal demand growth to 7–9% per year (The American Interest 2012).

With tightening supply, price goes up with increasing demand. Charles T. Maxwell, a highly respected oil analyst from Greenwich, Connecticut, told *Barron Magazine* (February 12, 2011) that continuing growing world demand would probably push oil to $300 a barrel by 2020.

Meanwhile, as more and more oil and other fossil fuels are burned, pollution increases and emissions—called greenhouse gases (GHGs) —grow, spreading around the world and causing climate change. The buildup of GHGs has marched in lock step with the expansion of fossil fuel use since the 1700s. While composed primarily of CO_2, greenhouse gases also include methane (CH_4) and nitrous oxide (N_2O). The gases float upward into the atmosphere and wrap themselves like a blanket around the Earth. As more and more are added, the blanket gets thicker and warmer.

Unlike empty beer cans, plastic bags, or the other garbage that piles up along our roads and rivers, GHGs pile up out of sight, in the Earth's atmosphere above our heads. Visualize all the CO_2 that is released from cars, from coal and gas-burning power generation, and from the burning and clearing of forests and the deforestation of regions such as Brazil or Indonesia. The results are devastating to the land areas and the atmosphere that is breathed daily around the world. Further, severe changes to weather patterns are created by the differences in temperatures and the impacts from evaporation as hot airstreams hit cold water.

As climate change raises the planet's temperature, one particular irony stands out. The Arctic region is warming twice as fast as the rest of the planet (Economist: Special Report 2012). While the ice melts, the world's major oil companies refuse to admit the role their products and activities play in this environmental death dance. The companies have poured millions of dollars into disinformation, politics, and public relations campaign arguing against climate change. At the same time, the companies are pushing forward with major explorations to harvest the vast natural resources of the Arctic region, even as the ice continues to recede at a faster rate. The companies are doing their best not to be seen as profiting from the environmental destruction contributed to by their activities.

THE GREEN INDUSTRIAL REVOLUTION

In the midst of this fossil-fueled, internal combustion, greenhouse gas-suffocating age, a new social and economic movement has emerged.

Social scientists argue that industrial revolutions are triggered by the confluences of a new energy-generating technology—steam in 1IR and internal combustion in 2IR—and a new form of communication technology—printing press and analog, respectively—that provides rapid dissemination of new ideas to accelerate the adoption of inventions. (Jeremy Rifkin 2004, *European Dream*).

Today, the world is on the verge of another industrial revolution. This one is driven by the nexus of the fast-as-light communication of the digital age, with its Internet access to almost all scientific knowledge, the Facebook and Twitter-led social networking that has truly created Marshall McLuhan's "global village," and the new energy generation sources of renewable energy, with its accompanied green smart grids, intelligent machines, and additive manufacturing. This worldwide Green Industrial Revolution is being led by the Asian nations of South Korea, Japan, and now China. The United States is lagging far behind.

In another major historical irony, the communications tools of this new Green Industrial Revolution helped overthrow the notorious despots who ruled the countries that controlled the oil supply of the 2IR. The Arab Spring, which has changed the political reality of the entire Middle East, was made possible by the instant communications of the social networks and Facebook, in particular.

This Green Industrial Revolution is potentially more significant and life changing than either the First or the Second Industrial Revolutions. It may also turn out to be the planet's only real chance for survival. With an estimated nine billion inhabitants by 2050, there is so much more at stake.

Today, the world is rapidly running out of fossil fuel, particularly oil. This alone threatens to shake the very foundation of human existence. Adding a heightened sense of urgency is the environmental degradation and the collapse of various parts of our planet's ecosystem, including the Brazilian watershed and the Arctic.

Fortunately, in many parts of the world, the Green Industrial Revolution has begun. Parts of Asia and Europe have been moving into it for over three decades, developing sustainable, energy-independent communities. South Korea has urban regions that are already energy independent and carbon neutral. Japan was heading in this direction as well, but got redirected toward nuclear power stations. Because of World War II, the Japanese were never proponents or supporters of nuclear power stations. However, American nuclear power companies and their financial backers were able to influence the acceptance of nuclear energy. Their influence is still strong in Japan today, despite the March 2011 nuclear disaster at Fukashima triggered by the earthquake and tsunami.

Meanwhile, a large-scale effort is also underway in China where the nation has "leapfrogged" other countries in the new green industrial revolution (Clark & Isherwood 2010). In 2008, the Climate Group, an international think tank, reported China's rapid gains in the race to become the leader in developing renewable energy technologies via its 12th 5-year plan (Climate Group 2008).

Since then, China's 13th 5-year plan, which started in March 2011, committed the nation to spending the equivalent of over three trillion dollars in funding for renewable energy (Lo 2011).

Germany, through its feed-in tariff (FiT) program, was the number one producer and installer of solar panels for homes, offices, and large open areas from 2006 to 2009. In 2010, Italy then copied the FiT and held that distinction of world leader in solar panel installation. China took the lead in 2011 and continues as the number one solar panel and PV manufacturer and installer. Japan is now leading in auto manufacturing, as it began to make vehicles that do not damage the environment and atmosphere.

Other nations in the European Union, such as the Nordic countries and Spain, have been aggressively implementing policies and programs to become energy independent through renewable energy by 2050. They are succeeding. Denmark has made extraordinary advances already. The Danish government has implemented a national policy that includes local plans and financing to develop distributed or onsite renewable energy power systems as the nation moves into the GIR (Lund & Østergaard 2009). By 2015, several Danish cities will be energy independent with renewable energy power and smart green grids, with the whole nation 100% using renewable energy by 2025 (Østergaard & Lund 2010).

SUSTAINABILITY IS THE KEY

Worldwide, natural resources are declining with increased demand. If global energy policies do not change, political and social tensions will mount over the supplies and locations of fossil fuels as they become scarcer and more expensive. The decline in fossil fuel and global warming will exacerbate the difficulties in feeding the world's expanding population.

The decline in natural resources and fossil fuels and the increase in climate change, plus an accelerating population, are pushing us toward the Green Industrial Revolution and its promise of sustainable communities, renewable and distributed energy, and smart green grids. Japan, South Korea, and western Europe have set the pace for sustainable and secure communities with their own renewable energy sources, storage devices, and emerging technologies.

Sustainable and smart agile communities represent an improved new design for how we live, particularly in urban areas. Nations, states, and cities want to control and centralize power and authority—that has been the historical pattern. However, today with the need to meet and address the global challenge of climate change, regional and local level solutions must be developed, along with policy and action for creating the GIR.

Sustainable communities integrate renewable energy generation and storage technologies with electric and hydrogen transportation, business development, job creation, and social activities. This is sustainable development or the interaction among a community's infrastructure requirements, economic needs, and social activities for the protection and preservation of the environment. This interaction stimulates business and provides compelling reasons for pursuing and creating sustainable communities (Clark & Isherwood 2010).

Most modern cities have the potential to implement some, if not all, sustainable activities. With a little guidance, our communities, colleges, shopping areas, towns, and cities can have locally distributed renewable energy, clean water, recycled garbage and waste, and efficient community transportation systems that run on renewable energy sources for power. We must create a sustainable lifestyle that is free from the carbon-intensive, fossil fuel-based, inefficient centralized energy generation of the 2IR.

RENEWABLE ENERGY: FROM CENTRAL GRID TO ON-SITE AND DISTRIBUTED POWER

Renewable energy generation is at the heart of the GIR. Yet, it is one of those terms that everyone thinks they understand until forced to use it in conversation. Basically, it is a source of energy that is not carbon based and will not diminish. For example, the sun shines during the day and the wind blows at night. Each needs some form of storage or feedback technology when the wind is not blowing or the sun is not shining. These forms of energy generation are called intermittent and need technologies to provide for "base load" or round-the-clock power generation. Renewable energy is the foundation for a sustainable community.

The most common renewable energy sources are systems that create power from wind, sun, or water; digestive processes that change waste into biomass; and systems that recycle waste for fuel generation. Other renewable sources include geothermal, "run of the river streams," and now, bacteria and algae.

Wind generation is fairly straightforward. Wind has been used as a power source for hundreds of years. Originally, windmills were used to power small machines for processing or for pushing water. Today, a large propeller is placed in the path of the wind. The force of the wind turns the propeller and a gear coupling interacts with a turbine, which generates electricity. The concept of wind generation may be ancient, but technological advances have transformed it. New-generation wind turbines are stronger, more efficient, quieter, and less expensive.

Wind farms harness the energy of dozens, even hundreds, of wind turbines. Turbines can be installed on land or offshore. They can be placed in small communities or even on rooftops to capture the natural flow of air.

Solar generation systems capture sunlight, including ultraviolet radiation, via solar cells (silicon). This process of passing sunlight through silicon creates a chemical reaction that generates a small amount of electricity. A photovoltaic reaction is at the core of solar panel systems. A second process uses sunlight to heat liquid (oil or water), which is then converted to electricity. A number of communities are now looking into solar "concentrated" systems in which the sun is captured in heat tubes and used for heating and cooling of homes, buildings, and central power plants. This is a great renewable energy technology for use in water systems and buildings that have swimming pools. Solar, like wind, can be installed on homes and building complexes closer to the end user instead of on distant farms. There is a one-third loss in efficiency transmitting the power generated over long distances. Hence, onsite renewable power generation is the key to becoming energy independent with solar and wind power as well as other renewable energy sources.

Biomass (biological material from living or recently living organisms such as plants) can be used to generate energy using a remarkable chemical process that converts plant sugars (such as corn) into gases (ethanol or methane). These are then burned or used to generate electricity. The process is referred to as "digestive" and it's not unlike an animal's digestive system. The appealing feature of this process is that abundant and seemingly unusable plant debris—rye grass, wood chips, weeds, grape sludge, almond hulls, and the like—can be used to generate energy.

Geothermal power is created from heat stored in the earth. This heat originates from the formation of the planet, from radioactive decay of minerals, and from solar energy absorbed at the Earth's surface. It has been used for space heating and bathing since ancient Roman times, but is now better known for generating electricity. In 2007, geothermal plants worldwide had the capacity to generate about 10 gigawatts (10 billion watts) of power and, in practice, generated enough power to meet 0.3% of the global electricity demand (Geothermal Energy Association www.geo-energy.org/data).

In the last few years, engineers have developed remarkable devices such as geothermal heat pumps, ground source heat pumps, and geoexchangers that gather ground heat to provide heating for buildings in cold climates. Through a similar process they can use ground sources for cooling buildings in hot climates. More and more communities with concentrations of buildings, such as colleges, government centers, and shopping malls, are turning to geothermal systems.

Ocean and tidal waves have power that can be harnessed to create usable energy. This concept is behind the revolutionary SeaGen tidal power system, which was pioneered by France and Ireland. The French have been generating power from the tides since 1966, and now Electricite de France has announced a large commercial-scale tidal power system that will generate 10 MW of electricity per year (Ocean Power Technologies info@ocean powertech.com).

America, particularly the Pacific coastline, is equally suitable for producing massive amounts of energy with the right technology. Ocean power technologies vary, but the primary types are **wave power** conversion devices, which bob up and down with passing swells; **tidal power** devices, which use strong tidal variations to produce power; **ocean current** devices, which look like wind turbines and are placed below the water surface to take advantage of the power of ocean currents; and **ocean thermal energy conversion devices**, which extract energy from the differences in temperature between the ocean's shallow and deep waters.

Run-of-the-river systems generate electricity without the large water storage required of traditional hydroelectric dams. Run-of-the-river systems are ideal for streams or rivers with consistent water levels or minimum loss of water flow during the dry season. In most cases, power turbines are mounted along the river and the turbines generate electricity as the water flows. This is being done in Europe and Asia, where flowing river water generates considerable amounts of energy without harming the surrounding land or changing the natural elements in the water. These systems do minimal destruction to pristine environments and could be adapted easily to large inland rivers.

Fuel cells are electrochemical cells that convert a source fuel into an electrical current. They generate electricity through reactions between a fuel and an oxidant, triggered in the presence of an electrolyte. The reactants flow into the cell, and the reaction products flow out of it, while the electrolyte remains within it. Fuel cells are energy storage devices that can operate continuously as long as the necessary reactant and oxidant flows are maintained. Fuel cells are different from conventional electrochemical cell batteries in that they consume reactants from an external source, which must be replaced. Many combinations of fuels and oxidants are possible. A hydrogen fuel cell uses hydrogen as its fuel and oxygen as its oxidant. Other fuels include hydrocarbons and alcohols. Other oxidants include chlorine and chlorine dioxide.

Bacterial, or microbial, fuel cells use living, nonhazardous microbial bacteria to generate electricity. British Petroleum has made a $500 million investment in this futuristic process, which is now being developed by researchers at the University of California, Berkeley (Renewable and Appropriate Energy Laboratory), and the University of Illinois, Urbana (Energy Agriculture Laboratory).

Researchers envision small household power generators that look like aquariums but are filled with water and microscopic bacteria instead of fish. When the bacteria inside are fed, the power generator—referred to as a "biogenerator"—produces electricity. Ironically, the funding for this technology comes from the same company that caused the April 2010 oil spill in the Gulf of Mexico that killed 11 people, damaged the Gulf waters, and polluted the coastland while destroying fishing and tourist businesses (www.nytimes.com/2010/04/29/us/29spill.html).

STORAGE AND INTERMITTENT TECHNOLOGIES

While all these power generation and storage systems produce electricity, none is as inexpensive today as the currently used fossil fuels—coal, oil, and natural gas—from the 2IR. However, fossil fuels were not inexpensive when their use escalated in the late 1890s and became the foundation for the 2IR.

To maximize renewable power efficiency, renewables need to be integrated as linked or bundled supply sources compatible with the natural physical characteristics of the locale. Further, these intermittent power generation resources must include storage devices because the sun is not always shining and the wind is not always blowing. Today, renewable energy technologies and their integrated systems need the same kind of policy and financing support that the fossil fuels received.

Storage devices can be either natural, such as salt formations, or artificial, such as batteries, flywheels, or fuel cells. Once the electricity is collected, these storage devices regulate the distribution so energy use is optimized.

Government support for fossil fuels (in terms of tax incentives, funds, and even land) must be repeated for the new renewable energy-generating sources. Incentives for the 2IR must be reduced and applied to the GIR. This tax shift has been very successful in other industries and businesses and can be designed so there is little or no additional tax burden on consumers.

In the 1990s, energy deregulation took hold in America, but was called "privatization" or "liberalization" in Europe. In both regions, the plan was a failure, as it was based on market economics, which turned into market manipulation, fraud, and loss of service. What has happened since then is that a number of communities and states around the world are promoting "distributed power generation" or the ability of local building owners to have renewable energy at their homes, offices, shopping malls, college campuses, and other locations where power is generated and then used. If there is any excess power, it can be sold to the power grid or large energy companies in the region.

Hence, a significant strategy of the GIR has been to modify deregulation. Central power plants are still the norm, partly owned by government, but far more regulations are needed to oversee supplies, costs, and delivery of energy. German and Danish central power plants, for example, have significant government involvement through either government ownership or appointed board members. The concept has begun in the United States and is critical in China where large and local renewable power systems are partly state owned.

The economic change from either extreme (government-controlled or market-central) power system generation has resulted in more local level energy generation, which is critical in the GIR. While some fossil fuels, such as coal, oil, and gas, are still inexpensive and used in central power plants, they are the major global atmospheric polluters (Natural Resources Defense Council www.nrdc.org). If a carbon tax or some other method that calculated the damage to the environment and public health was added, the costs would be higher than renewable power systems. That public strategy is beginning in Europe and China.

China and the United States rank at the top of emission polluters; in both cases, coal is the major problem (American Council on Renewable Energy, US-China Quarterly Review, March 2011). If the human and environmental impacts of coal were calculated into its true costs, then the real cost of coal energy generation for power would soar. To address this problem, the GIR needs the same sort of economic, tax, and funding support or incentives that the 2IR received over a century ago.

The result of the 2IR was the creation, operation, and maintenance of large, centralized, fossil-based power plants in the early part of the 20th century and then nuclear power plants in the last half of that century. The plants had to be powerful to withstand degradations over the vast distribution of a central-powered grid system. At each conversion from alternating current to direct current, electricity loses some power, but there is so much of it at the beginning that it does not matter several thousand miles away at the end. This results in the loss of efficiency in transmission over power lines as well as the constant need for repairs and upgrades.

Not so in the case of the environmentally friendly renewable systems of the GIR. For the best results, energy systems need local renewable power generation and distribution systems, "smart green" local and onsite grids, so electricity does not have to travel far and suffer losses from inefficiencies. An alternative is to hook into a transmission line. This way the local grid is added to the existing energy distribution system and the transmission line can act as a battery for the renewable energy that needs storage. Some have equated this to a model of the Internet where there is no single area for control over data or, in this case, power—rather it is spread out and localized.

The sooner the world evolves into the Green Industrial Revolution, the sooner the planet will begin to heal. How quickly the world adopts the GIR will be influenced heavily by the United States, the world's leading democracy. The United States can no longer afford oil wars, considering their financial impact and their costs in human lives and injuries. Nor can the country afford another environmentally crippling deep ocean oil spill, such as the April 2010 British Petroleum spill in the Gulf of Mexico. And it cannot continue to deny that the nation needs to take a new path.

ADDITIVE MANUFACTURING

Renewable energy and smart grid technologies are key components to the GIR. They have been evolving for decades and have received significant attention. However, millions of new "green" inventions are emerging from science and industrial laboratories. Smart appliances that can interact with utilities and other grid regulators are coming soon. Millions of other inventions for this new age are in design or development stages. Even more will be uncovered or dreamed of as the world begins to adopt more and more elements of the GIR.

One emerging process that may have a truly revolutionary impact is called additive manufacturing. Also referred to as three-dimensional printing, this remarkable process is now being refined in the science laboratories and clean rooms of some of the world's leading industrial companies.

The additive process is far more flexible and economical than conventional manufacturing. It requires fewer raw materials; because the design is done in the software, each product can be different or customized without retooling. Three-dimensional printing has the potential to transform manufacturing because it lower costs and reduces risks.

This new revolutionary technology will be a driving force in manufacturing and business development. Not only will it make wondrous new products possible, but it will also increase energy efficiency, maximize natural resources, and lower product cost. It will certainly be a critical component of the GIR.

THE NEXT ECONOMICS: SOCIAL CAPITALISM

It's not all about money—or is it?

In light of the October 2008 world financial meltdown, which even in 2012 continues with the monetary crisis in Europe, it seems silly to think that the supply-side, deregulated, free-market economics espoused so passionately by President Ronald Reagan and Prime Minister Margaret Thatcher in the 1980s will work for a 21st-century world threatened by irreversible environmental degradation.

The 2008 economic implosion from trillions of dollars in credit swaps, hedge funds, subprime mortgages, and related marginal derivatives (which nearly pushed the Western world's financial structure into the abyss) underlined what happens when governments ignore their responsibility to govern. Market economists and others had argued that there was no need for regulation. Government would act as "the invisible hand."

In the end, the worst financial disaster since the Great Depression was a testament to the venal side of free market capitalism—greed, stupidity, carelessness, and total disregard for risk management. These behaviors cannot be repeated if the planet is going to survive climate change and its impact on the earth and its inhabitants.

The Green Industrial Revolution must develop an economy that fits its social and political structures, similar to the First and Second Industrial Revolutions. The 1IR replaced an agrarian, draft animal-powered economy with one powered by steam engines and combustion machine-driven manufacturing, an evolution that was accelerated by colonial expansion. The 2IR created a fossil fuel-powered economy that extracted natural resources in an unregulated, consumer-fed, free-market capitalist society.

As the GIR emerges, the world is becoming much more interdependent. What happens in one part of the world, be it weather, pollution, politics, or economics, impacts other regions. For example, the dramatic change in the Egyptian government in early 2010 has affected the rest of the Middle East and will result in global changes of oil and gas supplies. The result might well be the forced end of the 2IR, so that as other nations strive for energy independence they implement the GIR.

There is historical precedence for a forced transition from the 2IR to the GIR. The Arab oil embargoes of the early 1970s pushed Europe and Asia toward social policies that eventually led to development of the GIR. Energy independence, climate change, and environmental protection became serious political issues. Both these regions have been developing economic forms of what has become known as "social capitalism," an economic view that includes sustainable growth, health and educational issues, environmental concerns, and climate change mitigation, along with interest in diverse populations, gender equality, and democratic processes (Clark 2012). The essence of social capitalism is that some social and political problems are so complex and overriding that free markets and deregulation cannot address them.

Social and environmental factors—sustainable communities, climate change mitigation, and environmental protection—are growing in importance and will soon demand far greater international cooperation and agreement. Rampant economic growth and individual accumulation of wealth are being replaced by social and environmental values that benefit the larger community.

Without a national policy and investment to fund these plans, countries cannot address their basic infrastructures, and there can be no action, no improvement, no resources, and certainly no response to environmental degradation. For example, the United States' inability to develop a national energy policy that addresses climate change is often cited as a monumental failure of its free market and deregulated economic model. Energy and infrastructure, the argument goes, are two extraordinarily important national issues. To address these basic systems for the greater good, a nation needs to have plans, which are outlined and offered by the central government.

The Peoples Republic of China, not the United States, is showing real global leadership as the world heads into the GIR. More than anything, China demonstrates how important a role the government plays in overseeing, directing, and supporting the economics of technologies and creation of employment. China's economic system is the prototype of social capitalism. Since the 1949 revolution, the Chinese have moved away from communism toward economic development through a series of 5-year plans, now being referred to as guidelines.

In the post-Mao era, China moved aggressively into a "market-capitalism" system, but one where state institutions were owned in part by the Chinese government and shared in joint ventures with foreign companies. Companies wanting to do business in China had to keep their "profits" there for reinvestment as well as have at least 49% of the company owned by the Chinese government.

Europeans adjusted their economies to fit the requirements of the GIR early on. Both the Scandinavians and the Germans realized that the move away from fossil fuels to renewable energy distribution would require more than the neoclassical free market economics could deliver. While the Danes and the other Scandinavians shifted national resources toward renewable energy power by national consensus, the Germans developed the innovative feed-in tariff process.

Germany's FiT was part of their 2000 Energy Renewable Sources Act, formally called the Act of Granting Priority to Renewable Energy Sources. This remarkable policy was designed to encourage the adoption of renewable energy sources and to help accelerate the move toward "grid parity," making renewable energy for the same price as existing power from the grid. Under a FiT, those generating eligible renewable energy, either homeowners or businesses, would be paid a premium price for the renewable electricity that they produced. Different tariff rates were set for different renewable energy technologies based on the development costs for each resource. By creating variable cost-based pricing, the Germans were able to encourage the use of new energy technologies such as wind power, biomass, hydropower, geothermal power, and solar photovoltaic, as well as to support the development of new technologies.

Creating an economy that can move the world into the GIR is an exceptionally complex process. Various nations and regions are approaching the problem differently. The European FiT program and China's direct government subsidies have been the most successful. Some U.S. states, such as California with its newly designed Renewable Auction Mechanism (RAM), have developed possible improvements over the European FiT, but the RAM is much more limited and available only in California.

So the Green Industrial Revolution is not all about the money. It is about climate change mitigation, renewable energy, smart grids, and health, education, and environmental sensitivity. But achieving the benefits of the

GIR—a wave of new technologies, business enterprises, and green jobs—will require substantial public and private financing. A new green economy will be needed to accelerate the necessary changes and stop climate change.

THE GREEN INDUSTRIAL REVOLUTION HAS STARTED

A Green Industrial Revolution is at the world's doorstep. The GIR is arriving as the Internet, with its social networks and digital electronics, intersects with renewable-source energy distributed by smart grids. Social and economic forces are coming together and pushing for an environmentally sustainable future.

Yet with global warming and climate change impacting daily lives, the United States is still mired in fossil fuel dependence without the political will to break loose and move forward.

As the world's leading democracy, the United States must also lead the Green Industrial Revolution. For the planet's sake and our children and grandchildren, the United States must establish a national energy policy with funding that makes sense. One that can move the entire country rapidly from the fossil fuels and wholesale resource extractions that dominated the 20th century to the GIR that will be the new world order of the 21st and 22nd centuries. The old, "dirty" economy was dependent on fossil fuels and internal combustion engines, as well as heavy manufacturing based on inexpensive oil and massive infrastructures to support energy and transportation. The GIR is about using renewable energy to power green local communities where renewable power and smart grids can monitor power and increase efficiencies.

Humanity's lust for fossil fuels may be our ruin. In the June 6 issue of *Nature,* Anthony Barnosky of the University of California, Berkeley, along with 21 coauthors, publically worried about the same thing. The planet may be on the verge of a "tipping point" they cautioned, which is a point of environmental decline that cannot be reversed.

The authors call for accelerated cooperation to reduce population growth and per-capita resource use. Fossil fuels need to be replaced with sustainable sources, the development of more efficient food production and distribution systems, and more protection for land and ocean. Barnosky's writes, "Humans may be forcing an irreversible, planetary-scale tipping point that could severely impact fisheries, agriculture, clean water and much of what Earth needs to sustain its inhabitants" (http://davis.patch.com/articles/are-humans-bringing-earth-to-an-irreversible-tipping-point).

In short, these scientists are among those urging the world to join the Green Industrial Revolution before it's too late.

REFERENCES

Asian Age, 2010. India car sales jump 21 percent. Available from: < www.asianage.com/business/india-car-sales-jump-21-cent-504. > [8 December 2010].

Clark 2012

Clark, W.W.II, Isherwood, W., 2010. "Report of Energy Strategies for the Inner Mongolia Autonomous Region." Utilities Policy , 1, 3–10, http://sciencedirect.com/science/article/pii/S0957178709000290.

Climate Group, 2008. China's clean revolution. Available from: < www.guardian.co.uk/environment/2008/aug/01/renewableenergy.climatechang. >

Economist: Special Report, 2012. The melting north. 16 June, p. 3.

Energy Sector Investing, 2011. China oil demand 9.2 barrels per day. Available from: < www.energysectorinvesting.com/2011/04/china-oil-demand-92m-barrels-per-day.html. >

Febvre, L., Martin, H.J., 1976. The coming of the book: the impact of printing 1450-1800. New Left Books, London.

Georgano, G.N., 1990. Cars: early and vintage, 1886-1930. Grange-Universal, London, p. 39.

Hubbert, M.K., 1956. 'Nuclear energy and the fossil fuels "drilling and production practice"' (PDF), Spring meeting of the Southern District, Division of Production. American Petroleum Institute, San Antonio TX, pp. 22–27.

Lo, 2011

Lund & Østergaard 2009

Ostergaard, P.A., Lund, H., 2010. "Climate Change Mitigation from the Bottom-Up Community Approach. In Sustainable Communities Design Handbook, Woodrow, W. ClarkII, (Ed.), Elsevier, New York, pp. 247–266.

The American Interest, 2012. The Folly of energy independence, Luft and Korin. July-August, p. 33.

CHAPTER 3
Public Policy and Leadership: Infrastructures and Integration

Michael "Hakeem" D. Johnson[1]
Acting Country Director, Adeso-African Development Solutions Juba, South Sudan

Contents

Introduction	41
Background	42
The Shift	43
Almere Zoneiland: The Sun Island, Kingdom of the Netherlands	45
Rizhao, China	46
Gaviotas, Colombia	47
The Birth of a Nation: The Republic of South Sudan	49

INTRODUCTION

The world has witnessed dramatic environmental changes since the early 1940s. With the dawn of the industrial age, nations have strived and achieved great technological advances in science and expanded the potential of societies to reach for even higher levels of productivity and efficiency. iPhones and Android platforms have replaced rudimentary forms of communications, and social media have brought together billions of people worldwide to exchange knowledge and ideas, as well as to conduct official business.

Certain pockets of the world have become transportation hubs, moving goods and people and providing services to an ever-increasing global and product-driven population. Technological progress, with industrialization, has brought untold wealth to millions and filled government coffers, but has also deepened marginalization and alienation for large parts of the Global South and poor communities in developed economies. Moreover, the plight of the environment versus technological advancement has been brought into sharper focus.

As more people move from rural to urban centers in search of opportunities, the more resources are required to provide goods and services to increasing metropolitan populations. The United Nations has recognized

---

[1] The views expressed herein are those of the author and do not reflect the views of the United Nations Development Programme.

the unsustainable nature of the cycle and in 1992 brought together leaders for its first-ever Conference on the Environment and Development to try and address the dichotomy and to agree on global, legally binding instruments to lay foundations from which the green economy has been able to gain traction in the early 21st century.

The author believes that absent sustained and active leadership from governments at all levels, the business community, and civil society organizations, attaining real achievement in protecting our natural environment and the valuable resources it contains, will be permanently lost. Attainment also means while ensuring technologies and economic growth do no harm to it, it provides opportunities to millions who are direct beneficiaries from the environment. It requires the development of and agreement on policy and legal frameworks at international and national levels to be linked up with the initiatives taking place at the municipal level to ensure not only sustainability, but also hold violators accountable. This chapter discusses what is current parlance in the international arena, particularly what is not working under the framework of the United Nations Conference on the Environment and Development and its ensuing multilateral environmental agreements, in its translation to guide governments, corporations, and civil society to build and sustain our cities. It highlights models of "sustainable communities" globally and the role the private sector must and is playing in ensuring sustainability, replication, and scaling up, without damaging our environment. The Republic of South Sudan will be a case study on where a new nation-state and other developing countries can capitalize on adopting policy and legal frameworks to create an enabling environment to build truly sustainable cities within their territories.

BACKGROUND

According to the United Nations Environment Programme, the 20th century saw more global environmental changes than in the previous 11,000 years combined. Prior to the 1930s, virtually no chlorofluorocarbons(CFCs) existed on Earth. The earth's protective layer of stratospheric ozone remained robust. However, by the late 20th century, large quantities of CFCs had escaped to the stratosphere, depleting the ozone shield significantly. Until the first self-sustaining nuclear reaction in 1942 there was no nuclear waste anywhere on earth, but the U.S. Department of Energy notes that within six decades there were millions of tons of it. What makes the last century such an era of unusual environmental turbulence is more the scale and scope of venerable human practices such as farming and forest clearance, the increasing use of fire and fossil fuels, fishing, and so forth. Little regulation and safeguards allowed such activity to continue, largely unabated.

Large-scale infrastructure projects throughout the industrial era have boosted economic production and output for countries such as the United States, western Europe, and parts of Asia and Latin America and lifted millions out of poverty. However, while such projects such as the Three Gorges Dam in China or high-speed rail/tunnel projects across the Alps in Europe (e.g., Lyon–Turin link) and in Queensland, Australia (the proposed Gold Coast Rapid Transit Project) have the propensity to expand irrigation systems in China and link major metropolitan centers throughout Europe and northeastern Australia, evidence suggests that the immediate environmental impact of the Three Gorges Dam has been enormous, but that the longer term effects are still unknown (Hvistendahl 2008). Millions of people have been uprooted with livelihoods lost, dozens of lives lost as a result of increasing landslides and other associated calamities, and entire communities relocated, without proper compensation, to make way for the dam (Gleick 2009). Hundreds of thousands of tons of earth has been stripped away, and many analysts have suggested that entire ecosystems, including valuable fish stocks, have been damaged irrevocably.

Many large-scale industrial projects, constructed not only in China, but around the world, lack fundamental safeguards and standards to reduce vulnerability to the environment and humans. Many are constructed without adequate planning frameworks and studies to determine economic, social, and environmental viability. Still many more projects lack proper consideration of community needs or include community-led consultation and mechanisms for local ownership.

THE SHIFT

At the global level, the shift occurred in 1972 in Stockholm. World leaders gathered in Sweden for the United Nations Conference on the Human Environment where they discussed the then state of the world's environment. It was well attended and is widely recognized as the beginning of modern political and public awareness on the plight of the global environment. Twenty-six principals were adopted by world leaders covering critical themes, such as national obligations for protecting the environment, technology development for environmental protection, wildlife and biodiversity safeguards, human settlements, and obligations for international organizations working to improve the environment.

The seminal issue to emerge from the Stockholm conference was the nexus drawn between poverty alleviation and protecting the environment. The Rio Declaration, coming nearly two decades following Stockholm in 1992, took that 1972 position a step further and states that the only way to achieve long-term economic progress is to link it to environmental protection. Nations were called to establish a new and equitable global partnership involving governments, populations, and key sectors of societies and to build international agreements that protect the integrity of the global environmental and developmental system. Three conventions, one legally binding, emanated from Rio in 1992: the United Nations Framework Convention on Climate Change, the United Nations Convention on Biological Diversity, and the United Nations Convention to Combat Desertification and the Effects of Drought, particularly in Africa, as well as an agreement on deforestation.

A significant product from Rio in 1992 was the adoption of Agenda 21. Agenda 21 was a blueprint on how to make the future development of the world sound and sustainable economically, socially, and environmentally. In addition, the Rio Declaration contains fundamental principles on which nations can base their future decisions and policies, considering the environmental implications of socioeconomic development. Rio underlined that thinking of environmental, economic, and social development as isolated fields is no longer possible. Most nations have ratified the Earth Conventions, with the U.S. Senate ratifying the United Nations Convention to Combat Desertification in 2000. The World Summit on Sustainable Development in Johannesburg in 2002 emphasized an urban dimension, and several paragraphs of the Johannesburg Plan of Implementation include commitments related to cities [e.g., paragraph 11, specifically dedicated to cities; paragraph 21 on transportation; paragraph 8 on water and sanitation, paragraph 39(a) and 56 on air pollution, and paragraph 167 on local authorities](Johannesburg Plan of Implementation 2002). The urban dimension is also present in the Millennium Development Goals. The United Nations Conference on Sustainable Development (Rio+20) convened in late 2012 was an action-oriented conference, where all stakeholders, including major groups, the UN system/IGOs, and member states were invited to make commitments focusing on delivering concrete results for sustainable development on a voluntary basis, while honoring prior commitments to protecting the environments, namely Rio 1992, the Millennium Summit, and the 2002 World Summit. Moreover, most nations have signed thematically based multilateral environmental agreements covering the protection of the atmosphere, hazardous substances, nuclear safety, freshwater resources, marine conservation, transboundary watercourses (rivers) and international lakes, and nature conservation and protection of terrestrial resources, among numerous other thematic agreements.

The policy environment to address ensuring sustainable development of urban centers while protecting our environment at the global level is rich, however; it is disjointed when translated at the local level, and many countries have adopted their own policies to respond to the threat posed by unsustainable development practices and environmental destruction. In order for cities of the future, as elaborated further with the world's newest country—the Republic of South Sudan as a case study—to match internationally binding agreements to sustainable and green practices on the ground requires a new orientation in how stakeholders cooperate, educate, legislate, and regulate. First, it is critical to highlight success stories from around the world where communities, together with government, partners, and industry, are building sustainable living spaces while reducing their carbon footprint and impact on the environment dramatically.

ALMERE ZONEILAND: THE SUN ISLAND, KINGDOM OF THE NETHERLANDS

Almere Zoneiland (Sun Island) is located in the Noorderplassen-West district in Almere, and it is the first project in The Netherlands where houses in a residential area are heated by a single solar energy system collectively. The project is supported financially by the European Union and implemented in cooperation between the Dutch energy company Nuon, Almere Municipality, and local citizens. The Netherlands is one of Europe's most densely populated countries, which has led to large artificial land reclamation projects. The province of Flevoland reclaimed from the former Zuiderzee Lake in the middle of The Netherlands is one of the largest land reclamation projects in the world. In the southern part of the province, east of Amsterdam, is the municipality and town of Almere.

In opposition to most historic European cities developed over centuries, Almere has developed from scratch to become a large city within 40 years. From having a population of zero 40 years ago, the city of Almere has approximately 180,000 inhabitants and it is estimated that by 2015 the number will rise to 250,000 inhabitants (www.nuon.com/corporatesocialresponsibility2009). Due to the population growth, the planning, and the location, Almere has become The Netherland's fastest growing city and

Europe's fastest growing "new town." Green plans for the sustainable development of the district of Almere ensured that the city was designed as a cohesive and well-connected city with bicycle lanes and bus and train links, in addition to being bound together by the city's heating system.

Another way identity was created for Almere was through experimental urban development such as the Zoneiland project. The Zoneiland contains 520 solar panels covering an area of approximately 7000 m², equivalent to 1.5 football fields (www.nuon.com/corporatesocialresponsibility2009). The scale of the project makes it the world's third largest solar panel project. As a supplement to the Zoneiland project, some of the ecohouses in Noorderplassen-West are built to be 10% more energy efficient than standard construction and contribute with their own private solar power systems (www.nuon.com/corporatesocialresponsibility2009).

The idea behind the solar island is simple: The Zoneiland system functions equivalent to having a water hose lying outside on a sunny day. The water in the tube is heated by the sun all year round, even on cloudy days. After being heated up, the water is pumped directly into the heating network of the city district, providing 2700 homes with heated water. Annually, the Zoneiland provides 9750 gigajoules of renewable, sustainable energy, equal to 10% of Almere city's total annual energy needs. The rest of the energy is provided by the local power plant in Almere.

By supplying the city with renewable sustainable energy, the Zoneiland has helped reduce CO_2 production by 50%. This reduction in CO_2 is equivalent to the amount of CO_2 saved if every household using the new energy source annually drove 12,000 km less in their cars. Another advantage of the solar energy system is that it has no rotating parts, which means that it is easier to maintain, unlike, for example, wind turbines. The estimated lifetime of the solar system is about 25 years.

RIZHAO, CHINA

Rizhao is a city of three million people in northern China. It is using solar energy to provide energy, heating, and lighting. Ninety-nine percent of Rizhao's households use solar water heaters, while almost all traffic lights, street lights, and park illuminations are powered by photovoltaic solar cells. The city of Rizhao combines incentives and legislative tools to encourage the large-scale, efficient use of renewable energy, especially solar energy. In the suburbs and surrounding villages, more than 30% of households use solar water heaters and over 6000 households have solar cooking facilities (www.greensolutionsmag.com/?p=1452).

Widespread use of solar energy has reduced the use of coal and helped improve the environmental quality of Rizhao. The vision of the government was to preserve the surrounding environment, thus helping the city's social, economic, and cultural development in the long run. Solar energy is seen as a starting point to trigger this positive cycle. After 15 years of promoting solar energy, the approach has proven effective. In 2007, the city began attracting a rapidly increasing amount of foreign direct investment.

The provincial government in Rizhao subsidized research and development activities in the solar water heater industry. This investment produced technological breakthroughs, which increased efficiency and lowered unit costs. The cost of a solar water heater was brought down to the same level as an electric system. The city has mandated all new buildings to incorporate solar panels and oversees the construction process to ensure proper installation (www.greensolutionsmag.com/?p=1452). To raise awareness, the city held open seminars and ran public media campaigns on television. Government buildings and the homes of city leaders were the first to have the panels installed as role models. Some government bodies and businesses even provided free installation for employees.

The achievement of Rizhao was the result of a convergence of three key factors: a government policy that encouraged solar energy use and supported research and development financially, local solar panel industries that seized the opportunity and improved their products, and the strong political will of the city's leadership to adopt it.

GAVIOTAS, COLOMBIA

Gaviotas is a village of about 200 people in the llanos region of eastern Colombia. Founded in 1971 by Paolo Lugari, for three decades Gaviotans—peasants, scientists, artists, and former street kids—have struggled to build an oasis in the remote, barren savannas of eastern Colombia. They planted millions of trees, thus regenerating an indigenous rainforest. They farm organically and use wind and solar power. Every family enjoys free housing, community meals, and free schooling. There are no weapons, no police, and no jail. There is no mayor. The United Nations has named the village a model of sustainable development. Many have labeled Paolo Lugari the "inventor of the world."

The community has produced a number of inventions and innovations over the years—notably including a children's seesaw that drives a water pump and a "distinctive 'sunflower' design" windmill that is well suited to the plains in Colombia (Gaviotas Rising 1994). As already-existing solutions are often very costly to adapt, Gaviotas' innovations are often simple changes to a means of production that make otherwise expensive products available at affordable prices. One of the most widespread Gaviotas' developments is a water pump that can tap aquifers six times as deep as conventional pumps with less effort being expended. While existing pumps in the region raised and lowered a heavy piston in a pipe, the Gaviotas' engineers created pumps that leave the piston in place and instead lift and lower an inexpensive, light PVC sleeve around the piston (Weisman 1995).

When leading solar hot water panel manufacturers explained the expenses and complicated manufacture of panels efficient enough to collect sunlight in the often overcast weather of the llanos, the Gaviotas' engineers crafted homemade solar water-heating panels out of inexpensive building materials that were better suited to the peculiar climate of the region. The ecovillage also creates some of its own building materials, such as a unique form of quick crete brick made with dirt of the region. The village is noted for the planting of over 1.5 million trees in the area. While the trees were originally part of an experiment to see if any significant growth could occur in the desiccated soil of the llanos region, they have become a significant feature of the grasslands. As a result of the shade these trees provide and the tropical climate of the area, the groundcover began hosting tropical rainforest species that were once native to the region. Resin harvested from the planted trees has provided Gaviotas with a sustainable source of income (Weisman 1999).

Replication of the Gaviotas model has taken place across Colombia. In the 1980s, the Las Gaviotas team was hired to install their innovative "appropriate technologies" in other parts of the country, including installation in many villages of water systems based on the Gaviotas windmills and pumps. The largest single effort was a solar hot water system for Ciudad Tunal, a 6000-apartment public housing project in Bogotá. The units still work perfectly, due greatly to the fact that they require no moving parts.

Gaviotas shows us that local communities hold keys to sustainable development, together with protecting the environment. The missing link with Gaviotas is sustained national involvement, support, and mechanisms designed to promote community-driven processes to sustain urbanization and environmental conservation. The community, itself, understood the value of conservation because products in llanos (resin from the replanted pine trees) are derived directly from the environment.

THE BIRTH OF A NATION: THE REPUBLIC OF SOUTH SUDAN

After nearly five decades of war with the Republic of Sudan, the Republic of South Sudan gained its independence on July 9, 2011. During the 6-year transition period leading to independence (2005–2011), South Sudan has presided over the fastest development of public sector institutions in modern statehood. Thirty-seven ministries, 19 commissions, 10 state governments, a national parliament, and 10 state legislatures were established. Rule of law institutions were created, and the first steps in transforming the liberation army into a professional force were taken. Despite these achievements, South Sudan is entering statehood with enormous challenges, most of them remnants of the long civil war. The 2012 South Sudan Peace Building Support Plan contextualizes key conflict drivers that, if left unchecked, have the propensity to further destabilize an already fragile situation and increase violence.

The overwhelming majority of South Sudanese remain distant from decision making and legitimate politics. Limited experience with democratic governance, intercommunal grievances, and unresolved issues with Sudan are significant factors driving violence and conflict. This is particularly the case with the youth, many of whom are highly militarized and frequently exploited by ethnic elites. With only a limited capacity to manage political diversity, state authorities are often unable to provide opportunities for dissent, discussion, and reconciliation. The lack of legitimate channels for expressing political aspirations and grievances exacerbates tensions among communities and is a major factor fuelling further tensions and, oftentimes, violence [Reference South Sudan Development Plan (SSDP), 2011: Section 2.2.1.1].

At the local level, communities frequently cite conflict over grazing land and access to water points, including between agriculturalists and pastoralists, as a major cause of violence. Sudanese pastoralists from neighboring regions migrating into South Sudanese territory to access watering points for their cattle, absent formal agreement between affected communities, have increased tensions in the volatile border area. Lack of government capacity to mediate between communities and regulate access to resources has played a significant role in escalating levels of intercommunal violence. In a context where rural areas will need to absorb high numbers of returnees and ex-combatants, pressure on access to land and water is expected to increase further. More than half of South Sudan's population relies on cattle as the currency for sociocultural interaction and exchange, as well as for socioeconomic status, rather than for more economically productive use. The primary use for cattle in this respect is payment of dowry. Marriage is seen as a rite of passage for both male and female youth, and young men are under severe pressure to meet escalating dowry costs. The consequence is the involvement of youth in raids of neighboring communities during which massive numbers of cattle are stolen and widespread civilian casualties occur, triggering reprisal attacks to recoup lost cattle and loss of life. The cycle is exacerbated by the absence of other livelihood or employment opportunities for youth (Reference SSDP, 2011: Sections 2.2.1.4; 2.2.1.2).

The results of the 2008 Sudan Census remain heavily contested in South Sudan due to perceptions of political bias. In the absence of agreed census data, authorities have struggled to distribute resources equitably (Reference SSDP, 2011: Section 6.5). Populous areas are chronically under-resourced, resulting in tensions between communities and ethnic groups. Where services reach local levels, they are largely provided through non-state entities with little state ownership. Absent tangible peace dividends, including education, health, and water/sanitation, and infrastructure visibly provided by the state fuel perceptions of exclusion from postindependence gains, contributing to resentment and diminishing confidence in government (Reference SSDP 2.2.4; 8.5.7).

The Republic of South Sudan is seizing opportunities to not only accede to the United Nations Conference on Environment and Development(UNCED) conventions and associated multilateral environmental agreements, but also to create a permissive environment for private industry to flourish and create jobs for the people. It takes 4 days to register a business in South Sudan, faster than many industrial nations. With abundant natural resources, land, water, livestock, and blessed with over 300 days of sunshine annually, the people of South Sudan stand to gain, provided the government enacts legislation designed to preserve the environment while opening the private sector to development around the country. In Juba, city authorities have mandated the installation of solar panels for city lights and other public structures. The model is being replicated in other cities and towns across South Sudan.

Pending and signed legislation covering the forest, water, agriculture, oil, and mining sectors offer the private sector, including local suppliers and service companies, to expand work in those areas, increasing employment and other tangible benefits greatly for an ever-increasing skilled workforce. Returnees, many of whom are highly educated or possess specific technical skills, have established import/export companies, hotels and guesthouse, mobile phone companies, law and accounting firms, and many other businesses. The government must ensure that the design of public and private structures is in adherence with international standards and position such structures to benefit from green technologies and materials. Today, many are not, which add to costs for operations and maintenance and reduce productivity over the longer term, providing virtually no returns to the business(es) therein. Economies of scale, even cost savings, are not achieved when buildings do not utilize proven green technologies and material designed to maximize the occupied space and increase efficiency.

The government of the Republic of South Sudan has many examples from around the world to draw from in designing its cities of the future. From China to India, Holland to the United States, experts in creating green economies and sustaining growth, working together and under the leadership of the government, can gift the people of South Sudan a prosperous future for generations to come. Translating action globally, through UNCED and other relevant fora, into dividends for the public and the environment locally requires sustained leadership and policies at all levels integrated with active participation from communities themselves. Working together, all of us stand to gain, while protecting our fragile environment.

REFERENCES

Gaviotas Rising, August 29, 1994. All Things Considered. National Public Radio: Segment #06. Transcript #1589.
Gleick, P., 2009. Three Gorges Dam project, Yangtze River, China world water. Available from: < www.johannesburgsummit.org/html/documents/summit_docs/2309_planfinal.htm. >
Hvistendahl, M., 2008. China's Three Gorges Dam: an environmental catastrophe? Sci. Am.
Weisman, A., 1995. A good harvest. In Context #42, Context Institute.
Weisman, A., 1999. Gaviotas: a village to reinvent the world. Chelsea Green Publishing Company.

CHAPTER 4

China: Resource Assessment of Offshore Wind Potential

Lixuan Hong
China Energy Group, Environmental, Energy Technologies Division, Lawrence Berkeley National Laboratory, Berkeley, CA, USA

Contents

Introduction	53
Methodology: Geographic Information System-Based Spatial Modeling	55
Offshore Wind Potential under Technical, Spatial, and Economic Constraints	57
Technical Potential	57
Spatial Constrained Potential	59
Economic Potential	61
Economic Costs of Tropical Cyclone Risks	63
National and Provincial Cost–Supply Curves	67
Conclusions	75

INTRODUCTION

The world's first offshore wind farm built in Denmark in 1991 marked the beginning of Europe's leadership in offshore wind power. Today, 4620 MW of offshore wind power capacity has been installed globally, and more than 90% of it is installed off northern Europe in the Baltic and Irish Seas, as well as the English Channel (Energy and Environmental Management Magazine 2012). While Europe has spent two decades to reach this level, China has announced targets that would surpass it within 5 years. In June 2010, the Chinese National Energy Administration (NEA) announced an ambitious target of 5 GW by 2015 and 30 GW by 2020, emerging as the second largest installation area after Europe for offshore wind development. At the same time, there are also great expectations placed for major offshore wind deployment elsewhere, including the United States, Canada, Japan, South Korea, Taiwan, and India. A total of 80 GW offshore wind capacity is projected to be installed by 2020 worldwide (Energy and Environmental Management Magazine 2012).

Offshore wind energy is generally being promoted for its enormous resource base and higher quality of resource compared to wind onshore, as there are vast available areas, far more consistent wind blowing, and less negative environmental impact, as well as close proximity to major demand centers. For China, it is particularly important to utilize its rich offshore wind resource because it is close to the coastal regions, which help enhance the security of the energy supply and mitigate greenhouse gas emissions. The coastal region is the most economically advanced part of China, with approximately 40% of the nation's population producing 60% of its gross domestic product. It is also the main driver of the nation's increasing energy use because of a high concentration of export industries, investment, and urbanization, accounting for 38% of growth in energy demands and 45% of growth in electricity consumption in 2010. Due to its lack of indigenous energy resources, the coastal region has become increasingly dependent on imported fuels, particularly coal, either from inland provinces or from international suppliers. However, as the center of domestic coal production continues to shift from middle China toward western China, congested railways in western China pose a major bottleneck for the coal supply chain (Fridley, Khanna & Hong 2012). Meanwhile, international coal prices are likely to soar in the years ahead because of growing demand and dwindling tradable supply globally (Heinberg & Fridley 2010). In addition, long-distance power transmission from west to east and from north to south not only causes great power losses, but also poses challenges to the capabilities of transmission lines and the stability of grids (Zhou et al. 2010). In order to address the issue of energy shortages, the central government has proposed an ambitious plan for nuclear power development along the coastal region in 2007, but the speed of implementation of this plan has decelerated after Japan's Fukushima accident in 2011. Offshore wind may be one option to ameliorate the current energy and climate dilemmas in the coastal region of China. Furthermore, wind can complement hydropower production in southwest China and offset the construction of new coal and nuclear power capacities in order to relieve stress on the nation's railway and grid systems.

At present, China has merely started its first step to assess its offshore wind potential. For instance, the China Meteorological Administration (CMA) organized the first and second national wind energy resource censuses in the late 1980s and during the period 2004–2005, drawing the conclusion that China's theoretically exploitable potential of offshore wind energy resources at a height of 10 meters above ground level was approximately 750 GW. From 2007 to 2009, a third national wind energy resource census was conducted using numerical simulation, concluding that the offshore wind potential (with water depth 5–25 m) reaches 200 GW calculated on the basis of 5 MW/km^2. This figure was adjusted by CMA to 400–500 GW recently. Another study by the World Wildlife Fund, China Wind Energy Association, and Sun Yet-sen University shows that the total technical offshore wind potential within the 100-km coastline of China is 11,580 TWh/year. However, these studies focus mainly on the assessment of technical potential, which are far from satisfying the needs for thorough and in-depth research on offshore wind energy in China.

METHODOLOGY: GEOGRAPHIC INFORMATION SYSTEM-BASED SPATIAL MODELING

The offshore model is built in a geographic information system (GIS), which is a system designed to capture, store, manipulate, analyze, manage, and present all types of geographical data. The idea of the offshore model is to reflect offshore potentials, economic costs, and tropical cyclone risks, as well as spatial constraints of ocean use across a region, therefore providing macroscopic information for decision makers and investors. A predecessor of this idea can be found in the technical and economic evaluation of the northern European offshore wind resource by Cockerill and colleagues (2001). The National Renewable Energy Laboratory (NREL) applied a GIS method for developing wind supply curves and provided a geographic and economic assessment of wind resources in the Zhangbei region of China (Kline, Heimiller & Cowlin 2008). The Danish Energy Agency used a GIS-based decision support system to identify future offshore wind turbine locations in 2025 (Danish Energy Agency 2007), but it is not mapping suitable areas continuously as in Möller (2011). The Energy Research Centre of The Netherlands published the project Windspeed for the purpose of analyzing the deployment of a wind energy plant in the North Sea, which suggests many valuable technical parameters and also offers a spatial decision support system (Jacquemin et al. 2009).

It is necessary to develop a comprehensive framework and methodology to assess offshore wind resources under China's condition. The offshore model consists of a series of models, including a power generation model, bottom-up cost model for offshore wind farms, probabilistic tropical cyclone event model, turbine damage model under tropical cyclones, and spatial constraints analysis (Figure 1). Outputs of the model can provide information, including a power generation potential, associated generation costs under

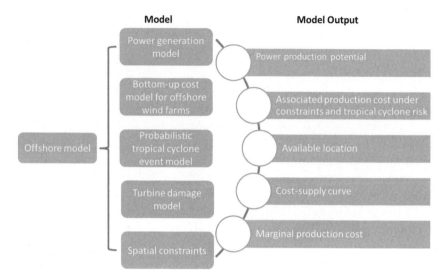

Figure 1 Schematic structure of a GIS-based offshore model. The GIS-based offshore model consists of five components, including power generation model, bottom-up cost model of offshore wind farms, probabilistic tropical cyclone event model, turbine damage model in the event of tropical cyclones, and spatial constraints for building offshore wind farms. Major outputs of the GIS-based offshore model are power production potential, associated production costs under spatial constraints and tropical cyclone risks, available locations for building offshore wind farms, and offshore wind cost–supply curves, as well as marginal production costs in different circumstances.

spatial constraints and tropical cyclone risks, available locations for offshore wind farms, and cost–supply curves, as well as marginal production costs in different circumstances. The ocean boundary of this analysis is the exclusive economic zone (EEZ)[1] of China, while initial spatial data inputs include average ocean wind speed, maximum wind speed during tropical cyclones, and a bathymetric map, as well as maps of spatial constraints such as shipping lanes, submarine cables, and conservation zones. All map layers in the GIS model have the unified spatial resolution of 1 km^2 in a geographical reference framework of the Universal Transverse Mercator system.

In this study, average ocean wind speed at a 10-meter height originates from the NASA SeaWinds scatterometer aboard the QuikSCAT satellite

---

[1] VLIZ (2009). Maritime Boundaries Geodatabase, version 5. Available online at http://www.vliz.be/vmdcdata/marbound. As there are still many disagreements regarding EEZ among countries worldwide, new treaties will be negotiated in the next years.

platform, which has a high spatial resolution of 12.5 km with a duration of measurement lasting more than 10 years. This data set has already been well applied for producing global ocean wind power density maps by NREL as part of the Solar and Wind Energy Resource Assessment project for the United Nations Environment Program. In addition, the CMA-STI Best Track Dataset for Tropical Cyclones in the Western North Pacific is used, which is compiled by the Shanghai Typhoon Institute (STI) of the China Meteorological Administration (CMA)(CMA-STI Best Track Dataset for Tropical Cyclones in the Western North Pacific 2010). This data set contains measurements of dates, intensities, latitudes, longitudes, minimum pressures at the storm center, and 2-minute mean maximum sustained wind speed near the storm center at 6-hour intervals for all tropical cyclones from 1949 to 2009 in the western north pacific (to the north of the equator and to the west of 180°E). The CMA-STI point data set is converted into spatially continuous map layers for modeling using the kernel density function of GIS. Since the mean size of tropical cyclone in the northwest Pacific Ocean is found to be 3.7 degrees of latitude (Liu & Chan 2010), a scan radius of this figure is used.

OFFSHORE WIND POTENTIAL UNDER TECHNICAL, SPATIAL, AND ECONOMIC CONSTRAINTS

Technical Potential

Wind power density, measured in watt per square meter of rotor area (W/m^2), indicates the flow of kinetic energy per unit swept area of a turbine. Power density is a useful way to evaluate the wind resource available at a potential site, as it is independent of turbine characteristics. Considering the constant air density of 1.225 kg/m^3, theoretical wind power density at a 90-meter height is shown in Figure 2. It ranges from 600 to 1300 W/m^2 in the southeastern domain, including Fujian, northern Guangdong, and southern Zhejiang, where are endowed with abundant offshore wind resources comparable to those in the Baltic and even the North Sea of Europe. Around the southern coast of Guangdong and Hainan, an average power density of 300–600 W/m^2 is expected, which is similar to that in southeastern Britain. Most areas of northern China, including Jiangsu, Shandong, and Bohai Rim, have an average power density of 200–400 W/m^2. These provinces are less windy regions where offshore wind power density is parallel to that in the Mediterranean.

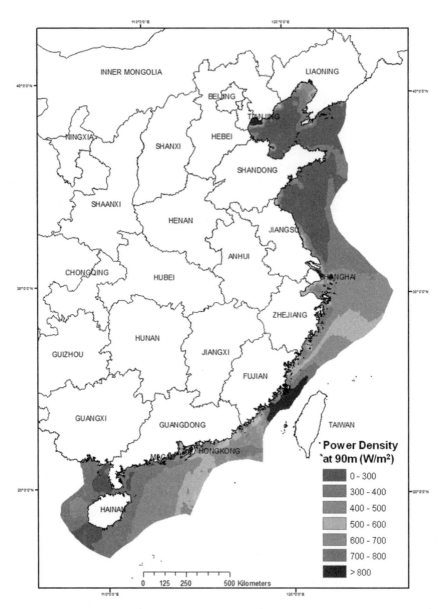

Figure 2 Offshore wind power density within the EEZ of China. Power density is a useful way to evaluate the wind resource available at a potential site, as it is independent of turbine characteristics. The accuracy of wind speed data decreases as it gets close to the coastline; however, it doesn't influence the overall spatial distribution of wind power density within the EEZ.

Table 1 Technical Potential at Different Sea Depths

Sea depth (m)	Technical potential (TWh)	Percentage (%)
0–20	211	12.3
20–50	627	36.5
50–100	521	30.4
>100	356	20.8
Total	1715	100

In addition to the resource condition, actual offshore wind power production is also influenced by the efficiency of a turbine rotor, turbine downtime for schedule maintenance, electrical conversion loss, layout efficiency of a wind farm, and so on. In this study, the technical potential of offshore wind energy generation refers to the maximum theoretical potential, based on overall resource availability and the maximum deployment density of turbines (in maximum water depths and maximum distance to coast), using existing technology or practice. Based on the assumptions of 5-MW offshore wind turbines and wind farm deployment described in the power generation model, the total technical potential of offshore wind energy is calculated to be 1715 TWh/year within the EEZ of China. Table 1 shows the distribution of technical potential in different sea depths. Around half of the technical potential locates in areas with sea depths less than 50 meters.

Spatial Constrained Potential

The spatial constrained potential refers to the fraction of the total technical potential that can be produced once spatial constraints within the extent of the EEZ have been taken into consideration. Due to the availability of spatial data, shipping lanes, submarine cables, bird migration routes, and visibility are so far considered as major spatial constraints for suitable areas of offshore wind development in this study (Figure 3). A reference scenario consisting of four spatial constraints is defined as follows: 1-km buffer for shipping lanes, 500-meter buffer for cables and pipelines, 3-km buffer for bird path, and 8-km buffer from all coasts for visibility. Table 2 compares the technical and spatial constrained potential in different sea depths. Overall, the considered spatial constraints would exclude around 8.7% of the total technical potential of offshore wind energy, leaving 1566 TWh/year available both technically and spatially. However, the impacts of spatial constraints vary greatly across the regions within different sea depths. Spatial constraints have the most severe impacts on shallow waters (below 20-m sea depths), excluding approximately 36.0% of the technical potential of offshore wind.

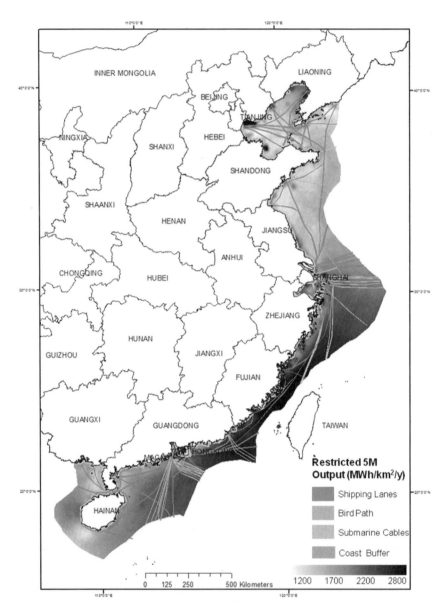

Figure 3 Major spatial constraints for offshore wind farms. This figure outlines a framework for marine spatial planning for offshore wind development. It is useful for planners and developers to identify available locations for offshore wind farms at an early stage. It could also be used for different stakeholders to make discussions and compromises on marine spatial use.

Table 2 Comparison of Technical and Spatial Constrained Potentials at Different Sea Depths

Sea depth (m)	Technical potential		Spatial potential		Excluded
	TWh	%	TWh	%	%
0–20	211	12.3	135	8.6	36.0
20–50	627	36.5	580	37.1	7.5
50–100	521	30.4	503	32.1	3.5
>100	356	20.8	348	22.2	2.2
Total	1715	100	1566	100	8.7

Table 3 Areas of Individual Exclusion (km^2)

Sea depth (m)	Shipping lanes	Submarine cables	Bird path	Visual exclusion
0–20	5,023	998	1765	63,769
20–50	28,976	2245	1455	0
50–100	7,867	3866	3380	0
>100	2,250	1224	0	0
Total	44,116	8333	6619	63,769

The individual exclusion areas are summarized quantitatively in Table 3, which can be used to examine the areas that were excluded for any one use. For example, designated shipping lanes are identified to exclude 44,116 km^2 or 5% of the total EEZ. The submarine cable and bird path would exclude 8333 and 6619 km^2 each, but more than half of them locate in deep sea regions (above 50-m water depths), which are not suitable for building offshore wind farms under current technological conditions. Visual exclusion has an overwhelming impact in shallow waters (below 20-m sea depths), which are considered technologically and economically viable locations for offshore wind farms. Considering the distances of 8 km, visual exclusion areas would reach a percentage as high as 34% of the total shallow waters.

Economic Potential

According to the aforementioned GIS-based cost model, the spatial distribution of levelized production cost (LPC) for offshore wind power is illustrated in Figure 4. The costs of offshore wind energy are highly correlated to sea depths and offshore wind resources. In other words, the lower sea depths and higher wind power density, the less production costs for offshore wind power generation. For example, regions located in the 0- to 50-meter sea depths of Fujian and 0- to 20-meter sea depths of Zhejiang, Shanghai, and partly

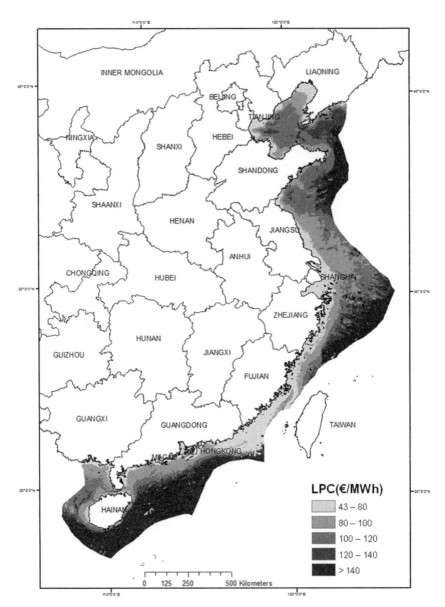

Figure 4 Spatial distribution of levelized production cost. Among a variety of meteorological and geographical parameters, the levelized production cost of offshore wind energy is highly dependent on sea depth. The cost becomes formidably high as a location's sea depth exceeds 50 meters.

Jiangsu are the least expensive areas for developing offshore wind farms, with an average LPC ranging from 43 to 80 €/MWh. The available energy in this category is 360 TWh, approximately 21% of the total technical potential of China. Available offshore wind energy, which costs between 80 and 100 €/MWh, is located mainly within the 0- to 20-meter sea depths of Guangdong and 20- to 50-meter sea depths of Zhejiang and Jiangsu. The potential of this category is 373 TWh, about 22% of the total available energy under the current technological level. Approximately 20% of the total available energy, that is, 348 TWh, costs between 100 and 120 €/MWh. It locates in the 20- to 50-meter sea depths of Jiangsu and southern Guangdong. Northern China, including the provinces around Bohai Rim and deeper sea depths (>50 m) of Zhejiang, Guangdong, and Hainan, have the most expensive offshore wind energy, which costs above 120 €/MWh. The annual amount of this category reaches 634 TWh, about 37% of the total available energy.

ECONOMIC COSTS OF TROPICAL CYCLONE RISKS

China has had an average of nine tropical cyclones striking landfall annually during the period from 1951 to 2008, being one of the most vulnerable countries that suffer from tropical cyclones worldwide. Tropical cyclones[2] are one of the most serious problems related to the economic risks of offshore wind farms globally. The design focus to date has been on offshore wind turbines intended for use in Europe, primarily in the North and Baltic Seas, where there is little risk of tropical cyclones historically. Current design guidelines relevant to offshore wind turbines and other offshore structures have not taken tropical cyclones into consideration. For example, the highest class of offshore wind turbines based on International Electro-technical Commission (IEC) standards has a reference wind speed lower than the actual maximum gust of a typhoon, while the extreme turbulence intensity during tropical cyclones is observed to be higher than the IEC requirement for offshore wind turbines. Therefore, it is necessary to assess possible economic risks of tropical cyclones on offshore wind farms and their impacts on cost–supply curves of offshore wind in tropical cyclone and hurricane prone regions such as the United States, China, Japan, and South Korea. Hurricane "Sandy" that

---

[2] A tropical cyclone is a generic term for a nonfrontal synoptic scale low-pressure system over tropical or subtropical waters with organized convection (i.e., thunderstorm activity) and definite cyclonic surface wind circulation. Depending on its location and strength, a tropical cyclone is referred to by names such as hurricane (in the western Atlantic and eastern Pacific), typhoon (in the western Pacific), and cyclone (in the Indian Ocean and South Pacific).

hit the east coast of the United States in late October 2012 is the case point for providing an offshore wind strategy along with back-up plans.

For offshore wind turbines, the most important external conditions are wind and waves. In the case of high wind speeds (above the cut out speed of a turbine, usually at 25 m/s), the mechanical brake will stop the turbine from rotating in order to reduce the loads. Otherwise, the blades will reach overspeed, creating extreme loads that the structure cannot withstand, eventually causing the blades to bend, be damaged, or collapse. Also, signals from wind vanes and other components of the turbine are sent to the control system, which help reduce extreme and fatigue loads from overspeed and turbulence intensity. The yaw system of the turbine uses electrical motors to turn the nacelle and rotor away from the prominent wind direction as to reduce the load, whereas its pitch mechanism uses hydraulics to control the angle of the blades relative to the wind. Failure of the yaw system can be caused by various reasons: grid failure (in that case it would be dependent on back-up power supply) or failure of the wind vane, which indicates the wind direction and failure in the electric motors controlling the system. Although offshore wind turbines in general are equipped with back-up power, these safety measures are not designed for use for long periods of time, and it is likely impossible to replace the back-up power or restore the grid system in a timely fashion in the event of a tropical cyclone. Therefore, offshore wind turbines may become vulnerable due to their inability to react appropriately to external conditions. Overall, once a tropical cyclone occurs, there could be a high risk of grid failure, which in turn could lead to the inability to adjust and stop the turbine, resulting in overspeed damage to the mechanical and electrical components. In more severe conditions, extreme high loads could cause a collapse of the turbine or a breaking blade might hit and induce its tower to collapse.

Based on a probabilistic tropical cyclone even model and turbine damage model, an annual expected economic loss is calculated for offshore wind farms conditional on occurrence of a 20-, 30-, 50-, and 100-year recurrence interval (Figure 5). Here it is assumed that there is no temporal change in turbine vulnerability and thus the annual expected economic loss is time invariant, considering that the developer is replacing the wearing components of a turbine continuously; in other words, it does not depend on the operation period and so the annual economic risk for year 1 is the same as the annual economic risk for year 20. An annual expected economic loss of a 20-year recurrence interval of tropical cyclones is much more severe than that of a longer recurrence interval due to its higher frequency. In regions of the northern Yangtze estuary, the annual expected economic loss of

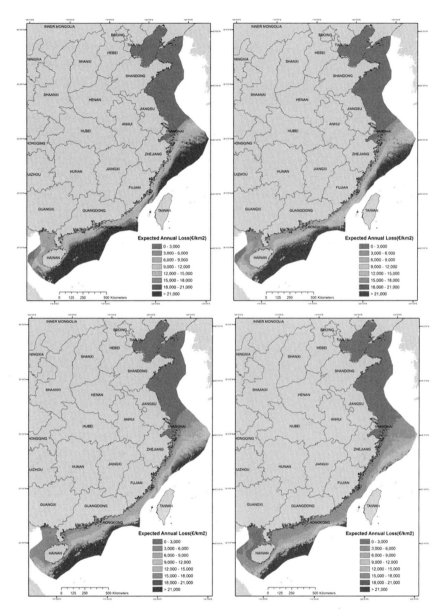

Figure 5 Spatial distribution of annual expected economic loss for offshore wind farms: (top left) a 20-year recurrence interval, (top right) a 30-year recurrence interval, (bottom left) a 50-year recurrence interval, and (bottom right) a 100-year recurrence interval.

tropical cyclones is in the range of 0–3000 €/km²/year. Southern Zhejiang, Fujian, Guangdong, and western Hainan provinces suffer from the most severe economic damages from tropical cyclones, with an annual expected economic loss of 6000–12,000 €/km²/year. Provinces located in the southern coast of the Yangtze estuary (including Shanghai, Zhejiang, Fujian, Guangdong, and Hainan) suffer two to seven times higher annual economic risks from tropical cyclones than that of the northern coast. Figure 6 shows the percentage of expected economic risks of offshore wind farms compared to their initial investment costs within a lifetime of 20 years. In

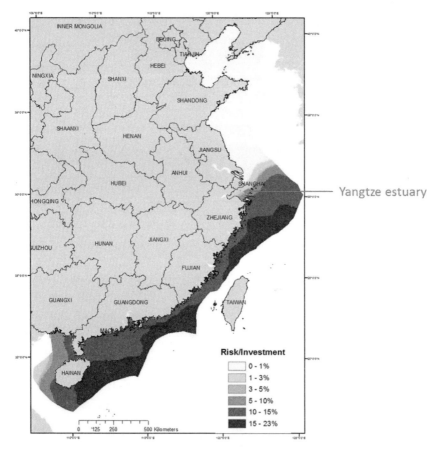

Figure 6 Percentage of economic risk in investment cost under a 20-year recurrence interval cyclone. Regions located north of the Yangtze estuary suffer from little tropical cyclone risks but have lower wind production potentials. In contrast, regions located south of the Yangtze estuary have higher tropical cyclone risks and higher wind production potentials as well.

general, regions located in the northern Yangtze estuary suffer little risks from tropical cyclones, less than 1% of total investment costs within 20 years. However, risks increase sharply to more than 10–15% for most regions of the southern Yangtze estuary.

Total economic losses of a 100-year recurrence interval cyclone are higher than that of a 20-year recurrence interval. However, annual economic losses of a 100-year recurrence interval cyclone are lower than a 20-year recurrence interval cyclone. A 100-year recurrence interval means the probability of this level of tropical cyclone occurring in a location in 1 year is merely 1%, while that for a 20-year recurrence interval cyclone is 5%.

Moreover, regions suffering from higher risks of tropical cyclones are usually endowed with better offshore wind resources. For example, the average wind power densities in regions including Fujian, northern Guangdong, and southern Zhejiang reach as high as 600–800 W/m^2 at a height of 90 meters compared to that of 200–400 W/m^2 in provinces located in the northern Yangtze estuary. Figure 7 shows the spatial distribution of LPC including economic losses of possible tropical cyclones and also reduced wind power productions for offshore wind farms during their lifetime. The least expensive sites for developing offshore wind energy are along the coasts of southern Jiangsu and Shanghai, with an average LPC of 47–80 €/MWh. In the 20- to 50-meter waters of Jiangsu and the 0- to 20-meter waters of Bohai Rim, Zhejiang, and Fujian, an average LPC is in the range of 80–100 €/MWh. As the distance being far away from the coasts, the average LPC of developing offshore wind farms would reach as high as 200 €/MWh.

NATIONAL AND PROVINCIAL COST–SUPPLY CURVES

In promoting wind power, the feed-in system has been used with some variations in Denmark, Germany, and Spain and has proved superior to other methods that have been tried in the European Union for promoting green electricity when evaluated in terms of installed electricity from renewable energy sources(RES-E) capacity (Meyer 2003). A nonhydro renewable is generally stimulated by various policies, including feed-in tariffs (FIT) and a renewable energy portfolio standard for grid and power companies in China (Buijs 2011). According to China's Renewable Energy Law and related price regulations, the FIT has been applied for wind, photovoltaic, and biomass, while prices for more nascent technologies such as offshore wind, concentrated solar thermal, and geothermal are determined through a public tender process in order to reduce their high generation costs. Offshore wind energy

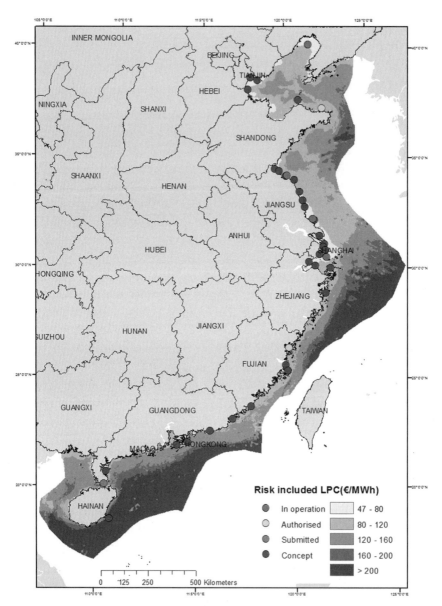

Figure 7 Spatial distribution of tropical cyclone risk included LPC for offshore wind farms. This figure considers both economic risks of tropical cyclones and wind production potentials in one location for the calculation of levelized production costs.

developers, mainly power companies combined with a wind turbine manufacturer and a construction and installation enterprise, are invited to bid for the development of a location. A low electricity price and a high equipment localization rate are two main factors used to decide the winning bidder. Advantages of this concession model include the combination of functions of government and power companies, the selection of suitable developers, and lower wind power prices (Han et al. 2009). Disadvantages include that the procedure excludes private and international companies and tends to produce extremely low bids. Most bidders for offshore concession projects are large state-owned power companies, whose motivation for low bids comes from the government's order to raise the proportion of renewable energy generation in their generation mix. Under the country's long-term renewable energy development strategy, which was launched by the National Development and Reform Commission (NDRC) in 2007, power companies that have more than a 5-GW power-generating capacity should have at least 3% of total installed capacity from nonhydroelectric renewable sources by 2010 and 8% by 2020. In order to win the concession project, some bidders underestimate investment and operation costs intentionally to get a lower price compared to other bidders. Once the bid is selected, it proves economically impossible to construct and operate the offshore wind farm.

Experience from the Shanghai Donghai Bridge suggests a price of 0.978 RMB/kWh (106 €/MWh) in 2009. According to the United Nations Framework Convention on Climate Changereport, the internal return rate of the project without certified emission reduction revenue is below the 8% benchmark even under the on-grid tariff (including VAT) of 1.1406 RMB/kWh (124 €/MWh)(United Nations Framework Convention on Climate Change 2008). In the first round of 1-GW offshore concession projects, the bid-winning prices are 0.6235 RMB/kWh (66.9 €/MWh) for the 200-MW Dongtai intertidal wind farm, 0.6396 RMB/kWh (68.6 €/MWh) for the 200-MW Dafeng intertidal wind farm, 0.737 RMB/kWh (79.1 €/MWh) for the 300-MW Binhai offshore wind farm, and 0.7047 RMB/kWh (75.6 €/MWh) for the 300-MW Sheyang offshore wind farm, respectively. These prices are much lower than offshore wind tariffs in other countries (Table 4), even approaching benchmark prices for inland wind farms.[3] The result of insufficient financial resources is the long-term delay of off-

---

[3] The NDRC issued notice on improving the price policy for wind power in July 2009. This establishes the principles for formulating the benchmark price for land-based wind power based on different resource areas, dividing the country into four categories of wind energy resource areas. The resulting four benchmark grid tariffs are correspondingly 0.51, 0.54, 0.58, and 0.61 RMB/kWh.

Table 4 Summary of Worldwide Offshore Wind Tariffs[a]

Jurisdiction	Years	€/MWh
Germany with Sprinter Bonus	20	190
Spain (maximum)	20	177
Germany	20	150
Ireland	15	140
France	15	130
Greece	20	97
Denmark (maximum)	20	83

[a]*Source:* Gipe (2012).

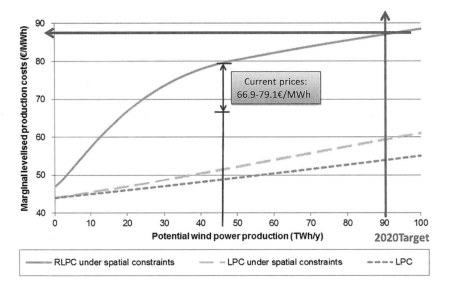

Figure 8 Comparison of marginal levelized production costs under spatial constraints and tropical cyclone risks. In order to achieve the 2020 offshore target, the marginal levelized production cost of offshore wind under the current technological level is estimated to be 87 €/MWh, much higher than the current bidding prices of 66.9–79.1 €/MWh.

shore wind farm constructions. The national 2020 target of 30 GW installed offshore capacity equals to approximately 90 TWh of electricity generation. Given the fact that spatial constraints and tropical cyclones have an influential impact on economically available offshore wind potential in China, Figure 8 shows that the increases of marginal levelized production costs for the 30-GW target caused by spatial constraints and tropical cyclone risk under a 20-year recurrence probability are 7 and 34 €/MWh, respectively. Moreover, Figure 9 indicates that the average marginal levelized production costs

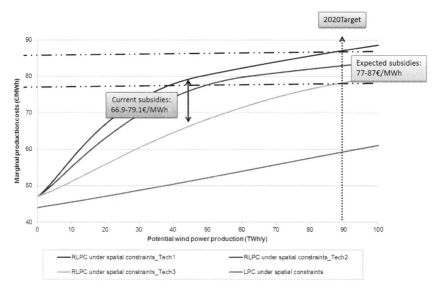

Figure 9 Comparison of marginal levelized production costs under different tropical cyclone-resistant technologies in the short term. Even if considering technological improvement in the aspect of resisting tropical cyclone risks, expected marginal levelized production costs of offshore wind are estimated in the range of 77–87 €/MWh.

for reaching this target are expected to be in the 77- to 87-€/MWh range under a tropical cyclone risk and spatial constraints even considering different levels of technological improvement and learning curves by 2020. It further implies that merely 40–70% of the national target in 2020 can be achieved under the present low bid price caused by the current pricing mechanism.

The main coastal provinces and municipalities interested in developing offshore wind energy in the near future include Shandong, Jiangsu, Shanghai, Zhejiang, and Fujian, where an estimated aggregate installed capacity of offshore wind energy will reach 22.8 GW (intertidal offshore 5.1 GW and near offshore 17.7 GW) in 2020 (Wang 2010). Currently, China's installed offshore capacity has reached 258.4 MW, which ranks it number three in the world after the United Kingdom (2362 MW) and Denmark (857 MW). The Shanghai Donghai Bridge project, totaling 102 MW installed in June 2010, was the first commercial offshore project outside Europe. By the end of 2011, the largest offshore project in China was the 150-MW demonstration project in Rudong, Jiangsu, with 99.3 MW installed and grid connected. Other installations are all small-scale demonstration projects, including the second phase of the 65.6-MW Shanghai Donghai Bridge project, with 8.6 MW installed in 2011. As a mechanism to achieve its offshore goals, China employs a concession tendering model, in which both developers and tariffs are determined by a tender. The first concession tendering of four projects totaling 1000 MW was finished in September 2010, but none of these projects has moved beyond the planning stage after 2 years, primarily due to planning and siting difficulties. The nation's second round concession tendering of 2000 MW for offshore wind, which was supposed to take place in 2011, has also been delayed currently at the end of 2012.

Offshore wind plans in various coastal provinces are at different stages (Figure 10). Generally speaking, Shanghai and Jiangsu have been the pioneers of developing offshore wind energy in China. The offshore wind plan of Shanghai has already been approved by the NEA. It proposes to develop offshore wind farms first in Donghai Bridge, Fengxian, and Nanhui with an

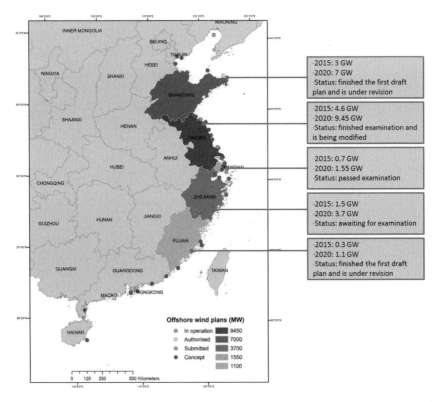

Figure 10 Provincial plans for offshore wind energy. Dots represent locations of planned offshore wind farms, and colors represent the total offshore target of that province by 2020.

installed capacity of 600 MW by 2015 (Shanghai investigation, design and research institute 2011). A final target of 6 GW offshore wind energy may be distributed to eight available locations of Shanghai by 2030 (Shanghai investigation, design and research institute 2011). Although Jiangsu's offshore wind plan remains to be examined, four offshore concession projects with a total installed capacity of 1 GW have already been fully commissioned. Two are offshore wind farms, located in Binhai and Sheyang, with an installed capacity of 300 MW each. Another two intertidal projects are located in Dongtai and Dafeng, sizing 200 MW each. Other provincial plans are at the initial stage, and the spatial distribution of offshore wind projects by the development stage in coastal provinces is shown in Figure 10.

Assessment of offshore wind potentials, their location associated costs, and economic risks of tropical cyclones under the existing technological level in each coastal province by means of the GIS model described are summarized in Table 5. Fujian is endowed with the highest wind density of 600–1300 W/m^2, but its total technical potential of offshore wind energy is much lower than that of Zhejiang and Guangdong. Even though the largest technical potential of offshore wind energy is located in Guangdong, the spatial constraints exclude as high as 14.5% of its exploitable potential. In addition, the economic risks of tropical cyclones are rather high in both

Table 5 Assessment of Offshore Wind Potential, Cost, and Risk in Coastal Provinces

Item	Shandong	Jiangsu	Shanghai	Zhejiang	Fujian	Guangdong
Wind density (W/m^2)	200–300	200–400	400–500	400–700	600–1300	400–700
Full-loaded hours (h)	2456	2604	3149	3301	3751	3303
2020 target (TWh)	19.6	28.35	4.96	12.95	4.4	—
Technical potential (TWh)	174	169	70	313	153	405
Spatial exclusion (%)	12.0	11.2	13.4	10.2	15.6	14.5
Risk/investment (%)	0	1	1–3	3–15	10–15	10–23
RLPC (€/MWh)	47–200	47–160	47–200	80–200	80–160	120–200

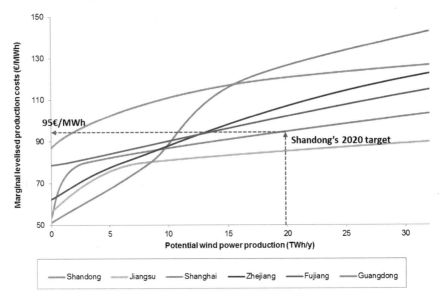

Figure 11 Marginal levelized production costs in coastal provinces. The provincial cost–supply curves of offshore wind energy could facilitate local governments to establish local feed-in tariffs. For example, the marginal levelized production cost is 95 €/MWh in Shandong in order to achieve its 2020 target.

Fujian and Guangdong, causing a much higher RLPC of 80–200 €/MWh compared to other provinces such as Jiangsu and Shanghai with less annual full-loaded hours and a lower risk of tropical cyclones.

The cost–supply curves under spatial constraints and tropical cyclone risk for each coastal province are displayed in Figure 11. Based on the annual full-loaded hour of each province calculated from the GIS model, the power generation is expected to be 19.6, 28.35, 4.96, 12.95, and 4.4 TWh in Shandong, Jiangsu, Shanghai, Zhejiang, and Fujian under their respective target by 2020. It shows that the marginal levelized production costs for reaching their respective target are approximately 95, 88, 67, 90, and 80 €/MWh in Shandong, Jiangsu, Shanghai, Zhejiang, and Fujian. In the short term, Shanghai is the most economically competitive location for developing offshore wind farms under spatial constraints and tropical cyclone risk. The marginal levelized production costs of Shanghai are the lowest among all the coastal provinces before the cumulative offshore wind power production reaching 8 TWh. However, costs will increase tremendously as more cumulative electricity is generated from offshore wind energy. The marginal levelized production costs of Shanghai exceed that of Jiangsu after the first

8 TWh of offshore wind generation and further increase to be higher than costs of Shandong and Zhejiang after the cumulative 10 TWh of generation. In the long term, Jiangsu is suitable for the large-scale development of offshore wind energy in terms of resource, cost, and tropical cyclone risk considerations.

CONCLUSIONS

A bottom-up GIS-based offshore model was applied to assess the technical potential of offshore wind energy under a multiple of spatial constraints and its location-specific economic costs. The total technical potential in China's EEZ is estimated to be 1715 TWh/year under the current technical level of offshore wind farms and turbines. However, spatial constraints, including designated shipping lanes, submarine cables, and bird migratory paths, as well as visual exclusion, would eliminate approximately 8.7% of the total technical potential, having the most significant impacts on shallower waters (below 20-m sea depths). Considering only the technical potential locates in areas with sea depths less than 50 meters to be economical, the available economic potential is merely 715 TWh with an average levelized production cost ranging from 0.419 to 0.975 RMB/kWh (43–100 €/MWh). Through a set of sensitivity analyses, it has been demonstrated that the GIS model may also provide a comprehensive framework for policy makers and investors to understand and strategize for the available offshore potential under different technical, spatial, and economic scenarios now and in the future; the GIS tool and analytical results are also flexible, upgradable, and updatable as more comprehensive and accurate data sets become accessible.

This study also pioneered the assessment of the economic risks of tropical cyclones on offshore wind farms through a GIS-based probabilistic tropical cyclone event model. It concluded that as much as 80% of the total offshore potential suffers from a certain degree of tropical cyclone risk, increasing the marginal levelized production costs of offshore wind by at least 0.293 RMB/kWh (30 €/MWh). Regions located north of the Yangtze estuary face little risk from tropical cyclones, equivalent to less than 1% of total investment costs. In contrast, risks increase sharply to more than 10–15% of total investment costs for most regions south of the Yangtze estuary. As a result, there is a potentially huge market demand for tropical cyclone-resistant wind turbine technologies. Currently, Minyang Wind Power of China has developed a tropical cyclone-resistant wind turbine with a single-unit capacity of 1.5 MW. Thirty-three of this type of turbine have been installed in the Yangqian wind farm of the Guangdong province and have succeeded in withstanding 15 tropical cyclones since July 2009. However, the quality of the turbine needs to be tested over a longer time period, and the safety standards of this kind of wind turbine need to be improved further prior to their utilization in locations with higher tropical cyclone risks. What's more, cost–supply curves indicate that much of the near future potential can be built with standard turbines and that the economic advantage of cyclone-resistant turbines materializes only after these potentials are utilized. This may give time to develop cyclone-resistant turbines.

Based on cost–supply curves derived from the tropical cyclone risk-included GIS-based model, the marginal levelized production cost of reaching China's 2020 offshore goal is estimated to be in the range of 0.75–0.85 RMB/kWh (77–87 €/MWh), depending on the progress of tropical cyclone-resistant turbine technology development. It implies that only 40–70% of the national target can be achieved by 2020 under the current low level of bidding prices. The marginal levelized production costs of reaching provincial targets have been investigated as well, reaching approximately 0.93 RMB/kWh (95 €/MWh) in Shangdong, 0.86 RMB/kWh (88 €/MWh) in Jiangsu, 0.65 RMB/kWh (67 €/MWh) in Shanghai, 0.88 RMB/kWh (90 €/MWh) in Zhejiang, and 0.78 RMB/kWh (80 €/MWh) in Fujian under spatial constraints and tropical cyclone risks. This analysis could serve as a basis for central and local governments to make offshore wind development plans and to set efficient incentive mechanisms to promote the implementation of plans.

REFERENCES

Buijs, B., 2011. Why China matters. Energy, sustainability and the environment: technology, incentives, behavior. Elsevier Inc., Burlington, pp. 445–476.

CMA-STI Best Track Dataset for Tropical Cyclones in the Western North Pacific, 2010. Shanghai Typhoon Institute, China Meteorological Administration. Available from: < http://www.typhoon.gov.cn. >

Cockerill, T.T., Kühn, M., Bussel, G.J.W., Bierbooms, W., Harrison, R., 2001. Combined technical and economic evaluation of the northern European offshore wind resource. J. Wind Eng. Ind. Aerodynamics 89, 689–711.

Danish Energy Agency, 2007. Future offshore wind power sites–2025.

Energy and Environmental Management Magazine, 2012. EAEM guide to the UK offshore wind industry. Available from: < http://viewer.zmags.com/publication/4e37c3e8#/4e37c3e8/1. >

Fridley, D., Khanna, N., Hong, L., 2012. Review of China's low-carbon city initiative and developments in the coal industry, China Energy Group, Environmental Energy Technologies Division. Lawrence Berkeley National Laboratory.

Gipe, P., 2012. Snapshot of feed-in tariffs around the world in 2011.
Han, J., Mol, A.P.J., Lu, Y., Zhang, L., 2009. Onshore wind power development in China: challenges behind a successful story. Energy Policy 37 (8), 2914–2951.
Heinberg, R., Fridley, D., 2010. The end of cheap coal. Nature 468, 367–369.
Jacquemin, J., Butterworth, D., Garret, C., Baldock, N., Henderson, A., 2009. Inventory of location specific wind energy cost. Garrad Hassan & Partners Ltd.
Kline, D., Heimiller, D., Cowlin, S., 2008. A GIS method for developing wind supply curves. National Renewable Energy Laboratory. Technical Report NREL/TP-670-43053.
Liu, K.S., Chan, J.C.L., 2010. Size of tropical cyclones as inferred from ERS-1 and ERS-2 data.
Meyer, N.I., 2003. European schemes for promoting renewables in liberalised markets. Energy Policy 31 (7), 665–676.
Möller, B., 2011. Continuous spatial modelling to analyse planning and economic consequences of offshore wind energy. Energy Policy 39 (2), 511–517.
Shanghai investigation, design and research institute, 2011. Offshore wind planning of Shanghai. 2011. Available from: < http://wenku.baidu.com/view/a2477e21aaea998fcc 220e9b.html?from=related. >
United Nations Framework Convention on Climate Change, 2008. Shanghai Donghai bridge offshore wind farm project.
Wang, M., 2010. Introduction of China's offshore wind planning. Offshore wind China 2010 conference and exhibition. China, Shanghai.
Zhou, X., Yi, J., Song, R., Yang, X., Li, Y., Tang, H., 2010. An overview of power transmission systems in China. Energy 35, 4302–4312.

CHAPTER 5

Sustainable Planning of Open Urban Areas in Developing Countries
A Lesson from a Case Study in Tel Aviv, Israel

Tali Hatuka, Hadas Saaroni
Geography and the Human Environment Department, Tel Aviv University, Ramat Aviv Tel Aviv, Israel

Contents

What is the Challenge? Quick Urbanization and Climate Change in Developing Countries	80
A lesson from the Design of Jaffa Slope Park: Contradictions and Gaps	84
Toward Establishing Design Codes for Outdoor Urban Spaces in Developing Countries	89
The Future of Publicness and its Meaning	91

Urban planning has long been challenged to address sustainability and, recently, the impact of climate change on cities. The task of increasing cities' resilience is multifaceted. One facet, unfortunately not a prominent one, is the fate of outdoor public spaces (e.g., streets, squares, parks). Urban microclimates and outdoor thermal comfort are generally given little importance in urban design and planning processes, especially in developing countries. To date, a growing body of research addresses the relationship between urban planning regulations and local microclimates (e.g., Erell, Pearlmutter & Williamson 2011). This relationship presents a compelling field of study given how modern society values outdoor public spaces as important assets, with a vast body of work addressing the significant social roles of these spaces as realms of contact and exchange among strangers. Socially, as a *common territory* that facilitates and regulates interpersonal relationships (Sennet 1976), outdoor spaces, and public spaces in particular, are where individuals display idealized selves adhering to (or challenging) cultural belief and behavior patterns. Contemporary criteria for defining the degree to which a space is public suggest that access, interest, and agency are the key dimensions of that space's social organization, with space being offered as an asset in exchanges. However, in this era of global warming, it can be argued that criteria for determining the degree to which a space is public must include environmental considerations if we are to support human comfort and the social role of public space.

Advocating design codes that consider human comfort in outdoor public spaces is relevant to cities worldwide, even more so for developing or underdeveloped countries (Farlex Financial Dictionary 2012). And yet, developing design codes is not a straightforward task. Although the threat of climate change is global, its implications are regional. Moreover, these changes' implications are subject to cultural, social, and ethnic differences, rendering a global manual of "how to design public outdoor spaces" improbable. We suggest expanding the research on interrelations among climate change, cities, culture, and the way climate change influences participants' thermal, emotional, and perceptual well-being in public spaces as a key step in developing contextual design codes for outdoor public spaces. We also suggest focusing on the developing world where outdoor spaces are extremely vulnerable and available studies are scarce. The remainder of this chapter elaborates on our argument, beginning with a brief discussion of climate conditions and change in developing countries. The subsequent section presents some lessons from a contemporary development of an urban park in Jaffa, Israel.

Using this case to illuminate some of the challenges in developing open urban areas in cities, this chapter suggests a general framework to address climate and environmental challenges in developing countries. The chapter concludes that in an era of global warming, design codes for outdoor public spaces are key means of supporting human comfort and social life.

WHAT IS THE CHALLENGE? QUICK URBANIZATION AND CLIMATE CHANGE IN DEVELOPING COUNTRIES

The past several decades have ushered in a growing number of pessimistic reports on megacities in developing countries. With an increasing number of people occupying squatter settlements in cities of the Global South, many states, nongovernmental organizations(NGOs), and academic consultants have returned to using language that presents slums as dirty, diseased, criminal, and depraved threats to society. Three well-known facts are often presented on slums in megacities: (a) around a billion people live in slum cities, mainly in the so-called Global South; (b) migration is increasing and is projected to double the slum population by 2030; and (c) current planning policies are often ineffective and slow to respond.

Why is this the case? In part, migration to cities is a dynamic phenomenon that is difficult to predict. Policy and urban development ensconced in rigid frameworks has difficulty tackling the large-scale, dynamic nature of the phenomenon. Moreover, the formal practices of urban development strategies tend to be abstract and often gloss over megacities' heterogeneity and informal natures. In response, increasing research on the phenomena surrounding migration could be categorized as Marxist, pragmatic, and humanistic in perspective. Focusing on spatial inequalities and housing problems, the Marxist perspective (Davis 2006) addresses slums as physical and spatial manifestations of urban poverty and intracity inequalities arising from the contemporary political economy. Advocates see slum residents as a potential positive force of resistance that may be able to change the contemporary economical framework. The pragmatic perspective, as taken by states, the World Bank, NGOs, and the United Nations (UN), is actively engaged with the question of what might be done and how. Because slums are perceived as a threat to the order of the state, national approaches to slums, and to informal settlements in particular, include forced eviction, benign neglect, and involuntary resettlement, as well as more positive policies, such as self-help and in situ upgrading, enabling, and rights-based policies (UN-Habitat 2008).

Exploring both the negative and the positive aspects of slums, the UN reports on intolerable conditions of urban housing, high concentrations of poverty, socioeconomic deprivation, and water-borne diseases; the UN also recognizes slums as entry points for immigrants, making for vibrant mixing of different cultures. The latter suggests how the humanistic perspective, developed as early as the 1970s, focuses on social networks of the poor as a means of survival. Instead of decrying the social disorganization of the poor, scholars addressed the ability of the poor to cope with extreme social and economical structural inequalities (e.g., Neuwirth 2005). Some of these writings address the black market of unregistered businesses and home construction as a spontaneous, creative, and popular response to the state's incapacity to spur production and distribution. Such informal systems are perceived as a way to resolve underdevelopment through reducing the state and expanding free enterprise (De Soto 1989).

This is not all. Beyond social challenges, megacities also face challenges from climates and climate change. Most developing countries lie in regions characterized by extreme climate conditions and severe weather events, including the tropic regions characterized by hot and humid conditions and the arid areas in subtropical regions. Climate significantly affects populations in tropical regions exposed to severe heat stress conditions from high temperatures and year-round humidity. Other difficulties include extreme rain events associated with floods, as well as the destructive effects of tropical storms, hurricanes, and typhoons. The monsoon system, particularly the Asian monsoon, is typical of some tropic regions and affects more than half of humanity worldwide, causing intensive summer rains and floods, as well as large interannual weather variability and severe droughts. Thus, developing countries are highly vulnerable to extreme weather conditions and high interannual and intra-annual variability in both tropic and arid regions. These extreme conditions and variations directly impact everyday life, water availability (both upper and ground), agricultural potential, and living and housing confidence. Stating it differently, these climate conditions significantly influence the daily life of a population highly dependent on the natural environment. Relatively low living standards, undeveloped industrial bases, and a low human development index increase this dependency further.

The vast, intensive process of urbanization enhances the difficulties of climate change, exposing populations to higher heat stress due to urban heat island (UHI) effects). Although UHI intensities in tropical and subtropical cities are generally lower than those of temperate cities with comparable populations (Roth 2007), noticeable UHIs are found in cities located in both tropic and arid regions. For example, a heat island intensity of up to 8.3°C was detected in Delhi, India, with the highest magnitude during the afternoon and night hours (Mohan et al. 2012). Moreover, the UHI shows a seasonal variation with lower (higher) intensities during the wet (dry) season (Roth 2007), indicating its larger effect in the dry season, which is characterized by the highest temperatures.

Global and regional warming is well documented (IPCC 2007), indicating that climate change is worldwide, including the low latitudes where most developing countries lie. Beyond the warming, tropical regions are becoming wetter, and the occurrence of severe rain events and floods are increasing. Conversely, the arid regions are drying, with increases in droughts and water shortages (IPCC 2007). Climatic models are predicting a continuation of the warming trend and intensification of the hydrological cycle that will cause more severe rain events in the tropics and further drying on the margins of arid regions due to the poleward shift of the subtropical high as part of the Hadley cell expansion (Seidel et al. 2008). The rapid urbanization of cities in developing countries is associated with sharp intensifications of the UHI in these areas, being on top of the regional warming in these warm areas that suffer from severe heat stress conditions. For example, the UHI of Beer Sheba, Israel, a city located in an arid region, increased by 2–3°C beyond the regional warming, which is already greater than the global rate, since the late 1970s (Saaroni & Ziv 2010).

The Stern review on the economics of climate change (Stern 2010) estimated that a 2°C rise in global temperature costs approximately 1% of the world's gross domestic product (GDP). The World Bank, in its World Development Report (World Bank 2012), estimated that the cost to Africa and India will be closer to 4 and 5% of the GDP, respectively, indicating the serious implications and higher vulnerability of developing countries to climate change. These countries' rapid urbanization, the absence of climate-adopted planning, and the lack of green construction and sustainable development further increase the severe effects of climate change on the population of fast-growing cities.

However, these challenges are not restricted only to the developed world. Israel, for example, as a young country, which was established in 1948, focused on growth and urban development over the last 6 decades, giving little attention to environmental aspects. It was only recently that planners have recognized the significance in developing sustainable urban spaces. One of the interesting cases in addressing this goal is Jaffa Slope Park (Figures 1 and 2), which is located on the eastern coast of the Mediterranean Sea, one of the key regions of high anthropogenic climate impact (IPCC 2007; Lionello 2012).

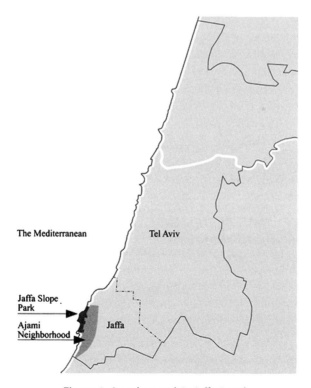

Figure 1 An urban park in Jaffa, Israel.

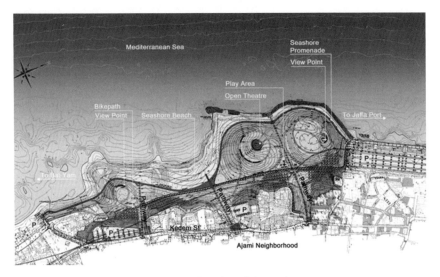

Figure 2 The plan of the park.

The park, a 50-acre (200 dunam) waterfront area, was established with public funding and resources rather than private donors, and it is regulated by the Tel Aviv-Jaffa city council. Analyzing the park's design, the basic hypothesis was that an increase in discomfort conditions, resulting from aggravation of heat stress, heat waves, and water shortages, would lead planners to respond comprehensively to this change, thus resulting in increased attention to human comfort conditions in the process of the park's design (Hatuka & Saaroni 2012). And yet, things are much more complex—a complexity that can illuminate the need to develop a better integration between scientific knowledge and planning practices.

A LESSON FROM THE DESIGN OF JAFFA SLOPE PARK: CONTRADICTIONS AND GAPS

Jaffa Slope Park, which served until 2004 as a construction debris waste site, was transformed into an urban park in 2010 and is now a popular place among residents and visitors alike. Located along the Mediterranean coast, the park is bounded by Jaffa port to the north, the Aliya Hill beach to the south, and the mixed (predominantly Arab) neighborhood Ajami to the east (Figures 3 and 4). In terms of design, the park consists of three major man-made hills, with three main axes drawn from the neighborhood down to the seafront, as well as a stretch of beach currently being rehabilitated, as shown in Figure 5.

Figure 3 The construction debris waste site and the Ajami neighborhood.

Figure 4 The new park: a view of the bike lanes.

The still incomplete park contains about 19.8 acres (80 dunams) of planned grass lawns, comprising 40% of the total 50 acres (200 dunams) of the park. The remaining area includes concrete paved roads and bike lanes, a playground, workout facilities, an amphitheater, and a fishing pier. The

Figure 5 A stretch of beach currently being rehabilitated at Jaffa Slope Park.

park's program and design should be seen as part of the Tel Aviv municipality's larger environmental and social vision, which aims to create a continuous waterfront boardwalk for pedestrians and cyclists stretching along the metropolitan area. The park represents a progressive approach both to the public (i.e., the planning included comprehensive public participation) and to the environment (i.e., construction included a large-scale recycling project; 1.275 tons of waste was recycled out of a total of 1.35 tons dug, with 200,000 m³ reused on-site).

Climatically, the area of the park is characterized by high temperatures and relative humidity during the summer, resulting in heat stress conditions along the entire day and night hours. Discomfort conditions are further aggravated by the UHI effect that has been documented in Tel Aviv (e.g., Saaroni et al. 2000). Climatic studies of the Mediterranean Basin over the last three decades point to a significant warming trend, suggesting that this area is warming faster than the global rate. Moreover, the highest warming trend is in the summer season, which is associated with the aggravation of heat waves and heat stress conditions. This has a severe environmental impact on human comfort, as well as on the hydrological cycle and water shortages (e.g., Ziv & Saaroni 2010). Climate models for the 21st century forecast intense warming, together with decreased precipitation over the area (e.g., IPCC 2007; Lionello 2012), that will further aggravate discomfort conditions and water shortages. Thus, from the perspective of environmentalists in general and climatologists in particular, there is an urgent need to incorporate climate considerations into planning and design practices in this area.

From the perspective of users and cultural norms, it is important to note that although summer in Israel is associated with heat waves and heat stress conditions (Harpaz et al. 2013), the use of outdoor space does not necessarily decline. With 14 hours of daylight, most outdoor activities take place during early mornings and late afternoons, which is why it is important to adjust public spaces climatically. This is particularly relevant to urban parks along the seashore, which, unlike parks within the city, attract users throughout the day (although less at midday).

Examining both the conceived space (as designed by planners) and the lived space (as experienced by residents), the strong emphasis given by planners and residents alike to the image of the place and its aesthetics has affected the design of the park as a whole. Among the various competing factors that influenced the design (i.e., social, environmental, and planning), climate considerations were perceived as one parameter to address among many, but not as a leading factor that might influence significant decisions in the design process.

As a result, Jaffa Slope Park, largely considered to be an environmentally friendly and social project, raises contradictory issues. Socially, despite the extensive and drawn-out public participation process, many of the facilities requested by the public were not integrated into the planning and design of the park. Environmentally, much of the building debris and waste were recycled on-site and used to construct the park, therefore increasing sustainability, but the final product is a water-wasting landscape where there is minimal shading and vast lawns, and presumably discomfort conditions prevail during the warm months (at least 6 months a year) in the coastal Mediterranean city. Still, these issues, along with the acknowledgment of the park's limitations, do not detract from its appreciation in terms of design and aesthetics. This case is one example out of many urban parks that have not incorporated climatic considerations into planning ontology and methodology. Even if the vast lawns found in public parks, and specifically recently built beach-front park in Tel Aviv, serve the general public, there are still measures that could be considered during the planning process in order to increase human comfort and sustainability, as shown in Figure 6 of a couple using the park and its surroundings.

So how can we explain this growing phenomenon of a dominant grass landscape in contemporary park planning, which seems to contradict accepted knowledge and governmental campaigns regarding environmentalism, climatic conditions, and the warming and drying trend?

Figure 6 A view of Jaffa Slope Park.

How can we explain extensive grass lawns in times of severe water shortage and national restrictions of water use, the lack of shade, and the park's sparse image, contradicting human comfort? The answer to this is not definitive and we suggest an explanation that is tied to the contemporary decentralization of planning processes in the neoliberal context. This context creates a competing situation on three interrelated aspects: (1) the actors, who decide on (2) the array of parameters, which influence the process of planning and (3) which in turn dictate the language of design. In other words, the array and interests of actors participating in the neoliberal contemporary planning process cause a shift. The planner becomes a mediator among a vast array of social and political interests being advocated during a public participation process, a method that gives power to the civil society and suspends scientific knowledge in favor of local wills.

Thus, in the process of planning, environmental and climate issues are often perceived as neutral, obvious, or barriers that attract significant parts of the budget and do not serve the immediate goals of community or political actors. Following the aforementioned, in a competition among social use, aesthetic considerations in the design process, and environmental and climate considerations, the first two parameters have gained higher priority. This influences the language of design and, in the Jaffa case, grass is used as an inexpensive solution that provides an immediate image. Grass is also a solution attached worldwide to cultivation; thus, culturally, it is associated with beauty and aesthetics, which have crucial values in raising property rates and rebranding an area. As such, transforming the park into a visual asset becomes an interest of both the municipality and the designers, who see it as the most significant component in the success of such a project.

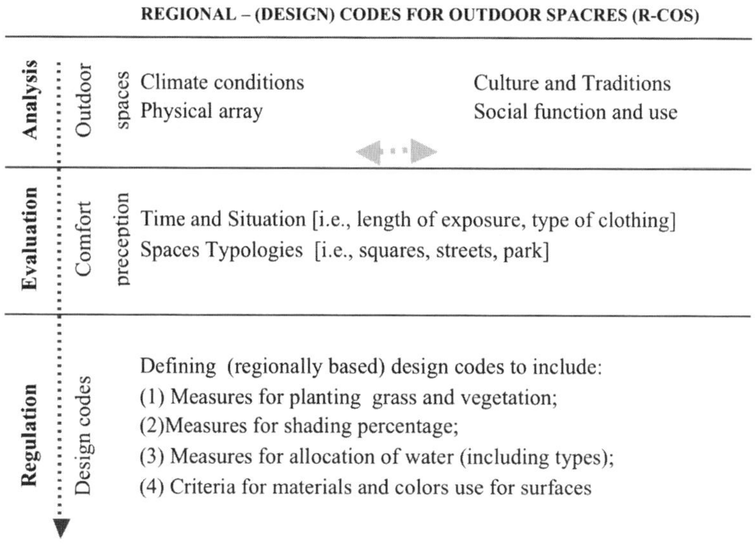

REGIONAL – (DESIGN) CODES FOR OUTDOOR SPACRES (R-COS)

Analysis — Outdoor spaces:
Climate conditions
Physical array
Culture and Traditions
Social function and use

Evaluation — Comfort preception:
Time and Situation [i.e., length of exposure, type of clothing]
Spaces Typologies [i.e., squares, streets, park]

Regulation — Design codes:
Defining (regionally based) design codes to include:
(1) Measures for planting grass and vegetation;
(2) Measures for shading percentage;
(3) Measures for allocation of water (including types);
(4) Criteria for materials and colors use for surfaces

Figure 7 Design codes for outdoor urban spaces.

While this case study explored a park in Israel, the conclusions and constraints of integrating go far beyond its geographical borders, raising the need for advocating human comfort design codes for public open spaces worldwide. How can we achieve this task while addressing the particular physical conditions and social dynamics of a place? How can environmental and climate considerations be integrated better into planning processes and what are the benefits of such integration? Figure 7 illustrates a possible path for such integration.

TOWARD ESTABLISHING DESIGN CODES FOR OUTDOOR URBAN SPACES IN DEVELOPING COUNTRIES

We suggest focusing on residents' daily lives and developing a microscale planning strategy that would enhance interventions in outdoor spaces with a focus on thermal comfort conditions and livability. The goal is to develop regional design codes for outdoor spaces to enrich residents' everyday life while managing thermal comfort conditions. This relatively modest strategy, thinking locally to impact globally, could be implemented quickly with relatively basic resources. Using this strategy frequently would assist greatly in tackling climate conditions and climate change.

Generally, when defining design codes, we suggest using a conceptual model: *regional codes for outdoor spaces* (R-COS), see Figure 7, which includes three key layers—analytic, evaluative, and regulative. The analytic layer suggests that the study of outdoor spaces entails an investigation of the relationships between the physical and the cultural. Acknowledging the interrelations between climactic/physical conditions and cultural/social norms and traditions is crucial, as noted by Knez and Thorsson (2006, 2008): "Different geographical/climatic zones can also be defined as different cultures…A member of a culture learns its rules and regulations, which she/he then shares with other members of the 'system.'" The perceived world is a social and psychological construct and significantly influences place-related identity processes and an environmental attitude toward outdoor spaces. Furthermore, as shown by Knez and Thorsson (2006, 2008), thermal, emotional, and perceptual assessments of a physical place may intertwine with psychological and cultural processes rather than remain fixed by general thermal indices.

The analytic step is followed by the evaluative step. This most crucial step focuses on estimating social groups' perception of comfort within a region or city. In this step, one should consider two parameters carefully: (1) length of stay and (2) space typologies; one should divide analytic results accordingly. For example, streets or urban corridors might have different functions and meanings than squares or open courts near religious buildings. Furthermore, length of stay and exposure can vary from one place to another, although they might be adjacent. These differences may affect perceptions of comfort.

Finally, based on comfort perception, the regulative step includes the development of design guidelines for outdoor spaces focusing on four key related aspects: *shading, vegetation, water use,* and *materials.* These four aspects have a direct effect on thermal comfort, defined as satisfaction with the thermal environment. The comfort sensation determined by the body's energy balance is an essential factor in open areas in regions known to suffer from discomfort conditions. Thus, it is clear that shading, vegetation, and materials have crucial impacts on the thermal comfort. Codes would incorporate percentages of shading, percentages of the area covered by water-intensive vegetation, and other factors, as well as a framework for implementation. As noted, this type of list should consider the conditions of the locale and should be based regionally. For example, the material and color of a particular surface, and the use of small artificial lakes, could have different impacts on different locales.

THE FUTURE OF PUBLICNESS AND ITS MEANING

Yahia (2007) describes the outdoor spaces of (planned) Damascus:

> The regulations for modern Damascus prescribe wide streets and pavements, large setbacks and relatively low building heights. This leads to a dispersed urban form where a large part of the buildings and streets are exposed to solar radiation. The existing planning regulations in Damascus have no requirements for shading for pedestrians, e.g., shading devices, arcades and projecting upper floors or shading trees.

Yet Damascus is one city among many suffering from discomfort climate conditions. Yahia's (2007) urging to establish regulations for improving comfort conditions is the right call. A city's outdoor spaces are its veins as well as the veins of a society's culture. In an era of global warming, of drying, and of increased water shortages, these types of codes in developing countries provide a feasible framework for supporting human comfort and sustainability. The underlying principle here is that design codes for outdoor urban spaces are not a luxury but a real step toward achieving socially just development.

Finally, the vulnerability of our daily life, as well as our behavior in outdoor spaces, should be seen in association with the gap between scientific knowledge (theory) and local landscape planning (practice) (e.g., Eliasson 2000). Although there is growing scientific research on the interrelationships between design and climate conditions from the individual unit to the city as a whole, most of these studies focus on analyses of buildings from a climatologic perspective, focusing on the effect of architectural and structural design features (e.g., layout, window orientation, shading, and ventilation) on human thermal comfort in indoor environments, and on the effects of city design and the spatial array (i.e., density, building height, street geometry) on the intensity and spatial distribution of the UHI that affects human thermal comfort in outdoor environments. However, while these studies have deeply explored relationships between the built environment and climates, it seems there are limited studies on microclimate open spaces in developing countries and even fewer on the impact of microclimate open spaces on social dynamics. In other words, we know very little about the spaces between buildings, the very places that we all inhabit and use every day.

REFERENCES

Davis, M., 2006. Planet of slums. Verso, New York.
De Soto, H., 1989. The other path: the invisible revolution in the third world. IB Taurus, London.
Eliasson, I., 2000. The use of climate knowledge in urban planning. Landscape Urban Plan. vol. 48, 31–44.

Erell, E., Pearlmutter, D., Williamson, T., 2011. Urban microclimate: designing the spaces between buildings. Earthscan, London/Washington DC.

Farlex Financial Dictionary, 2011. Financial definition of less-developed country, TheFreeDictionary.com. Available from: < http://financial-dictionary.thefreedictionary.com/less-developed+country, 2011. >. (15.09.12.).

Harpaz, T., Ziv, B., Saaroni, H., Beja, E., 2013. Extreme summer temperatures in the East Mediterranean - Dynamical analysis. Int. J. Climatol., http://dx.doi.org/10.1002/joc.3727 (in press).

Hatuka, T., Saaroni, H., 2012. The need for advocating regional human comfort design codes for public spaces: a case study of a Mediterranean urban park. Landscape Res.

Intergovernmental Panel on Climate Change (IPPC), 2007. The physical science basis, summary for policymakers (contribution of WG I to the 4th Assessment Report of the IPCC). Cambridge University Press, Cambridge/New York. Available from: < http://ipcc-wg1.ucar.edu/ >.

Knez, I., Thorsson, S., 2008. Thermal, emotional and perceptual evaluations of a park: cross-cultural and environmental attitude comparisons. Building Environ. 43 (9), 1483–1490.

Knez, I., Thorsson, S., 2006. Influences of culture and environmental attitude on thermal, emotional and perceptual evaluations of a public square. Int. J. Biometeorol. 50, 258–268.

Lionello, P. (Ed.), 2012. The climate of the Mediterranean region: from the past to the future. Elsevier, London.

Mohan, M., Kikegawa, Y., Gurjar, B., Bhati, S., Kandya, A., Ogawa, K., 2012. Urban heat island assessment for a tropical urban air shed in India. Atmospheric Climate Sci. 2 (2), 127–138.

Neuwirth, R., 2005. Shadow cities: a billion squatters, a new urban world. Routledge, New York.

Roth, M., 2007. Review of urban climate research in (sub)- tropical regions. Int. J. Climatol. 27 (14), 1859–1873.

Saaroni, H., Ben-Dor, E., Bitan, A., Potchter, O., 2000. Spatial distribution and microscale characteristics of the urban heat island in Tel-Aviv, Israel. Landscape Urban Plan. 48, 1–18.

Saaroni, H., Ziv, B., 2010. Isolating the urban heat island contribution in a complex terrain and its application for an arid city. J. Appl. Meteor. Climatol. 49, 2159–2166.

Seidel, D.J., Fu, Q., Randel, W.J., Reichler, T.J., 2008. Widening of the tropical belt in a changing climate. Nat. GeoScience 1, 21–24.

Sennet, R., 1976. The fall of the public man. Cambridge University Press, Cambridge.

Stern, N., 2010. Stern review on the economics of climate change. HM Treasury, Cambridge.

UN-Habitat, 2008. State of the world's cities 2010/2011: bridging the urban divide, UN-Habitat. Available from: http://www.unhabitat.org/content.asp?cid=8051&catid=7&typeid=46.

World Bank, 2012. World development report 2010: development and climate change. Available from: < http://wdronline.worldbank.org/worldbank/a/c.html/world_development_report_2010/abstract/WB.978-0-8213-7987-5.abstract. > (15.09.12.).

Yahia, M.W., 2007. Microclimate and thermal comfort of urban spaces in hot dry Damascus, influence of urban design and planning regulations. Ph.D thesis, Housing Development & Management Lund University, Sweden.

Ziv, B., Saaroni, H., 2010. The contribution of moisture to heat stress in a period of global warming: the case of the Mediterranean. Climatic Change 104 (2), 305–315.

CHAPTER 6

India: Issues for Sustainable Growth/Innovation for Sustainability

Namrita S. Heyden
University of Chapel Hill, NC, USA

Contents

Introduction	93
Society and Governance	94
Business/Corporate Role	96
Trends in Energy Use	99
Conventional Sources of Energy	100
Coal	100
Oil	100
Natural Gas	101
Renewable Sources of Energy	101
Wind Energy	101
Biomass	102
Hydropower	102
Nuclear Power Generation	102
Future Trends	103
Conclusion	103

INTRODUCTION

India's economy is developing rapidly along with its population. The growing economy and the growth in population are increasing the demand for energy and creating enormous pressure on the natural resources. The emerging technological revolution is giving rise to a middle class that is emulating Western behavior in consumerism (where more is better), foregoing the old Indian values of moderation and conservation. The cities are struggling with congestion, pollution, lack of basic services, and an increasing rich–poor divide.

    Prevalent socioeconomic issues such as illiteracy and poverty, along with the extreme diversity of cultures and languages and the presence of far-flung remote habitats, make it difficult for the government to create a comprehensive sustainable plan for the country. Moreover, the government is riddled with corruption, which makes regulation and enforcement of any plan more difficult.

But there's hope for India, as can be seen in grass root revolutions in sustainability and a culture of innovation emerging with hundreds of energy entrepreneurs. There are hundreds of success stories of rural areas using traditional methods to conserve and recycle natural resources. Many energy entrepreneurs have made it possible for millions of people in India to gain access to electricity through solar and wind energy. As more and more people are becoming aware of the dangers facing India in terms of climate change and environmental degradation, even businesses, along with government intervention, are playing important roles in bringing about green innovations.

These green innovations are necessary not just to solve the environmental problems in the country, but also as a way to bridge the divide between the "globally competitive India and India of the poor with acute inequalities and inefficiencies" (Innovation management). This chapter outlines the role of end users, communities, entrepreneurs, and the government in encouraging green innovations, some of the existing socioeconomic issues that encumber the success of such innovations, and potential for change and innovation in India.

SOCIETY AND GOVERNANCE

In the last two decades, India has witnessed rapid economic growth, which has changed the social structure of the country, resulting in a new "middle class." This emerging class has had a big influence in transforming India into one of the fastest-growing consumer markets in the world. This rapid economic growth has not only increased the purchasing power in urban and rural areas, but also changed the consumption habits of consumers. Traditionally, the country had a culture of need base consumption, but increased urbanization and influences of westernization are influencing consumer behavior, leading to unsustainable levels of consumption. Consequently, unsustainable levels of production and consumption are putting enormous pressures on the carrying capacity of the natural resource system, as well as the ability of the environment to absorb the waste by-products (Kapur Bakshi, 2012).

In emerging markets such as India, accommodating such rapid changes becomes a problem because of social and economic issues that are generally prevalent in such an economy. For example, a low adult literacy rate is one of the important factors inhibiting sustainable production and consumption in the country. The adult literacy rate in India for 2001–2004 was 61% (Kumar et al., 2011). A low level of education makes it difficult to influence consumption behavior because of the inability of individuals and households to understand the linkages between unsustainable levels of consumption and environmental degradation. Another example is the increasing social and economic divide between the rich and the poor. Rapid growth and urbanization have led to unequal income distribution, thus creating a stark contrast between the rich, with their conspicuous consumption habits, and the poor, living in slums and rural areas, where their livelihoods are threatened as they are unable to keep up with rapid technological innovations (Kapur Bakshi 2012).

The increasing levels of consumption in India provide a huge opportunity for the production and consumption of green goods. Although there is a rising movement in this direction, there are still important constraints in the path toward creating an efficient system of production and consumption of green goods. Some of the important limitations on the supply side are the lower level of research and development and technology, which leads to problems of availability and acquisition of green raw materials and technology, high cost of production of green goods, and lack of infrastructure to ensure access to consumers (Kumar et al., 2011). On the demand side, India still lacks a legal framework to incentivize the consumption of green goods or control the consumption of unsustainable goods.

Given the existing socioeconomic structure and the regulatory framework in India, it is important to promote sustainable consumption through intervention by the local government, nongovernmental organizations (NGOs), and local advocacy groups. Intervention mechanisms would include capacity building, aggressive policy measures, legislation, fiscal mechanisms, public investments, and public awareness campaigns (Martens & Spaargaren 2005).

Following a bottoms-up approach, local government capacity building should be emphasized through a provision of information, financial capacity building, and linkages with other local stakeholders, such as advocacy groups. This is important, as local governments are more attuned to the issues, needs, opportunities, and challenges of the population. Due to the extreme diversity in culture, norms, and traditions in different parts of the country, the role of local governments and grassroots networks becomes very important in influencing production and consumption behavior.

For example, if local governments include sustainability measures in their public procurement strategies, it will not only provide a boost to the production of green goods, but also generate reliability for such kinds of goods. Some progress has been made in India in terms of local governments

pushing sustainability measures through policy framework, fiscal incentives, and regulation. Some examples (Kapur Bakshi, 2012) are given here.
- The 11th 5-year plan introduced the Solar Cities Program of the Ministry of New and Renewable Energy that targeted 60 cities. The objective was to promote energy efficiency and renewable energy by driving local energy innovation and investment through cities.
- The city of Thane has committed to a 10% energy reduction over 5 years, promoting solar energy utilization at the same time, in an effort to become the first solar city of India.
- Thane Municipal Corporation introduced a 10% reduction in property tax for home owners who install solar water heating systems.
- Local governments are encouraging modern designs and systems to utilize biogas, derived from animal wastes and other biomass, as a cost-effective and climate-friendly renewable energy source in rural areas.

Apart from direct policy, legislation, and fiscal measures, local governments also play an important role in bringing together various stakeholders with different resources and skills to complement and support government efforts. For example, local governments can bring together important economic stakeholders such as businesses and small entrepreneurs, who can help in encouraging sustainable production and consumption through technology development, resource creation, and innovation, as well as incorporating sustainable practices in corporate strategies.

BUSINESS/CORPORATE ROLE

The businesses in a community in any country enjoy significant financial and intellectual assets and networks to be able to influence a change in society. Since the 1990s, the Indian economy has seen a rapid growth in its role in the global business environment. In recent years, an explosion of growth in the information technology sector has spurred economic progress and provided the Indian business sector with a global edge. The growing Indian business sector, composed of multinational companies, traditional Indian businesses, and small-scale entrepreneurs, has an important role to play in steering the country onto the path of sustainability.

There is an increasing global awareness in the business sector toward incorporating "corporate social responsibility" (CSR) in the overall strategy of businesses, with sustainability holding an important place in the overall CSR framework. Traditionally, the concept of social good has always been a part of the business community in India, although it took form mainly as a part of philanthropy, and the concept of CSR being part of a strategic business operation is a relatively new concept.

A survey conducted by KPMG in 2011, states that, only 16% of the top 100 listed firms in India reported having a corporate responsibility strategy in place. In contrast, 73% of the world's 250 largest companies have defined objectives related to corporate responsibility. However, 23% of top 100 Indian firms reported on the business risk of climate change, while 26% reported on the business opportunities related to it.

A number of recent initiatives signify that the Indian business community is taking a proactive role in incorporating sustainability goals in its business strategy. One of the more important changes has been adoption of the National Voluntary Guidelines for Social, Environmental and Economic Responsibilities of Business by the Ministry of Corporate Affairs in India. According to the drafting committee, sustainability is a significant factor in these guidelines. In another important step, "a parliamentary panel has sought a policy that makes it mandatory for banks, major public sector units, and companies in the public and private sectors to invest 50% of their CSR funds in afforestation initiatives " (Hindustan Times, 2010). The Confederation of Indian Industry (CII) has launched a green building rating system to encourage the design and construction of energy efficient buildings and to enable India to be one of the leaders in green buildings by 2015. CII is also involved in other initiatives geared towards building a responsible business leadership environment in India (CII, 2012).

Despite the Indian business sector's unprecedented success in the global arena and the effects of globalization making a positive economic impact on business growth, the intrinsic nature of India's problems sometimes makes it difficult to evaluate the net positive value of businesses and innovation. One such interesting example is introduction of the world's smallest and most affordable car, Tata Nano, in 2008 in India. The car was designed specifically such that it could be afforded by the millions of people in India, who could never dream of owning a car. With a fuel efficiency of approximately 50 mpg, such a car should have been considered as contributing toward economic and environmental sustainability. Although, interestingly, many leading environmentalists pointed out that even with high fuel efficiency, the fact that the car can be afforded by millions of people undermined its environmental friendly efficiency. Another factor that alarmed environmentalists was that the Indian infrastructure did not have the capacity to accommodate so many vehicles.

As the business sector in India is slowly becoming aware of its stake in promoting sustainability initiatives in India, small entrepreneurs, especially in rural communities, are using sustainability as a driver for innovation to protect and conserve natural resources and to provide real sustainable value to their communities. There are hundreds of examples of how small entrepreneurs are using their knowledge, skills, and a culture of innovation to make their lives more sustainable. Some examples (Green features, 2008) are given here.

- In Kerala, Biotech has developed an innovative biogas plant that runs on food waste. These plants are installed in indoor municipal fish markets, where they run on fish waste to produce enough power to light the indoor market. These plants not only produce electricity but also solve hygiene problems by eliminating the need to dispose of waste.
- In northern India, farmers are using treadle pumps to grow crops all year round. These pumps are inexpensive, simpler, affordable, and more environmentally friendly than diesel pumps. The NGO that developed them (NGO IDEI) has now formed a thriving network of energy entrepreneurs, including manufacturers, retailers, and installers, resulting in sales of over half a million pumps in total.
- People in rural India still rely on kerosene for household lighting. Matthew Scott and Amit Chugh, an Indian entrepreneur, came up with the idea of "Mighty Light." The innovation uses a combination of next-generation, light-emitting diodelighting and solar panels to provide a durable and portable light source. This light source lasts at least 100,000 hours. These light sources are being used by fishermen in south India for nighttime fishing.
- In 1995, Tulsi Tanti started Suzlon, which manufactures wind turbines and is now the *world's fifth largest manufacturer* of wind energy. He started Suzlon because a lack of reliable power supplies was hindering his family textile business.

India is currently the world's fourth largest economy in terms of real GDP. In recent years, India has emerged to be an important player in the global economic sector. With a rapidly growing economy, a significant pool of manpower, an abundance of entrepreneurial capabilities, and intellectual resources, India is positioned to be a leading innovator in the field of sustainability.

However, the pace of change in awareness is still slow in India, and the business sector needs to pool all of its resources toward making a difference in the community in which they operate and in their new risk management strategies that reflect the socioenvironmental aspects of their business operations. Businesses need to understand that their role is crucial in bringing the change toward a sustainable future.

TRENDS IN ENERGY USE

Energy is an important input to fuel a country's economic development. Especially for a developing country such as India, continued and reliable sources of energy are needed to meet the demands of a growing economic sector, as well as of a burgeoning population. According to the Energy Information Administration(EIA), in 2009, India ranked fifth in the world in terms of primary energy consumption. Despite having one of the fastest growing rates of energy consumption in the world, India's per capita energy consumption is still very low compared to other developing countries.

According to India's planning commission, to accommodate the growing population and economy, India needs to increase its primary energy use by four times over the next two decades.

Considering that more than half of the population in India doesn't even have access to electricity, and yet the country is the sixth largest consumer of energy in the world, India's future energy use choices will be critical in determining the country's commitment to sustainability issues.

With insufficient domestic energy sources and the existing supply insufficient to meet the increasing demand, India imports a large proportion of its energy requirements. Figure 1 shows a breakdown of the country's energy consumption by type in 2007.

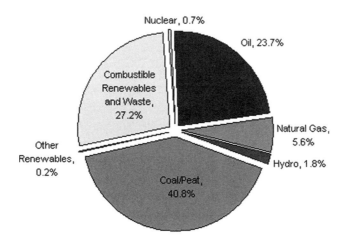

Figure 1 Total energy consumption in India, by type (2007). *Source: International Energy Agency (IEA).*

Conventional Sources of Energy

Coal

India is well endowed with exhaustible energy sources such as coal, which therefore accounts for 55% of the country's energy supply. India ranks third among the largest coal-producing as well as coal-consuming countries in the world. Due to a heavy reliance on coal as the primary energy source for so many years, India has started to face coal shortages and is now a net importer of coal. Despite shortages, coal continues to be a very important energy source in India, with 70% of power being generated by coal, and core infrastructure sectors such as power, steel, and cement still being heavily dependent on coal.

Oil

Currently, India has approximately 0.4% of the world's proven crude oil reserves. According to the EIA, India's consumption of crude oil has increased consistently; in 2009, India became the fourth largest consumer of crude oil, consuming nearly 3 billion barrels/day. Increasing oil consumption and limited production have resulted in India importing most of its crude oil requirements. In 2009, India was the sixth largest importer of crude oil, importing nearly 2.1 billion barrels/day, or about 70%, of its oil needs. According to the EIA, it is expected to become the fourth largest importer by 2025 (Figure 2).

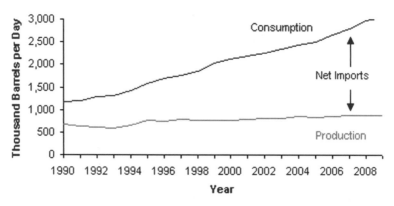

Figure 2 India's oil production and consumption 1990–2009 (2008–09 is forecast). *Source: U.S. Energy Information Administration.*

Natural Gas

With India's energy policy focusing on diversification and security, natural gas is expected to become an important energy source in the coming decade. The country's consumption of natural gas has been growing at a rate of about 6.5% during the last decade. The demand for natural gas has been increasing rapidly in the power sector over the last few years, due primarily to environmental considerations of using coal and its supply constraints. Industries such as fertilizer and petrochemical production are also shifting toward natural gas. India has very productive natural gas fields, especially in the western offshore regions, yet the demand has outstripped supply and India has been a net importer of natural gas since the year 2004 (Figure 3).

Renewable Sources of Energy

With a rapidly increasing population and a high rate of economic growth, India's growing reliance on traditional energy sources is a growing cause of concern due to increasing dependence on limited fossil fuel reserves and the environmental implications of their use. Renewable energy sources offer a viable option to address the energy security concerns of India. According to the India Energy Portal, "the country has one of the highest potentials for the effective use of renewable energy." The following are the statuses of renewable energy sources in India and their potential uses.

Wind Energy

Among all the renewable energy sources, perhaps wind energy is the most important and the most developed in India. At present, India is the fifth largest producer of wind power in the world. As of 2006, the total installed capacity

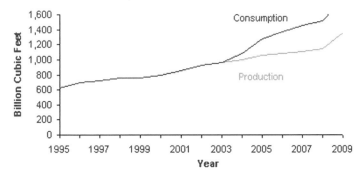

Figure 3 India's dry natural gas production and consumption 1995–2099. *Source: U.S. Energy Information Administration.*

of wind energy production in the country was 5340 MW, 46% of the total capacity of renewable resource-based power generation. According to the Centre for Wind Energy Technology, the total wind potential is estimated to be around 45,000 MW. With continuous research and development, extensions and improvements in the grid, and advances in wind turbine technology, the potential is expected to increase in the future.

Biomass

Biomass has always been an important source of energy in India, especially in rural areas. Even though the dependence on coal and oil as the primary energy source has been increasing, still 70% of the population of India relies on biomass for its primary energy needs. At present, biomass is used to meet approximately 30% of the country's total primary energy needs. India's agricultural sector is the primary producer of biomass material, some of which is used for fodder and fuel in the rural economy. According to estimates, 30 to 40% of the biomass produced is not used for energy use and has a potential of generating around 15,000 to 25,000 MW of electrical power. A more efficient system is needed to ensure that biomass is used more productively to generate power.

Hydropower

Currently, hydropower accounts for 26% of total power generation in India. It is estimated that the hydropower generated currently represents only 23% of the total potential existing in the country. To meet its target of providing reliable access to electricity by 2012 to all its citizens, the government of India has initiated several hydroelectric power projects in the country with financial support of the World Bank. India is aiming to reach an optimum ratio of 40% hydropower and 60% coal-based energy for its total power generation.

Nuclear Power Generation

India continues to focus on development of its nuclear power generation capacity to provide for the country's increasing power needs. According to the deal established in 2005 with the United States, India will increase its installed nuclear power generation capacity. Currently, the target is to increase the nuclear capacity to 40,000 MW by the year 2020. India currently has 14 nuclear reactors in commercial operation with 10 reactors recently purchased from France and Russia, which will add 11,000 MW of electric capacity to the country.

Future Trends

The Energy and Resources Institute of India conducted a 2-year study commissioned by the government that analyzed linkages between technological progress and availability of energy resources in the future. According to the study, if the country aims to meet its goals of a sustained growth rate of 8% in the next two decades, it will need aggressive policies and interventions both on the demand and on the supply side to ensure efficient utilization of resources. Some of the key areas emphasized in the report are hydro-based power generation, energy-efficient technologies in end-use sectors such as transport and residential, and efforts in research and development in the areas of deep-sea natural gas exploration. Renewable energy-based power generation is only emphasized to provide electricity in remote areas where the grid system cannot be extended. This will serve the dual purpose of reducing the demand for coal-based energy as well as making clean reliable energy to such regions.

CONCLUSION

Emerging economies such as India inevitably bring rapid transformation and a host of other issues, including increasing pressures on the natural ecosystem of the country. However, there is a huge potential for the country to strengthen sustainable production and consumption habits at the grass root level. The long tradition of innovation and entrepreneurship in India can play an important role in bringing about such a grass root level, but the intervention of the government in terms of policy measures and regulations is very important in integrating the efforts of diverse stakeholders in the country.

REFERENCES

Green features, 2008. Monsoons and miracles; India's search for a sustainable future.
Innovation Management, Unleashing innovations for sustainability: an Indian perspective. Retrieved from http://www.innovationmanagement.se/2010/06/09/unleashing-innovations-for-sustainability-an-indian-perspective/.
Kapur Bakshi, S., 2012. Role of local governments in fostering the transition to sustainable lifestyles and livelihoods and improved well-being. Proceedings: Global Research Forum on Sustainable Consumption and Production Workshop, Rio de Janiero, Brazil.
Kumar, D., Goyal, P., Rahman, Z., and Kumar, I. 2011. Sustainable consumption in India: challenges and opportunities. *International Journal of Management & Business Studies*, 1.3, 28-31.

Martens, S., Spaargaren, G., 2005. The politics of sustainable consumption: the case of the Netherlands. Sustainability: Sci. Pract. Policy 1 (1).

National Energy Map for India; technology Vision 2030, 2006. The Energy and Resources Institute. TERI Press.

PTI, 2010. 50% of CSR funds for afforestation. *Hindustan Times*. Retrieved from http://www.hindustantime.com.

Sathaye, J., Shukla, P.R., Ravindranath, N.H., 2006. Climate change, sustainable development and India: global and national concerns. Curr. Sci. 90 (3), 10.

The Confederation of Indian Industry (CII). Retrieved from: http://www.cii.com on October 20, 2013.

U.S. Energy Information Administration. Retrieved from http://www.eia.gov/countries/cab.cfm?fips=IN.

CHAPTER 7

Germany's Energiewende: Community-Driven Since the 1970s

Craig Morris
Director of Petite Planète, Krozingerstr., Freiburg, Germany

Contents

A Fledgling Movement from the 1970s	106
The Energiewende Today	109
Community Ownership Today	110
Conclusions	112

German Chancellor Angela Merkel turned a lot of heads both inside and outside Germany when she reversed her position on nuclear power in the wake of the Fukushima accident in March 2011. Only 7 months prior to the earthquake in Japan, Merkel's coalition had extended the service lives of the country's nuclear fleet by up to 14 years,[1] essentially extending the shutdown of the last nuclear plant from 2022 to the 2030s.

Then, in 2011, Merkel essentially returned the country to her predecessor Gerhard Schröder's original nuclear phase-out target, so the last nuclear plant once again gets shut down in 2022. But there are two major differences: (1) 8 of the country's 17 nuclear plants were switched off within a week and (2) dates are set for the remaining plants, with no wiggle room left for the firms (Schröder's original plan allowed plant operators to shift kilowatt-hour allotments from one plant to another so they could switch one off earlier and another one later).

Merkel's about-face has led many onlookers to associate the Energiewende with this second sudden nuclear phase out, but the original nuclear phase out

---

[1] In an official statement entitled "German Nuclear Power Plants Are Safe," Merkel's coalition defended the extension of nuclear plant commissions by saying that "An explosion or fire—like the one that happened in Chernobyl—is not possible" in any German nuclear plant (http://www.bundestag.de/presse/hib/2010_08/2010_265/02.html).

was implemented a decade earlier, and one could go even further back to three major policies to support renewables:

- The Feed-in Act of 1991, which gave generators of nuclear power the right to connect to the grid and guaranteed them payment as a share of the retail rate (such as 80% of the retail rate for wind power)
- The Renewable Energy Act (EEG) of 2000, which unlinked the calculation for these payments (called "feed-in tariffs") from the retail rate; now, the cost of a reference system (a wind turbine, say, or a biomass unit of a particular size) was estimated, and a small profit margin of 5–7% was added on to ensure that not only the least expensive source of renewable power (onshore wind) was built.
- The amendment of the EEG in 2004, which made photovoltaics eligible for feed-in tariffs without restriction for the first time.

Like the sudden phase out in 2011, the amendment in 2004 was spectacular and has therefore drawn the most attention, thereby skewing our understanding. While the feed-in tariffs for wind power were always below the retail rate and those for electricity from biomass were generally near it, roughly half a euro was paid for photovoltaics at the time—around three times the average retail rate in Germany at the time.

These feed-in tariffs for photovoltaics led people to conflate the policy with exorbitant subsidies,[2] and many prominent analysts have charged that feed-in tariffs can be done away with once grid parity (the point where power from a solar array costs the same as power from a wall socket) has been reached.[3] Clearly, however, such arguments overlook the origin of feed-in tariffs for all renewables, not just photovoltaics. With the exception of photovoltaics, German feed-in tariffs have always been below the retail rate and, as seen here, they now all are—even for solar.

One should think of feed-in tariffs not as a startup mechanism for the most expensive types of renewable energy, but rather as a way of protecting small investors in competition with corporations as a way of turning citizens into power producers.

A FLEDGLING MOVEMENT FROM THE 1970S

If we go back to the 1970s, we discover a widespread movement across all of Germany. In that decade, people of all walks of life came together

---

[2] See the author's rebuttal of an article in *Forbes* at http://www.renewablesinternational.net/a-spectacular-fail-by-command-economists/150/537/61648/.

[3] See the author's response to an article by Navigant's Paula Mints in 2011: 'Feed-in tariffs needed after grid parity' at http://www.renewableenergyworld.com/rea/news/article/2011/02/feed-in-tariffs-needed-after-grid-parity.

in "Energiewende groups" to discuss how to overcome corporate control of the power sector—and hence, of their communities—by making energy themselves. Pastors, farmers, and scientists all came together to squat on the land where plants were to be built. The government responded with authoritarian force, which only increased the number of protesters until the police were overwhelmed. In 1980, three of the researchers involved in the movement published a book entitled (in German) *Energiewende: Growth and Prosperity Without Petroleum and Uranium.* The book stresses that distributed energy—including biomass and coal in small cogeneration units—will be "hard for large corporations to monopolize."[4]

The movement focused initially on opposition to two nuclear plants, one in the south (Wyhl, which was never built) and the other in the north (Brokdorf, which was completed). From the outset, the Energiewende was thus linked to opposition to nuclear, but that opposition focused less on safety issues and more on who should get to decide what; specifically, should corporations and the government be able to tell communities they will have to accept construction of new industry? It was a democratic ground-roots uprising against authoritarian technocrats or, as one of the authors of the book from 1980 recently told the author in an interview, "our work did away with the claimed monopoly on technological competence that the energy industry used to assert: that there were no technically feasible alternatives to its ideas."

The movement led to a number of changes still visible today. For instance, German Environmental Minister Peter Altmaier approved carbon capture and storage (CCS) in 2012 contingent upon the consent of nearby communities. The international press reported the news as a compromise that would step up CCS "on a test basis,"[5] but in fact Altmaier himself saw the decision as the death knoll for CCS: "We cannot store carbon dioxide underground against the will of the population. And I do not see any political acceptance in a single German state for CCS."[6]

The experience has been the same with conventional coal power. As German environmental nongovernmental organization Klima-Allianz explains, there are no federal laws that limit coal plant construction in

---

[4] See the author's review of the book at http://energytransition.de/2013/03/time-for-a-coal-phase-out/.

[5] See *Reuters* http://articles.chicagotribune.com/2012-06-27/news/sns-rt-germany-energyco2l6e8h-rhgl-20120627_1_emissions-ccs-bundesrat.

[6] See http://www.renewablesinternational.net/germany-rejects-new-coal/150/537/39718/.

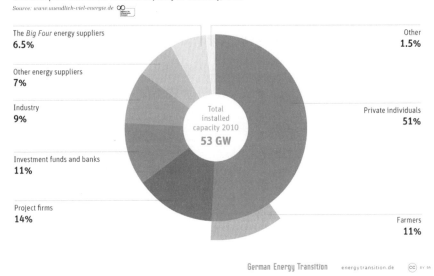

Figure 1 German citizens are behind most of the investments in new renewables, with the Big Four utilities only having made up less than 7% of the pie by the end of 2010.

Germany, but local community governments have the power to influence—and even reject—new coal plant projects.[7] Similar restrictions will apply when Germany attempts to select a site for its final nuclear waste repository.

Still, opposition to conventional energy monopolies will only get one so far—what should this electricity be replaced with? In the late 1980s, the first feed-in tariffs were implemented in three German towns, and when the policy slipped into the last day of the Bundestag's legislative session in 1991, the conventional energy sector did not think a little bit of renewables would hurt. Even in 1994, when Germany had 3% hydropower, they still claimed that renewables could never make up more than 4% of supply[8] (see Figure 1).

The rest, as they say, is history. The law led to a significant ground-roots movement in wind power and, to a lesser extent, in biomass. Most projects were community owned and "organic," as a representative of DEWI, the

---

[7] See http://www.die-klima-allianz.de/keine-neuen-kohlekraftwerke/hintergrund/.
[8] See "We've come a long way, baby" at http://energytransition.de/2013/03/weve-come-a-long-way-baby/.

German organization that tallies wind statistics, once told the author, who had noticed around 2005 that the American Wind Energy Association listed wind farms by size and owner, but DEWI did not. When the author called to ask why, they said ownership is too splintered in Germany, with dozens or even hundreds of local businesses and citizens having purchased shares of projects.[9]

Furthermore, wind farms in Germany start off with a small number of turbines, the DEWI representative explained. A community may put up a couple of turbines, and when they run well, those who were skeptical may realize what a good investment the project was and get on board for another round of turbines a few years later. The result is diverse ownership and wind farms that look like a museum of different technologies.

THE ENERGIEWENDE TODAY

By June 30, 2010, all European Union(EU) member states were required to publish their National Renewable Energy Action Plans (NREAP) for 2020.[10] Germany's current targets for its energy transition up to that point are essentially the same as those in the NREAP—further evidence for the Energiewende is not related to the sudden nuclear phase out of 2011 as shown in Figure 2 on German energy transition. For instance, the country aimed to have 38.6% of its electricity from renewables by the end of this decade in its NREAP. That figure has now been rounded down to 35%.

Whereas the original Energiewende book from 1980 actually considers all three sectors of energy use (electricity, heat, and motor fuel), the current energy transition has been criticized for being merely an "electricity transition." Nonetheless, Germany is well poised to reduce the demand for heat. In the early 1990s, the German-speaking world brought about Passive House, a style of architecture with ventilation heat recovery that allows homes even in the climate north of the Alps to make do with small backup heaters. Process heat can also come increasingly from waste heat, which will improve the overall efficiency of energy consumption.

Moreover, the heat sector is increasingly being seen as a way of storing excess renewable power. Denmark, which has a much more ambitious target of 100% renewable energy (not just electricity), is banking on electric

---

[9] See "Renewables from the bottom up" at http://archive.truthout.org/article/craig-morris-renewables-from-bottom-up.

[10] See http://ec.europa.eu/energy/renewables/transparency_platform/doc/dir_2009_0028_action_plan_germany.zip

German energy transition: high certainty with long-term targets

Long-term, comprehensive energy and climate targets set by the German government
Source: BMU

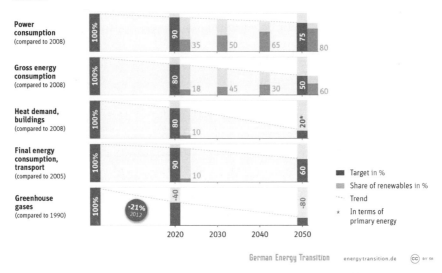

Figure 2 As this chart shows, Germany does not have a target of 100% renewables by 2050 in any area. In fact, its most ambitious target is for the electricity sector, with power consumption to drop by 25 percentage points and renewables to make up 80% of that lower demand. In contrast, the targets for heat and transport fuel are much more modest. *Source:* www.energytransition.de.

heating systems, which will cut on when power production exceeds power demand, thereby offsetting natural gas and oil used for heating purposes.[11] The main shortcomings are found in the transport sector. Germany has no specific roadmap for transitioning from motor fuels to some kind of renewable supply.

COMMUNITY OWNERSHIP TODAY

In all likelihood, Germany will therefore reach its targets for renewable power and heat, and it will continue to do so with great citizen participation and input. The result has been a marginalization of conventional power companies that have seen their profits and market shares shrink—an outcome that confuses foreign onlookers greatly.

---

[11] See "Electric heaters: a way of storing excess renewable power?" at http://www.renewablesinternational.net/electric-heaters-a-way-of-storing-excess-renewable-power/150/537/59292/.

For instance, Canada's *Financial Post* wrote[12] in 2013 that "subsidized" renewables are "getting ever nearer to destroying the business model of companies meant to invest in renewables, namely the big sector leaders such as Eon and RWE." In fact, the German renewables' sector has been telling conventional energy firms since the 1990s that they aim to marginalize them and offset conventional power—as mentioned previously, politicians and energy firms simply did not think it was possible.

The German tradition of citizen Energiewende meetings lives on today in places such as Jühnde, which became the country's first "bioenergy village," and Dardesheim, home of one of the largest community-owned wind farms in the country. As the project director in Dardesheim once told the author,[13] he held town meetings for years to get enough consent. Indeed, citizen participation is now considered a standard way of overcoming NIMBYism, as the German Wind Energy Association (BWE) puts it in its brochure on community wind power.[14]

The focus on community ownership is one reason why the BWE seems so tepid about offshore wind and so supportive of onshore wind.[15] The former consists mainly of gigantic projects that can only be financed by large corporations, whereas communities in Germany have always wanted to make their own power, which doesn't just mean not importing from abroad, but also not buying from an oligarchy of domestic corporations.

Going forward, Germany is even looking into ways of expanding community ownership of energy infrastructure. At present, there is a campaign among citizens of Berlin to buy back the city's power grid, and new power lines built across the country could also be partly opened up to citizen investors.[16] Unfortunately, an attempt by a citizens' group to own part of the Butendiek offshore wind farm did not work out, and the project has been sold to wind farm developer WPD—further evidence that offshore wind and community ownership do not go hand in hand.

---

[12] http://business.financialpost.com/2013/01/21/germany-grapples-with-switch-to-renewables/?__lsa=1085-837b.

[13] See "German energy" at http://www.boell.org/downloads/Morris_GermanEnergyFreedom.pdf.

[14] See http://www.wind-energie.de/sites/default/files/download/publication/community-wind-power/bwe_broschuere_buergerwindparks_engl_2012_l06_final.pdf. For another example in English of one of the country's many villages "going green," see the PBS video entitled "Germany's green revolution" http://video.pbs.org/video/2326679795/.

[15] See the interview with BWE head Hermann Albers at http://www.renewablesinternational.net/power-market-design-20/150/537/59421/.

[16] See "Berlin to buy back its grid" at http://www.renewablesinternational.net/berlin-to-buy-back-its-grid/150/537/38111/ and "German grid could open to citizen investors" at http://www.renewablesinternational.net/german-grid-could-open-to-citizen-investors/150/537/58612/.

CONCLUSIONS

While the task may seem daunting, Germany's energy transition actually appears a bit unambitious when viewed from Denmark, which has even stricter targets, and the German nuclear phase out also seems quite modest compared to Austria's decision not only to forgo domestic nuclear power, but also ban imports of nuclear power imports starting in 2015.[17]

But then, those who wonder whether Germany will succeed should consider how surprising the current outcome is. While many speak of exorbitant subsidies for renewables, cloudy Germany actually has the least expensive solar power in the world, with midsize solar roof arrays in Germany producing electricity at half the cost of utility-scale solar plants in sunny California—the difference is largely directly attributable to the corporate control of markets in the United States.[18] The highest solar feed-in tariff paid (as of June 1, 2013, with further reductions made for new systems each month) was just 0.15 euros per kilowatt-hour, with the average retail rate at around 0.27 euros.

Critics warned of industry leaving the country, but Germany continues to weather the current crisis with a healthy economy and the lowest unemployment rate since reunification two decades ago.[19] Critics warned of blackouts, but Germany had the lowest number of minutes of grid downtime of any EU country in 2011 (Figure 3; data for 2012 were not available as of this writing).

Critics spoke of a reliance on nuclear imports, but in 2012, Germany reached a record level of net power exports, and Germany is the only country with which nuclear-heavy France has a negative power trade balance.[20] Others warned that Germany would need to export its excess (and expensive) renewable power on the cheap while it imports inexpensive conventional power at a premium, but in 2012 the value of a kilowatt-hour of

---

[17] See http://www.renewablesinternational.net/austria-to-discontinue-imports-of-nuclear-power/150/537/38088/.
[18] See http://www.renewablesinternational.net/rfps-make-renewables-artificially-expensive/150/510/57958/.
[19] See http://www.reuters.com/article/2013/02/19/us-germany-employment-idUSBRE91 I0V720130219.
[20] See http://www.renewablesinternational.net/german-energy-consumption-up-adjusted-co2-emissions-down/150/537/59443/ and http://www.renewablesinternational.net/france-net-power-exporter-except-to-germany/150/537/59937/.

Grid reliability and renewable growth seem to go hand in hand
Minutes of power outages per year (excl. exceptional events), based on Saidi
Source: CEER and own calculations

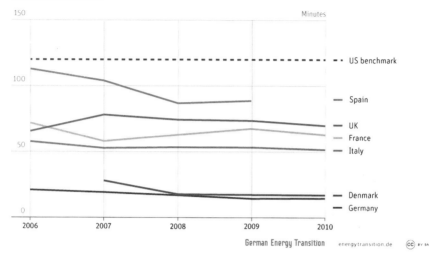

Figure 3 Germany had the most reliable grid in the EU in terms of minutes of downtime excluding natural disasters.

electricity Germany exported was roughly 7% above the value of the average kilowatt-hour imported.[21]

Indeed, it seems that just about all of the expectations that critics of the Energiewende had have turned out to be unfounded, at least as of this writing. Perhaps that is because these critics have not done one crucial thing that Germans do: pay attention to the macroeconomic impact of the switch from corporate-owned conventional energy to community-owned renewables on their national economy.

---

[21] See http://www.renewablesinternational.net/german-power-exports-more-valuable-than-imports/150/537/61663/.

CHAPTER 8

Multidisciplinary Integrated Development

M.K. Elahee
Faculty of Engineering, The University of Mauritius, Reduit, Mauritius

Contents

Sustainability	116
Energy	116
Limitations	119
Engineering	120
Toward Interdisciplinarity	121
Conclusion	121

Thus far, no project better depicts Maurice Ile Durable (MID) than the subsidy program on solar water heaters. Yet, there is a dark side to it as there has been abuse by a few suppliers. The second most successful MID-funded project is probably the compact fluorescent lamp project. It has changed customer attitudes to the extent that buying incandescent lighting is no more the obvious reflex. But who dares launch the scheme again in the shadow of the scandal of counterfeit lamps? Wind farms should have long been a reality had it not been for an invisible hand that procrastinates their advent. Even in the private sector, the few MID initiatives are proving difficult to sustain. The civil society response is all but proactive. The consultants appointed by the government to define the MID policy, strategy, and action plan are faced with a herculean task.

Yet, we have no other choice but to move ahead. MID is here to stay—it's much bigger than all the things that have gone wrong. What is at stake is our future. Either we believe in it or not. MID is for people of faith.

But there is no room for blind faith. We have to be critical about the process, rational about our expectations, and intelligent in our effort. Strategies must evolve, but our vision must not falter. Learning from our mistakes, particularly from those of others, is an enriching and necessary experience. Let us be modest and sincere above all, and recognize our weaknesses.

This chapter aims at focusing on what is going wrong according to the author's humble opinion. More importantly, it introduces constructive proposals toward turning hope into progress. The reference to energy and engineering issues reflects the author's own bias and in no way limits MID to the latter dimensions. The key message indeed is about making multidisciplinary integrated development a reality.

SUSTAINABILITY

The MID vision statement "to make Mauritius a model of sustainable development" is problematic. The statement should have captured the purpose and the values of MID. Is MID a global model? Is it not too pretentious to ask others to take us as a model? Is there just one model? Is a model static? Will the vision be true only if others accept us as their model?

The focus should have been on "sustainability" instead of "sustainable development," which is a worn-out and abused concept often limited to its economic dimension. Even the word "model" can have two meanings: referring to "an example" and, secondly as an adjective, it can mean "perfect." Considering the requirement to have the statement also in French, particularly in Creole, we do not have now the best vision statement for a concept, which, contrary to what many affirm, is well understood. Even children at school agree on what MID is, and also on what is not MID. But to find the right words to embody it is not easy.

It is humbly suggested, again, that the vision statement be revisited to integrate a statement of objectives (or even a mission list). In this context, our commitments should be made more explicit, as these have to be understood easily by one and all. To be fair to the consultants, clear signals from the policymakers will help the former show better clarity in the development of policy statements. Do they know what the government wants?

ENERGY

The energy aspect of MID should not dominate the concept, but it is inevitably at the heart of economic, social, and environmental integration, and hence a foremost sustainability issue. The reliance on imported fossil fuels, their rising prices, the geopolitical situation at the supply end, global climate change, existing potential for sustainable energy, and local environmental issues all add up to make energy and MID linked intimately. In fact, the world record oil prices of 2008 gave birth to MID.

Clarity of vision, shared commitment, and coherent strategies are needed to move ahead in terms of MID energy policy. Some progress has been made but we have not embarked on a paradigm shift. Many prefer a business-as-usual scenario or simply the law of the jungle where the strongest lobbies have it all at the end of the day. This is no blame-game but a reflection of the global situation when it comes to a common sustainable energy future for the planet.

The literature on sustainability provides essential lessons on the mistakes to avoid. For instance, national targets cannot be used as stand-alone reference points for local air quality strategies or for renewable energy policy. If evoked, these targets have to be linked as composite sets and assessed by referring to a normative target state for the whole system.

Concepts such as 24/7, two-million-tourists, treble-GDP, duty-free shopping paradise, oil terminal and refinery have to be assessed in terms of sustainability, including implications in terms of energy policy measures. It suffices to note, for example, that in the Niger Delta the life expectancy is only 40 years, and oil spills of the extent of the Exxon-Valdez catastrophe occur in 1 week. Neither Dubai nor even Singapore can be models of sustainability for Mauritius. This does not imply that innovative concepts cannot find their way in the MID vision. In fact, regional codevelopment, or still sustainable use of our exclusive economic zone, has a potential that has not been fully appreciated, particularly related to the energy dimension of MID.

The energy dimension is not just about energy as a resource. It has also engineering, economic, environmental, and ethical relevance, the latter including political and social aspects. In the context of a new more equitable economic framework, it can create jobs and adds value. Whatever power plant installed today will be still around in 20 years and will lock us in for at least 40 years strategically. The consultants should not only provide a process, but also identify the "red lines" not to be crossed and a "checklist" of MID criteria.

In the electricity sector, the current key choice seems to be, for some, to decide whether coal or liquid natural gas (LNG) should form part of the future energy mix to provide base load. Consideration of a technology that is MID compliant is as vital as the selection of the energy source. For instance, it should be decided whether coal or biomass with cogeneration can be treated at par with coal only using conventional technology. The end use is also important: is LNG meant only for power or is it for transport? These options are not mutually exclusive and should be prioritized.

As far as the target of renewable in our power mix is concerned, we have spent years fighting whether it should be 35, 60, or 100%. And we have been stuck at 20% for 12 years now, even decreasing this ratio since last year.

What is needed is more than an aspirational target setting a clear direction, but flexibility in terms of timeframe and coherent policies, strategies, and actions based on sustainability appraisal or criteria. For example, referring to a recommendation from the Working Group for Energy, we should be able to affirm, or deny, that "In the power sector, we aim at 100% renewable in the period 2040 to 2050." In the positive case, that would still be much less ambitious than Reunion and Maldives (both aiming at 100%, more than 15 years before us), but it has at least an unambiguous meaning!

The potential for conflicting interests exists and should be addressed. Whether it concerns the competing use of our scarce land resources for energy, food security, or housing or still the equity dimension of the so-called democratization of the energy sector, we should learn to avoid clashes and to manage them when they arise. Our fears can turn out to be self-fulfilled prophesies if we fail to plan properly. Again MID compliance criteria are needed, but dialogue and communication are just as essential.

Having the MID energy policy, strategy, and action plan is not an end in itself. We will have to apply it, together. It will be an awesome challenge. In order to keep a proportion between our thoughts and our actions, it is proposed that we do not delay any further with the MID process, even if we are not fully satisfied. The time for action has come and we have to leave our drawing boards.

Thus far, policy definition has been too much limited to the affirmation that "we have a plan." Time has come for decisiveness. The cost of inaction can be very high. We should be bold enough to adopt a concrete orientation of energy policy. And if it is to be MID relevant, it has better be a break from the business as usual, that is, innovative, daring, and forward looking.

For example, for the MID policy for power generation, we may agree to promote renewable energy to reach 35% (or indeed much higher if it were left to some) by 2025 in the power generation mix by:

1. Targeting an energy intensity of XXX by 2025 (to calculate XXX to attain the above as demand and supply are related and as there is a need to relate the target to the absolute demand projected in 2025).
2. Including YYY% of renewable capacity (in MW or electricity in GWh) to accompany any addition of generation capacity (to calculate YYY to attain the above).

Alternatively, (2) could be:

Reducing gradually by ZZZ% every 5 years the share of fossil fuel in the power generation mix by promoting in the following order:

dual-fuel generation including biomass, cogeneration, sustainable biofuels, renewable X, renewable Y, LNG (as a transition fuel ???) (where ZZZ is to be calculated to attain above target).
3. Subjecting all development projects, including demand-side requirements in terms of energy, to compliance to defined sustainability appraisal or criteria.

All policy statements should be elaborated more thoroughly (even if this means including specific strategic directions for the sake of clarity at an early stage). This should be done for transport, demand side, land use, industry, etc. For each sector, the policy statements, strategies, and outcomes (or targets) should be submitted to the compliance of sustainablity appraisal/criteria. When all sectors are taken together holistically, the totals should tally in terms of quantitative analysis.

LIMITATIONS

A critical review of the literature on sustainability reveals that a poor integration of economic, social, and environmental aspects leaves impacts to be balanced by decision makers, integrated through a political process or subjected to lobbies. Strong integration favors decision making through the technical process, whether multicriteria or economic, but needs comprehensive frameworks often laden with bureaucratic practice. In our context, this may be too heavy to be implemented in the short term. Another avenue is integration through community impact evaluation with expert definition of costs and benefits for different stakeholders, which again has "difficulty with cumulative effects, multiple values, and contested boundaries."

Experts have concluded that any sustainability appraisal/criteria are rarely a final or true answer, but a process of investigation. At best it will provide sound sensitivity analysis on different identified options, provided that sufficient raw data and information are available. Some have highlighted another condition: technical tools need to be embedded in a structured process of social debate and discourse.

Energy project multicriteria assessment with sustainability indicators, including those conducted in the case of renewable energy promotion in islands, converge to point out that the end result depends on which dimension is given more weight at the start. The solutions are pertinent only if they can be integrated in engineering practice. Nonnumeric information can also weigh significantly in decision making.

Thus at one end, sustainability assessment/criteria remain limited by the clarity of political orientation. And at the other end, engineering relevance is the litmus test before implementation can be envisaged. For these reasons, the success or failure of MID achieved at this stage should be taken with much humility: much remains to be done, which will depend on our sense of leadership and innovation, both in terms of decision making and technology development.

ENGINEERING

In this context, it is vital that we rethink our education, including the teaching and learning of the so-called sciences. A multidisciplinary approach is much warranted. This applies even to a specialized field such as engineering where an appreciation of the philosophy of ecological justice or of the economics of development would not be out of place. Thus far, we have rarely included sustainability as an engineering concern, even less considered engineering in direct relation to the specificity of our country's future. Students often enroll in mechanical engineering programs dreaming of fast cars, not really of solving our road traffic problems.

Research in the context of MID is also much overlooked. A list of projects in line with identified strategies should come along with the MID action plan. Among these projects should be energy topics such as gasification, remanufacturing, energy–waste linkage and management, ocean energy, cogeneration/trigeneration, cane and bioenergy research, sustainability analysis, and energy management, for example, related to smart-grid pilot projects, as well as behavioral research on sustainable consumption related to energy. The application of research in the local or regional situation should be a priority.

At a recent conference, a participant from the so-called developing world questioned those from the industrialized countries: "You have polluted the planet, including our country, and now you are doing research only for the sake of publishing papers on it." Indeed, particularly with the advent of the Internet, engineering publications have reached unimaginable proportions. However, most of it is not read, seldom cited and studied. Few are relevant to the problems of developing nations, for example, the case of small-island developing nations like ours. Issues such as the impacts of land grabbing in Africa for biofuel production or pollution-related cancer in poor countries are disregarded at the expense of commercially driven research on nanotechnology, for example. Access to energy is the key to development, but insufficient effort is made to ensure that developing countries tap their abundant potential of renewable energy.

TOWARD INTERDISCIPLINARITY

Another topic of crucial relevance for developing countries is energy efficiency because they will need more energy for sustaining development. From 2010 to 2011, the energy intensity of Mauritius dropped from 1.46 to 1.40 in terms of tons of oil equivalent per Rs 100,000 GDP. This implies that had the energy intensity remained the same as in 2010, the import bill of fossil fuels would have been at least Rs 1 billion higher with more than 200,000 tons of CO_2 emitted. A corresponding increase in the maximum peak demand of at least 20 MW and an extra power consumption of about 200 GWh would have been recorded had it not been for the drop of energy intensity.

Energy efficiency is an interesting example of a field where more than multidisciplinarity, there is a need to merge disciplines and integrate multiple realities. Emphasis should not be only on technology transfer, but also on that of know how through education, training, awareness, sensitization, and institutional capacity building. Targets and plans are not the ultimate ends of energy management programs. For a long time, energy saving has been considered a priority. But energy management cannot just be about saving energy—it is really about making *more* with *less* energy in the context of developing economies like ours that need energy for progress. It is about linking demand and supply in such a way so as to optimize resources. It is about being energy intelligent. It is about being energy conscious. It is about a new culture of generating, distributing, and consuming energy. Such a new interdisciplinary paradigm is essential if we want to exploit the potential of energy efficiency fully.

CONCLUSION

Ideas are abundant, the technology exists, the human resources can be trained to acquire the right skills, and the political will is unflinching, apparently at least. But this is not sufficient to make MID a reality, including the transformation of our "paysage énergétique" toward greater efficiency and sustainability reaping economic, environmental, and social benefits of significant measure. A new culture is needed. MID is calling for a real game changer, a shift to a vision different from the current business-as-usual scenario.

It calls for a new philosophy underpinning our very concept of energy production and use. This applies also to water, other resources, and waste. The distinction between producers and consumers will phase out gradually as we consider that all of us have the potential of being "prosumers." Decentralized energy (or resource) management based on a clear definition of individual rights and responsibilities will lead the way to a new framework where everyone is called upon to become resource conscious, whether at home or at work, while traveling or during leisure, as an investor or as a consumer. Addressing development requirements will be met as much by demand-side management as by installing new capacity. If the burden is shared, the former can be our solution.

We require adopting a multidisciplinary approach and turning to inter-disciplinarity. For instance, designers, architects, engineers, promoters, surveyors, planners, accountants, clients, and local authorities will have to work hand in hand to ensure that buildings are energy efficient. A holistic dimension is called for when we are designing new cities or transport systems. Integration of knowledge from different disciplines, including the behavioral sciences when it comes to assessing attitudes, will have to be considered if we want MID to happen.

Last but not least, MID is about education. However, it is more than just adding "ecoliteracy" modules to our curricula. From early childhood education to the Internet, teaching and learning will have to incorporate multiple realities. In other words, constructive new knowledge must be sought in view of having a transformative impact on the individual as well as on society. The outcome will be that people, or companies, will act in a conscientious way, not out of compulsion or guilt, not even for real financial benefits, but out of a sense of sheer responsibility and faith in a better world.

BIBLIOGRAPHY

Afghan, N., 2000. Energy system assessment with sustainability indicators. Energy Policy 28, 603–612.

Afgan, N., Calvalho, M., 2002. Multi-criteria assessment of new and renewable energy power plants. Energy (27), 739–755.

Buchholz, T., Luzadis, V.A., Volk, T.A., 2009. Sustainability criteria for bioenergy systems: results from an expert survey. J. Cleaner Prod. 17 (1), S86–S98.

Buytaert, V., Muys, B., Devriendt, N., Pelkmans, L., Kretzschmar, J.G., Samson, R., 2011. Towards integrated sustainability assessment for energetic use of biomass: a state of the art evaluation of assessment tools. Renewable and Sustainable Energy Rev. 15 (8), 3918–3933.

Hanegraaf, M.C., Biewinga, E.E., van derBijl, G., 1998. Assessing the ecological and economic sustainability of energy crops. Biomass and Bioenergy 15 (4-5), 345–355.

Kemmler, A., Spreng, D., 2007. Energy indicators for tracking sustainability in developing countries. Energy Policy 35 (4), 2466–2480.

Neves, A.R., Leal, V., 2010. Energy sustainability indicators for local energy planning: Review of current practices and derivation of a new framework. Renewable and Sustainable Energy Reviews 14 (9), 2723–2735.

NTA 8080 (en), 2009. Sustainability criteria for biomass for energy purposes. NEN, March.

Reijnders, L., 2006. Conditions for the sustainability of biomass based fuel use. Energy Policy 34 (7), 863–876.

Zupeng Zhou, Z., Jiang, H., Qin, L., 2007. Life cycle sustainability assessment of fuels. Fuel 86 (1-2), 256–263.

CHAPTER 9

Think Globally, Act Locally, and Plan Nationally An Evaluation of Sustainable Development in Indonesia at National, Regional, and Local Levels

Lacey M. Raak
Sustainability Director, University of California, Santa Cruz, CA, USA

Contents

Executive Summary	126
Policy and Budget Evaluation	126
Baseline Data	127
Recommendations	127
Methodology	129
Background	132
Indonesia Overview	132
East Kalimantan Overview	134
Balikpapan Overview	136
National Sustainable Development Strategies	138
Overview of National Sustainable Development Strategies	138
Indonesia's National Sustainable Development Strategy	140
Sectoral Strategy	140
Planning	141
Energy	141
Forestry[10]	142
Human Settlements	145
Mining	146
Tourism	148
Existing Planning Mechanisms	149
Development Planning Overview	149
Spatial Planning	152
Additional Methods of Sustainable Development Planning	154
Poverty Reduction Strategy Report	154
Millennium Development Goal Report	154
Climate Change Planning	154
Analysis of Sustainable Development Indicators	156
Results from Evaluation of Planning Priorities and Budget Allocations	157

Analysis of Planning Priorities and Budget Allocations 158
   National 158
   Provincial 160
   Local 162
   Overarching Issues 165
      Integration among Levels of Government 165
      Audience 166
      Legal Obstacles 167
      Private Sector 167
      Corruption and Illegal Activity 169
Recommendations 170
   Improving Sustainable Development in Indonesia 170
      National Recommendations 170
      Provincial Recommendations (East Kalimantan) 174
      Local Recommendations (Balikpapan) 174
   Improving Cooperation Between Local and National Governments 177
Conclusion 179
Acknowledgments 180
Translation Acknowledgment 180
Photo Acknowledgment 180

EXECUTIVE SUMMARY

This chapter offers an evaluation of the national sustainable development strategy (NSDS) for the Republic of Indonesia (Figure 1). The city of Balikpapan and East Kalimantan, the province in which it resides, were chosen as case study locations for further analysis of the relationship between the various levels of government in relation to sustainable development.

A NSDS is a country's coordinated efforts and process to achieve economic, environmental, and social objectives in a balanced and integrative manner. This effort should be replicated at the provincial and local governments.

The analysis includes an overview of existing sustainable development reports published in 2001 by the Ministry of Environment, as well as sustainable development data collected for each level of government to allow comparisons between different levels of government and to function as a baseline for existing sustainable development conditions. Finally, an evaluation of medium-term development plan policies and budget expenditures from a corresponding time frame were evaluated at each level of government.

Policy and Budget Evaluation

Each policy in the midterm development plan and every line item of the respective budget were analyzed and categorized into one of four categories,

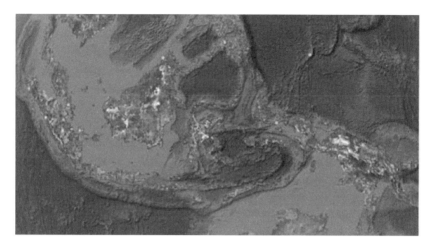

Figure 1 Google Earth image of Indonesia (2010).

three of which address sustainable development (environment, economic, and social). The fourth category covered administrative needs. By completing a comparative evaluation of policies and budgets it is possible to learn if policies are distributed evenly across the sustainable development sectors and if government spending reflects the policy goals.

Results of the policy evaluations show environmental policies to be included at least in medium-term development plans throughout every level of government, although in Balikpapan the economic sector is tied with the environment.

Results of the budget evaluation also indicate that the environment receives the smallest portion of budget allocation at each level of government. At each level, environmental projects and programs receive 5% or less of the budget.

Baseline Data

Baseline sustainable development data collected include information from multiple government agencies and international organizations. The result is the first comprehensive collection of sustainable development data across multiple levels of government in Indonesia. As data collection and government transparency improve, baseline data will become more available.

Recommendations

The primary recommendations for improving sustainable development in Indonesia are listed here. As a precursor, it is important to note that few of the recommendations will be achieved without reducing corruption and without political will. The recommendations account for corruption and transparency issues, but do not address it directly. Additional recommendations for each level of government are included in the body of the chapter.

Although sustainable development encompasses the environment, social needs, and economic growth, most recommendations are focused specifically on the environment because it is the aspect currently receiving the least attention.

1. **Strengthen the distinction between climate change planning and sustainable development planning (national level).** Currently, sustainable development planning has become eclipsed by climate change. Addressing climate change is an integral part to successful sustainable development, but it cannot replace it.
2. **Create an Indonesian civilian conservation corps (national level).** Create a well-paid and local team of people to work actively to protect the forest. This will provide a win–win situation by increasing the capacity of local citizens to ensure the protection of future resource while paying a livable wage and helping the country meet its carbon reduction targets.
3. **Enforce environmental laws and prosecute offenders (provincial level).** Enforcement and prosecution of laws should occur where violations occur. Prosecution should be quick and should incorporate not only a financial fine, but should shame the violator publically.
4. **Improve monitoring and evaluation of environmental indicators (local level).** Collect environmental data with the same consistency as economic and social indicators to provide current status updates and help direct future policy decision.

This research shows it is neither feasible nor desirable for the national sustainable development strategy to be recreated at the provincial and local level. However, the principle of balancing the needs of the three pillars of sustainable development should be transferred to each level of government. It is the responsibility of the national government to ensure that guidelines on achieving sustainable development reach the local government and are implemented. It is the responsibility of the local and provincial governments to implement policies and programs, then collect and monitor data to share with the national government to help guide policymaking.

The results of this evaluation are not static. Sustainable development is evolving continually. This chapter offers a snapshot of the current status of sustainable development in Indonesia by setting baseline data and evaluating planning documents and budgets. Recommended policies are a starting point and will need to evolve as progress is made and/or as new opportunities and needs arise.

METHODOLOGY

Field research was conducted in Balikpapan and Jakarta for a period of 8 months (February–October 2010). Research included literature review, key informant interviews, and field observation. Observation and literature review was ongoing as data became available. Interviews were conducted from April to June with follow-up interviews conducted in August.

The research location of East Kalimantan was chosen because of its natural resources and high contribution to Indonesia's gross domestic product (GDP). Although highly exploited for natural resources, it is an area that still has the ability to implement changes that would have a lasting effect. Balikpapan was chosen because it is an important district within East Kalimantan. It is relatively wealthy, technologically advanced, and diverse.

Interviews represented a range of local (Balikpapan), regional (East Kalimantan), and national (Indonesian) government institutions, including the Planning Ministry, Environmental Ministry, and National Statistics Bureau. Additional interviews were conducted with relevant nongovernment organizations (NGOs) and private businesses, including, The Nature Conservancy, Gesellschaft für Technische Zusammenarbeit (German Agency for Technical Cooperation), ICLEI–Local Governments for Sustainability, Wahana Lingkungan Hidup Indonesia, Pelangi, and Asosiasi Pemerintah Kota Seluruh Indonesia (Association of Indonesian city governments) (APEKSI). Many of these institutions also provided literature, data, and reports that are technically public, but are not widely available.

Evaluation included a review of planning documents and reports (medium-term and long-term development plans from each level of government). It also included budget data, environmental reports, statistical data reports, and academic reports on sustainable development.

Observations were completed primarily in Balikpapan and rural areas of East Kalimantan. Observation and information gathering were also conducted at conferences and community events. Because the research grant was intended to foster cross-cultural communication and understanding, examples and stories from experiences and conversations relevant to the research are included. These are intended to share the personal side of the issue addressed.

In all instances, the research was approached neutrally. By recognizing possible partiality based on past experience and education in the environmental field, the potential for bias was reduced.

A national sustainable development strategy is a nation's coordinated effort and process to achieve economic, social, and environmental objectives in a balanced and integrative manner. This definition holds true for both local and provincial governments, but operates at the appropriate scale (i.e., city efforts and provincial efforts and process to achieve the objectives).

Evaluating a NSDS is challenging. It is a nebulous process that may be partially described in a report or in many reports. It is also an evaluation of how the government works together to address the key development needs of a country, including coordinating efforts among different agencies, political parties, and personalities. The information in this chapter gives an overview of the NSDS in Indonesia while paying particular attention to the relationship between national government and local governments.

In order to provide structure to this unformulated process, an evaluation of the overall status of the NSDS was completed using a methodology developed by the University of Manchester and Central European University (Appendix 1). The methodology streamlines the existing principles for an effective NSDS offered by the Organization for Economic Cooperation and Development and the United Nations Division on Sustainable Development (Cherp 2004). Using these criteria allows the dynamic and nebulous NSDS to be evaluated using a comprehensive set of criteria that is applicable and comparable among different countries.

Quantitative materials used to evaluate the balance among the three main sustainable development sectors (environment, society, and economy) were gathered from medium-term development plans [Rencana Pembangunan Jangka Menengah (Medium-Term Development Plan)] (RPJM). This provided the policy goals and objectives. Policies were evaluated and categorized into one of four sectors. Three reflect the sectors of sustainable development and the fourth addresses administrative needs:

- **Environment**—Any policy that strongly addresses environmental issues related to conservation, protection, and monitoring. These distinctions were made based on how the policy influenced the environment positively or negatively. Some policies related to the environment that actually addressed economic policies were not included (such as "pursuing

potential exploitation of forest resources"). Other policies that were not related directly to the environment but had a strong impact on environment were included as an environmental policy, as well as a social policy ("family planning").
- **Social**—Addresses policies related directly to the benefit of the humans and their relationships to each other and the community. Social policies included those related to education, health, and religion. Infrastructure improvements were also included as a social policy (and an economic policy because of the contribution they make to improved welfare by access to markets and improved sanitation facilities).
- **Economic**—Any economic policy that addressed economic growth and stability directly or indirectly. For example, a policy related directly to economic growth is "developing centers of economic growth" and an indirect policy is "increasing nonoil exports."
- **Other/administrative**—This category included policies directed at administrative offices, record keeping, statistics, and bookkeeping.

Policies related to corruption eradication and certain disaster planning policies (such as fire prevention) were included in each category.

Following categorization, the percentage was then calculated. Because many of the policies were cross-sectoral, such as reducing corruption and population growth rate policies, policies were not mutually exclusive (it would go against the pretext of sustainability), which accounts for the total percentage being higher than 100%.

Evaluation of budget reports from 2006 (national) and 2007 (local) shows where the money was actually spent. The provincial budget is from 2009 to 2010 and is a projection. This dual evaluation provided information to determine if the policies in the RPJM were funded adequately. Budget analysis was conducted in the same manner as the policy document analysis; however, when percentages were calculated, the percentage is from the amount spent rather than the percentage of total budget line items.

The budget year analyzed was taken from the range of years included in the RPJM. For example, the national budget analysis was from 2006, which is the second year of the RPJM (2005–2009). The local budget analyzed was from 2007, which is the second year of the local RPJM (2006–2011). At the provincial level, the budget analyzed is from 2009—the first year of the current RPJM (2009–2013). Provincial data for past plans and budgets were unavailable. Therefore, the provincial analysis is based on proposed policies and budget expenditures.

Appendices 3–5 show complete results and analysis of the policy and budget documents for each level of government.

The criterion for establishing the baseline level of sustainable development is from the United Nations Commission on Sustainable Development (UNCSD) and published by the Department of Economic and Social Affairs of the United Nations. The indicators were developed following the World Summit on Sustainable Development as a tool for countries to monitor progress toward sustainable development. The indicators used for this report are the most recent indicator set (the third edition), which incorporates the Millennium Development Goals, as well as feedback from countries that have tested and applied the criteria extensively.

Data used in this baseline (shown in Appendix 1) were collected primarily from government reports. Additional data were collected from international organizations and academic reports. Ideally, data for establishing the baseline would come from the same year and same source across all of the indicators and levels of government; however, this was not always possible due to government capacity limitations.

BACKGROUND
Indonesia Overview

Indonesia is rich in natural resources. It has the world's highest marine diversity and the second largest remaining rainforest. Alfred Wallace (2008), an early explorer, described the region as

> Situated upon the equator and bathed by the tepid water of the great tropical oceans, this region enjoys a more uniformly hot and moist [climate] than almost any other part of the globe, with teems of natural productions that are elsewhere unknown. The richest of fruits and the most precious of spices are here indigenous. It produces the giant flowers of the Rafflesia, the great green-winged Ornithoptera (princes among the butterfly tribes), the great man-like Orangutans, and the gorgeous Birds of Paradise.[1]

Today, the natural resources that entranced Wallace face a plethora of threats. Some threats are man-made, including deforestation, rising global temperatures, water and air pollution, overfishing, and mining. Others are naturally occurring hazards such as forest fires, mudslides, earthquakes, tsunamis, and volcanic eruptions.[2]

---

[1] This quote was originally written in 1846.
[2] Forest fires and mudslides are also often a result of human actions.

Indonesia is a growing democracy. With a population of about 230,000,000 it is also the world's largest Muslim democracy.[3] (Central intelligence Agency, 2010). Although factions of extremism exist and the country has not been immune to the rise in global terrorism, the people of Indonesia are eager to meet newcomers. They are helpful, welcoming, and quick to return a smile. The diversity of the archipelago is evident not only in its flora and fauna, but also in the sundry human population (Figure 2).

Indonesia's economy is the strongest national economy in southeast Asia (Trading Economics 2010). Consumer spending continues to increase, and 2008 GDP per capita was 1087 USD, up from about 775 USD in 1998. The state manages various companies, including Pertamina (oil and gas), Bumi

Figure 2 Indonesia.

---

[3] Indonesia is currently conducting a census. Population estimates in 2010 are expected to increase significantly. Interestingly, during the census, demographers met a 157-year-old woman and her 118-year-old daughter, possibly the oldest living humans.

Resources (coal), and Perusahaan Listrik Negara (electricity). (Indonesian Statistic Bureau (BPS), 2009). Chinese companies continue to be significant investors in state companies, which feed China's growing need for energy and help Indonesia meet its need for infrastructure development (Jakarta Globe and Rueters 2010).

Although stymied by high corruption, Indonesia is hovering close to a foreign investment grade debt rating (which is based on a debt to GDP ratio). Indonesia's ratio of debt to GDP dropped by 29% between 2004 and 2009 (Antara News 2010). Indonesia joined CIVETS (Colombia, Indonesia, Vietnam, Egypt, Turkey, and South Africa), which identifies the next emerging market economies (Woods 2010).

Indonesia's economy is based largely on natural resource extraction, a significant portion of which is from the East Kalimantan province. (Ministry of Environment, 2002).

East Kalimantan Overview

East Kalimantan is located on the eastern side of the Island of Borneo (known locally as Kalimantan). Rich in natural resources, East Kalimantan contributes significantly to Indonesia's GDP, mainly through its timber, oil, natural gas, coal, and palm oil operations.[4] Due to their high economic and environmental value, forests are one of the province's greatest assets; as a result, East Kalimantan has experienced significant damage to its natural resources, affecting the health of the environment and the social well-being of many communities (Figure 3) (Provinsi Kalimantan Timur, 2008).

East Kalimantan has attempted action in the last 10 years to protect its forest. The 1999 spatial (or land use) plan showed 4,579,199.04 ha of protected forest, 25,786.38 ha of forest education and research area, and 9,761,197.79 ha of forest cultivation area (a current concession or potential concession). Projected estimates in the 2009 spatial plan showed a 2.4% increase in protected forest and an 11.6% decrease in forest cultivation area. This indicates a shift toward protection and away from concessions. (Bappeda/ East Kalimantan Ministry of Planning, 2009)

These numbers, however, do not account for illegal logging or acknowledge that extraction is still permitted in protected areas. Furthermore, forest laws in Indonesia are very porous and offer many opportunities for mismanagement.

---

[4] East Kalimantan provides the largest contribution to Kalimantan's regional GDP, which is the third highest in Indonesia. Indonesian Statistics Bureau (BPS 2009).

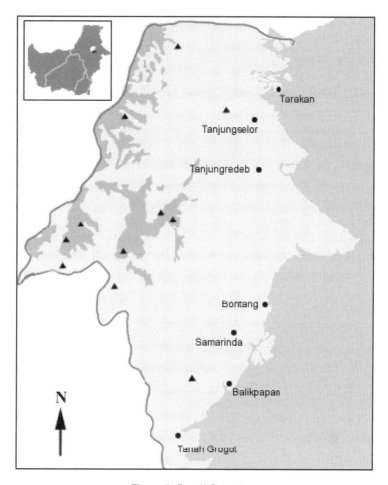

Figure 3 East Kalimantan.

In rural areas of East Kalimantan, many villages have strong connections to their indigenous past. The indigenous people (known as Dayak) have blended "modern" religions, including Islam and Christianity, with their traditional beliefs and still celebrate their past proudly. Local governments in rural areas often implement a community-based response to the threats and needs of the community. The needs of the local communities often contradict the needs of the resource extraction companies, leading to conflict and injustices against indigenous people.

East Kalimantan has a poverty rate of 9.5% and 19% of all children are underweight. The regional GDP per capita is also lower than the national average. These figures show the development challenges that face East

> ### Wehea Community
>
> **(Excerpt from The Nature Conservancy and personal interviews)**
> Along the Wehea River in the central north area of East Kalimantan is the small Wehea community. The indigenous people, known as Dyak, have taken a keen interest in the protection and preservation of the forest that sustains them.
>
> In response to threats to the forest, the Wehea Dayak people held a customary council meeting in 2004. Around 200 representatives from local Dayak communities, local government, and private sector representatives attended. They agreed to declare 38,000 hectares of an abandoned timber concession as "protected land" under their traditional law.
>
> The Wehea Dayak have used their traditional practices for conservation. With creation of the protected forest came a means of protecting it: the *Petkuq Mehuey*, or forest guardians. The forest guardians are 45 young people from the local community who protect the forest and have learned how to conduct surveys of the local flora and fauna. They are given a small salary from the East Kutai regency's budget.
>
> Under the mandate of the customary body of Wehea Dayak, the *Petkuq Mehuey* have accomplished some important results. They have captured illegal loggers and hunters and stopped people trying to take things from the protected forest. They actively save wild animals that stray in the settlement area and take them back to the local conservation office or forest. However, at the time of visiting (April 2010), only 10 boys and one manager were working to monitor the region. As a result, their efforts were focused on one area (the corner of the reserve most exposed to illegal logging), leaving the rest unmonitored.
>
> Although the community petitioned for protected status in the late 1990s and received verbal confirmation of the protected status from the national government, no formal declaration of the protected status has been made, posing a potential conflict if another entity declares their right to the land. The images here were taken from the Wehea Forest and from the community's Lom Plai festival, which celebrates their cultural heritage and gives thanks for the harvest.

Kalimantan. Despite these challenges, the people of East Kalimantan have high rates of technology use and education enrollment rates (when compared to national averages). (Badan Pusat Statistik, 2009).

Balikpapan Overview

Balikpapan is a bustling, industrial city situated on the Makassar Straight on the southern tip of East Kalimantan. It is about 2 hours south of the provincial capital, Samarinda. People come from all over the archipelago

Figure 4 Overview of development in Balikpapan.

to participate in its booming economy. It is the major port for many natural resource extraction companies working in East Kalimantan, including Chevron, Pertamina, and Total. Multiple coal barges float out of its port daily, and the flame from the refinery lights the western sky at night (Figure 4).

Balikpapan is a wealthy city and, when compared to other areas in the province and the nation, consistently ranks high on many socioeconomic indicators; 3.49% of the population lives below 1 USD per day, which is lower than both the provincial and the national level (9.5 and 7.5%, respectively)(Sulaiman 2010). Its unemployment rate is 4% and at 8,412 USD, the GPD per capita is well above the national rate (Badan Pusat Statistik, Balikpapan, 2009). Strong economic indicators also seem to boost the social indicators. The adult literacy rate is 98%, and 99% of elementary age students are enrolled in school. (Badan Pusat Statistik, Balikapapan, 2009).

It is rare to meet people born and raised in Balikpapan. For this reason, the city lacks much of the culture found in the surrounding areas, which still have strong connections to their indigenous (mostly Dayak) past. Consumerism is high, and local malls are popular "hang outs" for every demographic.

The majority of the population is Muslim and socially conservative. Although a number of foreign companies operate out of Balikpapan, most expatriates live in housing compounds provided by their companies and cross-cultural interaction is low.

Figure 5 View of Balikpapan Bay—a busy port with coal barges, imports, and exports.

Balikpapan and East Kalimantan are important parts of Indonesia's culture, economy, and environment. The nation's success in sustainable development is dependent on the actions that are taken on the regional and local levels (Figure 5).

NATIONAL SUSTAINABLE DEVELOPMENT STRATEGIES

Overview of National Sustainable Development Strategies

A national sustainable development strategy is a country's coordinated efforts and process to achieve economic, social, and environmental objectives in a balanced and integrative manner (United Nations, Department of Economic and Social Affairs 2007). There are five criteria associated with a NSDS:

1. Country ownership and commitment of the strategy
2. Integrated economic, environmental, and social policy across sectors
3. Broad participation and involved partners
4. Develop capacity and an enabling environment for implementing strategy
5. Outcome driven

National sustainable development strategies have evolved and matured since first introduced at the 1992 United Nations Conference on Environment and Development (UNCED) in Rio De Janeiro, Brazil. At UNCED, member parties produced a comprehensive action plan for governments and organizations at global, national, and local levels to better address the environment in development efforts. This plan became known as Agenda 21. Agenda 21 was further emphasized in 1997 at a special review session of the UN General Assembly and again in 2000, when NSDS were incorporated with the millennium development goals (MDGs).[5] In 2002, the world came together again in Johannesburg, South Africa, for the World Summit on Sustainable Development (WSSD). This UN-led conference was intended to move beyond rhetorical discussions to create actions and results. With policy-level commitment established, technical guidance continues to evolve (United Nations Department of Economic and Social Affairs, 2007).

The technical aspect of developing a NSDS is seemingly vague, but perhaps justifiably so. Sustainable development (and a NSDS) is not a project that is undertaken every few years, resulting in a paper or report. Rather, sustainable development is the constant application of its underlying principles. It involves a coordinated effort and encourages cooperation between government agencies that previously may not have interacted often. For example, it may require staff from the economics ministry to discuss a proposed project with the natural resources agency or the local social welfare institution. It is intended to change habits and linear practices to become more integrated and balanced.

Countries are free to develop a NSDS that meets the unique needs of their country. For example, although the basis remains the same—environment, economy, and society—some countries may choose to include faith and religious growth as aspects of sustainable development whereas other countries do not.

The International Institute for Sustainable Development (IISD) completed an analysis and developed lessons learned from countries with an existing NSDS. They concluded that four main types of strategies are used (Swanson 2004). The type of strategy used was determined by using the path of least resistance (i.e., that which incurred the least political resistance and capacity constraints).

Types of national sustainable development strategies:
- **Comprehensive and multidimensional (e.g., United Kingdom).** Often a compilation of existing economic, social, and environmental strategies and policy initiatives, this strategy approach provides a framework for articulating national strategic and coordinating action toward sustainable development.

---

[5] Specifically Goal 7: Ensure Environmental Sustainability and Target 9: Integrate the Principles of Sustainable Development into Country Policies and Programs and Reverse the Loss of Environmental Resources.

- **Cross-sectoral (e.g., Cameroon).** Sustainable development is addressed through existing cross-sectoral plans such as a poverty reduction strategy and AIDS strategy.
- **Sectoral (e.g., Canada).** Sectoral strategies address sustainable development by department or industry sector (mining, forestry, transportation, etc.). For example, in Canada, each ministry has a responsibility to report its sustainability strategy and actions to a coordinating body.
- **Integration into existing planning processes (e.g., Mexico).** Integrates sustainable development principles into the existing planning process.

Research by IISD also showed that NSDS strategies continue to function outside the national budgeting process and that sector and cross-sector plans need to overcome administration hurdles to improve coordination. Finally, national strategies often fail to connect to local strategies (Swanson 2004).

INDONESIA'S NATIONAL SUSTAINABLE DEVELOPMENT STRATEGY

Indonesia's NSDS fits into two of the categories listed previously. It is sectoral and is also integrated into the existing planning process. However, the extent to which the strategy has been successful in either category is debatable. A table detailing the results of the existing evaluation strategy is available in Appendix 1.

Sectoral Strategy

Indonesia's NSDS is outlined in eight short books written in response to Agenda 21.[6] Five books cover specific sectors on energy, forestry, human settlements, mining, and tourism. Three books relate specifically to planning for sustainable development and act as guidelines for planning agencies. The books offer insight into the history of sustainable development on the national level.

The Ministry of the Environment wrote the eight books prior to the WSSD in 2001 with feedback from additional government agencies and academic institutions. Local planning agencies (Bappeda) had not seen or heard of the books. The books are not referenced by any of the sectoral agency websites or in internally published reports. Additional research is needed to understand the extent to which the sectoral books have been integrated into their respective ministries.

---

[6] All of the books were in English, but translated from Indonesian. In many instances, the translation was unclear.

The book descriptions and recommendations in the following sections reflect the opinions of the Indonesian government and do not necessarily reflect the researchers view.

Planning

The first planning book gives a history of sustainable development from an academic and policy perspective. The second discusses the existing planning process in Indonesia and its ability to integrate sustainable development principles, as well as differences that exist for implementation between city and rural kabupatens (counties). It also addresses planning processes and methods. The third book addresses indicators.

Energy

Energy use in Indonesia is generated primarily through nonrenewable resources, including oil, gas, and coal. A much smaller (but much greater potential) energy source is from nonrenewable energy such as hydropower, geothermal power, and solar power.[7]

Indonesia currently holds 5.1% of the world's oil, gas, and coal reserves (combined) but has 40% of the world geothermal energy potential (U.S. Department of Commerce International Trade Administration 2010). With 3.5% of the world's population, Indonesia has the potential to meet the needs of its own growing population and also to export renewable energy.

The renewable energy sector continues to receive foreign investment, and the government projects an 8% increase in renewable energy by 2025. However, during the same time period, coal production is expected to increase by 14% (U.S. Department of Commerce International Trade Administration 2010).

The Agenda 21 sector book on energy includes charts listing the challenges for each type of energy. A large number of the problems stem from subsidies and monopolies within the energy industry. These problems are still relevant today.

Additional objectives to increase sustainability among the oil, gas and geothermal sector, coal sector and electricity sector include the following.
1. Eliminate subsidies
2. Improve environmental management
3. Local enforcement to reduce illegal mining
4. Improve business operations of state-owned companies

---

[7] Renewable energy (as described in the Agenda 21 sectoral book on energy) also includes nuclear and peat.

Figure 6 Arial view of refineries in central East Kalimantan.

5. Regional energy planning
6. Increase the use of coal[8]
7. Eliminate monopolies
8. Internalize environmental externalities[9]

The energy sector is one of the most powerful in Indonesia. It is also riddled with corruption and mismanagement. Like energy companies all over the world, they have a hard time shifting not only to renewable forms of energy, but also to more environmentally friendly practices within the nonrenewable industries. The books stated that in order to create lasting change in this sector, drastic changes will be needed from all actors.

Forestry[10]

Following a review of the legal framework regarding forestry in Indonesia and the global and national initiatives related to sustainable forest management (Figures 7 and 8), the sectoral Agenda 21 book on forestry notes the following important features from the global and national initiatives that need to be met:[11]

[8] Increasing the use of coal (although listed as a method for achieving sustainable development) is damaging to the environment and is not recommended as an effective sustainable development policy.
[9] An example of this would be a pollution tax. The negative externality (pollution), which causes damage that is not reflected in the market price, is internalized through a tax.
[10] The first three pages of this book were omitted; it appeared to be a publishing error.
[11] Global and national agreements include timber certification programs, Convention on International Trade in Endangered Species, climate change, and the International Tropical Timber Agreement.

Figure 7 Private protected forest outside Balikpapan managed by the Borneo Orangutan Survival Foundation.

Figure 8 Alternative forest products include honey from honeycombs in trees.

1. Reconciling environment and development—Recognizes that despite the nonbinding nature of Agenda 21, sustainable development efforts should still be pursued because "development that relies merely on economic development could not achieve the very meaning of development which is to improve the quality of life" (State Ministry for Environment and United Nations Development Program 2001).

2. Energy efficiency in all development sectors and the role of forest to sequester carbon dioxide—Recognizes the role that forests have in carbon sequestration.
3. Performance appraisal of forest management by community—A third-party review of forest sustainability should include local communities.
4. Interdepartment coordination in forest management—The current Interdepartmental Committee on Forestry has not been effective because it does not include all partners. A proposed alternative is to create an authoritative department responsible for managing all natural resources.
5. Indonesia's position and commitment on global initiatives—Indonesia should maintain its commitment to international agreements it has ratified, but it should review any initiatives that impair developing countries, particularly those affecting Indonesia negatively.

The book also recognizes conditions that will have to be met in order to achieve sustainable development in the forest sector (conditions that are closely related have been combined).

1. Good governance, consistent enforcement of existing regulations, law enforcement, illegal timber export stopped, abolition of illegal logging
2. Stop converting natural forest areas
3. Maintain a sustainable level of supply and demand of wood products
4. Proper use of regreening fund
5. Reforestation of critical land
6. Halt the allocation of critical land to estate (plantation) projects
7. Maintain healthy forest ecosystems, improve quality of life, biodiversity conservation; diversifying the products taken from the forest (such as resin and medicinal plants), incorporate ecosystem services into the value of the forest (clean water and erosion prevention)
8. Decentralize the authority of forest management[12]
9. Multistakeholder participation
10. Community-based forest management
11. Inter/sectoral coordination

Achieving these conditions will require four main methods; change the concept of how individuals and institutions should use a forest, improve institutions, as well as improve the spatial planning system and collection of data.

---

[12] Forest decentralization (as well as community-based forest management programs) has begun to be implemented since the writing of the Agenda 21 sectoral books (Barr 2006).

Human Settlements

Housing settlements refer to the growth and creation of communities and housing areas, allowing people a safe and healthy place to live and work. The Sectoral Agenda 21 Human Settlements published by Indonesia provides a history of human settlements and the steps needed to create future settlements sustainably.

During the New Order period of Indonesian history (late 1960s when President Suharto first came to power), settlements focused primarily on simply providing housing for all citizens. Today the need has developed to constructing the community, indicating a shift from the immediate need of shelter to providing the infrastructure to support a community, such as running water and sanitation (Figure 9).

Three overarching challenges dominate housing settlements in Indonesia: the transmigration program, the disparity between housing for wealthy and poor, and the lack of environmental consideration in planned developments.

Transmigration is a government program that relocates citizens from the heavily populated areas on Java to the less populated islands, mainly Kalimantan and Sumatra. Transmigration began in 1903 and still continues (The World Bank Group 2001). It receives about 1.5 trillion Rp. (1.7 million USD) from the annual budget. The Ministry for Manpower and Transmigration relocates about 10,000 families every year (Adhaiti 2001). The negative externalities generated by this program include conflict with indigenous

Figure 9 A model of a planned community in Balikpapan.

people, conflict over land rights, deforestation, and improper land use (The World Bank Group 2001). The plan was intended to increase the wealth of those moving, but in East Kalimantan, this has not been the case. Most transmigrants maintain the same level of poverty they had before moving (Adhaiti 2001; Anonymous Interview 2010).

Housing settlements exemplify the separation between the wealthy and the poor. With wealth comes the ability to buy a house and hold papers that recognize the ownership of that property legally. In rural communities, land or home ownership is based on community knowledge, and legal ownership of land can be a contentious issue. Infrastructure (water pipes, waste removal, and electricity) is also more likely to be designed into planned housing communities than fit retroactively into existing poor housing settlements. Additionally, little government assistance exists for low-income housing or home loans, which contributes to illegal housing settlements.

As planned communities continue to grow, little attention is given to the ecological impacts of the development. When they are used, environmental impact assessments (EIA) are often ineffective and subject to misrepresentation (see text box: Environmental Impact Assessment). New housing developments also favor the use of personal vehicles over public transportation.

Necessary actions identified for enhancing sustainable human settlements include the following.
1. Develop institutions and instruments for settlement that meet the housing needs of minorities and the poor.
2. Develop plans and review existing laws and regulations regarding housing (specifically spatial plans) to be effective and implemented at the regional level.
3. Develop a training system for governments and stakeholders to understand and consider alternative approaches to housing developments.
4. Create and disseminate information for the community and local governments to share with citizens on "how to be a good settler," which describe the role and responsibilities a settler needs to interact with neighbors and the community.

Mining

The discussion of sustainable development in mining is centered on changing the "paradigm of mining." The past paradigm was based on paragraph 33; article 3 of the 1945 constitution, which stated, "land and water and its natural resources are controlled by the country and used for maximum prosperity of the people" (State Ministry for Environment and United Nations Development Program 2001). This article has been interpreted to give reason for gross exploitation, as state-owned companies claim profits from the natural resource use benefit the local people and build the local economy. The authors of the Agenda 21 mining sector book argue that this same constitutional article could become a foundation for sustainable development by expanding the paradigm to promote prosperity among *generations*. However, this change is unlikely due to the strength of the existing interpretation of the law, the interpretation that the economic gains made from mining outweigh the effects on local communities or future generations (Figure 10).

In addition to changing the foundational perceptions of mining, the following goals need to be met to achieve sustainable development in the mining sector:
1. Social transformation
2. Effective decentralization
3. Acknowledgment of community rights and community involvement
4. Integrated environmental management
5. Good governance

Existing policies and regulations will need to be improved to reflect consideration for local indigenous communities and potential needs of future generations. Improvements include more strict land use policy, environmental

Figure 10 Open pit coal mining in East Kalimantan managed by Kaltim Prima Coal.

review, permit allocation, recycling programs, and instituting a mine closure policy that includes defining land rights after mining activities have ceased.[13]

Tourism

The tourism sector presents a duality between the economic gains and the potential for cultural loss. In order to meet these and other sustainable tourism goals, three main concepts as needs must be met: good governance, sustainable human development, and human rights. This approach is designed to develop the tourism market by meeting the national development needs.

Indonesian tourism has made a strong rebound following a drop in tourism after the Asian economic crisis of 1997 and the Bali bombings in 2002. Since 1997, tourism has been the third largest foreign exchange earner (State Ministry for Environment and United Nations Development Program 2001).

Five strategic issues are required to create a sustainable tourism industry:

1. Awareness of responsibility to the environment
2. Increase role of central government in regional tourism development
3. Improve the role of regional government
4. Stabilize and empower the tourism industry
5. Community involvement in tourism development

These strategies identify the economic, environmental, and social needs of sustainable tourism. Methods for achieving these goals include activities such as foreign partnerships, standardization of tourism services, law enforcement, waste management, energy conservation, measurement of tourism impact, and integrating tourism into regional development.

As noted in the book, many of these efforts can be coordinated by developing a professional tourism association to monitor and evaluate the progress of meeting sustainable tourism objectives and support cooperation between different actors in the tourism industry.

These books and reports are evidence of the sectoral approach Indonesia has adopted to address its NSDS. In addition to the information detailed in the eight Agenda 21 books, aspects of sustainable development are addressed in the MDG reports, poverty reduction strategy reports (PRSR), and the climate change plan (described further later).

---

[13] Current law (as of 2001) does not require mine closure procedures to be included in the initial feasibility study. Additionally, the mining company is released of its land rights after closure, leading to vast wasteland areas for which the company is not responsible.

The following overview offers a comprehensive description of existing planning mechanisms, including the RPJM. Understanding this process is important because it is recognized by the national government as a means to implement sustainable development and climate change policies.

Existing Planning Mechanisms

The second approach Indonesia has taken to address sustainable development is through the integration of policies into the current national planning documents. Of the three main planning documents (short, medium, and long term), the medium-term development plan offers the best opportunity for analyzing the extent to which sustainable development principles are being implemented at a planning level. Although each plan can provide insight into policy priorities, the RPJM offers an opportunity to evaluate trends and, unlike long-term development plans (RPJP), they can be altered periodically (every 5 years). They are also generally less vague in their policies than the long-term plan.

Development Planning Overview

Indonesia's long-, medium-, and short-term plans should (in theory) influence each other. The national long-term development plan is written every 20 years. This is then reviewed every 5 years, and immediate development plans are created on an annual basis. Figure 11 shows how the RPJP is

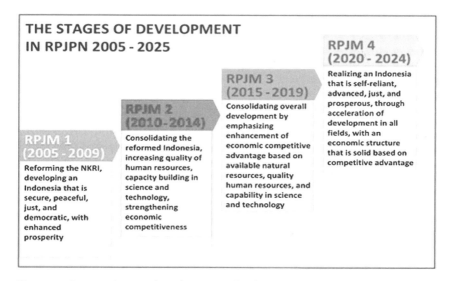

Figure 11 Stages of national medium-term development planning (Japanese Ministry of Land, Infrastructure, Transport and Tourism, 2010).

broken down into four stages, with unique priorities for each field, that evolve to reach the goals defined in the 20-year plan. The RPJM cycle begins whenever a new president is elected. The 5-year plan is then broken down further to an annual plan. This process is replicated at the provincial and local levels. At these levels, the 5-year plan is developed with the election of a new governor or mayor, respectively. Figure 11 further illustrates the national process as based on Law 25/2004.

The RPJM is based on the annual plan. Every January, meetings are held at the musranbang or neighborhood level, consisting of a small group of people living closely together (this would relate most closely to a block in the United States). The musranbang chooses a spokesperson to be their representative at the kelurahan level. The kelurahan consists of about 40 musranbangs. At this point, in addition to musranbang representatives, additional special interests are represented (women's groups, labor groups, etc.), as well as the Lembaga Peubedayan Masyarkat (LPM). The LPM is a community empowerment initiative. In February, kelurahan level (similar to a county) meetings advance to the kecamatan level (Figure 12).

At the kecamatan level, the LPM and kelurahan level staff meet. By this time they have received information from the neighborhoods and special

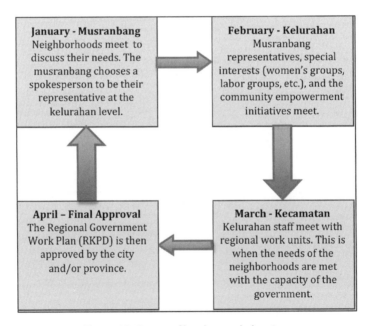

Figure 12 Stages of local annual planning.

interests. In addition, the government agency Satuan Kerja Perangkat Dearah (SKPD)[14] joins in the kecamatan meeting. This is when the needs of the neighborhoods are met with the capacity of the government. Contractors that may benefit from proposed projects also become involved at the kacamatan level. The interests of contractors may influence the policy direction by influencing government staff through kickbacks if certain projects are approved.

Then, the SKPD and additional planning staff develop the annual plan. The regional government work plan Rencana Kerja Pemerintah Daerah (Regional Government Work Plan) (RKPD) is then approved by the city or district if it is in a rural area. If the city's plan involves a project that affects the province (overlaps borders or uses provincial funds), the provincial government will also need to approve the RKPD (Figure 13).

The budget for annual plans is often directed by the medium-term development plan and current economic conditions. If the needs of the musrenbangs are the same (e.g., three areas may request improved sanitation facilities) and funds are insufficient to pay for all, the money will first go to the area that is the most lacking and then in subsequent years the other areas will receive the needed funding.

Three influential factors are used in determining if projects proposed in the annual plan are balanced among the sectors of sustainable development:

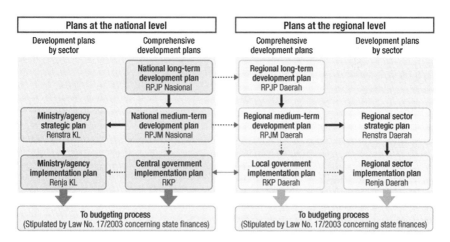

Figure 13 Development planning chart (Japanese Ministry of Land, Infrastructure, Transport, and Tourism, 2010).

---

[14] Satuan Kerja Perangkat Dearah translates in English to "regional work units."

(1) strong mayoral or governor commitment that is shared with government agencies, (2) funding sustainable development programs, and (3) neighborhood education to raise awareness of the need for developing programs that meet immediate wants of the community and support long-term sustainability.

Spatial Planning

In addition to long-, medium-, and short-term development plans, each level of government also completes a spatial plan, which is developed by the planning department and approved by the relevant elected official (president, governor, mayor). This plan is focused more precisely on land use than the development plans mentioned previously. The spatial plan is also a 20-year plan that is reviewed every 5 years. The review period allows changes to be made. In theory, any proposed change to the spatial plan must undergo an environmental review.

A builder must submit an environmental impact assessment before construction begins. The document is submitted to the local government, which then reviews and offers suggestions for improvement, if needed, to ensure compliance. Buildings must fit within the designated zone (residential, commercial, etc.) as defined in the spatial plan, as well as have wastewater treatment and meet other environmental standards as outlined in provincial and national laws. If the proposed construction meets these requirements, the local government will review it internally for approval. There is no public review process. Additionally, if a proposed project does not fit within the proposed spatial plan but is introduced as part of the local annual plan, it is eligible for construction.

> **Environmental Impact Assessment**
> The concept of environmental impact assessments was first introduced in 1986. Even then, the law was criticized for its weak enforcement powers and implementation. It was later strengthened in 1992, 1997, and 1999.
>
> The monitoring and coordination of EIA are managed in Jakarta but implementation occurs at each of the 27 provinces.
>
> In theory, an EIA is required for any project that will have an impact on the environment or land use.
>
> However, companies complete the EIA for their own project and submit it to the local government. Once the local government approves it, it is sent to the national level. National staff said no project has been denied at the national level.

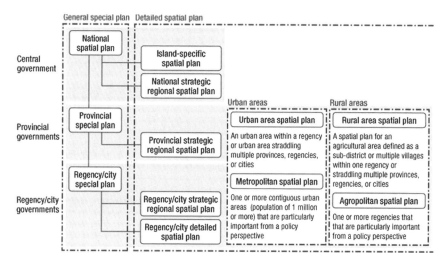

Figure 14 Spatial planning chart (Japanese Ministry of Land, Infrastructure, Transport, and Tourism, 2010).

After construction is complete, the site is monitored for environmental quality (mainly wastewater) three times a year. If they are found out of compliance, they are given a time period in which to improve the standards. Balikpapan does not currently have local regulation so they enforce the national and/or provincial standards (Kota Balikpapan, 1999).[15] Without having standards, the local government lacks its own ability for regulatory enforcement, leaving them unable to punish offenders. It is only if the national or provincial standards are violated that the local government will occasionally work with the national government to punish offenders. The national government also assists with regulation of large, state-owned or regulated institutions such as the airport or Pertamina refinery.

Environmental standards are currently enforced primarily for industrial sites and businesses. No regulation exists for housing in Balikpapan.[16]

The existing planning process (development plans and spatial plans) is widely recognized throughout each level of government and among multiple stakeholders. However, much work is needed to develop plans that adequately include environmental policies as an equal component of sustainable development (Figure 14).

---

[15] Standards are currently being developed and are expected to be complete by November 2010, with enforcement beginning in 2011. As of September 2010, no update or draft of the standards was available for review.

[16] Interviews with government officials were contradictory. Some officials said large housing developments were required to comply with all of the environmental review standards, but others said they did not. All staff interviewed stated that individual houses were not included in the regulation.

Additional Methods of Sustainable Development Planning
The national government also addresses sustainable development indirectly through existing programs.

Poverty Reduction Strategy Report
A poverty reduction strategy often incorporates many elements of a NSDS. An evaluation of how sustainable development goals are incorporated into Indonesia's PRSR was completed by the United Nations Environmental Program (UNEP) (United Nations Environmental Program). The findings of the study indicated that efforts to include an environmental or sustainable development perspective were not integrated in the PRSR, although they were recognized as an important element.

Millennium Development Goal Report
The MDGs are composed of eight goals:
- Goal 1: Eradicate extreme poverty and hunger
- Goal 2: Achieve universal primary education
- Goal 3: Promote gender equality and empower women
- Goal 4: Reduce child mortality
- Goal 5: Improve maternal health
- Goal 6: Combat HIV/AIDS, malaria, and other diseases
- Goal 7: Ensure environmental sustainability
- Goal 8: Promote global partnership for development

Indonesia's central government actively develops and monitors policies to meet the MDGs. The 2007 MDG report showed mixed results. Poverty reduction goals are being met, but the portion of the population suffering from hunger is increasing. Education and gender goals are being met, but AIDS/HIV rates and maternal mortality rates are increasing. A larger percentage of the population has access to running water and sanitation, but deforestation and energy use are increasing. Economic indicators are increasing steadily, as well as access to phones and the Internet (Stalker 2007).

Climate Change Planning
Climate change planning in Indonesia has fundamentally altered the issue of sustainable development within the country. Indonesia's goal to reduce emissions by 26% (41% with international support) by 2020 is one of the most aggressive in the world (Ministry of Finance 2009). The effort required to meet this goal will be demanding on the national budget and the technical capacity of the country.

Climate change is currently being integrated into the next national medium-term development plan (RPJMN 2010–2014). The RPJMN is divided into three books, which address climate change issues as follow.
1. **National priorities**—Addresses agriculture and food, energy, peat and forests, and climate adaptation
2. **Sectoral priorities**—In addition to national priorities, sectoral priorities also include waste, transport, and industry
3. **Regional priorities**—There is no specific direction on climate change policy included in regional priorities in the RPJMN.

In 2007, Indonesia published the first National Climate Action Plan (CAP). Its stated objective is "to address climate change for it to be used as guidance to various institutions in carrying out a coordinated and integrated effort to tackle climate change." By creating a foundation for a "coordinated and integrated effort," climate change policy has a solid chance of developing in a sustainable manner. (State Ministry of Environment, 2007).

The CAP includes mitigation and adaptation measures by sector. Some sectors overlap with those addressed in the Agenda 21 books (energy and forestry), whereas others do not (fisheries and marine, land use, water, infrastructure, health, and agriculture). Integrating climate change policy into all sectors is important to ensure a long-term and integrated approach to reducing greenhouse gas (GHG) emissions.

Addressing issues related to climate change also offers an opportunity to improve data collection and monitoring. The Indonesia Center on Climate Change, set to begin operation in 2010, will carry out research on forests,

---

Bersih, Hijah, Sehat

Clean, Green, and Healthy

In Balikpapan and other cities throughout the archipelago, local governments are participating in the "Bersih, Hijau, and Sehat" (Green, Clean, and Healthy) initiative. In Balikpapan, this includes a compost program, which reduces waste and also meets the need for improved sanitation. Many cities are developing waste management programs as part of this initiative.

This program is described as a sustainable development project because its objective is to improve "the air, the land, and the water." This program addresses the environmental aspect of sustainable development and helps raise public awareness for the environment.

These figures show the public awareness campaign for Balikpapan's "Clean, Green, and Healthy" campaign.

peatlands, oceans, renewable energy, and adaptation in order to improve the relationship between science and policy. One of the main responsibilities of this center is to monitor carbon dioxide and methane emissions (Simamora 2010).

Although current climate change plans have very few real consequences and little influence over local government, efforts are being made by organizations such as APEKSI to incorporate the interests of local governments in national planning, particularly in implementation.

ANALYSIS OF SUSTAINABLE DEVELOPMENT INDICATORS

Indicators are useful for evaluating progress on sustainable development. The United Nations Commission on Sustainable Development has developed a comprehensive set of indicators used to evaluate national sustainable development strategies. These indicators allow national governments to track economic, social, and environmental progress. Appendix 2 shows results from a collection of data corresponding to the UNCSD indicator set. In addition to national data, provincial data from East Kalimantan and local data from Balikpapan were also collected using the same UNCSD indicators.

The availability of data collection was dependent on the level of transparency, as well as the technical capacity of the staff. In general, data that are also part of another established indicator set (such as those that overlapped with MDG indicators) were more available. Also, the national government had more data available than other levels of government. A possible reason for this is the responsibility the national government has for meeting internationally established goals. International aid funds as well as foreign direct investment would demand transparency and information, particularly of economic and social indicators. This demand for transparency and increased access to environmental indicators is likely to increase for environmental indicators as Indonesia receives funds through the Reducing Emissions from Deforestation and Development (REDD) program.

In addition, the national government may have a disproportionately high amount of data available because the indicator set is created for NSDS. Therefore, data may inherently be data collected at a national level. For example, "investment share in GDP" and "proportion of marine protected area" are easier to measure at a national scale.

It is inappropriate to use the results of this indicator set to make comparisons between the levels of government. They each operate at a different scale and with different levels of capacity. It is interesting, however, to see the differences between certain indicator results. For example, results of the first indicator, "portion of population below $1 a day," show Balikpapan to have the lowest rate (3.49%), even though it is part of a province that has a higher rate (9.51%) than the national level (7.5%), although other poverty indicators, particularly those that reflect infrastructure, such as "population using improved water source" and "population using improved sanitation," show the province and the city to be higher than the national level. The author speculates that private companies operating in the province (particularly in rural areas) spur infrastructure development. In Balikpapan, a significant portion of the city's wealth (and therefore ability to provide infrastructure) is also raised from private companies through either taxes or corporate social responsibility (CSR) programs.

Additional results related to family planning were significant. The contraceptive prevalence rate for the nation is 61%; however, East Kalimantan and Balikpapan rates were about 48% lower (13 and 12% use rates, respectively). During interviews, no additional information was gained as to the reasons behind this results—one individual speculated that the people of Balikpapan have more education and therefore know that contraceptives are not needed to prevent pregnancy and can potentially be harmful to a woman's health. Data show no significant correlation, however, between education enrollment rates and literacy rates (both of which are in the 90th percentile for all levels of government) and contraceptive use. It would be helpful for future research to include how the contraceptive prevalence rate information is obtained. Some women and girls may be using contraceptives but may not feel comfortable sharing that information for cultural reasons.

These are just some examples of the information that can be gleaned from these data. Overall, sustainable development indicators are key to monitoring long-term trends toward meeting sustainable development goals. Results shown in Appendix 2 provide a baseline for sustainable development at each level of government, which will be useful for governments in coming years as policies continue to be implemented.

RESULTS FROM EVALUATION OF PLANNING PRIORITIES AND BUDGET ALLOCATIONS

A review of the budget and planning documents offers an opportunity to determine if budget provisions are supporting the policy goals defined in the planning documents. Table 1 shows results of the planning policies analysis, and Table 2 shows results of the budget analysis. The percentage

Table 1 Planning Policies by Sustainable Development Sector

Government level	Environmental	Social	Economic	Other
National[a]	6%	62%	32%	15%
Provincial[b]	20%	70%	43%	15%
City[c]	26%	66%	24%	32%

[a]RPJM 2004–2009.
[b]RPJM 2009–2014.
[c]RPJM 2006–2011.

Table 2 Budget Items by Sustainable Development Sector

Government level	Environmental	Social	Economic	Administrative
National[a]	5%	35%	41%	27%
Provincial[b]	3%	79%	24%	43%
City[c]	1%	56%	34%	8%

[a]2006 APBN.
[b]2009–2010 proposed APBD.
[c]2007 APBD.

indicates the portion of the total planning policies or total budget allocation dedicated to the respective sustainable development sector.

Each policy was categorized according to the sustainable development sector it addresses. For example, "increasing investment in oil and gas sector" was identified as meeting the economic needs, "improved quality of life and role of women and children welfare and protection" met a social need, and "improving management of natural resources and conservation of natural environment functions" is clearly an environmental-related policy. Many of the policies satisfied more than one sector: "increasing community access to quality education" addresses both social and economic sectors and therefore is counted in both sectors. An additional category was added to include the administrative and political sectors, such as "statistical archiving" and "arrangement of regional government institutions." Results, shown in Appendices 3–5, are summarized later.

Detailed descriptions of the policies and budget items are included in Appendices 3–5. The following section offers further analysis of the results.

ANALYSIS OF PLANNING PRIORITIES AND BUDGET ALLOCATIONS

National

The national midterm development plan used for this analysis was from the years 2004–2009; the budget year used was 2006 (Beppenas, 2004; Department

Keuangan Republic Indonesia). Policies at the national level were generally vague and unquantifiable; they were listed more like goals than policies. These goals then directed policy.

The social category received the greatest policy attention and funding, followed by economic policies. Policies related to the environment had the lowest percentage of all policies. Even the "admin" category (which included administrative policies, such as "consolidating foreign policy and enhancing international cooperation") was higher than environmental policies (Figure 15).

When analyzing results of policy programs against budget funding, results show that the economic sector has a 9% difference (41% minus 32%) between policies directed at economic activities (32%) and the amount of budget funding allocated to economic institutions and programs (41%). This is in direct contrast to social programs, which constitute a large portion of the planning policies (62%), but appear to be underfunded by the government, with low budget allocations (35%). The environmental sector receives the least policy attention and has the smallest budget allocation. The other category receives a significant portion of the budget, most of which (about 18%) is spent on interest payments. Interest payments are not included in the economic category because they do not contribute significantly to economic growth and the high percentage would skew the results of the economic policy category.

Of the national budget, 49.74% is spent on subsidies and debt interest payments, as shown in Appendix 3. Debt interest is categorized into the "admin" category, and subsidies are included in the economic category. Although the argument could be made that subsidies also benefit society, it was included in the economic category for this analysis because a large portion of national subsidies are given to the energy sector, very little of which benefits the energy users in Indonesia. Most subsidies benefit the companies.

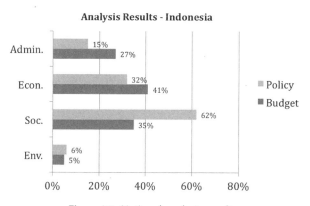

Figure 15 National analysis results.

Additional research is needed to determine the breakdown of subsidies in Indonesia and whom they benefit.

Further research is suggested to ascertain how future policy will be influenced by actions taken to address climate change. The national planning ministry has begun to incorporate climate change into the RPJMN, which is expected to increase the percentage of policies focused on the environment (Beppenas – National Ministry of Planning, 2005). It will be equally important to monitor national funding. The Indonesian president has committed to a 26% reduction in GHG emissions. This portion will need to be funded independently by the government. The president also agreed to an additional 15% reduction if the international community offered support. Indonesia has already received millions of dollars in international support, funding that is transparent and has accountability measures preset. It is not clear, however, where the domestic portion of the funding contribution will come from.

Provincial

The 2009–2014 RPJM was used for analysis. This is the current RPJM (local and national RPJMs are from the previous 5-year cycle) and therefore are more difficult to evaluate. In addition, the budget is projected (2009–2010) and subject to change (again, local and national budgets show actual expenditures rather than projections) (Bappeda/East Kalimantan Ministry of Planning, 2009; Kalimantan Timur, 2009).

In addition to the policy document, environmental/sustainable development goals are expressed in the "Kaltim Hijau" (Green Kaltim) declaration (Province of East Kalimantan, 2010). Less than one page long, this declaration was a response to the threat of climate change. It declares the following:

> We, the Government of East Kalimantan, together with the Government District in East Kalimantan, communities, universities, private circles in East Kalimantan, deem it necessary to do actions that aim to:
> 1. Improving quality of life of people of East Kalimantan as a whole and balanced, both in economic, social, cultural and environmental quality of life.
> 2. Reducing the threat of ecological disasters such as flooding, landslides, droughts, forest fires and land in the entire region of East Kalimantan.
> 3. Reduce pollution and destruction of the quality of terrestrial ecosystems, water and air in East Kalimantan
> 4. Increased knowledge and awareness institutionalize all parties, both governmental, private, and the people of East Kalimantan, against the interests of natural resource conservation and wise use of renewable natural resources [that] are not renewable

For that, we, the Government of East Kalimantan, together with the Government District in East Kalimantan, communities, universities, private sectors in East Kalimantan, declared a "Green East Kalimantan."

This declaration (like the RPJM) gives no quantifiable goals or specific policy changes needed to reach the goals. Additionally, the 2009–2014 RPJM still shows the environment receiving the least policy attention.

The actions of the provincial government often seem to be in direct contradiction to the goals outlined earlier. Natural resources are being exploited in East Kalimantan, and the lack of oversight by the government provides little repercussions for offenders. Illegal logging is one of the most important sustainable development issues for East Kalimantan, but action on this issue is not encouraging. For example, between 2007 and 2008, the number of forest crimes prosecuted dropped 17%, although the number of crimes reported increased (The Jakarta Post 2010).

Large sections of once healthy forest lay as wasteland from mining and forest concessions. The provincial government does not work to rehabilitate these areas. Instead it supports infrastructure projects that would continue to expand access to natural resources (legal and illegal) further into the central and northern parts of the province under the guise of poverty reduction. In reality, areas with the highest contribution to national GDP (from resource extraction) remain the poorest in the region, as Appendix 2 shows.

Results of analysis show, unlike the national government, that social policies are well funded in relation. The environment and economic sectors, however, appear to be underfunded. Additional funding sources from NGOs and international organizations may help narrow this gap for the environmental sector (Figure 16).

The RPJM also includes many policies directed at developing the border region (the northern part of East Kalimantan borders Malaysia). It

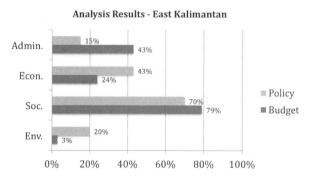

Figure 16 Provincial analysis results.

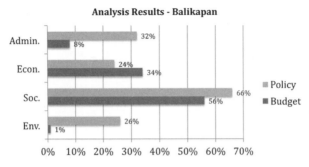

Figure 17 Local analysis results.

is possible that policies regarding this are included in the RPJM, but not the budget with the hope that it will increase the chances of receiving national funds, as these projects are of national interest. This could partially account for the economic policy being higher that the economic budget allocation.

Local

Policy priorities in Balikpapan are weighted toward social policies. Economic and environmental policies were addressed in 24% and 26%, respectively, of all policies in the 2006–2011 RPJM (Balikpapan Ministry of Planning, 2007). When policy priorities are evaluated against actual budgetary provisions, it is clear that spending is generally aligned with policy goals for the social sector (Badan Pusat Statistik, Balikpapan, 2009). However, the largest negative discrepancy is related to the environmental policy. The policies constitute 26% of all policies listed in the midterm plan, yet just 1% of the budget is spent on environmental offices or projects, a difference of 25%. While the economic planning policies also constitute 24% of all policies, the budgetary provisions actually increase to 34% (Figure 17).

This does not indicate that environmental policies are not being met, however, just that the local government is not paying for them. Environmental programs are often funded by outside aid organizations or by corporate social responsibility programs. CSR programs are active in Balikpapan, but are mainly limited to small-scale wastewater treatment and solid waste programs.

Like East Kalimantan, Balikpapan has a statement of its environmental policy (Kota Balikpapan, 2009).

Environmental Commitment Balikpapan City up to 2030

Long-Term Goals
Protect at least 52% of Balikpapan as "Green Area" to be achieved through:
1. Committed according to established land use planning
2. Forest and land rehabilitation program
   a. Increase reforestation and planting activities
   b. Increase forest fire prevention management
3. Management and development of city cleanliness program
4. Blue sky program: clean development mechanism, methane gas energy utilization
5. Integrated coastal and marine management program
6. Drinking water and wastewater management program
7. Botanical garden development to conserve native plants (such as *Dipterocarps*), to improve public awareness, and to promote botanical research and ecotourism.

Short-Term Target 2010–2012
1. Reduce amount of solid waste to landfill by 10% within 3 years.
2. Involve NGO and citizen groups in all environmental activities to contribute to local environment budget by 10%.
3. Increase the number of schools that enclose "Adiwiyata" Program by 30%.
4. Increase the quantity of existing green area by at least 15 ha in 3 years.
5. Increase the number of access to clean water supply by 72% of Balikpapan society until 2012.
6. Increase the number of access to wastewater treatment facility by 7% until 2012.
7. Develop conservation plan for Balikpapan Bay Biodiversity protected area (20% of total area).
8. Monitor air quality by installing additional air quality monitoring devises.
9. Sustainable biodiversity by conserving Sungai Wain and Manggar Protected Forest Balikpapan, including:
   a. Maintaining "screen burn" area at least 10 km each year.
   b. Preventing both illegal logging and illegal mining.
   c. Constructing botanical garden.

The Measurement of Activities to Achieve
1. Composting
   a. Four units of composting will be located in traditional markets.
   b. Fifteen composting centers will be located in five districts (3 units for each district), 10 units in Islamic boarding schools, and 1377 aerobic composters will be distributed within Balikpapan city neighborhoods.

c. Establish three composting centers as pilot projects.
   d. Apply concept of fund sharing in funding composting process by 10% in 3 years.
2. Fund support from NGOs and sharing each 5% of CSR budget/multi-stakeholders.
3. Existence of Center for Environmental Education at sanitary landfill in Manggar.
4. The green movement: "one man one tree" through a variety of activities; for instance, promotion, wedding, school program, celebration of anniversary.
5. Increase housing connection to water supply to 81,600 units.
6. Conservation of world endangered species, especially animals in Balikpapan Bay (freshwater dolphin, bekantan monkey, dugong, green turtle).
7. Add two units of air quality monitoring devices.
8. Sustainable biodiversity by conserving Sungai Wain and Manggar Protected Forest by:
   a. Organizing training and capacity building twice a year.
   b. Extending "screen burn" by 10 km each year.
   c. Decreasing square area of illegal logging to 0.1% (from 1000–3500 to 1–3 Ha)
   d. Establishing a botanical garden in Balikpapan within 5% of the total green area.

This plan, if implemented, will address the policy gap that currently exists in the RPJM for environmental policies. The policies given here address some of the current deficiencies in data collection, as shown in the sustainable development indicators chart (Appendix 2), such as solid waste management and wastewater treatment. It also sets measurable goals. It will be important for an evaluation to be completed in 2012 to monitor progress.

The two largest economic-related budget provisions in Balikpapan are public works and transportation. The categories are also included in the social sector because they improve social well-being, as well as contributing to a functioning, robust economy. This categorization matches the other levels of government, but Balikpapan seems to be spending a larger percentage on these items. The rate of spending is likely a result of the fast-paced growth the city is experiencing.

The existing corporations in Balikpapan make it the hub of economic growth in East Kalimantan, which is also increasing the social well-being of its citizens through improved infrastructure. This mutually beneficial situation shows how different aspects of development can works together; however, it is incomplete and unbalanced because it does not include the environment.

> **Balikpapan Bay**
> The city of Balikpapan's efforts to preserve 52% of its area as green space are currently threatened by a road project that would cut directly throughout the city's forested area, including mangroves.
> The city would like to build a bridge across a narrow stretch of the Bay, which would connect to South Kalimantan with minimal impact to the environment. However, the Province of East Kalimantan is instead proposing a road that would extend around the Bay and disrupt critical habitats.

Overarching Issues

The following issues are relevant across various levels of government.

Integration among Levels of Government

There are significant challenges with communication among the various levels of government. Much of the breakdown happens at the provincial level. This is true across all sectors of government and is not unique to sustainable development. This breakdown has broad implications for land use and regulation enforcement. New environmental laws and regulations are usually passed from the national government to the provincial government and then distributed to regencies and cities within the province.

Through interviews, it was revealed that this information is often not shared with the relevant staff members throughout the provincial government. The information is also disseminated poorly to rural areas that would be influenced by the new regulations. This often means that those with the most access to information (through direct and indirect sources) are most aware of new regulation, leaving the rural communities without access vulnerable to exploitation. For example, if a small community in central East Kalimantan is not informed of a new law that requires companies to consult with communities before submitting an EIA, there is a possibility that the company will not fulfill this obligation, despite potential negative effects to the community. Conversely, if a law is passed that bans the removal of certain forest products (including removal for traditional purposes), individuals could be punished for violating a law they did not know existed.

The effective implementation of national laws by the local government has been an ongoing struggle since decentralization began in 2001. Growing pains are expected after 30 years of centralized government, but the realization of good governance today seems far from reality (Syaikhu 2001).

Although local and provincial government officials often had a difficult time defining what sustainable development meant, in practice they

> **Logging Moratorium**
> A moratorium on logging permits was announced in May 2010 with a proposed start date of January 2011.
> The moratorium does not currently include existing concessions, only future concessions. This could increase the possibility of concessions being postdated to ensure they will be grandfathered in.

understand the need to achieve economic stability and growth, environmental health, and social improvement. It is clear, however, that despite a cognitive recognition of this need, policy development and implementation are lacking. For example, the current mayor of Balikpapan has taken significant steps to preserve the forested areas of the city and is a strong proponent of "green" initiatives. He has designated 52% of the city to remain in its current forested state. However, this is not included in the long-term development plan and because the short-term (and medium-term) development plans are altered with the election of a new mayor, this 52% conservation area could change.[17] In addition, there are no other policies to ensure that as the community continues to grow the development will be "in-fill development" (essentially referring to growing "up" rather than "out"). Without strong policies, as the population grows, development will expand and contribute to urban sprawl.

Audience

Audiences influence the amount of commitment each level of government makes to sustainable development. The national government caters much more to international pressure and regulations than the local or provincial government. Local and provincial governments report to the national government and to their local constituents. One of the difficulties with the provincial government and the national government is the distance between policymakers and constituents.

Individual participation in governance is not ingrained in the Indonesian culture as it is in Western-style democracies. One of the possible reasons for low involvement is due to the high rates of corruption and a general lack of perceived influence. The level of active involvement by the relevant audience is crucial to the implementation of sustainable development programs. If there is no oversight, governments can easily produce rhetoric that appeals to the masses but does not reflect reality.

---

[17] Protection of the 52% green area is being integrated into the spatial plan.

Legal Obstacles

There are three main problems associated with laws and regulations in Indonesia. First, current laws and regulations are not entirely reflective of future policy desires. Second, laws are not harmonious between different ministries. Finally, there is ineffective enforcement of laws and regulations.

Past laws allow protected forest areas to be used for certain types of natural resource extraction or to be eligible for concession. However, this contradicts future efforts to preserve forested areas for carbon sequestration as part of climate change legislation.

Different ministries act consistently in their own interest. Occasionally, even different offices within the same ministry will work on contradicting interests. For example, the Forest Ministry has an office for production, which encourages the economic development of forest resources, and an office for conservation, which advocates for the preservation of protected areas. This issue has received a lot of attention recently when it was announced that palm oil plantations would be eligible for carbon credit. This is clearly contradictory to environmental goals because palm oil plantations are one of the most notorious causes of deforestation. Yet, palm oil is a major economic interest with a lot of influence in the national government. Most often, when there are contradictory interests the one with the most funding is the one that achieves its policy objective.

Most environmental laws have loopholes that act against public interest. Although this can happen at any level of government, it is particularly common at the provincial level. For example, if a private company wants to log in a forested area not designated as a production forest it will need a release granted by the national authority. The company can go to a local government official and use whatever means they deem necessary to convince that government official to recommend the area to be released, which will then lead to a concession. The local authority will send the concession request to the national government, which will grant a release. This process leaves obvious room for corruption to occur. Additionally, the national government does not have the resources to investigate each application to ensure due diligence (Christopher Barr, 2006).

Private Sector

Corporate social responsibility is a popular phrase in Indonesia; however, its interpretation varies. On the very first day of the author's research in Balikpapan, she attended a meeting of a NGO consortium in Balikpapan. One of the members gave a presentation on corporate social responsibility.

> **The Pretty Ones Win**
> In Indonesia, companies are rewarded for their sustainability reports. Sadly, the award is just given to the best looking and most complete report, not to the actual CSR actions the company performed.

It was interesting to learn that the popular consensus of most members of the group had been that CSR was defined as money given to the community by a company with little to no oversight or regulation. It was a foreign concept that corporations have a responsibility for their own internal actions as well as the need for an ongoing commitment to whatever project they fund. This is understandable because that belief reflects what companies have done historically.

Through CSR, corporations are taking on many responsibilities that normally fall to government. Chevron has been operating in Indonesia since the 1970s, but its CSR project began in 2001. Much of their work focuses on education and capacity building because it has long-term implications. In their rural work areas, their goal of capacity building often goes against the local desire for infrastructure development. Although both are needed, as one Chevron employee noted, the difference is teaching man to fish vs giving a man fish.

> **Corporate UN-Social Responsibility**
> An employee of a coal company working in East Kalimantan spoke anonymously of his work as a CSR program coordinator. At the time the author and he met (April 2010), he was involved in negotiations with a local community to settle on a "CSR" payment amount. The company was offering the community money to relocate so they could use the site to dispose of mining waste.

In addition, in Balikpapan (and many other major cities), private companies have taken a strong role in solid waste management. Chevron introduced waste separation (organic and inorganic) to the neighborhoods directly surrounding its company compound in Balikpapan. The success of the project led to additional facilities being set up in other neighborhoods. With additional funding from a Japanese aid organization, neighborhood leaders will receive training on how to use the new tools (mainly compost boxes). These are two examples of services that normally fall to government, being supported and funded by a private company, which provides a needed service, but may also create a dependency situation that may not be manageable in the long term.

Corruption and Illegal Activity

Corruption damages the credibility and ability for Indonesia to meet its development goals. Acts of corruption and nepotism have done irreparable harm to the country for hundreds of years through lost revenue and services (Alberts, 2010). Today, the government has established a government institution that carries the power and anonymity of the Supreme Court. The Komisi Pemberantasan Korupsi (Corruption Eradication Committee) (KPK) is charged with finding and sentencing individuals and institutions suspected of corruption. This is an important first step, but stronger enforcement is needed. Scandal seems to follow every case. Those convicted are found to be living in luxury prison suites or move to Singapore prior to arrest. At other times, when those arrested share information about companies or individuals that are also corrupt, the information is disregarded. In an ironic twist, there is even speculation of corruption within the KPK. The Transparency International's Corruption Perception Index shows an increase in perceived corruption in Indonesia since KPK was introduced (National Coalition of Indonesia for Anticorruption 2009).

> **Corruption over Coffee**
>
> In Jakarta, the author had a couple of hours between interviews. She went to a local Dunkin Doughnuts to use the Internet and have a snack. Sitting near her were four men. Two were from the Indonesian government (in uniform) and two were from China (they were in business attire but made several phone calls in Chinese). Their shared language was English.
>
> The discussion revolved around the location of a ship and the ship's capacity (the number of logs it could hold). The ship would be waiting off the coast of a city in the northern region of East Kalimantan.
>
> Discussion and phone calls were made when deciding the price of the logs. After an agreement was made, they discussed what would be changed on the official documents (names removed, etc.). After the decisions were made, two men stayed behind to create the documents. When the author walked past them it was clear that the documents were official Indonesian government documents (official documents in Indonesia have a similar heading with a circular stamp and signature on the bottom).

RECOMMENDATIONS

Improving Sustainable Development in Indonesia

Recommendations are intended to create more balance among the three pillars of sustainable development (Central Intelligence Agency, 2010). In order to achieve this, more policies and budget allocations are needed to address the environment. As a developing country, Indonesia will need to achieve this increase without reducing social and economic programs. Most of the policy recommendations achieve this; however, some recommendations, such as "enforcing market incentives for sustainable development," would mean removing existing subsidies. These recommendations may cause temporary economic hardship for some industries, but the long-term benefits would outweigh any short-term loss. Additionally, these recommendations are consistent with the Indonesian economic policy of reducing energy subsidies.

Recommendations are based on the review of the current national sustainable development plans (Agenda 21 books) and the current planning process, as well as baseline indicators. Special consideration is given to the issue of climate change because it currently has the most potential as an avenue for strengthening sustainable development in Indonesia. Climate change planning is not a strong factor in the RPJM evaluated in this analysis because they were written before climate change became a strong national priority.

Many of the following suggestions complement the recommendations set forth in the Ministry of Environment's report "From Crisis to Sustainability; Paving the Way for Sustainable Development in Indonesia: Overview of the Implementation of Agenda 21 (Ministry of Environment, Republic of Indonesia, 2002)." Select recommendations are highlighted (noted with an asterisk). Logistical constraints prevented in-depth details on all recommendations.

National Recommendations

a. **Strengthen provincial government.** Following decentralization, many provincial governments remain weak and lack the capacity to implement national regulations or to develop their own effective regulations. The central government and citizens from the province would benefit from a stronger provincial government—specifically, a stronger ability to monitor and verify resource-use permit applications and environmental regulations. The national government has a responsibility and an interest to help provincial governments improve.

b. **Strengthen and enforce existing conservation laws and regulations.** Current laws and regulations are contradictory, ill-enforced, and/or blatantly disregarded. Without laws and regulations that force natural resource companies to be accountable for their actions (or inactions), laws will continue to be ineffective. Current laws and regulations are inconsistent and/or unclear with many opportunities for loopholes. This is especially true of environmental laws. As a result, many actors (loggers, conservationists, government officials) are able to bypass laws intended to protect the environment. Before continuing to advance similar, ineffective legal mechanisms, an analysis should be completed to evaluate existing laws.[18] (This is especially relevant to legal issues related to climate change and deforestation.)

c. **Establish a system for monitoring sustainable development.** The reports completed in 2001 offer a strong foundation for establishing sustainable development in Indonesia. Without monitoring of goals, strengthening the ability of sustainable development sectors, and strengthening local institutions, they have little value.

d. ***Create an Indonesian Civilian Conservation Corps.** With support of the international community, much attention has been directed to creating forest protection areas. One of the main challenges is ensuring that illegal logging does not continue in protected areas. Creating a well-paid and local team of people to work actively to protect the forest will provide a win–win situation of increasing the capacity of local citizens to ensure the protection of future resources. At the same time it will provide a livable wage and help the country meet its carbon reduction targets.

Modeled after the civilian conservation corps implemented in the United States, this program creates public works jobs for unemployed persons. By offering a fair wage (that could be higher than most civilian wages if the program uses REDD or other international funding) to rural citizens on forest protection and monitoring, individuals have a stronger economic incentive to work for an industry that helps the environment. Training programs could be offered in tourism development, forest ecology, and monitoring.

One of the main challenges associated with resource extraction in rural areas is finding another way to meet the same level of economic prosperity it brings to the region. However, as was shown earlier, this prosperity does not often reach the people. By training and paying the people directly it is possible to break the current cycle of funds being used by the government. Additionally, by increasing their income, the

---

[18] The Indonesian Center for Environmental Law has begun an evaluation to determine how new laws would impact existing laws.

locals will have increased purchasing power, which has the potential to increase the health and education of their children.

The Wehea Forest (described previously) provides an example of citizens who may benefit from this type of program. The men and women currently monitoring the forest are recruited from the local community schools and are paid a meager wage from provincial government funds. Most of the workers are teenagers and not yet married. Their wage (about $100 USD per month) could not support a family and, when the time comes, most leave to work in the palm oil industry. The manager of the forest has ambitions to increase the staff in order to monitor a larger section of the forest and to offer training for his staff, as well as increasing their wages. Doubling the wages per month and the number of staff would cost approximately $6000 USD per year. Training and additional monitoring equipment may cost an additional $20,000 USD, which would be a one-time cost. REDD program funding in Indonesia is expected to reach more than $1 billion USD (Governments of Norway and Indonesia 2010).

e. **Increase funding for APEKSI to initiate programs with ICLEI.** APEKSI is an existing government organization that coordinates information sharing between cities and helps foster communication between local and national governments. Increased funding for APEKSI would allow new programs to develop, such as a pilot for cities to join ICLEI. ICLEI offers climate change adaptation and mitigation programs and support that allow cities to monitor their GHG emissions and develop mitigation measures. By combining the expertise of these two programs it is possible to improve the technical capacity of local governments and the communication between local governments.

f. **Enforce market incentives for conservation.** Remove or reduce subsidies on polluting sectors that distort the price of energy and limit the growth of nonrenewable energy industries.

g. **Clarify the relationship between climate change and sustainable development.** Currently, these two aspects seem to be aligned so closely that it is difficult to distinguish between the two. While climate change policies should be sustainable and sustainable development policies should address climate change, replacing a comprehensive sustainable development strategy with a climate change policy is not recommended. Climate change does not address poverty linkages, education, toxic run-off, etc.

Both climate change and sustainable development require a coordinated effort. Clarity should be given to who manages each process. Currently, the Planning Ministry is managing the majority of the climate change policy, leaving an opportunity for the Environmental Ministry to address sustainable development.

Additionally, the sectoral approach to climate change mentioned previously (agriculture and food, energy, peat and forests, climate adaptation, waste, transportation, and industry) is similar, but not the same as the current sustainable development sectors (energy, forestry, human settlements, mining, and tourism). The two sets of sectoral analysis should be coordinated to not only complement each other but to also give a more holistic approach to each program.

h. ***Increase environmental policies in RPJM.** Indonesia is spread across different climates and ecosystems. Each region has a unique set of needs and abilities to address their needs. The next national RPJM will include a sectoral approach to climate change but little direction is provided for specific regions. Working with regional governments to address needs will increase regional government awareness of the actions they will need to undertake to help the country meet its GHG emission reduction goals. The regional (or provincial) government can also include those policies in their RPJMs.

This recommendation does not guarantee that because policies are included they will be implemented and funded. However, this basic change in an important and established planning document, which is used at all levels of government, will show a fundamental shift in the value of the environment as equal to the economic and social needs of the country. This is also important for disseminating national policies to the local levels, as local governments often base RPJM goals on the national government. If local governments see an increase in environmental policies, their policies are likely to reflect the same increase.

This recommendation does not require a large amount of financial investment but it does require a significant amount of education and shift in norms. The development plans are often derived from past plans and from national plans; in order to change the type of policies (or increase environmental policies), both knowledge of the need for change and the willingness to do so are needed. This can be achieved mainly through education and through making changes to the national plan that will then filter down to the local levels.

Provincial Recommendations (East Kalimantan)

a. **Improve transparency.** The provincial government continues to adapt to decentralization. Transparency will require both technical capacity (electronic data resources, improved communication mechanisms, etc.) and a staff commitment to sharing public information with the public quickly and easily.

b. ***Enforce environmental laws and prosecute offenders.** Provincial governments have the ability to enact stricter environmental laws and a responsibility to prosecute violators of existing laws.

 As a region rich in natural resources, East Kalimantan has a responsibility to ensure that its resources are used in a legal and sustainable manner. Increasing transparency and making companies more accountable are key, but enacting punishments that fit the crime are also essential. Currently, known violators are often released or left unpunished.

 Creating a dedicated environmental crimes division is one possible avenue for achieving this. The division would need to operate in conjunction with the existing police force, but would be autonomous and report to a separate agency, such as the environmental ministry. It would be unwise to use the existing police force because of its notorious level of corruption. The effort would also require a legal system with a quick turn-around.

 It would also be necessary to create a public shame aspect of punishment, which would create more long-term repercussions to a company than paying a fee. The intent is not to damage a company permanently, but to encourage it to shift its practices to be more sustainable. Because saving face is an important part of the Indonesian culture, ensuring that violators lose face when they violate environmental laws (in addition to fines and monetary damage) could achieve results without demanding a large amount of financial resources and government capacity.

c. **Implement education/media campaign.** Raising public awareness of the need for stronger environmental action will reach across multiple generations and create a significant impact if the campaign focuses on shared interest by all citizens instead of a "green-washing" campaign that overemphasizes action by the government. In addition, developing a curriculum for schools will change practices for future generations.

Local Recommendations (Balikpapan)

a. ***Increase monitoring of environmental indicators.** Monitoring of environmental indicators such as water pollution, GHG emissions, and land use change will provide the city with a better understanding of trends that can influence policy direction. Completing a local MDG report would provide a good basis for evaluation.

Local governments are important for tracking environmental indicators because they are the closest to the source and are the first line of defense. The environmental office should coordinate efforts with the government statistics department.

b. **Implement smart growth and development practices.** Balikpapan, like other cities in Indonesia, is growing quickly. In order to accommodate the burgeoning population, public transportation and effective land use are both crucial to ensure growth does not continue to expand past its limits.

c. **Reduce solid waste and litter.** The city of Balikpapan has already started to develop a program for solid waste. However, waste (mainly plastic) still clogs drainage canals, which contributes to flooding, and littering is common. By enforcing litter laws, citizens may develop more responsibility toward environmental protection while the city earns income from fines paid. That income can then be used to fund additional environmental programs (Figures 18 and 19).

d. **Consider program-based budgeting.** Program-based budgeting allows funds to be directed to a particular program instead of to individual offices. This encourages collaborative work on projects that involve multiple actors. It also creates a better mechanism for monitoring the use of public funds. Although this type of budgeting could also be used at the provincial level, Balikpapan could act as a pilot project.

e. ***Improve and expand public transportation.** Current public transportation is by an angkot (a shared minivan that runs select routes throughout the city). Although useful to many people, the angkot does not appeal to a broad base of possible users. By improving the vehicle (safer, cleaner, cooler, etc.) and expanding the routes traveled, people may be discouraged from driving personal vehicles.

If government funding is unavailable for this program, it would be helpful to partner with local companies for CSR funding. In Balikpapan, two routes could be selected as trail routes along the major thoroughfares for bus rapid transit. These buses (larger than angkots) would be air-conditioned, in a dedicated lane, and stop at select locations. Road construction and expansion continue to occur as the city grows. It will be less expensive and easier to incorporate public transport at inception rather than in the future.

Figure 18 Trash and pollution from drainage systems flow to rivers and directly to the ocean (Balikpapan).

f. **Renew participation in ICLEI.** Balikpapan has been a member of ICLEI in the past. However, there was a lack of active involvement. When staff from ICLEI–Southeast Asia would call Balikpapan, the staff in Balikpapan would often hang up (mainly as a result of language barriers). These challenges can be overcome by identifying a staff member able to communicate with ICLEI staff and then share information with other city personnel.

g. ***Expand "clean, green, and healthy" campaign.** Like the provincial recommendation, raising public awareness of environmental issues will increase the collaborative approach to protecting the environment and increasing citizen participation. Balikpapan should use the existing "clean, green, and healthy" campaign as a base for beginning a more broadly focused program that includes education and encourages personal responsibility.

Improving environmental health is a personal responsibility as much as a government responsibility. It requires education and often a changing of

Figure 19 Trash, including plastic bags and plastic water bottles, often clogs drainage systems, contributing to flooding (Balikpapan).

norms. By highlighting specific actions individuals can take, such as reducing the use of plastic bags, the campaign becomes more of an action than a statement.

Improving Cooperation Between Local and National Governments

The relationship between local governments and the central government changed dramatically in 2001 and 2002 following decentralization. The actions by the national government compared to local governments have in many ways taken two separate roads in response to sustainable development, albeit with the same destination.

The national government has been directed primarily by international commitments and pressures to halt its rampant deforestation, while the local government has responded to the local needs of constituents as well as to the problems that arise from a lack of environmental regulation (pollution, health problems, etc.). Both influence each other and would benefit from increased communication. The central government will be able to better develop policies (particularly regional policies) by understanding what issues are of concern to the local communities. Local governments will also benefit from understanding the national sustainable development strategy, particularly objectives related to regional concerns.

The role of provincial governments in Indonesia has grown in importance since decentralization. However, the provincial government is not succeeding in enforcing environmental laws, investing in its future adequately, or supporting the social needs of its population. The balance of sustainable development is weighted too heavily toward the economy in East Kalimantan, and the argument that economic growth is used to lift people out of poverty is not supported by reality.

The Indonesian economy relies heavily on natural resource extraction, which occurs at the regional level. If this income source is reduced in an effort to meet sustainable development objectives, an alternative must be found. One of the most effective ways to meet this need is by reducing corruption and improving the rate of return from products harvested.

Recovering the loss of direct income from product sales as well as lost tax revenue from illegally harvested forest products that currently fill the pockets of corrupt officials could improve education and be reinvested into the economy. Also, expanding the downstream manufacturing of natural resources will increase the revenue from resources used. Currently, most of the resources are primary forest products that are exported to foreign countries. Developing manufacturing and secondary resource processes will create jobs and increase the value of the products being exported.

If less polluting energy sources, including geothermal and solar power, can be developed to a level that will allow exportation, the country could maintain (or possibly increase) its foreign income, as well as meet its own expanding resource use. Foreign direct investment in renewable energy infrastructures has been increasing, but corruption often deters many potential investors.

In addition to a more effective use of natural resource revenues, development of other revenue-generating industries such as tourism and service industries will create a more diversified economy that relies less on natural resource extraction.

Local governments are often forced to address the issues of sustainable development because they are affected more directly by environmental damage. Local governments around the world (including Indonesia) are responding to the need for sustainable development.

Through programs like ICLEI, which spawned from international inaction by member states following the Rio conference, many local governments have initiated actions to address climate change. While this spurs a larger debate on the role of climate change amidst the sustainable development field, addressing climate change does lead to increased action on addressing environmental issues.

It is neither feasible nor desirable for the NSDS to be recreated at the provincial and local levels. The needs of the nation are of a different scale and reflect the various ecosystems and resources throughout the archipelago. However, the principle of balancing the needs of the three pillars of sustainable development should be transferred to each level of government. In order to accomplish this, each level of government needs to not only incorporate but implement and fund more environmental policies.

CONCLUSION

Sustainable development is a continuum. It is an ongoing process that responds to changing technology and capacity. An important question to ask regularly would be, "is sustainable development improving?" However, because there are few, if any, baselines dedicated specifically to sustainable development in Indonesia, it is difficult to give an overall rate of sustainability improvements. This research is intended to begin filling this gap.

It is clear from the results of this study that of the three sustainable development sectors, the environment is receiving the least policy attention and budget allocation. In order to improve its sustainability, each level of government will need to recognize the environment as equally as economic growth and social well-being. Programs that address climate change and GHG reductions are becoming the modus operandi for increasing the effectiveness of environmental policies. However, these programs are nascent and will require strong implementation and monitoring at the provincial and local levels. Environmental policies will require a broader focus than just climate change in order to meet sustainable development objectives.

Each level of Indonesian governance must begin to recognize the importance of the environment as part of its future well-being. They must also actively implement strategies that impact those most responsible for achieving the desired outcome of sustainable development and dedicate the financial resources to ensure success.

Climate change has begun to attempt this by coordinating between ministry departments and stakeholders. However, it has yet to adequately incorporate local governments in climate change planning. Without each level of government working in concert with each other, the success of climate change policies will be put in jeopardy. This will make it less useful for expanding to become a model for other environmental concerns that need to be addressed to advance sustainable development in Indonesia.

Indonesia may be at the beginning of the sustainable development continuum, but it has the incentives and ability to begin addressing this important issue in a more coordinated and integrative manner.

ACKNOWLEDGMENTS

Numerous institutions and individuals were integral to the success of this research, including the Fulbright Program and the American Indonesian Exchange Foundation for funding and logistical support and the Indonesian Institute of Science and Technology, particularly Bapak Herry Yogyaswara and Ibu Aswatini. Thanks to the Natural Resources Law Institute, with special thanks to Ibu Rhamina and Diah Sembering, who were kind, patient, and helpful while I learned Indonesian and tried to navigate my way through the government systems. The NGO Consortium in Balikpapan, coordinated by Ibu Yulita and government staff, particularly Ibu Murni in Balikpapan and Bapak Dana in Jakarta, were helpful and supportive. Additional thanks are given to people who were helpful in many small ways by sharing their time and skills during interviews, translation, and personal stories.

TRANSLATION ACKNOWLEDGMENT

Thank you to Mbak Eka from Puri Indonesia in Yogyakarta for providing translation of the executive summary.

PHOTO ACKNOWLEDGMENT

Unless otherwise noted, all photographs were taken or collected by the author during the research period.

REFERENCES

Adhaiti, M.A., 2001. In: Bobsien, A. (Ed.), Indonesia's transmigration programme - an update. Available from: < http://dte.gn.apc.org/ctrans.htm >(21.08.10.).

Alberts, H.A., 2010. March 10, Asia's most corrupt countries. Available from: < http://www.forbes.com/2010/03/10/indonesia-yudhoyono-bailout-business-asia-most-corrupt-countries.html >(21.08.10.).

Antara News, 2010. July 15, Japan upgrades Indonesia's credit rating to investment grade. Available from: < http://www.antaranews.com/en/news/1279176961/japan-upgrades-indonesias-credit-rating-to-investment-grade >(21.08.10.).
Badan Pusat Statistik, 2009. Population of Indonesia by province 1971, 1980, 1990, 1995 and 2000. Available from: <http://dds.bps.go.id/eng/tab_sub/view.php?tabel=1&daftar=1&id_subyek=12&not >(25.08.10.).
Badan Pusat Statistik, Balikpapan, 2009. Balikpapam Dalam Angka Final.
Badan Pusat Statistik, Balikpapan, 2009. Indikator Ekonomi Kota Balikpapan, Badan Pusat Statistik, Balikpapan: City of Balikpapan.
Balikpapan Ministry of Planning, 2007. Rencana Pembangunan Jangka Menengah Daerah (RPJMD)/medium term development plan, Kota Balikpapan Tahun 2006–2011, Balikpapan: City of Balikpapan.
Bappeda/East Kalimantan Ministry of Planning, 2009. RPJM - medium term development plan 2009-2014. Bappeda, Samarinda.
Bappenas - National Ministry of Planning, 2005. The national long term development plan 2005-2025. Bappenas, Jakarta.
Bappenas, 2004. The national medium-term development plan 2004-2009. Bappenas, Jakarta.
Central Intelligence Agency, 2010. August 19, CIA world factbook. Available from: < www.cia.gov: https://www.cia.gov/library/publications/the-world-factbook/geos/id.html >(25.08.10.).
Cherp, A.C., 2004. A methodology for assessing national sustainable development strategies. Institute for Development Policy & Management, University of Manchester, Manchester.
Christopher Barr, I.A., 2006. Decentralization of forest administration in Indonesia, Center for International Forestry Research (CIFOR), Bogor.
Department Keuangan Republik Indonesia, Data Pokok APBN, 2007-2008. Department Keuangan Republik Indonesia. Jakarta.
Governments of Norway and Indonesia 2010, May 26, http://www.norway.or.id/PageFiles/404362/Letter_of_Intent_Norway_Indonesia_26_May_2010.pdf. [2010 September 30].
Indonesian Statistics Bureau (BPS), 2009. Gross regional domestic product at 2000 constant market prices by provinces, 2004-2008 (million rupiahs). Available from: <http://dds.bps.go.id/eng/tab_sub/view.php?tabel=1&daftar=1&id_subyek=52&notab=2 >(21.08.10.).
Jakarta Globe and Rueters 2010, August 3, http://www.thejakartaglobe.com/business/china-investment-corp-eyes-indonesian-state-owned-enterprises-minister/389290. [2010 August 21].
Japanese Ministry of Land, Infrastructure, Transport and Tourism n.d., Overview of spatial planning policies in Asian and European countries. Available from: <http://www.mlit.go.jp/kokudokeikaku/international/spw/general/indonesia/index_e.html#top>. [2010 August 29].
Kalimantan Timur, 2009. Strategic Planning Budget Program and Activity (2009-10 proposed APBD).
Kota Balikpapan, 2009. Buku data status Lingkungan Hidup Daerah Kota Balikpapan, Badan Lingkungan Hidup, Balikpapan: City of Balikpapan.
Kota Balikpapan, 1999. East Kalimantan spatial plan. City of Balikpapan, Balikpapan.
Ministry of Environment, Republic of Indonesia, 2002. From crisis to sustainability: paving the way for sustainable development in Indonesia, overview of the implementation of Agenda 21. Ministry of Environment, Jakarta.
Ministry of Finance, 2009. Ministry of Finance Green Paper: economic and fiscal policy strategies for climate change mitigation in Indonesia. Ministry of Finance and Australia Indonesia Partnership, Jakarta.
National Coalition of Indonesia For Anticorruption, 2009. Weakening of corruption eradication commission in Indonesia. Available from: <http://www.antikorupsi.org/docs/independentreportofUNCACimplementationinindonesiaqatar09.pdf >(21.08.10.).

Province of East Kalimantan, 2010. Kaltim Hijau - Green Kaltim declaration. Government of East Kalimantan, Samarinda.

Provinsi Kalimantan Timur, 2008. Profile Kesehatan Provinsi Kalimantan Timur. Government of East Kalimantan, Samarinda.

Provinsi Kalimantan Timur, 2008. Status Lingkungan Hidup Daerah Provinsi, Bapedalda. Samarinda: Government of East Kalimantan.

Simamora, A.P., 2010. New center to supply govt with credible data. Available from: <http://www.thejakartapost.com/news/2010/09/06/new-center-supply-govt-with-credible-data.html >(06.09.10.).

Stalker, P., 2007. Millenium development goals, United Nations Development Programme and Indonesian Planning Ministry. UNDP and BAPPENAS, Jakarta.

State Ministry for Development Planning, 2010. Climate mitigation and adaptation considerations in Indonesia's medium and long term development plans. International Climate Change Workshop on Research Priorities and Policy Development, Depak.

State Ministry for Environment and United Nations Development Program, 2001. Sectoral Agenda 21 energy agenda for developing sustainable quality of life. Sectoral Agenda 21 Project, Jakarta.

State Ministry for Environment and United Nations Development Program, 2001. Sectoral Agenda 21 forestry agenda for seveloping sustainable quality of life. Sectoral Agenda 21 Project, Jakarta.

State Ministry for Environment and United Nations Development Program, 2001. Sectoral Agenda 21 human settlements agenda for developing sustainable quality of life. Sectoral Agenda 21 Project, Jakarta.

State Ministry for Environment and United Nations Development Program, 2001. Sectoral Agenda 21 mining agenda for developing sustainable quality of life. Sectoral Agenda 21 Project, Jakarta.

State Ministry for Environment and United Nations Development Programme, 2001. Sectoral Agenda 21 tourism agenda for developing sustainable quality of life. Sectoral Agenda 21 Project, Jakart.

State Ministry of Environment, 2007. National action plan addressing climate change. State Ministry of Environment, Jakarta.

Sulaiman, N., 2010. E. Kalimantan seeks economic equality by 2010. Available from: < http://www.thejakartapost.com/news/2010/03/17/e-kalimantan-seeks-economic-equality-2013.html >(21.08.10.).

Swanson, D.A., 2004. National strategies for sustainable development: challenges, approaches and innovations in strategic and co-ordinated action. International Institute for Sustainable Development, International Institute for Sustainable Development and Deutsche Gesellschaft für Technische Zusammenarbeit (GTZ) GmbH, Winnipeg.

Syaikhu, U., 2001. Indonesia's decentralization policy: initial experiences and emerging problems. Available from: < http://www.smeru.or.id/report/workpaper/euroseasdecentral/euroseasexperience.pdf >(25.08.10.).

The Jakarta Post, 2010. May 26, Archipelago: 13 illegal logging suspects arrested in East Kalimantan. Available from: http://www.thejakartapost.com/news/2010/05/26/13-illegal-logging-suspects-arrested-east-kalimantan.html (21.08.10.).

The World Bank Group, 2001. Independant evaluations: publications. Available from: < http://lnweb90.worldbank.org/oed/oeddoclib.nsf/DocUNIDViewForJavaSearch/4b8b0e01445d8351852567f5005d87b8?OpenDocument&Click= >(05.09.10.).

Trading Economics, 2010. Indonesia GDP per capita (constant prices since 2000). Available from: < http://www.tradingeconomics.com/Economics/GDP-Per-Capita.aspx?Symbol=IDR >(21.08.10.).

United Nations Department of Economic and Social Affairs, 2007. Indicators of sustainable development: guidelines and methodologies. United Nations, New York.

United Nations Environmental Program n.d., Indonesia: integrated assessment of the poverty reduction strategy paper. Available from: <http://www.unep.ch/etb/publications/FINALIndonesianReport.pdf> [2010August 25].

U.S. Department of Commerce International Trade Administration, 2010. Renewable energy market assessment report: Indonesia. Available from: http://ita.doc.gov/td/energy/Indonesia%20Renewable%20Energy%20Assessment%20(FINA).pdf.

Wallace, A.R., 2008. The Malay Archipelago. Periplus Editions, Jakarta.

Woods, J., 2010. Stock picks: CIVETS: the new BRIC in the wall of international investing. Available from: < http://www.investorplace.com/stock-picks/emerging-markets/civets-the-new-bric-in-the-wall-of-international-investing.html >(06.09.10.).

APPENDIX 1: NSDS PROCESS EVALUATION

Indonesia NSDS Process Evaluation

Principles and criteria	Score (A–D)	Remarks
A. Integration of economic, social, and environmental objective		
a.1. Integration Strategic planning in the country is based on a comprehensive and integrated analysis of economic, social, and environmental issues, which clarifies links among the three spheres, resolves, and negotiates conflicts.	B	Previous national medium-term development plans set policy goals that were weighted more heavily toward social and economic issues. However, with the emergence of climate change planning being integrated into the RPJM, environmental policy development is expected to rise.
a.2. Social and poverty issues Strategic planning in the country integrates poverty eradication, gender issues, and the short-and long-term needs of disadvantages and marginalized groups into economic policy.	B	Each RPJM includes multiple policies on social and provincial issues. Gender issues are also addressed in planning documents but in most cases it is included as a part of family welfare. Strategic poverty reduction planning is described mainly in the PRSR. Social and poverty issues are addressed in planning policies but it is unclear how they are incorporated into economic policy.
a.3. Environment and resource issues Strategic planning in the country integrates the maintenance of sustainable levels of resource extraction and control of pollutants to maintain a healthy environment into economic policy.	D	Although economic considerations are often included in the analysis of various social and environmental policies, the opposite is less true. Most economic policies (including the annual budget) do not give adequate attention to the environmental or social aspects of development. This is often most clear in the natural resource extraction industries. While an EIA is required for various development projects, they are ineffective or not enforced.

Indonesia NSDS Process Evaluation—cont'd

Principles and criteria	Score (A–D)	Remarks
a.4. International commitments Measures are in place to ensure compliance with international agreements, which the country has entered into for environmental and social issues.	B	The most high-profile international agreements currently being implemented are related to climate change. Government actions are contradictory with new policies favoring forest conservation. The most effective measures used to ensure compliance with agreements are funding and international pressure, both of which are present with climate change but not sustainable development.
B. Participation and consensus		
b.1. Involvement of stakeholders The country's process of strategic planning, implementation, monitoring, and review includes the participation of stakeholders. (Government, decentralized authorities, elected bodies, nongovernment organizations, sectoral institutions, and marginalized groups.)	C	Efforts are made to include stakeholders; however, interviews with various NGOs and community members indicated that efforts extended to getting a name on a list to show that there was involvement of stakeholders. Suggestions offered by community members rarely impact the decisions. Involvement of stakeholders appeared to be more of a formality.
b.2. Transparency and accountability Management of the country's strategic planning process is transparent, with accountability for decisions.	C	Transparency at every level of government is poor. Planning documents at the national level are available and most are accessible on-line. The process of accountability is unclear, however, and the policies are vague. At the provincial level in East Kalimantan, planning documents are almost impossible to view. Although they are creating a new website that has links to all documents, the documents are dead links and an actual hard copy of the document must be obtained.

Continued

Indonesia NSDS Process Evaluation—cont'd

Principles and criteria	Score (A–D)	Remarks
b.3. Communication and public awareness Measures are taken to increase public awareness of sustainable development, to communicate relevant information, and to encourage the development of stakeholder involvement in the strategic planning process.	D	Little public awareness exists regarding sustainable development. Minor public campaigns encourage environmental awareness. Stakeholder involvement is limited primarily to the annual development plans and does not ensure that program options are sustainable in nature.
b.4. Long-term visions and consensus The country's strategic planning processes are based on a long-term vision for the country's development, which is consistent with the country's capabilities, allows for short- and medium-term necessities, and has wide political and stakeholder support.	B	The long-term vision is articulated in the long-term development plan at each level of government. However, the vision is vague and lofty. Although actions are taken to boost the country's capabilities, current capabilities do not match the goals. Infrastructure, such as e-mail use by government officials, is not widespread. Medium- and short-term goals are more measurable, but are not necessarily more achievable (such as reducing population growth).
C. Country ownership and commitment		
c.1. High-level government commitment The process of formulating and implementing the national strategy is led by government, with high-level commitment.	A	The current strategy (of using the existing planning process) is well established and has high-level commitment. The Planning Ministry (Bappenas/Bappeda) coordinates development planning documents and the highest government official (president, governor, mayor) approves them.
c.2. Broad-based political support The country's strategic planning process has broad-based political support.	B	The process of planning has broad political support, but the actual policies may not have broad support. This is due to the fact that medium- and short-term development plans are developed and implemented by the current leader. If the political power shifts, the policies will shift accordingly.

Indonesia NSDS Process Evaluation—cont'd

Principles and criteria	Score (A–D)	Remarks
c.3. Responsibility for implementation Responsibility for implementing strategies is clearly assigned to bodies with the appropriate authority.	C	Responsibility for implementation is unclear. The sectoral books on sustainable development written in 2001 do not appear to have been shared with the relevant agencies. Sustainable development remains in the environmental realm; however, climate change is becoming integrated into other fields. In general, the policies in the RPJM are managed by the appropriate agency, but without an evaluation of each department's budget, it is impossible to confirm which agency manages specific programs.
c.4. Coordination with donors The country's strategic planning process is coordinated with donor programs.	—	Insufficient information available

D. Comprehensive and coordinated policy process

d.1. Build on existing processes The NSDS is based on existing strategic planning processes, with coordination between them and the mechanisms to identify and resolve potential conflicts.	C	Sustainable development is part of the country's planning process through the RPJM, but little conflict resolution is needed because there is little conflict (mainly due to inaction). However, as the country addresses climate change, more conflict between competing interests arises. These conflicts are being addressed without a formal process
d.2. Analysis and information Strategic planning in the country is based on a comprehensive analysis of the present situation and of forecasted trends and risks using reliable information on changing environmental, social, and economic conditions.	D	Analysis and information are incorporated adequately into planning mechanisms to the extent possible. This is more true at the national level. At the local level, plans change little from one year to the next, as data gathering is insufficient and/or inaccurate. The lack of data and the evaluation of risks and trends. The lack of data is most obvious in environmental and social data, economic indicators are monitored closely, and economic policies in the RPJM are adjusted if needed (e.g., Asian economic crisis).

Continued

Indonesia NSDS Process Evaluation—cont'd

Principles and criteria	Score (A–D)	Remarks
d.3. Realistic goals The national strategy is based on a realistic analysis of national resources and capacities in the economic, social, and environmental spheres, taking into account external pressures.	D	The goals of the RPJM are vague and lofty, often without quantitative targets. Additionally, without eliminating the amount of corruption, mismanagement of environmental resources and social funds will continue, making all goals unrealistic.
d.4. Decentralization The country's strategic planning process embraces both national and decentralized levels, with two-way communication between these levels.	C−	Communication between the levels of government is poor. The national government receives data from regional statistics offices. This has the ability to offer a snapshot of the current status of sustainable development, but this is dependent on quality data and consistent gathering of data at the local and provincial levels. The extent of communication differs among regions. In East Kalimantan, there is a high amount of communication breakdown at the provincial level. When information on new national regulations is passed down, it is not disseminated effectively between relevant offices or to local governments that may be affected by the new regulation.
E. Targeting, resourcing, and monitoring		
e.1. Budgetary provision The NSDS is integrated into the budget process, such that plans have sufficient resources to achieve their goals.	D	There is no budgetary provision for sustainable development in national or local budgets. Additionally, budget allocations are not balanced among the three pillars of sustainable development.

Indonesia NSDS Process Evaluation—cont'd

Principles and criteria	Score (A–D)	Remarks
e.2. Capacity for implementation The NSDS includes realistic mechanisms to develop and the capacity required to implement it.	C	The national plan includes mechanisms to develop capacity, but some of them may not be realistic due to funding constrains. For example, the integration of climate action policies into the national budget will require significant resources—not all of which have been identified. In addition, there is not a clear indication of how to educate local governments on incorporating sustainable development strategy into local plans. Much of the capacity for instituting environmental policies will require strong monitoring and evaluation, which are already woefully inadequate.
e.3. Targets and indicators Targets have been defined for key strategic economic, social, and environmental objectives with indicators for monitoring.	C	Indicators for monitoring are stronger for economic and social goals than for environmental goals. Millennium development goals (MDGs) remain the most applicable quantifiable targets for sustainable development goals, as they are established and widely accepted.
e.4. Monitoring and feedback Systems are in place for monitoring the implementation of strategies and achievement of the defined objective, as well as for recording and reviewing the results, with effective mechanisms for feedback and revisions.	D	Monitoring is one of the greatest areas for improvement. The national MDGs offer some of the best examples of successful monitoring and evaluation.

APPENDIX 2

Indicators			Results		
Indicator - 96 original	Indicators - 50 revised	MDG	National: Indonesia	Provincial: East Kalimantan	Local: Balikpapan
Theme: Poverty					
Portion of population below $1 a day		Goal 1; Target 1	7.50%	9.51%	3.49%
Ratio of share in national income (Gini coefficient)	Ratio of share in national income (Gini coefficient)		2.77%	N/A	N/A
Proportion of population using improved sanitation facilities	Proportion of population using improved sanitation facilities		69.30%	N/A	97%
	Percentage of houses with a bathroom and septic tank		53.33%	59%	94%
Proportion of population using an improved water source	Proportion of population using an improved water source	Goal 7; Target 10	55.07%	67.51%	70%
Proportion of urban population living in slums	Proportion of urban population living in slums	Goal 7; Target 11	23.10%	N/A	N/A
Theme: Governance					
Percentage of population having paid bribes	Percentage of population having paid bribes		31%	N/A	N/A

—cont'd

Indicators			Results		
Indicator - 96 original	Indicators - 50 revised	MDG	National: Indonesia	Provincial: East Kalimantan	Local: Balikpapan
Theme: Health					
Under-five mortality rate	Under-five mortality rate (deaths per 1000 children)	Goal 4; Target 5	44	3	4
Life expectancy at birth	Life expectancy at birth (years)		70.76	70.5	71.73
Contraceptive prevalence rate		Goal 5; Target 6	61%	13%	12%
Immunization against infectious childhood diseases	Immunization against infectious childhood diseases		–	–	–
	DPT		68.77%	71.60%	N/A
	Polio		72.21%	72.44%	N/A
	Hepatitis B		60.79%	67.17%	N/A
Nutritional status of children	Nutritional status of children (shows unhealthy/underweight percentage of all children)	Goal 7; Target 8	28%	19%	0.02%
Morbidity of major diseases (AIDS, malaria, TB, etc.)	Morbidity of major diseases (total persons died)	Goal 7; Target 8 and Target 7	N/A	388	77
	AIDS (total recorded cases)		170,000	183	52
	Malaria (confirmed cases)		411,979 (863,213 est.)	2979	30
	TB (cases per 100,000)		39	–	–

Continued

—cont'd

Indicators			Results		
Indicator - 96 original	Indicators - 50 revised	MDG	National: Indonesia	Provincial: East Kalimantan	Local: Balikpapan
Prevalence of tobacco use[19]			Males: 63.1% Females: 4.5%	N/A	N/A
Theme: Education					
Net enrollment rate in primary education	Net enrollment rate in primary education	Goal 2; Target 3	94.70%	96.60%	99.16
Adult (age 15–24) literacy rate	Adult (age 15–24) literacy rate	Goal 2; Target 3	99.40%	96.71%	98.32%
Theme: Demographics					
Population growth rate	Population growth rate (2008)		1.175%	2.81%	1.46%
Dependency ratio	Dependency ratio (2010 est.)		53.73%	N/A	13%
Theme: Atmosphere					
Carbon dioxide emissions	Carbon dioxide emissions (MT per capita based on World Bank CO_2 estimates and BPS population data)	41% reduction	391,902,731	4,664,728	1,001,300
Ambient concentration of air pollutants in urban areas	Ambient concentration of air pollutants in urban areas (TSP)	Goal 7	N/A	114.25 Nm^{320}	127.6 Nm^{321}
	PM10 (mg/Nm^3)	50 m^3	102 m^3	N/A	42 m^{322}

—cont'd

Indicators			Results		
Indicator - 96 original	Indicators - 50 revised	MDG	National: Indonesia	Provincial: East Kalimantan	Local: Balikpapan
Theme: Land					
Proportion of land area covered by forests	Proportion of land covered by forest (national and provincial data show proportion of land area under forest concession)		26,169,813 ha	6,581,712 ha	5,044 ha
Area of forest under sustainable forest management	Area of protected forest		N/A	23%	34.30%
Theme: Oceans, seas, and coasts					
Proportion of marine-protected areas	Proportion of marine-protected areas		11%	N/A	N/A
Theme: Fresh water					
	Quantity of clean water distributed to customers (m^3)	Goal 7; Target 10	2,296,055	57,665	N/A
Presence of fecal coliforms in fresh water	Presence of fecal coliforms in fresh water (MPN averages shown)	Goal 7; Target 10	N/A	565[23]	443[24]
Theme: Biodiversity					
Proportion of terrestrial area protected (total and by ecological region)	Proportion of terrestrial area protected (total and by ecological region)	Goal 7; Target 9	29.50%	23%	52%

Continued

—cont'd

Indicators			Results		
Indicator - 96 original	Indicators - 50 revised	MDG	National: Indonesia	Provincial: East Kalimantan	Local: Balikpapan
Change in threat status of species	Change in threat status of species (EX, EW, CR, EN, and VU)[25] (shows total threatened species)	Goal 7; Target 10	1,127	N/A	N/A
Theme: Economic development					
Gross domestic product per capita	Gross domestic product per capita (2008)		3900 USD	2351 USD	8412 USD
Investment share in GDP	Investment share in GDP (2009 est.)		27.10%	N/A	N/A
Employment–population ratio	Employment–population ratio (percentage unemployed)		8.40%	N/A	4%
Share of women in wage employment in nonagricultural sector	Share of women in wage employment in nonagricultural sector	Goal 3; Target 4	30.60%	N/A	N/A
Internet users per 100 population	Internet users (percentage of population)	Goal 8; Target 18	4.20%	15%	
Mobile cellular telephone subscribers per 100 population (used per % of population)		Goal 8; Target 18	52.00%	76.57%	

—cont'd

Indicators			Results		
Indicator - 96 original	Indicators - 50 revised	MDG	National: Indonesia	Provincial: East Kalimantan	Local: Balikpapan
Tourism contribution to GDP	Tourism contribution to GDP		7.7%	N/A	N/A
Theme: Global economic partnerships					
Current account deficit as percentage of GDP	Current account deficit as percentage of GDP (current account balance; 2008)	Goal 8; Target 15	604 million USD	N/A	N/A
Net official development assistance given or received as a percentage of gross national income	Net official development assistance given or received as a percentage of gross national income	Goal 8; Target 12	44.90%	N/A	N/A
Theme: Consumption and production patterns					
Annual energy consumption (total and by main user category)	Annual energy consumption (total and by main user category)	Goal 7; Target 8	95.3 kg oil-eq/ 1.000$	N/A	496,600,470 kWh
Modal split of passenger transport	Modal split of passenger transport		—	—	—
	Motor bike		73%	N/A	78.50%
	Car		15%	N/A	13%
	Bus		4%	N/A	8%
	Truck		8%	N/A	0.50%

[19] "Prevalence of tobacco use" is not a core United Nations Commission on Sustainable Development indicator but is included as an "other indicator" because it is relevant for Indonesia due to the high rate of smoking.

—cont'd

[20] Status Lingkungan Hidup Daerah Provinsi Kalimantan Timur November 2008, Samarinda East Kalimantan, p. 33-58. Data shown are the average of four cities where data were available: Balikpapan, Samarinda, Pasar, Bontang, and Kutai Kartanegara.

[21] Status Lingkungan Hidup Daerah Provinsi Kalimantan Timur November 2008, Samarinda East Kalimantan, p. 33-38. Data are average of three locations in Balikpapan (Simpang Maura Rapak, Simpang Tiga Plaza, and Jl Soekarno-Hatta Km 4,5). Data collected by Bapedalda Provinsi.

[22] Buku data status Lingkungan Hidup Daerah Kota Balikpapan Tahun 2009, Fecal coliform data represent the average of six downstream river samplings: Klandasan, Klandasan Kecil, Manggar, Wain, Sepinggan, and Somber.

[23] Kumpulan data status Lingkungan Hidup Daerah Provinsi Kalimantan Timur 2007, Fecal cloriform data showed great discrepancy between the two testing dates with sites 5 (palaran) and 6 (anggana/sungai merlam) highest.

[24] Buku data status Lingkungan Hidup Daerah Kota Balikpapan Tahun 2009, Fecal coliform data represent the average of six downstream river samplings: Klandasan, Klandasan Kecil, Manggar, Wain, Sepinggan, and Somber.

[25] EX, extinct; EW, extinct in the wild; CR, critically endangered; EN, endangered; VU, vulnerable. Additional notes: The original 96 indicators as well as the 50 revised indicators are included because data in some instances were available for an original indicator but not the revised indicator. For example, data on "portion of population living below poverty line" were only available for the national level, but each level of government has data for "portion of population below $1 a day." Indicators not included due to lack of data are listed here. Some data were available for one level of government (usually national government). Also, data results often varied significantly between sources.

Poverty:
- Share of household without electricity or other modern energy services
- Portion of population living below poverty line

Governance:
- Number of intentional homicides per 100,000 population

Health:
- Percentage of population with access to primary health care facilities

Education:
- gross intake ratio to last grade of primary education
- Adult secondary school attainment level

Natural hazards:
- Percentage of population living in hazard-prone areas

Atmosphere:
- Consumption of ozone-depleting substances

Land:
- Arable and permanent cropland area

Oceans, seas, and coasts:
- Percentage of total population living in coastal area
- Proportion of fish stocks within their safe biological limits

Fresh water:
- Proportion of total water resources used
- Water use intensity by economic activity

Economic development:
- Labor productivity and unit labor cost

Consumption and production patterns:
- Material intensity of the economy
- Intensity of energy use
- Generation of hazardous waste
- Wastewater treatment and disposal

APPENDIX 3: INDONESIA PLANNING AND BUDGET ANALYSIS CHARTS

Evaluation of National Midterm Development Plan 2004–2009

Policy—as indicated by chapter description	Sustainable development sector			Other
	Env.	Soc.	Eco.	
Enhancing mutual trust and harmonization among social groups		1		
Development of culture based on noble ancestral values		1		
Enhancing security and order and overcoming crime		1		
Preventing and overcoming separatism		1		
Preventing and overcoming terrorism		1		
Enhancing capacity of state defense				1
Consolidating foreign policy and enhancing international cooperation				1
Improving the legal and political system		1		
Elimination of discrimination in various forms		1		
Respect, recognition, and enforcement of law and human rights		1		
Improved quality of life and role of women and children welfare and protection		1		
Revitalization process of decentralization and regional autonomy				1
Creating a clean and credible government	1	1	1	
Realization of increasingly solid democratic institutions				1
Poverty reduction		1	1	
Increasing investment and export nonoil/gas			1	
Improved competitiveness of manufacturing industry			1	
Revitalization of agriculture			1	
Empowerment cooperative and micro, small, and medium enterprises		1	1	
Improved management of state-owned enterprises			1	

Continued

Evaluation of National Midterm Development Plan 2004–2009—cont'd

Policy—as indicated by chapter description	Sustainable development sector			Other
	Env.	Soc.	Eco.	
Increasing the capability in science and technology				1
Improving manpower		1		
Consolidation of macroeconomic stability			1	
Rural development		1	1	
Reducing inequalities in regional development		1		
Increasing community access to quality education		1		
Increased public access to quality health care		1		
Increased social security and welfare		1		
Development of population, quality small families, youth, and sports		1		
Improving the quality of religious life		1		
Improving management of natural resources and conservation of natural environment functions	1			
Acceleration of infrastructure development		1	1	
Macroeconomic framework and development financing			1	
Total	2	21	11	5
Total percentage (34 policies)	6%	62%	32%	15%

Central Government Expenditure by Organization—2006 APBN

Description of organization/project	Expenditure (trillions Rp.)	Percent of total budget	Sustainable development categorization			
			Env.	Soc.	Econ.	Admin.
People's Consultative Assembly	130.5	0.03%				0.03
Legislative Council	939.9	0.21%				0.21
Audit Board	566.6	0.13%			0.13	
Justice Court	1,948.20	0.44%		0.44		
Attorney General	1,401.10	0.32%				0.32
State Secretariat	729.9	0.17%				0.17
Vice President	157.1	0.04%				0.04
Department of State	1,158.00	0.26%	0.26	0.26	0.26	
Foreign Ministry	3,152.80	0.72%	0.72	0.72	0.72	
Defense Department	23,922.80	5.44%		5.44		
Ministry of Law	2,875.90	0.65%		0.65		
Financial Department	5,167.00	1.17%			1.17	
Agriculture Department	5,551.20	1.26%	1.26		1.26	
Industry Department	1,126.50	0.26%			0.26	
Department of Energy and Mineral Resources	4,657.60	1.06%	1.06		1.06	
Transportation Department	6,769.70	1.54%		1.54	1.54	
National Education Ministry	37,095.10	8.43%		8.43		
Department of Health	12,260.60	2.79%		2.79		
Department of Religion	10,023.30	2.28%		2.28		
Manpower and Transmigration Ministry	2,069.40	0.47%		0.47		
Social Department	2,221.40	0.50%		0.5		

Continued

Central Government Expenditure by Organization—2006 APBN —cont'd

Description of organization/project	Expenditure (trillions Rp.)	Percent of total budget	Sustainable development categorization			
			Env.	Soc.	Econ.	Admin.
Forestry Department	1,485.20	0.34%	0.34		0.34	
Marine and Fisheries Department	2,566.30	0.58%	0.58		0.58	
Public Works Department	19,186.70	4.36%		4.36		
Coordinating Ministry for Political, Legal, and Security	76.3	0.02%				0.02
Ministry of Economic Affairs	65.6	0.01%			0.01	
People's Welfare Ministry	68.1	0.02%		0.02		
Ministry of Culture and Tourism	609.7	0.14%			0.14	
Ministry for State-Owned Businesses	155.1	0.04%			0.04	
Research and Technology Ministry	342.6	0.08%				0.08
Environmental Ministry	300.9	0.07%	0.07			
Ministry of Cooperative and SMEs	930.2	0.21%			0.21	
Ministry of Women's Empowerment	116.9	0.03%		0.03	0.03	
State Apparatus Ministry	169.8	0.04%				0.04
State Intelligence Agency	1,012.40	0.23%		0.23		
Pass-State Institution	690.3	0.16%				0.16
National Board of Resilience	29.6	0.01%				0.01
Statistics Board	912.1	0.21%				0.21
Ministry of State—Planning	198.1	0.05%	0.05	0.05	0.05	
National Land Agency	1,211.50	0.28%	0.28		0.28	
National Library	138.7	0.03%		0.03		0.03
Communication and Information Department	1,235.70	0.28%				0.28

State Police	16,449.90	3.74%			3.74
Food and Drug Supervisory Board	302.3	0.07%			0.07
National Resilience Institute	72.3	0.02%		0.04		0.02
Investment Coordination Agency	183.2	0.04%				0.04
National Narcotics Board	285.7	0.06%		0.05	0.06	
Ministry of State Development of Disadvantaged Regions	230.2	0.05%			0.05	
National Family Planning Coordinating Agency	637.5	0.14%	0.14		0.14	
National Human Rights Commission	36.6	0.01%			0.01	
Meteorological and Geophysical Agency	521.8	0.12%	0.12			
Election Commission	318.1	0.07%				0.07
Constitutional Court	204.6	0.05%			0.05	0.05
Center for Analysis and Reporting Financial Transaction	33	0.01%		0.01		
LIPI (Indonesian Institute of Science)	396.6	0.09%			0.09	
National Nuclear Energy Agency	250.9	0.06%			0.06	0.06
Tax Assessment and Technology Application	413.4	0.09%		0.09		0.09
Institute of Aeronautics and Space	162.5	0.04%			0.04	0.04
Survey and Mapping Agency	144.9	0.03%				0.03
National Standardization Bodies	31.1	0.01%				0.01
Nuclear Power Supervisory Board	46.9	0.01%				0.01
State Administrative Institution	126	0.03%				0.03
National Archives	83.9	0.02%			0.02	
National Civil Service Agency	228.5	0.05%			0.05	0.02

Continued

Central Government Expenditure by Organization—2006 APBN —cont'd

Description of organization/project	Expenditure (trillions Rp.)	Percent of total budget	Sustainable development categorization			
			Env.	Soc.	Econ.	Admin.
Financial Supervision and Development Agency	437.1	0.10%			0.1	
Department of Trade	1,128.70	0.26%			0.26	
State Housing Ministry for the People	369.2	0.08%		0.08		
State Ministry of Youth and Sports	457.4	0.10%		0.1		
Corruption Eradication Commission	221.7	0.05%	0.05	0.05	0.05	0.05
Rehabilitation and Reconstruction Agency	9,976.70	2.27%		2.27		
Regional Legislative Council	149.2	0.03%				0.03
Judiciary Commission	34.9	0.01%		0.01		0.01
Agency Coordination of National Disaster Response	—			0		0
National Agency for Placement and Protection	—					0
Ministry and State Institutions (I)	189,361.20	43.03%				
Debt interest	78,828.10	17.91%				17.91
Subsidy	140,058.50	31.83%			31.83	
Other expenditure	31,784.30	7.22%				7.22
	250,670.90	56.97%	4.93	35.13	40.51	27.42
Total	440,032.10	100%	5%	35%	41%	27%

APPENDIX 4: EAST KALIMANTAN PLANNING AND BUDGET ANALYSIS CHARTS

Evaluation of Provincial Midterm Development Plan 2009–2014

Policy description (as indicated by policy listing in chapter 7 "Public Policy and Regional Development Program")	Sustainable development sector			
	Env.	Soc.	Eco.	Other
A. Legal affairs, politics, and government				
1. Enhance good governance through supervisory and control governance and regulatory improvement efforts supporting legislation.				1
2. Improving implementation of the principles of good governance local governance and prevention, eradication, and acceleration prosecution of corruption cases.				1
3. Rule of law and transparency and improving service system public.				1
4. Increasing community participation in prevention efforts and drug abuse prevention.		1		
5. Enhance security and public order for a smooth election and activities in the area through community participation and enhance community-policing functions. Create a tiered direction from the village level to provincial level, also participation of police, armed forces, commands, member linmas, and youth organizations.		1		
6. Increasing consolidation and understanding of nationality in the territory of homeland and community-based behavioral ideology Pancasila and Constitution 1945.		1		
7. Increasing political education to the community and improving the system, methods, and materials in order to increase alertness.		1		
8. Improving the development of border regions in an effort to accelerate development to improve accessibility and growth economics.		1	1	

Continued

Evaluation of Provincial Midterm Development Plan 2009–2014—cont'd

Policy description (as indicated by policy listing in chapter 7 "Public Policy and Regional Development Program")	Sustainable development sector			
	Env.	Soc.	Eco.	Other
9. Improve the performance and coordination of governance and development of regional governance through capacity building and coordination.	1	1	1	
10. Increasing domestic and international cooperation, as well as between government and institutions.				1
11. Develop and increase regulation of government administrative area through dispute settlement procedures between administrative boundaries, including districts/cities, interprovincial, and interstate areas in East Kalimantan.				1
12. Improving the efficiency of local and regional organizations, organizational structure, and working procedures of the institution through technical improvements, training and personnel development and education service, bureaucratic reform, and equitable distribution of staff to remote areas and inland.		1		1
13. Improving infrastructure and services through preservation of local archives and documents to improve archival information services.		1		
14. Increasing research and development and utilizing science and technology research results as bases for formulating policies and decision making by government and society.				1
15. Improving the quality of handling and disaster relief through early prevention and risk reduction, including disaster preparedness, early warning detection and mitigation, and postdisaster rehabilitation and reconstruction.	1	1	1	

Evaluation of Provincial Midterm Development Plan 2009–2014—cont'd

Policy description (as indicated by policy listing in chapter 7 "Public Policy and Regional Development Program")	Sustainable development sector			
	Env.	Soc.	Eco.	Other
16. Improving quality and information quality through the development of information and communications technology networks within the provincial and district/city, as well as dissemination and range of information to all areas in East Kalimantan.		1		
17. Improving the quality of local development planning to ensure sustainable development and implement good governance	1	1	1	
B. Economic affairs				
1. Enhance regional economic growth through the development of structures in a balanced economy between sectors of the capital-intensive economy and economic sectors in labor intensive to spur economic sectors and potential labor-intensive agricultural sector.			1	
2. Develop centers of economic growth that can encourage regional economic growth, adjusted for comparative advantage in each region.			1	
3. Building a strong agricultural sector, which consists of food crops, plantation, fishery, and animal husbandry, to develop comparative advantages and a competitive agro-industry.			1	
4. Developing small and medium industrial sectors, which can result in intermediate goods in an effort to encourage industrial development upstream.			1	
C. Infrastructure sector				
1. Improving basic infrastructure development, including transportation and communications, electricity, and clean water, as a prerequisite for entry of both foreign and domestic direct investment.		1	1	

Continued

Evaluation of Provincial Midterm Development Plan 2009–2014—cont'd

Policy description (as indicated by policy listing in chapter 7 "Public Policy and Regional Development Program")	Sustainable development sector			
	Env.	Soc.	Eco.	Other
2. Developing border regions and rural and disadvantaged areas to organize and explore various potential economic and infrastructure developments at the border.		1	1	
3. Construction of highway.		1	1	
4. Development and road improvements to drive the economy.		1	1	
5. Construction and development of airports and ports.			1	
6. Building and developing energy sources and power plant potential alternatives, such as hydroelectric power generation, plant microhydro power, gas power plant, and steam or coal power plant.	1		1	
7. Procurement of a generator for a village.		1	1	
8. Development of water resource facilities and infrastructure in order to support food self-sufficiency.	1	1		
9. Construction of flood control infrastructure in some districts.		1		
D. Social area				
1. Improving the quality of education through increased development of facilities for primary and secondary education that leads to national standards education.		1		
2. Develop a winning school in each district/city in order to recruit students who excel in every area.		1		
3. Improving community health service facilities, especially for low-income communities in rural areas, border areas, backward, and remote areas.		1		
4. Free health service, health center 24 hours with a second doctor.		1		
5. Increasing participation and empowerment of women in developing communities and villages.		1		
6. Increasing employment opportunities.		1	1	

Evaluation of Provincial Midterm Development Plan 2009–2014—cont'd

Policy description (as indicated by policy listing in chapter 7 "Public Policy and Regional Development Program")	Sustainable development sector			
	Env.	Soc.	Eco.	Other
7. Increasing the participation of youth in development and improve performance sports		1		
8. Improve management to control the growth of population	1	1		
9. Improving security and social protection and the fulfillment of basic rights for the poor.		1		
E. Environmental field				
1. Protecting and maintaining protected areas, national parks, water catchment areas, and greenery on critical lands in the context of reducing the occurrence of natural disasters, especially floods and landslides.	1			
Total	8	28	17	6
Total percentage (40 total policies)	20%	70%	43%	15%

East Kalimantan Expenditure by Sector

Description of organization/project	Expenditure (USD)a	Expenditure (trillions Rp.)	Percentage of total budget	Env.	Soc.	Econ.	Admin.
Family planning and family welfare	368,684	3,845,005,000	0.05%	0.05	0.05		
Cooperative and UKM	525,837	5,483,950,000	0.08%				0.08
Investment	1,132,971	11,815,750,000	0.16%			0.16	
College	1,241,398	12,946,541,375	0.18%		0.18		
Agency of Indonesian Broadcasting Commission	1,399,223	14,592,500,000	0.20%				0.2
Women's empowerment	1,401,506	14,616,306,000	0.20%		0.2		
Border area management	1,412,887	14,735,000,000	0.20%				0.2
Sekretariat DPP Korpri	1,446,843	15,089,127,500	0.21%				0.21
Environment	2,073,585	21,625,421,000	0.30%	0.3			
Filing	2,232,339	23,281,067,000	0.32%				0.32
Drugs	2,257,168	23,540,000,000	0.33%		0.33		
Industry	2,269,811	23,671,857,550	0.33%			0.33	
Empowerment of rural communities	2,777,831	28,970,000,000	0.40%		0.4		
Tourism	3,353,420	34,972,813,728	0.48%		0.48	0.48	

Sector	USD[a]	IDR	%				
Development planning	3,595,839	37,501,000,000	0.52%				0.52
Forestry	4,156,133	43,344,314,000	0.60%	0.6			
Social	4,729,915	49,328,280,300	0.68%		0.68		
Unity of national and domestic police	5,041,327	52,576,004,000	0.73%		0.73		0.73
Marine/fisheries	5,139,344	53,598,214,000	0.74%	0.74		0.74	
Employment	5,207,346	54,307,406,750	0.75%			0.75	
Communication and information	5,241,439	54,662,970,000	0.75%		0.75		0.75
Trade	5,439,141	56,724,805,000	0.78%			0.78	
Transmigration	6,659,315	69,450,000,000	0.96%				0.96
Energy and mineral resources	6,709,117	69,969,379,150	0.97%	0.97	0.97		
Youth sports	11,967,529	124,809,360,400	1.72%		1.72		
Employment	12,602,672	131,433,263,400	1.82%		1.82	1.82	
Library	30,076,003	313,662,635,250	4.33%		4.33		4.33
Agriculture	38,120,067	397,554,181,687	5.49%			5.49	
Regional autonomy, general government	57,015,877	594,618,577,901	8.21%				8.21
Health	71,854,763	749,373,322,211	10.35%		10.35		
Transportation	85,826,671	895,086,348,200	12.36%		12.36	12.36	
Education	124,243,152	1,295,731,831,677	17.90%		17.9		
Public works	186,734,123	1,947,450,171,385	26.90%		26.9		26.9
Total	694,253,275	7,240,367,404,464	100.00%	3%	79%	24%	43%

[a] USD based on average exchange from 1/1/2009 to 12/31/2009 = 10,429 IDR.

APPENDIX 5: BALIKPAPAN PLANNING AND BUDGET ANALYSIS CHARTS

Evaluation of Local Midterm Development Plan 2004–2009

Policy—as indicated by chapter description	Sustainable development sector			Other
	Env.	Soc.	Eco.	
Realization of human resources and sound mind and competitiveness in science and technology				1
Achieving the availability of city infrastructure able to meet the needs of communities and support functions of the city in the future		1	1	
Realization of natural resources in a controlled, livable, and environmentally sound urban condition	1			
City oriented to the development of democracy and development potential of the economic base of cities in the future		1	1	
Realization of the rule of law and good governance that guarantees justice and legal certainty for the community	1	1	1	
Increase the faith and devotion to God Almighty		1		
Increasing the degree and quality of public health services		1		
Controlling the amount and rate of population growth	1	1		
Increasing youth activities and sports		1		
Increasing the role of women and children		1		
Improving the quality and coverage of public education in the community for every level of education, including a library and museum		1		
Reduce the unemployment rate, improve the quality of the labor force, and implement the employment information system		1		

Evaluation of Local Midterm Development Plan 2004–2009—cont'd

Policy—as indicated by chapter description	Sustainable development sector			
	Env.	Soc.	Eco.	Other
Reduce poverty		1		
Improving the social welfare of society		1		
Build city infrastructure able to meet the needs of the community and function in the future city (duplicate policy)	–	–	–	–
Develop a road and transport infrastructure (transportation)		1		
Develop a harmonious arrangement of settlements and provide livable and affordable housing		1		
Develop a quality water supply	1	1		
Develop sustainable energy resources	1	1		
Develop spatial planning able to control future development of Balikpapan				1
Develop tourism			1	
Achieve natural resources, a sustainable and livable urban condition and environment (duplicate policy)	–	–	–	–
Integration and harmonization of policies of natural resource management and environment with other sectors	1			
Mainstreaming of sustainable development principles into all areas of development	1	1	1	
Increasing the capacity of environmental management institutions	1			
Increase public awareness on environmental care	1			
Realizing the potential of economy-oriented economic development and development of the economic base of cities in the future (duplicate policy)	–	–	–	–
Development of economic potential to the community, namely, agriculture, plantation, animal husbandry, fisheries and marine, and small and medium enterprises and cooperatives		1	1	

Continued

Evaluation of Local Midterm Development Plan 2004–2009—cont'd

Policy—as indicated by chapter description	Sustainable development sector			Other
	Env.	Soc.	Eco.	
Development of Balikpapan in the future economic base, particularly for industry, commerce and services, and tourism			1	
Investment and business development			1	
To realize the implementation of good governance (duplicate policy)	–	–	–	–
Arrangement of regional government institutions				1
Development of management information systems of corporate governance				1
Improving the quality of partnerships between regions and/or abroad				1
Increased capacity of Balikpapan city government				1
Development of public participation in public policy making		1		1
Coaching and supervision				1
To realize the rule of law that guarantees justice, expediency, and legal certainty and safety for the community (duplicate policy)	–	–	–	–
Update bylaws				1
Consistent enforcement of local regulations		1		1
Development of regional law enforcement code of ethics		1		1
Community empowerment in the legal field		1		
Achieve a sense of security and public order		1		
Total	10	24	10	12
Total percentage (38 policies)	26%	66%	24%	32%

Balikpapan Government Expenditure by Department/Project—2007 APBD

Description of office/project	Total expenditure (Rp.)	Percent	Sustainable development criteria			
			Env.	Soc.	Econ.	Admin.
Education	164,037,826,044.73	12.67		12.67		
Health	52,948,894,653.40	4.09		4.09		
Public works	362,556,331,697.70	28.00		28.00	28.00	
Public housing	4,219,954,793.60	0.33		0.33		
Spatial planning	7,921,749,222.40	0.61				0.61
Development planning	14,282,515,676.80	1.10				1.1
Transportation department	19,087,394,428.20	1.47		1.47	1.47	
Environment	40,630,757,079.60	3.14	3.14			
Regional Environmental Impact Management Agency	15,262,779,123.20	1.18	1.18			
Department of Hygiene and Cemetery	25,367,977,956.40	1.96		1.96		
Land	0.00	0.00	0.00			
Population and civil records	4,597,886,037.80	0.36		0.36		
Women's empowerment	0.00	0.00		0.00		
Family planning and family welfare	3,364,207,340.00	0.26	0.26	0.26		
Social	0.00	0.00				
Manpower	7,843,904,295.00	0.61		0.61		
Cooperatives and small and medium enterprises	6,274,026,212.00	0.48		0.48	0.48	
Department of Industry, Trade, and Cooperation	6,274,026,212.00	0.48		0.48		

Continued

Balikpapan Government Expenditure by Department/Project—2007 APBD—cont'd

Description of office/project	Total expenditure (Rp.)	Percent	Sustainable development criteria			
			Env.	Soc.	Econ.	Admin.
Capital investment	0.00	0.00			0.00	
Culture	0.00	0.00			0.00	
Youth and sports	0.00	0.00			0.00	
National unity and politics in city	9,699,682,321.76	0.75				0.75
Kesbang Office and Wellness	2,638,808,528.80	0.20				0.2
Police unit office of civil service	7,060,873,792.96	0.55		0.55		0.55
General government	529,754,031,597.52	40.91				
Regional Representatives Council	11,968,795,640.00	0.92				0.92
Mayor and deputy mayor	593,388,910.00	0.05				0.05
Regional secretariat	431,220,871,201.79	33.30				2.64
Legislative secretariat	34,173,097,271.20	2.64				2.64
Regional revenue office	22,448,828,337.69	1.73				1.73
Local supervisory board	5,578,982,487.80	0.43				0.43
Neighborhoods and villages (33 total)	23,770,067,749.04	1.79		1.79		
Civil service	17,168,131,917.60	1.33		1.33		
Rural community empowerment	2,325,274,013.20	0.18		0.18		
Statistical	0.00	0.00				0
Filing	972,157,455.80	0.08				0.08
Office of archives and library	972,157,455.80	0.08				0.08
Communications and information	0.00	0.00				

Subtotal I. Mandatory affairs	1,247,684,724,607.11	96.35		
Agriculture	24,496,067,382.60	1.89	1.89	1.89
Office of food crops and horticulture	4,560,819,610.40	0.35		0.35
Plantation office	15,985,730,399.80	1.23		1.23
Livestock office	3,949,517,372.40	0.31		0.31
Forestry	0.00	0.00		
Energy and mineral resources	0.00	0.00		
Tourism	2,957,513,926.80	0.23		0.23
Maritime affairs and fisheries	3,025,379,900.80	0.23		0.23
Trade	16,751,799,617.20	1.29		1.29
Industry	0.00	0.00		
Transmigration	0.00	0.00		
Subtotal II. Optional affairs	47,230,760,827.40	3.65		
Total	1,294,915,485,433.51	100.00	1%	56%	34%	8%

CHAPTER 10

Financial Investments for Zero Energy Houses: The Case of Near-Zero Energy Buildings

Natalija Lepkova[1], Domantas Zubka[2], Rasmus Lund Jensen[3]

[1]Department of Construction Economics and Property Management, Vilnius Gediminas Technical University, Vilnius, Lithuania
[2]Master of Management and Business Administration, Vilnius Gediminas Technical University, Vilnius, Lithuania
[3]Department of Civil Engineering, Head of Indoor Environmental Engineering Laboratory, Aalborg University, Aalborg, Denmark

Contents

Introduction	217
Concept of a Zero Energy Building	219
Importance of Renovating Residential Buildings	223
Evaluating Investments for Renovation of a Detached House into a Zero Energy Building in Denmark	225
Background	225
General Information about the Existing Building and its Renovation	226
Building Simulation Data	226
Assessment Methods Used to Measure the Investment Efficiency of Residential House Renovation	228
Results of Investment Estimations	231
A Survey on Possibilities to Implement Zero Energy Buildings	236
Background	236
Survey Results	238
Conclusions	245
Appendix	248

INTRODUCTION

Climate change is already affecting wildlife and human lives and their future. Due to natural causes, such as slight changes of solar radiation, volcanic eruptions, and natural climate variability, the climate is constantly changing. However, natural causes can explain only a small part of such a climate change. The main cause is the effect on the environment by greenhouse gases related to human activity. People use fossil fuels, which formed from oxygen millions of years ago and now release CO_2. One of the main goals for today's humans is to protect the planet's environment and find ways to reduce pollution and energy consumption, as well as increase the use of sustainable energy sources.

The construction sector has become energy-intensive and a serious source of pollution. Today, houses use more energy and fossil fuels than ever before. In fact, buildings currently consume more than one-third of all energy and two-thirds of all electricity used in the United States and the European Union (EU). Ideas of efficient energy use in buildings are being developed to solve these problems, noticed almost three decades ago. Many ideas have already been validated and are widely used in the construction of low-energy, superinsulated, passive, zero energy, and other types of eco-friendly buildings. However, construction of new houses alone cannot solve the problems: many buildings—detached houses, blocks of flats, multistory houses, commercial and public buildings, and so on—need to be renovated.

Renovation of detached houses is a topic less discussed than renovation of apartment buildings, but single-family houses should also be included in the renovation process and turned into zero energy buildings, which will be the standard in the near future of today's construction sector. One of the main factors that affect decisions to start renovation is the money needed to turn an existing house into a zero energy building. It is important to analyze whether renovation of a detached house by turning it into a near-zero energy building is economically sensible and what is the expected investment payback time. The results may affect the amount of renovations in the future: promising and good results of first projects will encourage more and more people to start reconstruction of their houses by turning them into near- or net-zero energy buildings.

There are no instances of zero energy buildings in Lithuania. It is important to analyze possible future perspectives of zero energy houses and see which type of building is interesting to construction professionals. New construction of zero energy houses and favorable results of their monitoring can be a decent example for people and encourage them not only to build new buildings of this type, but also to renovate their old houses to achieve a zero energy level.

Research aims to estimate the investments required to turn a detached house into a near-zero energy building and to determine the future perspectives of zero energy houses in general.

The part dealing with investment estimation considers a detached house in Denmark and its possible renovation models. The information from simulations and investments made was presented by the research group of Aalborg University and its leader R.L. Jensen (2010). The part dealing with the survey determined the opinions of Lithuanian and foreign construction professionals on zero energy buildings and their perspectives. The objectives of this chapter have been achieved using descriptive and numerical presentation methods based on the analysis of literature and official documents, measurement of financial indicators, and social surveys. The main information came from academic publications by Lithuanian and foreign authors, scientific articles from scientific journals, and statistical data published by EU, Lithuanian, and Danish institutions.

CONCEPT OF A ZERO ENERGY BUILDING

The construction industry is one of the biggest consumers of fossil fuels and producers of carbon dioxide. Energy—consumption of nonrenewable energy sources in particular—is one of the key issues in the 21st century. It is therefore very important to come up with, and examine, ideas of an energy-efficient house, which, compared to a conventional house, reduces heating needs dramatically. There is a huge diversity of energy-efficient buildings by energy consumption and CO_2 emissions—zero energy houses are among them. Zero energy houses are the vision of modern reality in the construction industry.

Hammon et al., (2009) state that in attempts to achieve a zero energy level, detached houses should be focused on first. The variety of definitions of a zero energy house is huge. Zero energy houses are sometimes referred to as zero energy buildings (ZEB). Despite sharing the name, zero energy buildings in practice may be defined in several ways. Torcellini et al., (2006) distinguish such definitions of different types of zero energy buildings:

- "Net Zero Site Energy: a site ZEB produces at least as much energy as it uses in a year, when accounted for at the site."
- "Net Zero Source Energy: a source ZEB produces at least as much energy as it uses in a year, when accounted for at the source. Source energy refers to the primary energy used to generate and deliver the energy to the site. To calculate a building's total source energy, imported and exported energy is multiplied by the appropriate site-to-source conversion multipliers."
- "Net Zero Energy Costs: in a cost ZEB, the amount of money the utility pays the building owner for the energy the building exports to the grid is at least equal to the amount the owner pays the utility for the energy services and energy used over the year."
- "Net Zero Energy Emissions: a net-zero emissions building produces at least as much emissions-free renewable energy as it uses from emissions-producing energy sources."

In the Energy Performance of Buildings Directive (2010), the European parliament defines zero energy buildings as follows: "Net zero energy building means a building where, as a result of the very high level of energy efficiency of the building, the overall annual primary energy consumption is equal to or less than the energy production from renewable energy sources on site." A more specific definition was released in 2011. The directive also defines a near-zero energy building, which may sometimes be referred to as a zero energy building: "nearly zero energy building means a building that has a very high energy performance, the nearly zero or very low amount of energy required should be covered to a very significant extent by energy from renewable sources, including energy from renewable sources produced on-site or nearby."

Hernandez et al., (2010) say that "historical definitions of zero energy are based mainly on annual energy use for the building's operation (heating, cooling, ventilation, lighting, etc.). The term net zero energy is frequently used to present the annual energy balance of a grid connected building."

Iqbal (2004) defines a zero energy house as follows: "zero energy home annual energy consumption is equal to the annual energy production using one or more available renewable energy resources."

Laustsen (2008) provides a definition of a net-zero energy building and its difference from zero carbon buildings:

> Zero net energy buildings are buildings that over a year are neutral, meaning that they deliver as much energy to the supply grids as they use from the grids. Seen in these terms they do not need any fossil fuel for heating, cooling, lighting or other energy uses although they sometimes draw energy from the grid. Zero carbon buildings differ from zero energy building in the way that they can use for instance electricity produced by CO_2 free sources, such as large windmills, nuclear power and PV solar systems which are not integrated in the buildings or at the construction site.

In this chapter, the term zero energy house corresponds to the "net-zero energy building" as defined by the European parliament.

Zero energy houses produce as much energy as they consume. There are several ways to achieve zero energy levels in a building. Energy conversion: the house requires very little energy for its maintenance and has renewable energy systems for energy production on-site or is connected to renewable energy sources on a grid. If a zero energy house generates energy on-site and is not connected to a grid, it may be called an autonomous house. If a house produces more energy than it consumes, it may be called a "plus energy" house. A zero energy house can be connected to a grid and has an option to purchase energy generated off-site by renewable energy sources.

A house connected to a grid may also supply its surplus energy to utilities. After renovation, when detached houses have been turned into zero energy buildings, they can produce electricity by photovoltaic panels and wind turbines. Some EU countries and the United States already offer an option to sell excess electricity generated on-site. The Lithuanian government is considering approval of an option to sell excess generated electricity, but the intention is to give such permissions only to a limited number of subjects.

A clear standard of a zero energy house does not exist: there are passive building standards in Germany and classes of low energy buildings in Denmark. Because zero energy houses have a very low demand for energy and can sometimes reach the level of a passive house, the first step of house renovation is to reduce the energy consumption of the building. But ways to reach the standards in one climate zone cannot be copied and used in another climate zone. Venckus et al., (2009) compare the differences of energy consumption in the same type of a passive house in several European countries. They describe a passive house designed on the basis of Lithuanian climatic data (the city of Vilnius) and equivalent to the German standard of a passive house, which consumes 15 kWh/m² of heating. The building's heat demand has been evaluated using climatic data from a number of European cities. Results show that the same house built in Austria would use half the energy for heating than that in Lithuania, but in Helsinki the energy demand would increase by 60%. According to the results, such a building would have the most similar energy demands in Warsaw. Data presented by Venckus et al., (2010) suggest that, among the reviewed countries, the best example of a passive house for Lithuania is in Poland.

Reyes (2008) states that the general characteristics of a zero energy house aim to meet the following demands:
- average annual energy production equal to energy consumption
- minimum consumption of water from the main's supply
- maximum use of environment-friendly materials
- standard or improved comfort of living

Zero energy houses are constructed mostly from the same materials as conventional houses. But energy demand may be reduced in the entire building. The best chance to save energy is to reduce the energy consumed for heating, ventilation, and hot water. This means a zero energy building needs more insulation than a standard house. The efficiency of the main energy generation equipment must be as high as possible; it also must be installed properly. Solar collectors, photovoltaic systems, thermal pumps, and wind turbines may be sources of energy in a zero energy house.

The design of a zero energy house may include the following solutions:
- specific, climate-adapted design
- passive solar design
- active renewable energy systems
- thermal mass
- superinsulation
- tightly sealed envelope
- low-emission windows
- energy-saving appliances

Zero energy buildings are very high energy performance buildings and many of them meet the standard of a low energy house. According to Harvey (2009) and Hamada and colleagues (2003), the standard of a low energy house gives energy savings from 50 to 75%. Also, Harvey (2009) analyzed the studies by Holton (2002), Gamble and colleagues (2004), and Rudd and colleagues (2006), which show that a series of modest insulation and window improvements can lead to energy savings of 30–75% in a wide variety of U.S. climates.

Indoor thermal comfort must be guaranteed, and EU countries have their own regulations and standards defining the quality of indoor comfort. Johnston and Gibson (2010) say that, compared to conventional houses, zero energy buildings have higher thermal comfort because of better insulation and modern ventilation.

According to Balaras et al., (2007), better energy performance of a building can help countries achieve the requirements of the Kyoto Protocol. Zero energy houses use only either renewable energy sources or together with a tiny amount of fossil fuels; consequently, they produce very low CO_2 emissions compared to a standard detached house. A renovated single-family house with properties of a zero energy building becomes more eco-friendly and pollutes the environment less.

In the wake of the global economic crisis, investments into energy-efficient construction decreased—into renovation of existing buildings as well. If the global economic situation remains unchanged and the crisis continues, many renovation projects might be suspended or will not start. Nistorescu and Ploscaru (2009) present development statistics of the European construction industry since 2008. Data presented by Nistorescu and Ploscaru (2009) show that Lithuania's construction industry between 2005 and 2007 was growing, but the ensuing economic crisis cut the growth by almost 21%. Before 2006, Denmark recorded a growth of construction, but the rates dropped in 2008 and 2009.

A stagnating construction industry may have a negative effect on the progress of renovation.

The next section discusses the importance of renovating residential buildings.

IMPORTANCE OF RENOVATING RESIDENTIAL BUILDINGS

Research by Itard et al., (2008) reveals that the main reason to renovate the stock of residential buildings is a combination of needing to reduce energy consumption and needing to replace structural members that have finished their service life.

One of the major problems today is drying nonrenewable energy sources and the negative effect on the environment. There are tons of old buildings that not only lack energy efficiency, but also are a big source of pollution; moreover, a growing energy demand increases the dependency of buildings on fossil fuels. Balaras et al., (2005) present statistical data showing that over 50% of existing residential buildings in 25 EU member states were built before 1970 and one-third of dwellings were built between 1970 and 1990. Hence, structural members of more than half of the buildings in Europe are old and energy loss can be prevented by their renovation. In the EU, buildings consume 40% of end energy, whereas residential buildings account for 63% of total energy consumption in the buildings sector. Statistical data presented by Balaras and colleagues (2005) show that residential buildings in Germany, France, Italy, and the United Kingdom are the biggest energy consumers in the EU. In Latvia, Estonia, and Hungary, residential buildings consume more than 60% of all energy used by these countries. These and other statistical data show broad possibilities to reduce energy demands in residential buildings, which can make a positive impact on energy consumption issues worldwide.

Reducing energy consumption is one of the key goals of the EU. Here is a selection of the requirements issued by EU institutions:

> *the current directive of the European Parliament and the Council on energy performance of buildings sets forth reduction of energy consumption in buildings and claims that the use of energy from renewable sources in the buildings sector constitutes important measures needed to reduce the EU energy dependency and greenhouse gas emissions; the Council of European Union said all new buildings must be nearly zero-energy buildings by 31 December 2020, that an intermediate target must be set for 2015, and buildings occupied and owned by public authorities have to be nearly zero-energy buildings after 31 December 2018, in line with the leading role that the public sector should play in this field; by 31 December 2018 at the latest*

> *EU Member States must ensure that all newly-constructed buildings produce as much energy as they consume on-site, e.g., via solar panels or heat pumps. The Industry Committee also wants Member States to set intermediate national targets for existing buildings, i.e., to fix minimum percentages of buildings that should be zero energy by 2015 and by 2020, respectively.*

Furthermore, member states should develop policies for the transformation of existing buildings into nearly zero energy buildings.

Zavadskas et al., (2004) say that the condition of the majority of residential buildings in Vilnius, Lithuanian's capital, is less than satisfactory and need renovation. Ustinovičius (2004) claims the choice of a right object for reconstruction investments is the most important. In 2004, Lithuania's government approved the program of buildings' modernization. It states that 70% of the current housing stock must be renovated by 2020. This program covers not only apartment buildings, but also detached houses constructed before 1993. The majority of buildings in Lithuania were built in the Soviet era and need renovation to improve their energy efficiency. Such buildings are eligible for financial support from the program.

By year of construction, Denmark and Lithuania have similar building stock. Just like in Lithuania, the majority of buildings in Denmark are old and were built between 1946 and 1980. Energy consumption in Denmark is similar to the EU average: about 40% of all energy. Tommerup et al., (2005) say that "a large potential for energy savings exists in the Danish residential building stock." Three-fourths of all buildings in Denmark were built before introduction of the first regulations of energy performance for buildings in 1979.

Meyer (2010) lists policy examples of the Danish building sector: renovation of existing buildings; energy labeling of all buildings combined with graduated green tax; investment subsidies for renovation of buildings and installation of solar collectors, photovoltaic panels, heat pumps, etc.; and investment subsidies for the replacement of selected old houses by passive houses. New construction of passive houses may also cut energy consumption in Danish buildings by over 40% from its current levels.

Tommerup et al., (2006) states that "profitable savings potential of energy used for space heating of about 80% is identified over 45 years (until 2050) within the residential building stock if the energy performances are upgraded when buildings are renovated."

Danish buildings constructed between 1998 and 2003 lose less energy than buildings built before that period. Energy losses through windows are rather similar in all periods. Obviously, the biggest heat losses occur through exterior walls of buildings built between 1930 and 1970. Buildings from this period also lose more energy through floors, roofs, and ventilation than buildings from other periods. Hence, renovation of buildings built between 1930 and 1970 will result in the best energy savings and outcomes.

A report by the Danish Climate Commission (2010) suggests that Denmark could be fossil fuel free by 2050. The report shows that the country could cover all of its energy demands by energy from renewable sources such as wind, sun, waves, geothermal power, and biomass power.

It also means that all buildings will only use energy produced by sustainable sources, which will replace fossil fuels such as natural gas, coal, and oil, now accounting for more than two-thirds of all energy in Denmark. Buildings will have to produce their own energy on-site or will be connected to renewable energy suppliers on grid. Furthermore, greenhouse gas emissions will be cut to very low levels.

The Danish Climate Commission (2010) states that "there are generally greater opportunities to reduce energy consumption in existing than in new buildings. In many cases, efficiency initiatives can be profitable if energy improvements are implemented in connection with renovation." As a result, there is a necessity to reduce energy demands in existing buildings, thus achieving the standard of a low energy building. Some cases make it possible for existing buildings to meet the standard of a zero energy building and to produce their own energy on-site.

Tons of new zero energy houses have been constructed worldwide, but only a few detached houses have been renovated into zero energy buildings. In line with the aims and requirements of the EU parliament and individual countries to cut fossil fuel consumption for heating and electricity demands of buildings to very low levels or stop altogether, all existing buildings will have to become energy efficient, or even zero energy, in the next 20 years.

EVALUATING INVESTMENTS FOR RENOVATION OF A DETACHED HOUSE INTO A ZERO ENERGY BUILDING IN DENMARK

Background

To estimate the investments, data of eight simulated buildings were used. The energy consumption and production values for all buildings were generated considering Danish climatic conditions. First, spreadsheets were prepared for calculation of the investment indicators of renovation to achieve the simulated levels in existing buildings. The measures of energy policy were chosen to win over public support for the renovation. These measures were evaluated separately. Having made all calculations and analyzed their results, conclusions were drawn.

Danish krones were used in calculations and later converted to Lithuanian litas at the exchange rate of the SEB Bank on November 2, 2011, which was LTL 2.43 for USD 1. All results and prices are in U.S. dollars. Some errors are possible in decimal places.

General Information about the Existing Building and its Renovation

The research deals with a detached house near Aarhus, Denmark. The building has been renovated into a first-class low energy building in line with the Danish building standards. The house was renovated as part of the EnergiParcel project. The EnergiParcel is a pilot project by Realea, Ltd.; its scope was renovation of houses constructed in the 1960s and 1970s. Houses selected for the EnergiParcel project were the types of single-family houses that were the biggest energy consumers in Denmark. The prototype house is an experiment that tests the limits of energy cuts in a conventional house.

The building is in a neighborhood of detached houses. Originally, it was a white two-story house built in 1974, with its facing wall from silicate bricks and its internal bearing walls from lightweight concrete. The 176-m² house has a conventional layout: rooms above and the stay and kitchen below. The house originally had white elements in its façade and a black asbestos cement roof. The exterior walls and the roof (or ceiling) had minimum insulation, while the doors and windows dated to the year of construction. The house was poorly maintained, which had a major negative impact on overall energy consumption. Before the renovation, the house had the energy label E. Its reconstruction budget made up USD 370,000.

Building Simulation Data

To simulate the building with renewable energy technologies installed, the research group of Aalborg University used Bsim and Mathcad software tools. The simulation returned 2880 results. Results illustrate energy consumption, energy production, economic values, CO_2 emissions, and the operating temperature for different values of parameters. In calculations, the study used seven different combinations of solar collectors, photovoltaic systems, and heat pumps (Table 1). The simulations of buildings for further

Table 1 Data of Simulations of Houses (Done by the Authors in Reference to the Research Group of Aalborg University)

Building	Renewable energy technologies	Investment costs, USD	Energy demand for heating, kWh/m²	Energy demand for electricity, kWh/m²	CO_2 emissions, t
"2010"-1	Good heat pump	20,000	0	39	3.754
"2010"-2	Large solar collectors, good heat pump	27,500	0	41	3.988
"2015"	Small solar collectors, good heat pump, small PV panels	46,000	0	30	2.840
"2020"	Good heat pump, medium solar collector, medium PV panels	48,900	0	20	1.959
"2030"-1	Good heat pump, large PV panels	51,400	0	13	1.238
"2030"-2	Large solar collectors, large PV panels	38,700	62	−9	0.598
"2030"-3	Large solar collectors, large PV panels, good heat pump	58,900	0	10	1.004

calculation were selected considering these parameters: south orientation of renewable energy systems; two people living in the building and using its appliances; medium level of energy consumption in the building; and the building has been renovated and has better infiltration and good mechanical ventilation installed. Table 1 presents energy consumption, CO_2 emissions, installed renewable energy systems, and the cost of investments in selected buildings.

Table 2 Data of Building in Question Before Renovation and Energy Prices

Name	Unit	Value
Floor area	m²	176
Heating demand before renovation	kWh/m²	161
Electricity demand before renovation	kWh/m²	20
Type of heating	—	District heating
CO_2 emissions	T	5.621
Electricity price	USD/kWh	0.33
District heating price	USD/kWh	0.15

Data in Table 2 were used in further investment estimations. The buildings have been labeled by the year of the standard to which they corresponded. Data show that the difference between annual heating energy consumption and production amounts is 0 kWh/m² in six of seven buildings. In one of the buildings, power generation exceeds consumption by 9 kWh/m²; hence, this building produces more electricity than consumes. None of the buildings (in all simulation results) has 0 kWh/m² annual energy demand for both heating and electricity. The buildings need more powerful photovoltaic systems to achieve an annual electricity demand of 0 kWh/m²; then the buildings will become net zero energy houses. Data show that none of the buildings are real net-zero energy buildings, but three of seven may be considered near-zero energy buildings.

Assessment Methods Used to Measure the Investment Efficiency of Residential House Renovation

An ideal assessment method has to be simple, has to be easy to comprehend, and has to avoid questionable assumptions. Several methods and criteria may be used in evaluations of building renovation. Investments may be estimated with help of the multiple criteria method when project benefits are evaluated by several criteria. The multiple criteria method evaluates social, environmental, improved comfort, new jobs, and other criteria. But to make this method useful, it demands a lot of human effort and time. This survey chose the financial criterion as one of the most important methods in the project analysis.

Rapcevičienė (2010) says that financial project analysis is a system of financial calculations, which reveals expenses, evaluates income and outcomes throughout project implementation, and shows the project's benefits to an organization. The project's economic benefit has been considered only from the perspective of energy-cost savings.

This survey chose the following financial criteria for the evaluation of building renovation:
- simple payback time (SPT)
- return of investment (ROI)
- net present value (NPV)
- internal rate of return (IRR)

Martinaitis et al., (2004) claim that "one of the most popular criteria used is a simple payback time because it is readily comprehensible for non-economists." The criterion is defined as a time period necessary to recover the initial investment. Simple payback time is defined as the number of years when money saved after the renovation will cover the investment.

When annual savings remain the same throughout the project period, a simple payback period is calculated as follows:

$$SPT = I/P,\qquad [1]$$

where I is investments, USD; and P is annual savings, USD (Rapcevičienė 2010).

However, Martinaitis et al., (2004) state that

simple payback time of an individual energy saving measure or a package of them are to be used only for the superficial evaluation of the cost effectiveness. The major limitation is that the lifetime of an "energy" saving measure is absolutely not taken into account. But simple payback time is that it does not value the cost of borrowing money. The payback time indicates if savings, resulting from the implementation of a measure, are sufficient to repay a loan. However, it also does not show if the savings are enough to pay the interest.

This tool is, however, useful for project evaluation before turning to a more serious analysis. If the simple payback time is shorter than a building's life cycle, project evaluation can proceed.

Return of investments is variously interpreted as an indicator depending on its purpose. ROI shows the efficiency of investments. ROI is calculated using the formula:

$$ROI = P/I * 100\%,\qquad [2]$$

where I is investments, USD; and P is annual savings, USD.

The investment profitability indicator ROI shows the return rate of investments. The ROI indicator is less important in evaluating a building's renovation because it is used more in analyzing a company's finances.

Net present value and internal rate of return are among the most important criteria evaluating the financial efficiency of investments. Martinaitis et al., (2007) say that *these criteria estimate benefits from investments over a certain period of time. The cost of borrowing is reflected by using discounted cash flows. A discount rate is usually equated to the market's interest rate. Nevertheless, like the simple payback period method, both the NPV and the IRR depend on estimates of future energy prices.*

Nikolaidis and colleagues (2009) state that "NPV sums the discounted cash flows; it integrates and converts at the same time amounts of money of various time periods." The formula to determine the NPV when all investments have been made at the beginning of the period in question is as follows:

$$NPV = -CF_0 + \sum_{t=1}^{T} \frac{CF_t}{(1+i)^t},$$ [3]

where CF_0 is the present value of investments, USD; CF_t is the net cash flow (the amount of cash, inflow minus outflow) at time t, USD; t is time of cash flow; and i is the discount rate, % (Nikolaidis, Pilavachi & Chletsis 2009).

The indicator is used mainly in calculations of the net present value: to subtract discounted cash outflows for the investment period from discounted cash inflows at the beginning of the investment period. Because cash flows are distributed in time periods, they are discounted by the percentage rate.

When NPV > 0, the project is acceptable and pays off its investments; when NPV < 0, the project is unprofitable and has to be rejected; and when NPV = 0, the project is neither profitable nor unprofitable.

Negative NPV values do not mean there is no benefit from the investment. Negative NPV values signal that it is better to opt for an alternative decision or to improve the investment project, not that the decision to invest is generally bad. NPV values show the best alternative among investment projects.

Nikolaidis and colleagues (2009) state that the criterion of the internal rate of return shows the investment return rate. It is used for comparing alternatives. IRR is equal to the rate of discount, which makes the value of future savings equal to the value of investments—IRR is a rate of discount that makes NPV equal to 0 (Nikolaidis, Pilavachi & Chletsis 2009). IRR can be calculated using the following formula:

$$0 = -CF_0 + \sum_{t=1}^{T} \frac{CF_t}{(1+IRR)^t},$$ [4]

where CF_0 is the present value of investments, USD; CF_t is the net cash flow (the amount of cash, inflow minus outflow) at time t, USD; and t is time of cash flow.

The internal rate of return is understood as the discount rate, which ensures equal present values of expected cash outflows (expenses) and expected cash inflows. The internal rate of return indicator shows the maximum possible relative level of expenditure. A negative IRR value indicates that an investment is likely to lose money and should be ruled out. A positive IRR value indicates viable future returns and should be maximized.

This study of renovating a house into a near-zero energy level building used SPT, ROI, NPV, and IRR indicators in investment estimations.

Results of Investment Estimations

Data about the selected prototype house and the heating and electricity prices used in the calculations are presented in Table 2. The invested amount, the floor area, and the type of heating were given in a brochure by Realea, Ltd (2010). The heating and electricity prices applicable in the area of the house in question were determined using information from NRG, Ltd. The house's demands for electricity and heating and CO_2 emissions were taken from simulation results considering the following facts: two people live in the building, the energy consumption is medium, the infiltration has not been improved, the building lacks mechanical ventilation, and the building's structural members have not been renovated.

Data about the building after its renovation but before installing renewable energy systems are presented in Table 3. Data about energy consumption and CO_2 emissions are taken from simulation results considering the following facts: two people live in the building, the energy consumption is medium, the infiltration has been improved, mechanical ventilation has been installed, and structural members have been renovated.

According to simulation results, data in Tables 2 and 3 show that renovation cut the heating demand but the demand for electricity remains the same.

Estimated investment amounts and savings after the renovation of simulated buildings are shown in Figure 1. Despite the fact that one of the

Table 3 Data About the Building After Improvement but Before Installing Renewable Energy Systems

Name	Unit	Value
Floor area	m^2	176
Heating demand after renovation	kWh/m^2	71
Electricity demand after renovation	kWh/m^2	20
Type of heating	—	District heating
CO_2 emissions	T	3.550

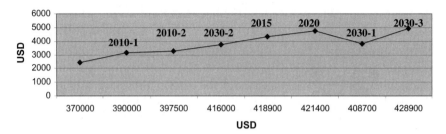

Figure 1 Dependence between investments and annual cost savings in the simulated building.

simulated buildings produces more energy than it consumes, its energy consumption is 0 kWh/m² because Denmark has no energy policy instruments that enable selling of surplus energy.

The amount of investments was calculated by summing up the investments into energy conservation and improvement of the quality of building and the investments into renewable energy systems. The final investments were calculated considering that all investments will be made at once without a bank loan. The reduction of CO_2 emissions is the difference between CO_2 emissions before and after renovation.

As seen in Figure 1, buildings "2030"-1 and "2030"-2 required the biggest investments. They also have the biggest savings after their renovation. The cuts of energy costs after renovation are also the biggest in buildings "2030"-1 and "2030"-2. The results of calculations show that higher amounts invested into renewable energy sources give more savings of energy costs. Building "2030"-2 has the highest cut of CO_2 emissions with the lowest (0) annual electricity demand. Buildings with heat pumps and without PV panels have lower CO_2 cuts than buildings with PV panels.

The calculated values of renovation's simple payback time and its ROI, NPV, and IRR are presented in Table 4. The indicators were calculated assuming that energy prices (thus the benefits after renovation also) will be the same every year in the future. NPV and IRR were calculated using a period of 40 years. A 2% annual inflation rate was assumed as the discount rate. Maintenance expenses have not been evaluated.

The financial indicators of investments presented in Table 4 show that the investments are inefficient: the simple payback time is very long, ROI is very close to 0%, the NPV and IRR values are negative.

The results of simple payback time calculations (Table 4) show that the payback period of the investments into renovation of these buildings is not reasonable. Buildings "2030"-1 and "2030"-2 have the best simple payback time of about 87–89 years, which more than twice exceeds the possible maximum bank loan period and thus is unacceptable. Moreover, the simple payback time of all simulated models exceeds the life cycle of renovated items. Results also show that the simple payback time depends directly on the investments into renewable energy installations; the evaluation of simulated buildings shows that more investments into renewable energy systems give a better simple payback time. Buildings with PV panels installed have better simple payback time than buildings without them. Consequently, renovation of buildings aiming at near-zero energy levels requires installation of PV panels to reduce the electricity demand after the heat pump has been installed.

The best ROI results (Table 4) were recorded in buildings that have shorter simple payback periods. The ROI for all buildings is above 0, which means the return of investments is positive; the investment efficiency, however, is very low. The best ROI is when investments aim to achieve the near-zero energy level. The values of ROI confirm that more investments into renewable energy technologies improve the renovation investment efficiency for the houses in question.

Relying on the NPV rule, these investments into simulated houses should be rejected because the NPV values are negative. However, these NPV values (Table 4) can also be interpreted as a sign that the best alternative among these models is investment into renovation aiming at near-zero energy levels because houses "2030"-1 and "2030"-3 have the highest NPV. The calculated NPV values also confirm that more investments into renewable energy systems give better renovation investment efficiency for the houses in question because the house models with bigger investments into renewable energy systems have higher NPV values. NPV values show that the worst alternative is to renovate a building as the "2005" house model without any renewable energy technologies.

The IRR values that achieve 0 value of NPV are negative. Consequently, all cases of renovation lead to loss of money and the projects have to be rejected.

The IRR and NPV values were calculated considering a 40-year period and a low discount rate. A higher discount rate would give an even lower value of NPV. Moreover, a 40-year period, which is rather long, is insufficient to get positive NPV and IRR values.

One must consider the fact that, in this renovation of buildings, not all investments sought energy savings. Investments may be divided in two groups: aimed at energy saving and at improvement of the quality of the

Table 4 Calculated Values of Simple Payback Time, ROI, NPV, and IRR

Indicator	Unit	"2005"	"2010"-1	"2010"-2	"2015"	"2020"	"2030"-1	"2030"-2	"2030"-3
SPT	Years	153	125	123	109	97	89	107	87
ROI	%	0.65	0.80	0.82	0.92	1.03	1.12	0.93	1.14
NPV	USD	−313,827	−300,608	−304,684	−301,854	−297,154	−289,158	−301,567	−291,909
IRR	%	−6	−5	−5	−4	−4	−4	−4	−4

Table 5 Estimated Investments into Renewable Energy Systems of Simulated House Models

Indicator	Unit	"2010"-1	"2010"-2	"2015"	"2020"	"2030"-1	"2030"-2	"2030"-3
SPT	Years	28	33	29	26	22	28	24
ROI	%	3.52	3.01	3.49	3.90	4.48	3.57	4.21
NPV	USD	−1028	−5104	−2274	2425	10,421	−1476	7669
IRR	%	2	1	2	2	3	2	3

house and its grounds. During renovation, investments into better quality of a building cannot be avoided, but if the main reason of renovation is to save energy in the building's use, the investment into the building's quality can be reduced. Then, a better payback time and other indicators can be expected.

Because the house has already been renovated into a low energy building by investing USD 370,000, and this fact cannot be changed, investments into renewable energy equipment have been estimated in order to choose the best combination for further improvement of the building in question. To determine the combination of renewable energy systems with the best results, selected simulated buildings were assessed, excluding investments into energy saving and quality improvement. The investment estimations of renewable energy systems in the simulated buildings are presented in Table 5. The values of buildings after renovation presented in Table 3 were used as the starting values of energy demand for heating and electricity. As in previous calculations, it was assumed that the annual savings remain the same every year and all investments are made at once at the beginning of the period; the period in question is 40 years and the discount rate is 2%. The operating expenses have not been evaluated.

The financial indicators of investments into the houses in question presented in Table 5 show that the investment efficiency is different when only the investments into renewable energy technologies are considered. The best values are for the investments into a combination of a good heat pump and large photovoltaic panels.

As in the case of estimated investments into energy conservation and improving the quality of buildings, investments into buildings "2030"-1 and "2030"-3 have the best simple payback time. Hence, despite the largest amount of investment required, the combination of large PV panels, a good heat pump, and solar collectors gives the best simple payback time. The investment into a good heat pump alone has a better simple payback time than installation of a good heat pump and a solar collector. Thus the price and effectiveness ratio of solar collectors fail to give a reasonable effect. The evaluation results also confirm the previous statement that a decision to heat a building by geothermal sources makes installation of PV panels very important when a house is to be renovated into a near-zero energy building.

The highest ROI, which shows investment effectiveness, was recorded in buildings that satisfy the majority of their energy demands by using renewable energy systems. The highest ROI was recorded in models with the biggest investments. Hence, more investments give a better return.

Investments into combinations used in buildings "2020," "2030"-1, and "2030"-2 have positive NPV values in a 40-year period when the annual savings remain constant. But a 40-year period is rather long.

All IRR values are positive. Hence, in a 40-year period with a discount rate of 2%, a profit may be expected.

A SURVEY ON POSSIBILITIES TO IMPLEMENT ZERO ENERGY BUILDINGS

Background

The calculations reviewed earlier show that renovation of a house into a zero energy building is sometimes economically imprudent. The possibilities of constructing new zero energy houses, therefore, need to be researched. Overall perspectives of zero energy houses affect the progress of renovation of single-family houses into zero energy buildings. This part of the chapter presents an opinion survey of construction and real estate professionals, politicians, and academic circles on the perspectives of zero energy buildings, their advantages, disadvantages, and implementation possibilities in Lithuania.

Surveys are a research tool widely used in research papers and articles. The basic survey procedure is that people are asked a number of questions on the aspects of interest to the researcher. A scientific survey must get as much objective information as possible.

This research aims to expose and compare the opinions of Lithuanian and foreign professionals about zero energy houses, their perspectives, and implementation possibilities in Lithuania and abroad. The questionnaire aimed to show opinions of respondents on such issues as

- EU requirements to build near-zero energy buildings from 2019
- barriers and drivers for a zero energy house
- climate suitability for a zero energy house
- perspectives of zero energy houses
- advantages and disadvantages of zero energy houses

The questionnaire was prepared in two languages: Lithuanian and English. It included 25 closed questions. Five questions gave the respondents an option to provide their own answer if they did not agree with the ones given. The questionnaire can be found in the appendix. The survey continued from August 2010 until December 2010. The choice of respondents fell on professionals working in industries related to zero energy buildings. A criterion to choose the respondents was their interest in renewable energy and energy efficiency. The respondents filled in a live form of the questionnaire online. Google spreadsheets were used to collect answers.

Analysis and interpretation of the survey results were used as methods of result description. This part reviews the answers to the questionnaire of Lithuanian and foreign specialists and compares the differences.

The sample in this research included 50 respondents. They were divided by their nationality and occupation. The distribution of their nationalities is shown in Figure 2.

Figure 2 shows that 50% of the professionals surveyed were from Lithuania. The other half came from different EU members: Germany, Denmark, Belgium, Spain, and Finland, which are countries with very fast development of construction of energy-efficient buildings, while the reduction of energy consumption in buildings is a priority in the agenda of their governments. The comparison of the opinions of Lithuanian and foreign professionals might show differences of experience in this field and different approaches to the problem.

One of the research aims was therefore to predict the future and evaluate the perspectives of zero energy houses, and it was important to have respondents who had jobs related to the construction industry. Figure 3 shows the distribution of the respondents by their occupation.

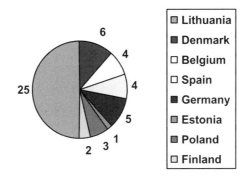

Figure 2 Distribution of respondents by nationality.

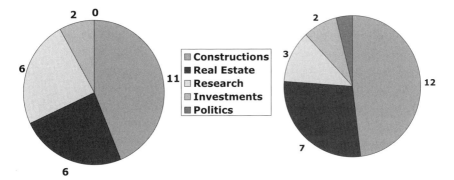

Figure 3 Lithuanian and foreign respondents by their occupation.

Most respondents worked in construction (23) and real estate (13). A survey of university researchers (9) was also important because they participate in many projects related to energy-efficient buildings by doing measurements and calculations.

All respondents answered "yes" to the question whether they were worried about energy consumption and global warming. It means these issues were familiar to them. It was also important to find out whether the respondents were interested in the topic of zero energy buildings. Figure 4 shows the self-assessment of the respondents' knowledge about zero energy buildings.

There are differences in the knowledge of Lithuanian and foreign respondents. Compared to the respondents from Lithuania, the proportion of foreign respondents who gave 100 points to their knowledge is larger—they have much interest, work, and invest in this field. It shows that the topic of zero energy buildings is closer to foreign respondents than to Lithuanian. The overall result shows that the respondents were professionals really interested in the topic of zero energy buildings and thus their opinion can be important in predicting the perspectives of the development of zero energy houses.

Answers of the respondents to the survey questions and interpretations of the answers are presented in the next section.

Survey Results

To determine the position of zero energy houses in today's construction industry, the respondents were asked whether the idea of a zero energy house is the reality, the future, or has no future. The distribution of answers is shown in Figure 5.

None of the respondents believed that the idea of a zero energy building had "no future"; it means the idea of such buildings has perspectives.

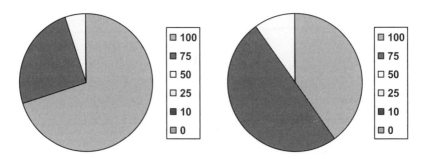

Figure 4 Assessment of knowledge about zero energy buildings.

In contrast to Lithuanian respondents, more foreign professionals believe that the idea of a zero energy building is a reality of the modern construction industry. It can be attributed to the fact that countries such as Germany, Belgium, and Britain already have many new zero energy buildings, and professionals from these countries are working in projects on the development of zero energy buildings. Results show that the idea of zero energy buildings is the future in the Lithuanian construction industry.

As mentioned before, from 2019 on, the EU intends to construct only near-zero energy buildings. This requirement sparked numerous debates among construction professionals and politicians about the feasibility to fulfil this requirement across the EU. This survey determined the real future of this requirement and the difference of opinions among foreign and Lithuanian professionals on this issue. All respondents confirmed they were aware of the intention to construct only near-zero energy buildings in the EU from 2019, but their opinions about the fulfilment of this requirement were not unanimous. The opinions are shown in Figure 6.

The majority of the respondents (60% foreign and 72% Lithuanian) agreed that this requirement would not be fulfilled and the European parliament would set a new date. It means this decision does not have enough support from the professionals of this field. Foreign professionals were more optimistic than Lithuanian respondents, not only because fewer among

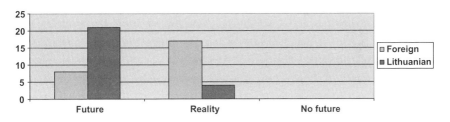

Figure 5 The idea of a zero energy house.

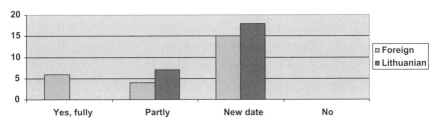

Figure 6 Will the requirement to build only near-zero energy buildings from 2019 be implemented across the EU?

them believed in the failure to fulfil this requirement, but six of them also stated that the requirement would be implemented across the EU.

Then the respondents were asked about the general types of buildings (zero energy—ZEB, passive, new model, or conventional) and the types of zero energy buildings with better perspectives in the next 10 years. Results are shown in Figure 7.

Only one respondent believed that conventional standard buildings have better perspectives in the next 10 years than energy-efficient buildings. Some professionals (20% foreign and one Lithuanian) think a new model of buildings will be invented. In contrast to Lithuanians, foreign professionals state that zero energy buildings have better perspectives than passive buildings in the next 10 years.

Both foreign and Lithuanian professionals believe that zero energy houses and zero energy commercial buildings clearly have better perspectives than zero energy industrial buildings. Results show that Lithuanian respondents see better development perspectives for zero energy houses than commercial buildings. The opinion of foreign professionals differs: they mostly (44%) specified zero energy commercial buildings. It shows that the development of zero energy buildings in Lithuania now concerns more zero energy houses.

It is very important to determine the main pluses and minuses of zero energy buildings. The main advantages can be used for marketing, while the disadvantages should be eliminated to make zero energy houses more attractive to developers and the public. Respondents were given options and asked to pick the biggest advantage and the biggest disadvantage of zero energy houses. Results are shown in Figure 8.

Foreign professionals believe that the main disadvantage of zero energy houses is their expensive construction. Lithuanian and foreign professionals agreed on the advantages of zero energy houses and specified financial

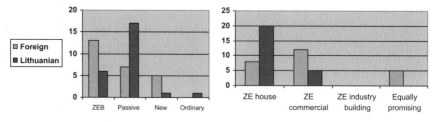

Figure 7 Which type of buildings (left) and which type of ZEBs (right) have better perspectives in the next 10 years?

benefits as the main advantages: economy, energy savings, and independence from centralized power grids, plus property appreciation.

Respondents were asked to choose the main barriers for zero energy buildings and possible key drivers of this sector's development. Results are shown in Figure 9.

Lithuanian and foreign professionals share their opinion. According to the results, a lack of public interest is one of the main barriers for zero energy buildings, thus government support and the possibility of selling produced energy may be among the main measures to make construction of such buildings attractive to the public and developers.

Respondents were asked about which aspects they considered to be the most important in construction of a detached house—about criteria relevant to the intention to build/buy a house (they were asked to choose the three most important criteria). Results are presented in Figure 10, which shows only those criteria selected by the respondents.

Both Lithuanian and foreign respondents said financial criteria (cost and stable value in the future) were the most important when intending to build/buy a house. In contrast to Lithuanian professionals, foreign respondents mentioned eco-friendliness of a house as an important criterion.

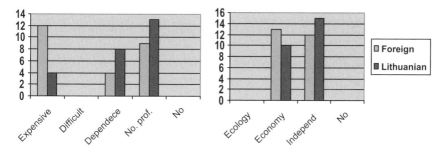

Figure 8 Main advantages (right) and disadvantages (left) of ZEHs.

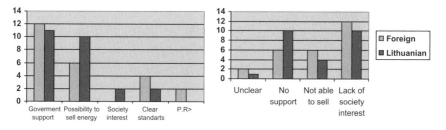

Figure 9 Key barriers (right) and drivers (left) of zero energy buildings.

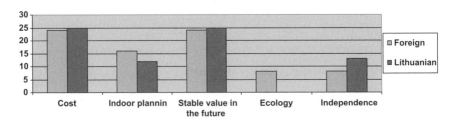

Figure 10 What aspects are the most important in the construction of a detached house and which criteria are relevant to the intention to build/buy a house?

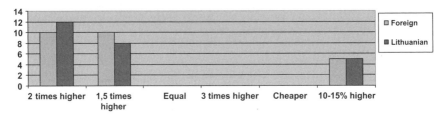

Figure 11 Rate construction costs of a ZEH and a standard house of the same floor area.

Indoor layout and independence from utilities were two other criteria pointed out by the respondents.

Respondents were asked to share their opinion on differences between construction costs of a zero energy house and a conventional standard house when their floor area was the same. The cost of construction depends on many different factors: materials, design, construction method, house location, and others. It is very difficult to judge the cost of construction without any project analysis. But this question aimed to evaluate the basic experience and knowledge of Lithuanian and foreign professionals. Results are shown in Figure 11.

Some respondents chose not to answer this question. Those who answered believed that construction of a new zero energy house would be more expensive than construction of a standard house with the same floor area. Lithuanian respondents were more willing to evaluate construction of a new zero energy house than their foreign peers.

As mentioned before, it is important to reduce a building's heating needs before installing active renewable energy systems in zero energy houses. Respondents were given options and asked to choose three solutions that they believed were the best for the construction of zero energy houses and had the biggest potential to reduce loss of heat compared to standard houses. Results are presented in Figure 12. Evaluations of heat loss through

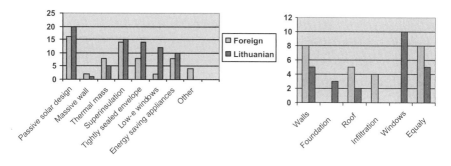

Figure 12 Solutions that make the biggest impact on the reduction of heating demands (left) and possible places to cut the loss of heat in a detached house (right).

structural members require many calculations and experiments and cannot be determined by a social survey. But the question aimed to establish opinions and knowledge of professionals.

Results showed that both Lithuanian and foreign professionals considered passive solar design and superinsulation to be the most important means to reduce a building's heating needs, while a thick wall was not a very popular option among given technologies. Ten Lithuanian professionals thought windows of high standards make a significant impact, but only two foreigners believed so. Among other important solutions, the surveyed professionals mentioned tightly sealed envelopes and energy-saving appliances. Respondents were not in accord about the possible ways to reduce heat losses. In contrast to the foreign respondents, the majority of Lithuanian professionals stated that windows had the best potential to reduce heat loss in detached houses. More foreign professionals believed that all structural members of a house may contribute to the reduction of heat loss equally.

A conclusion follows that, in contrast to their foreign peers, Lithuanian professionals highlight high efficiency windows as an important aspect in improvement of a house's energy performance.

Before construction of a zero energy house starts, one of the most important factors to be evaluated in planning is the climate. Local climate conditions determine optimal renewable energy technologies and also the best passive technologies to cut building's energy demands. Respondents were asked to select the part of Europe that, in their opinion, had the best climatic conditions for the construction of zero energy houses. Results are shown in Figure 13.

Lithuanian and foreign professionals answered differently. Foreign respondents (48%) believed central Europe (Germany, Austria) had the best

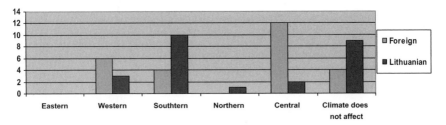

Figure 13 Parts of Europe with the best climatic conditions for ZEHs.

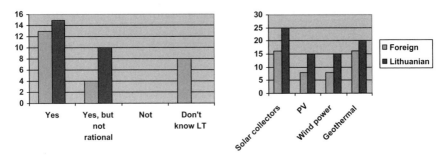

Figure 14 Do you think Lithuania's climate is suitable for ZEHs (left)? Which active zero-emission systems are the best for Lithuania (right)?

conditions, while the majority of Lithuanians (40%) stated that southern Europe (Spain, Italy, Greece) had the best climatic conditions for zero energy houses. More Lithuanian professionals than foreign believed that climate had no impact.

Lithuania is in the eastern part of Europe and none of the professionals stated it as the best location for such types of buildings. The next survey questions asked whether Lithuania's climate is suitable for zero energy buildings. Respondents were also asked to point out the renewable energy technologies that were the most suitable for Lithuania's climate (results are shown in Figure 14).

According to the results, most of the respondents (52% foreign, 60% Lithuanian) stated that Lithuania's climate is good enough for zero energy houses. However, 40% of Lithuanian professionals also believe that zero energy houses are not rational for Lithuania. Results show that all active renewable energy systems are suitable for Lithuania's climate. Thus all these systems, such as solar collectors, PV systems, wind power turbines, and geothermal pumps, can be used in the construction of zero energy houses in Lithuania. Lithuanian professionals, however, exclude solar collectors and geothermal pumps as the best systems for Lithuania's climate.

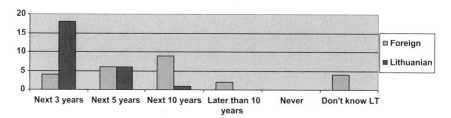

Figure 15 When, in your opinion, will construction of first ZEHs start in Lithuania?

At the time the research had been done, Lithuania had no net-zero energy houses. Respondents thus were given options and asked which period, in their opinion, was the most probable for Lithuania to start the construction of first zero energy houses. Results are shown in Figure 15.

None of the respondents thought Lithuania would never build zero energy houses, but foreign and Lithuanian professionals had different opinions about the starting period of the construction of zero energy houses in Lithuania. Results show that the majority of foreign respondents believe that the construction of first zero energy houses will start in the next 10 years, but not earlier than 5 years. Lithuanian professionals are more optimistic, and most Lithuanian respondents (68%) forecast that the construction of first zero energy houses in Lithuania will start in the next 3 years.

CONCLUSIONS

After practical investigations of renovation of a detached house into a near-zero energy building and an analysis of the social survey results, the following conclusions can be drawn.

1. In view of the financial aspects, investment into renovation of a prototype house is inexpedient. Further investments into renewable energy technologies do not make the renovation financially more expedient but they do improve simple payback time, NPV, IRR, and ROI. The best renovation alternative for a house with medium energy consumption and two residents is to turn it into a near-zero energy house with a good heat pump and large photovoltaic panels.
2. A combination of a heat pump and photovoltaic panels is the best alternative of renewable energy equipment considering the financial aspects. In view of the financial aspects, photovoltaic panels are recommended as the most important renewable energy system when the house is to be renovated into an active building.

3. The evaluation of investments for renovation of the house in question into simulated model buildings showed that in case of this house full costs cannot be covered by saved costs coming from energy savings in a normal period. The models show that the best solution is to renovate the house in question by turning it into a near-zero energy house. The best investment alternative is to renovate the house by turning it into a near-zero energy building with large photovoltaic panels and a good heat pump. When a house is heated by a heat pump, it must also have photovoltaic panels installed to satisfy the increased demand for electricity. Photovoltaic panels also give an opportunity to sell excess electricity in case such policy instruments are available. Investments into renewable energy technologies of this building help reduce the payback time of investments. In contrast to renovation without renewable energy technologies installed, investments into a heat pump, photovoltaic panels, and solar collectors help save more energy costs and improve the simple payback time.
4. The social survey confirms that zero energy buildings are a perspective branch in modern construction. Nevertheless, the EU requirement for new construction to be only near-zero energy buildings from 2019 has no support among construction professionals and they forecast that a new date will be set. Foreign professionals believe that central Europe has better climatic conditions than other regions of Europe. Financial criteria, rather than advantages and disadvantages of zero energy houses, were chosen as the most important aspects determining the intentions to build a new house. This means that respondents would decide to invest into construction of a zero energy house only seeing economic benefits and the evaluation of their investments would make an impact on their decision. A stable value in the future is another criterion pointed out by both Lithuanian and foreign professionals. Respondents believe that government support and the possibility of selling produced energy are the most important factors in motivating the public to invest into, and build, zero energy houses.
5. The social survey shows that Lithuania's climate is considered good enough for the construction of zero energy houses and that all kinds of renewable energy technologies can be used. Respondents claim that zero energy houses have better perspectives in Lithuania than other types of zero energy buildings, and construction professionals forecast that construction of the first zero energy houses will begin in the next 3 years. Although Lithuania's climate was said to be suitable for

renewable energy technologies, some Lithuanian professionals still doubt that zero energy buildings are rational for Lithuania, but most of them forecast that construction of the first zero energy houses will begin in the next 3 years.

REFERENCES

Balaras, C.A., Gaglia, A., Georgopoulou, E., Mirasgedis, S., Sarafidis, Y., Lalas, D., 2007. European residential buildings and empirical assessment of the Hellenic building stock, energy consumption, emissions and potential energy savings. Building Environ. 42, 1298–1314.

Balaras, C.A., Droutsa, K., Dascalaki, E., Kontoyiannidis, S., 2005. Heating energy consumption and resulting environmental impact of European apartment buildings. Energy Buildings 37, 429–442.

Directive of the European Parliament and the Council of 19 May 2010 on the energy performance of buildings. 2010. Official J. Eur. Union. Available from: <http://www.europarl.europa.eu-/sides/getDoc.do?pubRef=-//EP//TEXT+REPORT+A7-2010-0124+0+DOC+XML+V0//EN.> (01.10.10.).

Harvey, D., 2009. Reducing energy use in the buildings sector: measures, costs, and examples. Energy Efficiency 2, 139–163.

Hamada, Y., Nakamura, M., Ochifuji, K., Yokoyama, S., Nagano, K., 2003. Development of a database of low energy homes around the world and analysis of their trends. Renewable Energy 28, 321–328.

Hammon, R., Neugebauer, A., 2009. Applications for large residential communities: what is net-zero energy? Strategic Plan. Energy Environ. 29, 26–55.

Hernandez, P., Kenny, P., 2010. From net energy to zero energy buildings: defining life cycle zero energy buildings. Energy Buildings (42), 815–821.

Iqbal, M.T., 2004. A feasibility study of a zero energy home in Newfoundland. Renewable Energy (29), 277–289.

Itard, L., Meijer, F., Vrins, E., Hoiting, H., 2008. Building renovation and modernisation in Europe: state of the art review, OTB Research Institute for Housing. Available from: <www.buildup.eu/publications/12198>. (30.11.10.).

Johnston, D., Gibson, S., 2010. Towards zero energy home: a complete guide to energy self-sufficiency at home. 250.

Laustsen, J., 2008. Energy efficiency requirements in building codes, energy efficiency policies for new buildings, International Energy Agency (IEA). Available from: <http://www.iea.org/efficiency/CD-EnergyEfficiencyPolicy2009/2-Buildings/2-Building%20Codes%20for%20COP%202009.pdf>. (23.11.10.).

Martinaitis, V., Kazakevičius, E., Vitkauskas, A., 2007. A two-factor method for appraising building renovation and energy efficiency improvement projects. Energy Policy 35, 192–201.

Martinaitis, V., Ragoûa, A., Bikmanienė, I., 2004. Criterion to evaluate the ìtwofold benefitî of the renovation of buildings and their elements. Energy Buildings 36, 3–8.

Meyer, N.I., 2010. Out pashing of fossil fuels before 2050, Danish case study. Available from: <http://www.folkecenter.net/mediafiles/folkecenter/Niels-I.pdf>. (21.09.10.).

Nikolaidis, Y., Pilavachi, P., Chletsis, A., 2009. Economic evaluation of energy saving measures in a common type of Greek building. Applied Energy 86, 255002559.

Nistorescu, T., Ploscaru, C., 2009. Impact of economic and financial crisis in the construction industry. University Craiova. Available from: <http://wwww.mnmk.ro/documents/2010/3NistorescuFFF.pdf.>. (16.12.10.).

NRG, Ltd electricity and district heating prices information in Orum region. Available from: <http://www.nrginet.dk/priser+og+gebyrer/el-+og+varmepriser>. (16.11.10.).
Rapcevičienė, D., 2010. Evaluation of Multifamily buildings renovation efficiency (In Lithuanian). Business in XXI Century 2 (2), 83–89.
Realea Ltd, 2010. Energi Parcel, Projektmateriale. Available from: <http://www.realea.dk/pages_from_321_energiparcel_-_projektmappe_februar_2010_(a4_lo)_1–5.pdf.> (23.10.10.).
Reyes, I., 2008. Zero energy home. Available from: <http://www4.architektur.tu-darmstadt.de/powerhouse/db/248, id_25, s_Papers.fb15.>. (12.03.10.).
Tommerup, H., Svendsen, S., 2006. Energy savings in Danish residential building stock. Energy Buildings 38, 618–626.
Torcellini, P., Pless, S., Deru, M., 2006. Zero energy buildings: a critical look at the definition, National Renewable Energy Laboratory (NREL), USA. Available from: <http://www.nrel.gov/docs/fy06osti/39833.pdf>. (28.05.10.).
Ustinovičius, L., 2004. Determination of efficiency of investments in construction. Int. J. Strategic Property Manag. 8, 25–43.
Venckus, N., Bliudzius, R., Endriukaityte, A., Parasonis, J., 2010. Research of low energy design and construction opportunities in Lithuania. Technological and economic development of economy, 16 (3), 541–554.
Zavadskas, E.K., Kaklauskas, A., Raslanas, S., 2004. Evaluation of investments into housing renovation. Int. J. Strategic Property Manag. 8 (3), 177–190.

APPENDIX

Survey questionnaire

Zero energy building concept, perspectives and application possibilities

Your occupation field

Choose most suitable

O Constructions
O Real Estate
O Investment
O Politics
O University research
O Other

Your nationality

O Polish
O Spanish
O German
O Dutch
O Danish
O Other

Are you worried about energy consumption, global warming problems?

O Yes
O A little
O No

Are you interested in the use of renewable energy, ecological architecture?
- ○ (100%) Yes. I am working, investing in this field. Using for my personal needs
- ○ (75%) Yes I am very interested
- ○ (50%) I know a lot, but I don't have any relation with it
- ○ (25%) Just a little.
- ○ (10%) Just heard something but not really interested
- ○ (0%) No, I am not

Do you know anything about zero-energy buildings?
- ○ Yes, a lot
- ○ Yes, a little
- ○ No

How would you rate your knowledge of the zero-energy Buildings?
- ○ 100%
- ○ 75%
- ○ 50%
- ○ 25%
- ○ 10%
- ○ 0%

Do you know that from 1 January 2019 In EU will be Built only net-zero energy homes?
- ○ Yes
- ○ No

Do you think this requirement will be implemented by that data (01 01 2019)?
- ○ Yes, in all EU
- ○ Yes, but it won't be implemented in new EU countries.
- ○ No, but they will set a new date
- ○ No, never

If you'll built your own house, do you prefer zero energy house?
- ○ Yes
- ○ No, but it will be other low e-building
- ○ No, traditional house
- ○ I do not think it is rational to live in an individual house.

Would you invest or advice to others to invest in zero energy buildings constructions for commercial reasons?
eg.: to build zero energy homes for sale
- ○ Yes
- ○ No

Idea of zero energy house:
- ○ Reality
- ○ Future
- ○ Has no future

Better perspectives In next 10 years have:
- ○ Zero-energy buildings
- ○ Passive buildings
- ○ Invented new model buildings
- ○ Traditional buildings.

Which type of zero-energy buildings has better perspectives?
- ○ Zero energy house
- ○ Zero energy commercial building
- ○ Zero energy industrial building
- ○ Equally promising
- ○ Equally viable
- ○ Other

What is the main disadvantage of zero energy building?
- ○ Expensive to realize
- ○ Difficult to realize
- ○ Dependence on climate
- ○ There are not many professionals who can design this type of building
- ○ No disadvantages
- ○ Other.

What is the main advantage of zero energy building?
- ○ Ecology
- ○ Economy
- ○ Independence from centralized power networks and its price appreciation
- ○ No advantages
- ○ Other

What are the key drivers for zero energy building in Your country?
- ○ Government support
- ○ Possibility to sell produced energy
- ○ Society interest
- ○ Clear standards for such type of buildings
- ○ Other

What are the key barriers for zero energy buildings in Your country?
- ○ Unclear definition
- ○ No government support

- ○ No possibility to sell produced energy
- ○ Lack of society interest
- ○ There are no barriers
- ○ Other

Evaluate the criteria which are relevant to the intention to build/buy a house. Choose 3 most important

- ☐ Construction/building cost
- ☐ Existing communications
- ☐ House exterior (outdoor design)
- ☐ Indoor planning/design
- ☐ Low maintenance costs
- ☐ House ecology
- ☐ Security
- ☐ Stable value in the future, longevity
- ☐ Independence from utilities, usage of renewable energy resource
- ☐ Quality of the materials

Which decisions make biggest impact to reducing home heating needs? Choose three

- ☐ Passive solar design
- ☐ Massive Wall
- ☐ Thermal mass
- ☐ Super insulation
- ☐ Tightly sealed envelope
- ☐ Low emission windows.
- ☐ Energy-saving appliances
- ☐ Other.

Where is possible mostly reduces the loss of heat in an individual house?

- ○ Windows
- ○ Walls
- ○ Foundation
- ○ Roof
- ○ Infiltration
- ○ Equally
- ○ Other

What is the most important in construction of individual house?

- ○ Construction cost
- ○ Construction time
- ○ Construction method
- ○ Technical supervision of construction
- ○ Other.

Rate zero energy home construction cost difference with the same area traditional house construction cost.
- ○ Zero-energy house construction cost is two times higher
- ○ Good Zero-energy house construction cost is one and a half time higher
- ○ Cost is equal
- ○ Zero-energy house construction cost is three times higher
- ○ There is cheaper to built zero energy house
- ○ Other

Which part of Europe are the best conditions for the construction of zero energy buildings?
- ○ Eastern Europe
- ○ Western Europe
- ○ South Europe
- ○ North Europe
- ○ Central Europe
- ○ Climate does not affect

Do you think the climate in Lithuania is suitable for zero-energy house construction?
- ○ Yes
- ○ Yes, but this type of houses is not rational for Lithuania
- ○ Not suitable
- ○ I do not know where Lithuania is

Which active zero emission systems are the most suitable for Lithuania?
- ☐ Solar collectors
- ☐ Photovoltanic elements
- ☐ Wind power
- ☐ Geothermal energy
- ☐ No one is good for Lithuania climate
- ☐ I do not know where Lithuania is
- ☐ Other

Do you know any zero-energy buildings in the neighbor countries of Lithuania?
- ☐ Latvia
- ☐ Estonia
- ☐ Belarus
- ☐ Poland
- ☐ There are no zero energy buildings in these countries
- ☐ I do not know where Lithuania is
- ☐ Other.

When you are prognozing first zero energy house/building construction in Lithuania?
- ○ In next 3 years
- ○ In next 5 years
- ○ In next 10 years
- ○ Later then next 10 years
- ○ It will never happen
- ○ I do not know where Lithuania is

Any other comments on zero energy buildings?

You can describe zero energy buildings situation in your country.

CHAPTER 11

The Canadian Context: Energy

Kartik Sameer Madiraju
Department of Bioresource Engineering, McGill University, Montreal, Quebec, Canada

Contents

Introduction	256
Energy Conservation and Efficiency	258
Federal Commitments to Energy Efficiency and Conservation	259
The Kyoto Accord	259
Passive Decentralization	259
Federal Energy Efficiency Regulations	262
EnerGuide	262
R-2000 Standard	263
ENERGY STAR	264
FleetSmart	264
Provincial Commitments to Energy Efficiency	265
Quebec	265
British Columbia	266
Northwest Territories, Nunavut, and the Yukon	266
Atlantic Canada	266
Appraisal of Current Federal and Provincial Efforts	267
Renewable Energy Research and Development	270
Renewable Energy Production in Canada by Type	271
Solar	271
Hydroelectricity	272
Wind	272
Biomass and Biofuels	273
Major Federal Renewable Energy Regulations and Policies	273
ecoENERGY for Biofuels	274
NextGEN Biofuels and Sustainable Development Tech Fund	274
National Strategy on Renewable Energy	275
Major Renewable Energy Technology Research Initiatives	275
CanmetENERGY	275
Office of Energy Research and Development	276
Canadian Biomass Innovation Network	276
Appraisal of Federal Regulations, Renewable Energy Commitments, and Research	276
Agile, Sustainable Communities	279
Success Stories in Sustainable Communities	280
Drake Solar Landing Community	280
Earthship Biotecture and Similar Sustainable Housing	280

Microcosmic Initiatives	282
Appraisal of Sustainable Community Efforts	282
Benefits of Sustainable, Agile Communities	283
Resilience	283
Balance	284
Decentralization	284
Cooperation	285
The Future of Sustainable Design in Canada	286
Conclusion	287

INTRODUCTION

Canada has in its possession a veritable "treasure trove" of natural resources and energy potential. Recent estimates suggest that in the energy sector, Canada has nearly twice its current hydroelectric capacity in unexplored potential (163 GW), of which only 15% is currently under development (Irving, 2010). Canada also has a vast untapped resource in solar energy, whether through thermal or photovoltaic applications (Figure 1)(Solar photovoltaic map). In the nuclear sector, the World Nuclear Association has listed pending projects to develop new reactors in the province of Ontario; Canada holds the world's third largest reserves of uranium and is currently the largest producer of uranium ore (21% of global production)(Nuclear Power in Canada; Whitlock). Where wind energy is concerned, the scenario is equally promising (Figure 2)—with only 5000 MW installed capacity to date, many wind energy agencies such as Canadian Wind Energy Association (CanWEA) estimate up to 20% of electricity requirements in Canada could be met by wind power alone as early as 2025 (Wind energy).

While resources and potential in renewable energy are undoubtedly significant, proven reserves in natural gas and oil are equally vast. Natural gas reserves amount to 58 trillion cubic feet if only proven reserves are counted (Atlas of Canada). Similarly, the recoverable quantities of oil from the Athabasca tar sands are close to 173 billion barrels (Alberta's oil sands, 2006). The growing demand for these nonrenewable fossil fuels has led to the approval of massive extraction and transportation projects. While the tar sands and the related Keystone XL pipeline are well-known examples of this drive to profit off of fossil fuel extraction, several other projects are either underway or have been proposed (Keystone XL pipeline project).

If development strategies must be guided by principles of sustainable design, then greater focus must be placed on capitalizing Canada's renewable energy potential, efficiency measures, and various policy initiatives to incentivize sustainable lifestyles.

Figure 1 Solar potential map of Canada. Warmer colors indicate a greater potential (kWh/kW), where yellow is ~1000 kWh/kW, orange is 1300 kWh/kW, and green is 900 kWh/kW. *(courtesy of NRCan)*.

This chapter outlines the state of renewable energy research and development, as well as energy sustainability initiatives across government, industrial, and societal sectors in Canada. First, it examines the conservation and energy efficiency initiatives already in place in Canada, whether as a result of government policy or through widespread nongovernmental organizations campaigns; in the context of these initiatives, possible improvements and areas where current approaches are lacking are discussed. Second, in light of these required improvements, this chapter reviews the renewable energy resources developed in Canada and ongoing scientific research in green technology. Whenever possible, case studies are used to contextualize the benefits and shortcomings of these technologies. Third, this chapter looks to the future and endeavors to give an outlook for completely

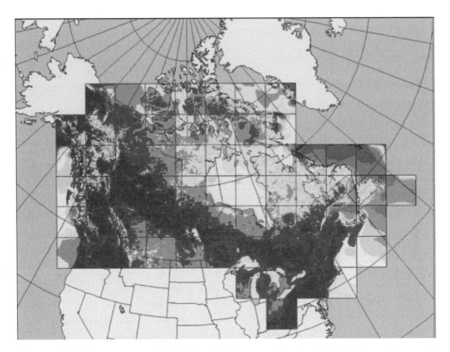

Figure 2 Wind energy potential in Canada. Warmer colors indicate higher wind speeds at a 50-meter altitude. *(courtesy of NRCan).*

sustainable communities that are adaptable, versatile, and "off-grid." Once again, through the use of case studies of existing communities that emulate these principles, the relationship among local government, federal institutions, industries, and citizens must be to achieve the sustainable communities' goal is examined.

ENERGY CONSERVATION AND EFFICIENCY

The design of sustainable communities can be thought of as a three-pronged approach—social, technological, and governmental. The last two approaches often dominate the public discourse with respect to clean energy, mainly because the government holds the most power to implement clean energy technology, and advances in the performance of renewable energy technology are thought to be prerequisites for its widespread use. However, the societal aspect of sustainable design, especially with respect to energy, is a crucial step that can be taken with minimal investment, government intervention, and without any long-term research.

The social approach to energy sustainability in Canada is intuitive—end users, whether they are industrial, public, or private, can make dramatic contributions to achieving a sustainable use of energy by reducing consumption, making efficient lifestyle choices, or incorporating energy-conscientious strategies into their business models. To this end, the Canadian government has laid out several initiatives and policies, which aim to reduce energy use or incentivize the purchase of more efficient technologies. The bicameral governmental system of Canada allows for decentralization of energy conservation strategies to the provincial level. As such, certain policies and programs are implemented across the nation by the federal government, while others are specific to each province's demographics and energy use trends.

Federal Commitments to Energy Efficiency and Conservation
The Kyoto Accord
In 1997 the Kyoto Accord was ratified by over 100 member states of the United Nations. The goal of the agreement was to implement policies to reduce greenhouse gas emissions 5% below 1990 levels by the year 2012 (Kyoto Protocol). Although Canada, by way of erstwhile Prime Minister Jean Chretien's ratification (Chretien signs Kyoto agreement), was a signatory party to this accord, the current government, led by Stephen Harper, has withdrawn from the commitments as early as 2008. The official withdrawal came just after the COP18 summit in Doha, Qatar (Canada to withdraw from Kyoto Protocol). While the Kyoto Accord was considered binding among signatory nations, the lack of any mechanism to enforce the commitments meant that no party was obliged to achieve its emissions reduction target.

The withdrawal of Canada from the Kyoto Accord was met with widespread disapproval, but the Canadian government has not abandoned all commitments to reducing greenhouse gas emissions. Through its National Action Plan, the Canadian government aims to reduce emissions 17% below 2000 levels by the year 2020 (Canada's action on climate change). In early 2012, it was announced that through several initiatives, not the least of which are energy efficiency and conservation projects, the nation had achieved half its target (Canada's emissions trends, 2012). It is clear that energy efficiency and conservation are major components of the National Action Plan, despite its separation from international agreements.

Passive Decentralization
The federal branch of sustainable energy policy is under the jurisdiction of Natural Resources Canada (NRCan). The Office of Energy Efficiency was

established in 1998 with the specific mandate to devise incentives and policies that would encourage conservation of energy (Office of Energy Efficiency), as well as the responsibility to coordinate with provincial governments to achieve their efficiency standards. It must be noted that in Canada, regulations and standards with respect to the environment usually only prescribe maximum limits and merely suggest standards through guidelines. Certain provinces may choose to implement more stringent laws to conserve energy. This creates two sets of regulations, which carry their own benefits and shortcomings. We can denote the process of enabling diverse laws to come about with only guiding principles from higher government as "passive decentralization." Passive decentralization can be described by three major characteristics:

1. it recognizes the autonomy of lower government in achieving national goals and so does not require streamlined policies across all provinces (or states)
2. it does not actively commit to decentralization but through the absence of laws requiring the use certain methods, allows lower levels of government to implement a wide range of projects
3. it serves as a tool of guidance in most cases, and as a baseline in other cases—this means that in some cases, educational and informative resources are provided for users to decide to what *extent* they wish to pursue the national goal, and in other cases a minimum standard must be met.

There are two major benefits of a passive decentralization approach to energy efficiency. First, because provinces are not constrained to follow streamlined methods of energy conservation, there is a greater incentive to invest in research and development and a greater incentive for citizens to voice their opinions with their local municipal governments. Energy efficiency is an inherently social approach to energy policy, as mentioned earlier, and relies on the choices of industries and consumers. Therefore, any incentive to increase the role of these actors in the formulation of policies will create a more cooperative atmosphere between decision makers and citizens—it follows that industry actors and consumers will be more likely to adhere to regulations and incorporate initiatives that they formulated. If policies were determined centrally, the distance between the decision-making body and the affected actors would increase dramatically; likewise the incentive to innovate of one's own accord would decrease.

The second major benefit is implicit recognition of the autonomy of lower echelons of government. While many sectors of governance are relegated to the provinces explicitly (e.g., education and public transport), the absence of a national energy policy has largely allowed elements specific to certain provinces to define their respective energy conservation strategies. For example, in Ontario, the feed-in tariff mechanism to increase renewable energy production was introduced recently, despite no national policy aimed at phasing in feed-in tariffs (Feed-in tariff program). Similarly, certain provinces are able to implement energy efficiency projects, which include specific provisions for First Nations who play a stronger role in governance as compared to other provinces (Power Smart First Nations Program). The implicit acknowledgment of provincial autonomy also prevents the loss of investment dollars in unnecessary projects—funds earmarked for the regulation of certain products in a given province, for example, would be better spent to reduce energy consumption in another sector if the products in question already meet the standard. A centralized efficiency initiative may be less likely to recognize these particularities.

It is important to understand that passive decentralization is not incompatible with national policies or strategies. Indeed, without the guidance and direction provided by a national strategy, negative consequences can also occur. One major disadvantage of the passive decentralization approach is that for provinces who have largely met all standards, there is no incentive to go further—this is especially problematic if the national standards on energy efficiency *themselves* are insufficient to combat climate change or reduce energy consumption markedly. If a national strategy were in place that advanced its goals in energy efficiency continuously, such as the nation-wide feed-in tariff schemes implemented by European Union member states, local and provincial governments would be under constant pressure to achieve those national goals. In Germany, the national energy policy has declared a goal of complete fossil fuel independence by 2050 (100% renewable electricity supply by 2050). Since then, every intermediate goal has been exceeded. In contrast, the extremely invasive methods used to develop the tar sands in Alberta face no federal opposition since investments are not funneled into renewable energy via a national energy policy. A strict rule such as a ban or high efficiency requirement could yield changes in consumer choices far quicker than a passive decentralization approach. Quicker and broader changes are required because they speak to the urgency of environmental crises. For example, in the U.S. state of California, plastic bags were banned altogether in both San Jose and San Francisco in light of their environmental hazards (California local ordinances). Countries such as China and South Africa have instituted nation-wide bans of plastic bags and even fine stores that still provide them (Taylor, 2012; South Africa - plastic bags regulation). While the Canadian government, through NRCan, amends its regulations and standards for energy use and product energy efficiency routinely, the language used in these policies is aligned very closely with intent to guide and inform, not to oblige.

The government has earmarked nearly 200 million dollars to be spent between 2011 and 2016 on its umbrella program ecoENERGY Efficiency (ecoENERGY efficiency), which comprises several initiatives geared toward the residential, transport, consumer, and industrial sectors. Some key programs receiving funding under this mandate are listed here.

Federal Energy Efficiency Regulations

Efficiency regulations in Canada are amended regularly to reflect the changing list of products available to consumers—the list is very comprehensive and includes products such as air conditioners, gas furnaces, lighting and bulbs, televisions, and other appliances. These regulations apply to dealers who must comply by submitting efficiency reports and product information so that adherence to standards can be verified. Noncompliance is penalized with fines, although the upper limit on these fines, as of the 2002 Administrative Monetary Penalty System, is 1000$ CAD. The standards will denote minimum efficiency or maximum energy use in units relevant to the product in question (wattage for lighting, cycles or kWh for dryers, etc.). In conjunction with these regulations, labeling such as the EnerGuide, Energy Star Rating, and the R-2000 housing standard is used to promote, inform, and encourage sustainable design (Canada Customs and Revenue Agency's Administrative Monetary Penalty System).

EnerGuide

The EnerGuide rating offers two services—first, labels are placed on household appliances that rate their energy consumption on a scale (Figure 3), allowing consumers to assess where their appliances rank in their class with respect to energy use. The second and major service provided by the EnerGuide is a full household audit by an energy advisor, who provides an efficiency rating for a household *at the request of the owner*. The rating is on a 0–100 scale, with 100 being a perfect efficiency household (maximum energy savings based on available products). The energy advisor not only conducts the audit but also offers mechanisms for household owners to reduce energy use by making appropriate sustainable choices. The ranking system is not compulsory, but provides a very strong visual aid for

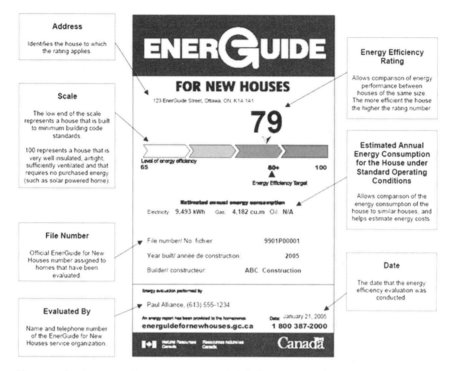

Figure 3 A schematic of common EnerGuide labels. *(courtesy of EnerBuild Homes).*

consumers when making purchasing decisions and serves as marketing tool for companies that provide comparatively more efficient products (EnerGuide home rating system).

R-2000 Standard

The R-2000 standard is specific to home construction and is also administered by NRCan. R-2000 is an opt-in regulation as well and is more informative and educational than functional. Homebuilders and contractors who wish to adhere to R-2000 standards are provided training and options to choose materials and equipment that will reduce the overall energy footprint of the household. The important factor here is that an R-2000-compliant building will *exceed* currently Canadian home construction laws—therefore a new house built lawfully need not comply with the additional R-2000 efficiency requirements. According to NRCan, a typical house compliant to the building codes will have an EnerGuide rating of 60–70, while an R-2000-compliant home will be rated closer to 80. The Office of Energy Efficiency estimates that

R-2000-compliant houses consume 30% less energy per annum to perform the same functions as any typical household (with respect to insulation, stability, durability, etc.). If a builder adopts the R-2000 standard, they must submit to independent verification of the household's compliance (What is R-2000?).

ENERGY STAR

ENERGY STAR can be thought of as an intensified set of efficiency regulations that are stipulated by Canadian law—where as these standards are to be achieved competitively by corporations and retailers, the baseline efficiency requirements are required. ENERGY STAR certification is achieved when companies have a product that exceeds these baseline efficiency requirements, which are more stringent than existing regulations (What is ENERGY STAR?). For example, to receive a Tier 1 ENERGY STAR rating, a clothes washer must achieve less than 1 liter of water use per cycle; but to comply with efficiency regulations, a clothes washer of equal volume can use up to 18 liters per cycle (Clothes washers). While NRCan notes that appliances have improved in efficiency in the last 20 years, it is unclear whether this is due to increasingly ambitious standards and regulations or as a result of unrelated factors. The ENERGY STAR rating is also an effective marketing tool for companies since its strict requirements limit ENERGY STAR-certified products to a select few.

FleetSmart

According to the International Transport Forum, fossil fuel consumption by transport vehicles contributes at least one-quarter of global CO_2 emissions from fossil fuels and nearly 15% of global greenhouse gas emissions (Reducing transport greenhouse gas emissions 2010). The Canadian government has responded to the impact this sector is having on climate change primarily through the FleetSmart program, which provides education and incentives for vehicle operators (of government transport vehicles such as forestry trucks and transit fleets) to make energy-conscious decisions when upgrading or selecting fleet models and organizing urban planning to minimize the expenditure of fossil fuels. While less of a marketing tool for fleet operators, the FleetSmart program showcases the potential savings in fuel costs from investing in the federal training programs. The Fuel Consumption Guide 2012 also lists consumer vehicles by class and fuel economy, and annually awards models in each class with the best fuel economy (FleetSmart).

Provincial Commitments to Energy Efficiency

It is clear that the steps taken by the Canadian government to improve energy efficiency are extensive; however, the approach is evidently one of recommendation rather than requirement. The concept of passive decentralization described earlier is granted new importance when considering the individual efforts of provinces and territories with respect to environmental efficiency and comparing those efforts to the federal government overall. One such study is conducted approximately every 2 years by the Canadian Energy Efficiency Alliance (CEEA). The CEEA (Canadian Energy Efficiency Alliance) releases its energy efficiency report card periodically, which, in their words, aims to evaluate:

- how the jurisdiction supported activities such as energy efficiency and public outreach
- the existence of public/private partnerships to support energy efficiency
- responsiveness to energy efficiency issues in key legislation, such as building codes and energy efficiency acts.

On a scale of F to A+, in 2005, the federal government was awarded a grade of A, with only Manitoba equaling that grade provincially; in 2007, the federal grade dropped to a B, with every province except New Brunswick scoring higher (Energy efficiency improving in Canada). The 2009 evaluation elevated the federal government to a B+ rating but three provinces scored an A+ rating. It must also be noted that six provinces and territories scored lower than the federal government in 2009. Although the "report card" system is slightly vague, the diversity of approaches among the provinces is unmistakable (2009 energy efficiency report card). Some of the unique programs and incentive structures spearheaded at the provincial level are given here.

Quebec

Quebec has an intriguing history with the federal government because of existing tensions relating to the nationalist movement in the province. The ideology that Quebec is sufficiently distinct in culture, views, and goals forms the basis for the argument that the province should ultimately secede from the Confederation and become an autonomous nation. While by no means a condoning of separatist activity, the idea that Quebec was a distinct "nation within Canada" was not only entertained but tabled as a motion and passed in the House of Commons by the federal government in 2006 under current Prime Minister Stephen Harper (39th parliament 1st session,

2006). The bifurcation between federal and Quebecois stances on various issues can be related to these inherent differences. While the federal government officially withdrew its commitment to the Kyoto Protocol, Quebec as a province still set targets, which largely mimic the Kyoto Protocol (20% reduction below 1990 levels of greenhouse gases by the year 2020). Today, the province boasts the second smallest per capita emissions of all provinces, a rate that is 50% below the national average (Harper vs Kyoto).

The province offers a range of initiatives, ranging from a refrigerator recycling initiative (which in turn benefits low-income households) called RECYC-FRIGO (RECYC-FRIGO); residential geothermal energy grants; and rebates offered for the purchase of drain water heat recovery technology from Gazifere (Recuperateur douche).

British Columbia

British Columbia's provincial government also offers a refrigerator buy-back program as well as the Scrap-It BC initiative, which encourages drivers of older and less efficient vehicles to trade them in for rebates on hybrid or fuel-efficient cars (Scrap-it). The provincial Ministry of Environment has also set up a special sustainability fund that provides grants to citizens who wish to initiate their own conservation and efficiency projects in the local community (Sustainable endowment fund); a separate fund of 35 million dollars was set up with the mandate to provide financial assistance to homeowners who opt into energy audits or retrofitting projects (LiveSmart BC home efficiency program extended).

Northwest Territories, Nunavut, and the Yukon

Under the auspices of the Arctic Energy Alliance, the territories in Canada have set aside funds that provide rebates to businesses that invest in technologies that save fuel or use more efficiency appliances in their offices; incentives are also offered to homeowners, who can apply for rebates after having purchased specific products with energy-efficient features (low-flow toilets, ENERGY STAR-rated kitchen appliances, etc.)(Energy conservation program).

Atlantic Canada

A number of initiatives have been proposed and funded in the provinces of Prince Edward Island (PEI), New Brunswick, Nova Scotia, and Newfoundland and Labrador. PEI has offered up to 3000 dollars in tax rebates for consumers who purchase a hybrid car and maintains a granting agency for low-income households that still wish to invest in energy-efficient home upgrades (Hybrid vehicle tax incentive). On the academic front, the commitment to sustainability research overall has been thorough and consistent in Atlantic Canada, where an independent study found that of 18 campuses in Atlantic Canada, 84% addressed the issue of sustainability to some extent in their curricula (Beringer et al., 2008). In Newfoundland, the Take-CHARGE program offers rebates for households purchasing ENERGY STAR products such as lighting, windows, and home appliances. In New Brunswick, several incentive structures exist for those constructing new buildings to promote energy efficiency construction and building operation in both residential and commercial sectors (TakeCHARGE!).

Appraisal of Current Federal and Provincial Efforts

This section outlined many programs, regulations, and financial rewards that promote energy-efficient decision making in all sectors. However, while bulk emissions of carbon dioxide have remained largely static in most provinces since 1990, dramatic increases have been observed in the oil-rich province of Alberta and the agricultural hub of Canada, Saskatchewan (Greenhouse gas emissions by province). Per capita emissions overall have also remained at 1990 levels or slightly above (Greenhouse gas emissions per person)—while this might suggest that efficiency measures put in place are having a significantly positive impact, Canada's emissions have actually increased by nearly 25% relative to 1990 (the Kyoto base year), when their Kyoto target was to reduce emissions *below* 1990 levels by at least 6% (Report of the individual review of the annual submission of Canada, 2010). In a 2008 study, only two provinces showed a decreased in emissions relative to 1990 levels (Quebec and PEI)(National inventory report greenhouse gas sources and sinks in Canada 1990-2007, 2009). Interestingly, in the European Union (EU), nearly all member states have achieved reductions in emissions relative to the 1990 Kyoto Protocol base year, amounting to a 15% reduction by the EU as a whole and close to a 60% reduction in some member states (Petz, 2012).

The willingness or interest to participate in energy efficiency and emissions reduction targets is prevalent in Canada—the 2012 Energy Efficiency Indicator results for the United States and Canada showed that in 2011, 66% of business executives were very interested in energy efficiency and energy management with respect to their organizations and, in 2012, that figure rose to 86%. However, the study also found that only 25% of businesses surveyed actually invested in renewable energy technology, a crucial component of energy efficiency and sustainable design. Of note is that among all surveyed regions, the United States and Canada demonstrated the highest level of interest *and* the lowest level of renewable energy investment. In Europe, only 53% of business executives indicated high interest in energy efficiency investment, but 43% were already investing in renewable energy technology (energy efficiency indicators US and Canada, 2012). It can be inferred that the presence of strong European energy policies, such as feed-in tariffs and maintaining Kyoto obligations, has made energy efficiency a routine and daily item of business rather than a novel and cutting-edge feature that would pique the interest of executives.

Recent scholarship on energy policy in Canada has indicated that a strong national policy and obligatory standards would be required, *in addition* to existing regulations, for a marked improvement in Canada's environmental record. A study published in 2009 by Dusyk and colleagues concluded that while sustainable development strategy in British Columbia had made appreciable gains in energy efficiency and environmental protection, factors that would ultimately play the deciding role in continued progress would be "large, mandated…reduction targets" and "integrated and coherent action… across sectors" (Dusyk et al., 2009). At the national level, the major programs detailed in this chapter do not fit that model. In Ontario, however, the provincial feed-in tariff scheme is a perfect example of such an integrated and long-term target—the program has branches for power projects of 10 kW and higher and for "micro-FIT" projects of less than 10 kW (Feed-in tariff program). An earlier study on the province of Nova Scotia concluded that while interest in eco-efficiency was high among businesses, the initiatives actually taken by enterprises were minimal. Of 35 potential actions that a business could take toward energy efficiency, the study found that, on average, most companies only undertook 9 (Cote et al., 2006). Once again, in comparison to the level of interest and corresponding renewable energy investment trends seen in Europe, the existence of a national energy policy seems to be the missing catalyst.

Hopper and colleagues (Hopper et al., 2009) highlight the importance that state and provincial initiatives to meet emissions targets or improve energy efficiency have on energy savings. The programs studied were mainly in western North America and had the potential to conserve up to 5% of projected electricity demand over a 10-year period. However, Hopper and colleagues (Hopper et al., 2009) note the lack of a national climate policy in the United States as one hindrance to achieving uniform progress—indeed, while many states are signatories to the "Western Carbon Initiative," a market-based approach to mitigating emissions, there are still others who have not opted in. And in those states and provinces that have signed the initiative, regulatory policies to enforce compliance have not been implemented (Hopper et al., 2009). On a more political note, the issue of national identity and citizenship responsibility can be of special importance when a national policy is put in place—Gamtessa (Gamtessa, 2013) notes that the propensity of household owners to request EnerGuide housing audits was significantly determined by financial incentives. While incentives are a necessary component of any energy efficiency program, the fact that economic reward governs sustainability decisions in Canada is problematic. For those who face a return on their investment that is less than the expected return on any other investment, the allure of the incentive disappears and the will to make energy-efficient household upgrades along with it. As such, it appears that in current programs in Canada, the idea that such practices are a part of one's duty as citizen or part of a nation's identity is not well communicated. Even working groups commissioned by the Canadian government have reached similar conclusions, recommending stronger targets and regulations and more legal tools for local governments to encourage sustainable energy use.

However, it cannot be concluded from this that energy efficiency measures put in place in Canada have been without impact. A study that tracked the trends in energy use in Canada from 1990 to 2009 found that energy use projections without efficiency regulations in place were 15% higher than those actually observed for the year 2009. In the residential sector, a comparison of appliance energy use in 1990 and in 2009 showed dramatic decreases across all major appliances (refrigerator, washer, dryer, dishwasher, and range). Similarly, incandescent light bulb use in households plunged from an average of 29.4 in 1990 to 19.2 in 2009, with fluorescent, compact fluorescent light, and halogen bulbs making up the remainder. Overall energy intensity (denoted as gigajoules/household) decreased from nearly 130 GJ/household to just under 110 GJ/household. In commercial and industrial sectors, however, energy intensity did not vary appreciably nor did energy savings due to efficiency measures (Improving energy performance in Canada, 2010-2011).

This section dealt primarily with a description and evaluation of the energy efficiency initiatives and regulations that have been put forward by the federal and provincial governments in Canada. We have seen how the passive decentralization mechanism leads to a diverse set of programs at the provincial level, but it also leads to diversity in emissions reductions and achievements in energy efficiency. Finally, we established that the diversity in approach could be maintained, but under the guidance of stricter national policies and regulations—this would call for, for example, a national energy policy equivalent to those proposed by many EU nations. Within the context of sustainable design, examining renewable energy is crucial, as energy efficiency can only limit the amount of emissions from a fossil fuel-driven economy. As long as nations are pushed to grow on the basis of a fossil fuel economy, emissions will necessarily increase. Renewable energy technology (RET) aims to take energy efficiency to the next level, aiming ultimately for a "carbon-free" economy.

The next section examines renewable energy investment trends in Canada, the development of renewable energy technology by the federal government, and once again provides a critical analysis of Canadian efforts in RET as compared with other nations.

RENEWABLE ENERGY RESEARCH AND DEVELOPMENT

A 2009 federal report on improving energy performance in Canada states that 17% of Canada's "primary energy sources" are from renewable technologies such as hydroelectricity, biomass, wind, solar, and geothermal. The report differentiates "primary energy" from electricity, in which sector renewable technologies make up 62% of the supply. However, it must be noted that the bulk of renewable energy technologies are large and small hydropower plants, whose environmental impacts and life-cycle greenhouse gas emissions have been criticized by environmental interest groups around the world (Improving energy performance in Canada, 2010-2011). Only 3% of electricity generated in Canada comes from nonhydro renewable sources (Nyboer & Lutes, 2009), and although wind energy installed capacity increased nearly 40-fold from 2000 to 2011 (Canadian wind energy industry), France achieved 60-fold increases and China achieved nearly 150-fold increases over the same time period. Similarly, wind energy alone accounts for 18% of electricity produced in Portugal, 21% in Denmark, and 9% in Germany (International Energy Statistics).

The Canadian government has also made significant progress in the production of biofuels. Between 2005 and 2010, the production of ethanol had increased to 1.5 billion liters and in 2010 ethanol use accounted for approximately 4% of gasoline sales. Although to a lesser degree, solar capacity has also increased threefold, but only stands at 290 MW as of 2010 (Nyboer & Lutes, 2009). Figure 4 gives a percentage breakdown of renewable contributions to Canada's energy sector.

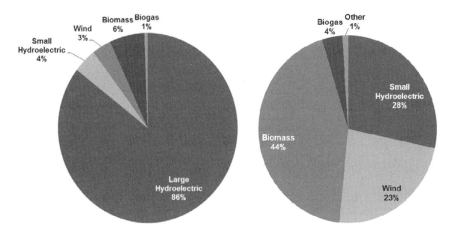

Figure 4 Contribution of renewables to electricity, including large hydro (left) and excluding (right). *(Courtesy of the Canadian Industrial Energy End-Use Data and Analysis Centre renewables report).*

Renewable Energy Production in Canada by Type
Solar

As mentioned previously, solar photovoltaic capacity increased from just under 100 MW to just under 300 MW by 2010. Photovoltaics provide electricity directly, while solar thermal uses dissipated solar energy to provide heating for various applications in residential, commercial, and industrial sectors. The total installed capacity for solar thermal systems as of 2010 was 712 MWe (megawatts equivalent). The Canadian Solar Industries Association reports that the public funds budget (i.e., government spending) on solar photovoltaic installations increased to approximately 75 million dollars. As trends in investment have increased, trends in installation and production costs have also decreased—in 2000 the average price per watt of solar photovoltaic system was around 10 CAD and subsequently decreased to 2.27 CAD in 2010 (National survey report of PV power applications, 2010).

Thus far, only Ontario has capitalized on its solar potential, with both a feed-in tariff program initiated in 2006 (and later expanded in 2009) and several large-scale demonstration projects. Of note is the Sarnia solar demonstration project, which provides electricity for over 10,000 households, created 800 jobs during construction, and required a 400 million dollar investment. Smaller demonstration projects have been explored in Alberta. The feed-in tariff program in Ontario was also a success—upon advertising the initial offer, the government received more than 20,000 applications to supply power to the grid, amounting to an installed capacity of 5000 MW. If all applications are accepted, Canada's installed capacity would increase by more than 1000% due to the feed-in tariff program alone. Contracts last 20 years and guarantee a fixed price of up to 0.80 CAD/kWh (Feed-in tariff program).

Hydroelectricity

The 2011 Canadian Industrial Energy End-Use Data and Analysis Centre report pegged Canada's hydroelectricity contribution to renewable energy generation at 89%, including large and small systems. In terms of capacity, of the 77.5-GW installed renewable capacity in the nation, 94% of it is hydro-based power (Nyboer & Lutes, 2009). Although already an impressive statistic, hydroelectric companies in Canada are seeking to increase the installed capacity of hydroelectric dams in the next decade. Clean Technica reports that an expected investment of 80 billion dollars over the next decade has been set aside to add 4.5 GW in Quebec, 3.3 GW in British Columbia, 2.4 GW in Manitoba, and another 3 GW in Newfoundland and Labrador. The large majority of this investment is aimed at export to the United States, where coal still provides 45% of the national energy requirements and nearly all the energy requirements in eight states (Canada boosting hydro power to 88.5 GW). Provincially, Quebec produces more hydroelectricity than all other provinces and territories combined—this also means that it has exploited nearly half of its commercially viable hydroelectric potential. In British Columbia, the province with the second highest hydroelectric output, only 25% of potential has been developed. Little to none of the hydroelectric potential has been developed in the territories (Irving, 2010; Nyboer & Lutes, 2009).

Wind

Wind power accounts for only about 2% of energy production in Canada and 4% of renewable electricity. Ontario contributes more than one-third of total national wind energy capacity, with Quebec and Alberta coming in second and third place, respectively (Nyboer & Lutes, 2009). Installed capacity according to Industry Canada stands at 3.4 GW as of 2010, increasing from just 26 MW in 1998 (Canadian wind energy industry). This growth in wind energy capacity over that time period is approximately 300 MW per year—despite this, the Canadian Wind Energy Association estimates that wind energy capacity could exceed 55 GW by 2025, satisfying 20% of national energy requirements. Those estimates would require an annual added capacity of more than 4000 MW per year starting in 2013 (Wind vision). Indeed, the required investment to achieve these targets would be 79 billion CAD. Interestingly, both CanWEA and the Pembina Institute expressed dismay at the relative lack of federal funding increases for wind energy in Canada since 2010, despite these prospects (Federal budget fails to extend support for new wind energy development; Budget, 2012). However, provinces have once again taken on the responsibility of developing renewable energy technology in the absence of federal policy—Ontario is once again the leader and is the only province with a feed-in tariff program for wind energy producers. The province has financed 1 billion CAD worth of projects in 2011 and expects 7 GW of wind energy projects to be approved by 2018 (Baillie). Although still in its infancy, offshore wind energy is also being explored in Canada. A contract has been approved by British Columbia to allow for a large-scale offshore wind project on the northwestern coast of the province. The company, NaiKun Wind Energy, estimates that the project would supply power to more than 100,000 households upon complete (Offshore wind energy blows into Canada).

Biomass and Biofuels

As mentioned earlier in this section, biofuel production has increased to over 1 billion liters per year. Close to 2 billion CAD is earmarked by the federal government over the 2008–2017 time period to increase biofuel production in Canada (ecoENERGY for Biofuels Program). Currently, Ontario and British Columbia are the leading provinces in biomass energy production (more than half of the 5-GW installed capacity), although biomass only accounts for 2% of electricity production (Nyboer & Lutes, 2009). The Canadian Bioenergy Association has chronicled several "bioenergy success stories" that detail collaborations among waste facilities, pulp and paper mills, and lumber industries to convert waste and unused organic matter to energy in biomass reactors. Biomass energy as whole is the second largest renewable energy resource in Canada in terms of capacity, due primarily to Canada's extensive pulp and paper, forestry, and agricultural industries (Bioenergy success stories).

Major Federal Renewable Energy Regulations and Policies

The state of production of renewable energy in Canada has progressed dramatically in all types, with hydroelectricity making up the large majority of renewable electricity generated yearly. Although federal investment in renewable energy technology has increased appreciably, the major advances in wind, solar, and biomass have come from provincial commitments and programs. We have already elaborated on the energy plans in provinces such as Quebec, British Columbia, and Ontario—provincially there is a strong commitment to renewable energy but the variation in commitments and achievements is a result of the nature of federal commitments to RET. This section goes over some of the major federal initiatives in the renewable energy sector.

ecoENERGY for Biofuels

The Harper government announced in 2007 a 9-year incentive scheme to increase biofuel production in Canada. The incentive would provide declining rates (starting at 0.10 CAD per liter produced) over the 9-year period starting in 2008 and ending in 2017. Although similar to a feed-in tariff (since the rates decline as cost of production is projected to decrease) in principle, the scheme does not guarantee a market for producers. The project investment was valued at 1.5 billion CAD. There are a few regulatory burdens on this incentive, however; the Office of Energy Efficiency admissibility criteria for the incentive range from a cap on production volume eligible for the payoff (600 million liters), a minimum production rate (3 million liters per year for alternative diesel and 5 million liters per year for alternative gasoline), and a construction deadline for all production facilities of September 2012. Under this program, NRCan has signed 32 agreements with producers, which by December of 2012 was projected to result in more than 2 billion liters per year of ethanol or biodiesel (ecoENERGY for Biofuels Program).

NextGEN Biofuels and Sustainable Development Tech Fund

Sustainable Development Technology Canada (SDTC) is a nonprofit granting agency funded entirely by the Canadian federal government. Its mandate is to the fund the research, development, and implementation of clean energy technology through the formation of public–private partnerships. Two major funds operated under the SDTC to fulfill this mandate: (1) the SD Tech Fund, with an operating budget of 590 million CAD, and (2) the NextGEN Biofuels Fund, with an operating budget of 500 million CAD. The funds have two distinct aims: while the SD Tech Fund aims at helping commercially viable, "ready-for-market" clean energy technology, the NextGEN Biofuels Fund, as the name suggests, provides grants for demonstration projects and other initiatives to increase the commercial potential and visibility of emerging alternative fuel systems. The SDTC claims to have allocated 560 million CAD to over 200 projects and estimates that industry partners in these projects have contributed an additional 1.4 billion CAD. The 2000 statements of interest received by the agency during its regular calls for applications have an estimated potential of 19.2 billion CAD (SDTC funds).

National Strategy on Renewable Energy

As of 2012, there was no national policy or framework for reducing the dependency of Canadians on fossil fuels and nonrenewable energy sources. Hydroelectric dams also have not been evaluated for decommissioning as they have been elsewhere in the world. While provinces have taken on initiatives to diversify their energy portfolio, no commitment is in place in Canada to increase renewable electricity sources. An unsolicited "national strategy" was released by the Canadian Renewable Energy Association, which comprehensively recommends energy efficiency targets to meet all new energy demand, community power project approval infrastructure, decentralized power generation, level playing fields through feed-in tariff programs, and many other policy and fiscal tools. This "model" strategy is, surprisingly, the only document available to Canadians that emulates a real federal energy strategy (National energy strategy).

Major Renewable Energy Technology Research Initiatives

While the development of existing renewable resources, facilitating their introduction into the grid, forms a crucial component of sustainable design in Canada, a view to the future must also pay special attention to research and exploration of emerging technologies that may outpace and outperform current technologies. What follows is a list of major research programs funded by the federal government whose mandate is the research of existing RET and developing emerging technology.

CanmetENERGY

CanmetENERGY is the research branch of Natural Resources Canada, tasked specifically with the innovation and exploration of new energy technologies. These initiatives are not limited to renewable energy, but extend to buildings, clean fossil fuels, oil sands, and transportation. A few examples of nonhydro marine energy research projects are the Fundy Ocean Research Centre for Energy, which aims to invest at least 100 million CAD in its first phase and envisions 65 MW of tidal energy systems in Nova Scotia by 2015; the Offshore Energy Research group also conducts policy and technology research to increase the output of tidal energy systems (Technology roadmap). CanmetENERGY has also been instrumental in increasing the effectiveness of solar thermal systems, most notably commercializing the transpired solar air collector through its Solarwall product line, which was installed in 40 countries. Other initiatives and research collaborations are collected under the CanmetENERGY umbrella group (CanmetENERGY).

Office of Energy Research and Development

The Office of Energy Research and Development funds several projects under the mandate of the Clean Energy Fund—as of now, there are no projects or funding grants available. Past demonstration projects have been concentrated in the carbon capture discipline, specifically the Shell Canada Energy Quest Project, which, in collaboration with the Athabasca Oil Sands Project, aimed to siphon 1 million tons of CO_2 per year to be stored under layers of impermeable rock. The Enhance Energy-Alberta Carbon Trunk Line project also has similar aims (OERD programs).

Canadian Biomass Innovation Network

The Canadian Biomass Innovation Network (CBIN) also operates under the auspices of NRCan and funds research projects with goals of improving biofuel generation and purification technology. Between 2009 and 2013, projects funded by the CBIN under the Program of Energy Research and Development Bio-Based Energy Systems and Technologies initiative include efforts to improve biofuel inventory and feedstock supply models, harvest and storage of agricultural waste for biofuel conversion, scaling up biomass production, and developing refinery systems (PERD BEST).

Appraisal of Federal Regulations, Renewable Energy Commitments, and Research

Although provincial commitments to renewable energy are growing, especially in Ontario, there is currently a power and guidance vacuum with respect to renewable energy strategy at the federal level. It is not that Canada's economic output would be insufficient to develop strong renewable energy commitments and adhere to them. For example, the federal government in Spain has initiated several policies that form the basis for an aggressive renewable energy strategy. In 2010, nearly one-quarter of Spain's electricity was generated from renewable sources and it is en route to achieving a more ambitious target of 20% renewable energy contribution (Renewable theology vs. economic reality). As of 2005, a policy was implemented obliging all buildings to install photovoltaic systems and is only the second nation to make solar water heating systems mandatory. Spain is also the third most prolific producer of wind energy (REN21 global status report, 2006). Interestingly, the International Monetary Fund lists Canada and Spain as 13th and 14th in terms of gross domestic product (adjusted for purchasing power parity), separated only by about 10 billion USD (World economic outlook database, 2012). And while South Korea, also economically equivalent to Canada, only derives 2% of its energy needs from renewable sources, it currently has a "Green Growth" national policy initiative with a clear goal of increasing that fraction to 10% by 2022 (South Korea's drive for renewable energy). These examples indicate that investment in a strong renewable energy vision across the nation is not predicated on a budget not in Canada's possession.

This section identified a clear potential, especially in the solar and wind energy sectors, for increased electricity and energy supply from renewable resources. From the previous section, it could be inferred that the lack of federal guidance on energy efficiency led to two outcomes: (1) that the large-scale changes required to combat climate change and environmental damage were being delayed and (2) that the shortcomings at the federal level did not prevent lower levels of governance in their efforts to develop provincial energy efficiency goals. Similarly, the absence of directives from the upper echelons of decision making has created a power vacuum to be filled by scattered efforts at the provincial level. While this has resulted in achievements of significant impact in Ontario, British Columbia, and Quebec, it has not pushed the whole nation toward a renewable energy economy as observed in most European Union states.

According to the Renewable Energy Network (REN21) 2012 status report, the European Union draws nearly one-third of its electricity from renewable resources, and that three-quarters of added electricity capacity was completely renewable in 2012. Even China boasts 70 GW of nonhydro renewable power, although a staggering more than 200 GW comes from immense dam projects that threaten the stability of its riverine ecosystems. However, the point to be made is that 118 nations have renewable energy targets of some sort at the federal level. As of now, Canada does not feature in that list. Even from a strict economic growth perspective, investments in renewable energy have created over five million jobs around the world. Especially intriguing is the summary table produced in the REN21 status report, in which regulatory policies, fiscal incentives, and public financing initiatives are listed by nation and further divided into federal or provincial level policies. While nearly every other nation had significant federal or national level policies *and* incentives, the Canadian context had only one national regulation for biofuel production (REN21 global status report, 2012). When juxtaposed against the relative added capacity in renewable energy, it appears that the national energy goals of many nations have resulted in progress in green technology.

Finally, we must understand that the absence of a national renewable energy goal has not been the status quo for a long period of time—in fact, as early as 2005 the Renewable Power Production Incentive (RPPI) was proposed by the federal government, which earmarked 886 million CAD over 15 years (starting in 2006) to provide 0.01 CAD/kWh of renewable energy produced by eligible projects. The goal was to add 1.5 GW of new renewable electricity capacity from nonwind sources such as solar, biomass, and geothermal. However, according to Human Resources and Skills Development Canada, the RPPI has been shelved indefinitely. This incentive would have been the closest Canada has come to a national feed-in tariff policy, a staple feature of EU member energy policies (Renewable power production incentive). Liming and colleagues (Liming et al., 2008) extolled the virtues of having national sustainable energy goals and expounded on the importance of developing such a strategy in Canada, but have admitted that even a basic framework for such a policy is missing.

In conclusion, it can be inferred that while research initiatives and incentives are existent, the potential for renewable energy production in Canada has barely been explored due to the lack of federal initiative. This section highlighted the national energy policies of many other nations, but also indicated the strength of lower level governance in pushing a renewable energy agenda forward. Therefore, it can be said that Canada is at a veritable crossroads—on the one hand, initiation of a national renewable energy strategy requires both political will and immense popular pressure, and it cannot be ascertained that the latter will beget the former; on the other hand, a grassroots ambition to do away with reliance on carbon-based energy resources may be tougher to coordinate, but holds a great deal of promise. The next section explores this second path in depth. It examines the feasibility of developing agile, sustainable communities, which are not only self-sufficient and self-reliant, but also capable of interaction and technology transfer with other communities to promulgate sustainable design and ecological living principles *even in the absence of federal input.*

AGILE, SUSTAINABLE COMMUNITIES

As a response to Rawlsian thought, an initial reading of the philosophy of community suggests that the practice of community membership, at its core, is a practice of validating and debating beliefs, values, morals, and objectives. Communitarianism, or the ideology of communal living, offers the notion that the individual does not exist or develop in a vacuum of reflective rationalization, but rather is molded, shaped, influenced, and ultimately enhanced by his or her community experience. The connection between communities, individual membership in that community, and conceptions of environmental ethics are common to known scholars of environmental justice such as Aldo Leopold, Henry David Thoreau, and Rachel Carson (Sandler, 2013). The basic tenet of this school of thought is that community includes the natural world in which it exists, and therefore the values and morals that extend to human members of the community naturally extend toward the nonhuman members of that community.

If we acknowledge this ideology as the basis for one sort of approach to environmental sustainability, we might quickly realize that urbanization, globalization, and removal of the individual from the natural environment ultimately serve to sever the communal ties that may have existed between nature and humans. By extension, we seek to discipline and control nature, and ourselves, through a systematic bureaucratization and mechanization of nature and human activity—a concept known as biopower or biopolitics put forth by Foucault. Prior to Foucault, Cartesian dualism promulgated a similar binary and dissected view of the universe, through which machinery and industrial practice could flourish, but the essential relationship among life, ecosystems, and species interaction could not. Our understanding of natural complexity, of which we are an inextricable component, has become quite robust, yet the major institutions such as government, law, finance, economics, and policy have not adapted to this understanding. This means that the disconnect between humans and nature is no longer valid, but its influence persists in nearly all sectors that affect our daily lives and that dictate the identity of nations.

In the previous sections, we came to understand that policy in Canada was insufficient in the face of growing climate chaos threats and pollution impacts—the question raised was how might societies reorganize to better coexist within a complex natural environment to promote an ecologically balanced lifestyle? How could we truly achieve energy efficiency and promote renewable energy use, even in the absence of national strategy?

This section attempts to answer these questions by achieving the following three objectives.

1. Provide an overview of "experimental" communities in Canada, and around the world, that have indeed formed a greater harmony with the natural environment, abandoned the mechanized view of the universe, and reduced their impact on their surroundings.
2. Detail the major benefits such practices could have on sustaining communities, increasing their resilience, promoting cultural enrichment, and ultimately alleviating discontent with democratic representation.
3. Offer a projection of sustainability initiatives in Canada in the future and what contribution we can expect from federal governance structures in relation to sustainable community living.

Success Stories in Sustainable Communities
Drake Solar Landing Community
The Drake Solar Landing Community consists of 52 households all built to the R-2000 standard mentioned earlier, meaning that overall the energy efficiency of each household is 30% more than its conventional counterpart. The project, completed in 2007, derives 100% of its heating energy from the 800 solar panels installed in the community. The community project was initiated in 2005 by NRCan and today runs an independent Web site where visitors can track the operation of the energy collection system in real time via an animated schematic (Drake solar landing community). The initiative represents, at a superficial level, the idea of sustainable living where energy sources are decentralized (i.e., off grid), but localized to provide for the community.

Earthship Biotecture and Similar Sustainable Housing
Possibly the most advanced and comprehensive architectural initiative to achieve ecologically balanced living, engineer and architect Michael Reynolds founded Earthship Biotecture to construct "earthships" or off-grid households that not only ran on self-produced electricity, but also treated wastewater, grew food, and provided thermal insulation. While initial experimental versions of the structures led to a prolonged battle with the state of New Mexico, resulting in the loss of Reynolds' architecture license, Earthship households have since been constructed all over the world—South Africa, United Kingdom, Canada (Manitoba), Germany, France, the Netherlands, Argentina, and in several cities of the United States. The black water is treated in a multistep process, whose effluent can be used for irrigation or

nonconsumptive uses. Greywater and rainwater are collected, purified, and used for drinking. Electricity is usually provided with a combination of solar (dominant source), geothermal, and wind energy. Figures 5 and 6 illustrate some of the household operating principles. However, the completed Earthships range from over 200,000 USD to 1.5 million USD in cost—it is unclear how long it would take to pay the mortgage on an Earthship (Earthships for sale).

Model sustainable villages, where housing is not structured as intensely around off-grid principles but overall village energy use is targeted for

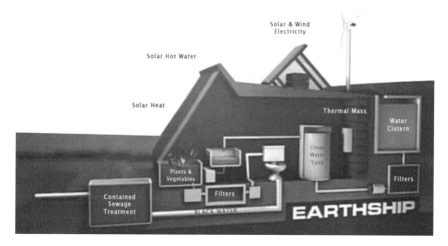

Figure 5 Basic operating systems in an Earthship. *(courtesy of Earthship Biotecture).*

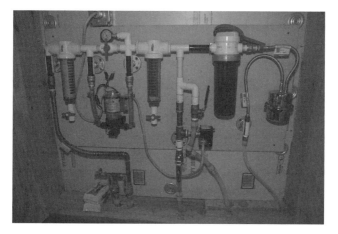

Figure 6 Real greywater recycling system in an Earthship. *(courtesy of Earthship Biotecture).*

reduction, exist in New Zealand and in China. The latter village, Huangbaiyu, has come under great criticism for the lack of consistency in construction and energy sources and did not receive positive reviews from anthropologist Shannon May, who lived in the village for over a year (China's first ecovillage proves a hard sell).

Microcosmic Initiatives

Microcosmic initiatives are efforts at introducing components of sustainable communities into urban centers to help reduce the overall impact the city has on the environment. One example is the Solar Schools project in the United Kingdom, founded by the 10:10 organization, whose sole aim is the reduction of carbon emissions by 10% every year. Solar Schools has a very simple concept—funding targets are set by elementary and secondary schools around the United Kingdom, put toward the purchase and installation of solar panels over the school buildings' roofs. The solar panels offset the energy use of the school by a significant amount and serve the dual purpose of educating children in community sharing and environmental awareness. The added communal element of the project is that the solar panels are funded entirely through donation. The webpage divides each school's roof into virtual solar "tiles," which can be purchased by parents, local businesses, and other supporters. The offset costs also help boost the school budgets, and children can benefit from an increased quality of education and after-school programs (Solar Schools).

Another microcosmic initiative in Vancouver is the hydrogen fuel cell-powered bus fleet, which has also been used in Whistler, British Columbia. The fuel cell company Ballard has provided 20 buses to Whistler (Bus case studies), which recently completed 1.5 million miles of transit (Hydrogen buses pass 1.5 million mile mark). While the transport of hydrogen fuel from Quebec has come under fire for diminishing the life cycle reduction in greenhouse gases, efforts are underway to provide hydrogen fuel in British Columbia itself.

Appraisal of Sustainable Community Efforts

The examples just given, although few, illustrate the various characteristics of an agile, sustainable community. First, a sustainable, agile community is defined to be one that does not depend on streams of material inputs, or that is self-sufficient; one that predicates governance on community-based decision making, knowledge and skills sharing, community-funded projects, decentralization of energy use, and cohabitation with nature; respect, reverence, and legal defense of ecosystems; divestment from polluting industries; diversified material and energy inputs to adapt to changing seasons and unforeseen events; real-time monitoring of energy and resource consumption; individual awareness; accessibility to decision-making bodies; and healthy debate. These characteristics combine generally under themes of resilience, decentralization, cooperation, and balance. The next section examines the benefits of these characteristics.

Benefits of Sustainable, Agile Communities
Resilience

Resilience as an ideology is the notion that as we overextend our boundaries and draw from natural sources more than what natural processes can replenish, our resilience to shock events or long-term consequences is diminished. For example, the climate change fueled delay of the monsoon in India led to overextraction of groundwater via electric pumps; the highly centralized grid of invasive thermal power plants and hydroelectric dams could not handle the sudden increase in electricity use. The result was a massive power outage in which half a billion people were left without electricity (India blackout). Folke and associates (Folke et al., 2002) have provided the seminal argumentation in favor of resilience as a tenet of sustainable development, criticizing free market models of economics and unlimited growth as inherently counter to natural ecosystems that pursue a, although dynamic, certain stable and balanced existence. A sustainable community would benefit from pursuing resilience because sudden environmental changes, outbreaks of disease, or similar unpredictable events can set off highly structured systems even if one component is compromised. Another example was the 2008 mortgage crisis in which the inability of household owners to pay mortgage loans that had been given to them without stringent credit verification set off a cascade of toxic collateralized debt obligations being written off, resulting in over 4 trillion USD lost in home equity and over 8 trillion USD lost by the American public in total (Altman, 2009). As a principle, resilience pushes for less central structure and more diversity. The resilience model would have prescribed household or village-level energy resources and low-cost water treatment to prevent overextraction in India; it would have also issued stringent mortgage regulations to prevent predatory lending and would have incentivized energy efficiency measures via mortgage payment reductions. Briefly, investing in energy efficiency invariably helps increase the value of a household, while allowing mortgage payments to accelerate (through saved energy costs) or be lowered (via incentives).

Balance

The concept of balance draws from the symbiotic and stable relationships species form in their ecosystems—resources are not consumed such that the species population is decimated. In addition, living space and territory are not expanded such that other species are unable to survive. The intricacy of the food web is preserved through coexistence and balanced living—and so while there is a high level of ordering and structure, it is not centralized or hinged on one single component. All members of the ecosystem act in concert, and no boundaries are overstepped to preserve the integrity of the system. By achieving this balance, species also achieve resilience—their ability to use only what resources are needed and maintain stable populations secures them from utter collapse even when a certain food source or abiotic requirement is suddenly reduced in availability. Communities can achieve balance through downsizing their levels of consumption, divesting from corporations, which are polluting or engaging in extractive industry, and pursuing self-sufficiency. As the Earthship model suggests, growing food locally, conserving water, and investing in renewable energy are all simple methods that help achieve balance. At the governance level, communities must provide incentives to encourage these practices, such as feed-in tariffs, skills building workshops, mandatory rainwater harvesting, and small-scale farming. Of importance is to note that these initiatives are *not* novel and in that lies their value. Feed-in tariffs were explained earlier in this chapter, but rainwater harvesting policies were implemented in the Indian state of Tamil Nadu with great effect (Madhavan), and the Brooklyn Grange Farm produces over 40,000 pounds of food annually through the cultivation of its two rooftop farms in the heart of New York City (About the Grange).

Decentralization

First, the ideology behind decentralization is that policy decisions, and similarly provision of resources, cannot adapt quickly to dynamic environments if the structure is widespread, rigid, and uniform. The repercussions of centralized power are evident in India's power outage, and the repercussions of centralized policy are evident in the 2008 financial crisis in the United States. Decentralizing power generation not only reduces energy costs for the household or building in question, but also increases resilience in the face of an outage or inclement weather and reduces dependence on nonrenewables. The proximity of the electricity-generating device to the household can also be thought to increase awareness of energy use; we can think of running water from the tap as analogous, in that our lack of knowledge of that water's source and treatment pathway ultimately reduces our ability to preempt contamination or track consumption. Such is the case with centralized electricity—by removing ourselves from the process by which that power reaches us, we absolve ourselves of guilt for the generation mechanism and abdicate our responsibility to conserve. Decentralization in diversified communities, which are prevalent in Canada, and in communities of distinct cultural heritage, such as the First Nations in northern Canada, has been shown to be an effective policy tool for inclusion, cooperation, and reconciliation. Studies in India showed that devolving policymaking to the village level led to increased participation of marginalized caste groups, women, and conflict resolution (Heller et al., 2007). With respect to the environment, the different bioregions and ecosystems in Canada are conducive to a less homogenized approach to environmental sustainability. The design of buildings and their respective energy requirements should be tailored to the prevailing climate and ecosystem—high elevation correlating with wind power, insolation with solar panels, nearby farms with biomass reactors, etc. Decentralization does not, however, diminish the value and necessity for national strategies. National policies help fund, audit, and set targets for community-based sustainable design, but do not dictate the means by which that goal is achieved. Decentralization can reinforce resilience and balance.

Cooperation

Cooperation at all levels of society is, naturally, a goal to pursue. However, in reference to developing sustainable, agile communities, cooperative innovation is required among citizens of varying backgrounds and expertise. Therefore, cooperation for sustainable design requires interdisciplinary education, combining unlikely fields to produce innovative projects, and including various viewpoints to respect cultural diversity. For example, the creation of rooftop gardens would require the input of agricultural engineers, public policy professionals, architects, and contributions of all the building tenants. Similarly, the use of nearby rivers or forest lands requires the advice of local First Nations' groups regarding their significance, how they can be used in harmony with other species, and how negative impacts can be reduced. Sustainable design incorporates cultural, technical, and political factors because if one of these is poorly addressed, the design or initiative cannot last for long and, by consequence, cannot be sustainable. Cooperation is achieved first through education and awareness; second through discourse, inclusive decision making and consultation; third through evaluation and revalidation of existing projects; and finally through evolving and transforming initiatives to respect the dynamics and complexity of the natural world. It is clear that decentralization facilitates these steps, and so it is inferred that decentralization can breed cooperation.

The Future of Sustainable Design in Canada

Throughout this chapter we have juxtaposed the promise of local movements toward renewable energy and efficiency, global examples of initiatives, which encourage participation, diversity, and adaptability, with the predominant approach of central authority, top-down policies, and profit-oriented governance. The scenario in Canada can be summarized as follows: provincial and local willingness to achieve greater ecological balance is present but scattered and in need of guidance and direction from federal authorities, which is absent. If Canada is to develop in a sustainable way, and become "agile" enough to deal with changing ecosystems and complexity, the decentralization of energy and sustainability will need to become a matter of policy, while the targets for reducing emissions, eliminating fossil fuel use, and reducing consumption overall will need to become matters of national strategy. The example provided by Tamil Nadu showcases decentralization with policy guidance—rainwater harvesting must be done in every household, but the methods or devices used are not controlled. As a nation, Germany exemplifies national strategy—renewable energy must become the primary source of electricity in the nation by a certain date, and feed-in tariffs diversify the sources of that energy.

It should not be assumed that the path to this future will be easy—the economic value of the tar sands, timber, unexplored mineral deposits, and Arctic trade routes have largely dominated the political discourse in Canada. Several initiatives and avenues for empowering citizens to practice democratic life have been curtailed or shut down (Yuen & Hien, 2005)—while we cannot place the blame on any government, party, or individual, we must question the validity of democratic representation if the most fundamental of requirements—a healthy environment—has been largely ignored for the past several years. Public awareness campaigns to inform citizens must continue; those who would demonstrate and protest in favor of stronger environmental policies should exercise their rights to do so; but most importantly, as discussed in this chapter, the future of sustainable design in Canada, and arguably across the world, lies in the hands of community initiatives to promote renewable energy and reduced resource consumption. At the core, environmental degradation increases because the products consumers purchase and the *scale* at which they are consumed require such a massive extraction of resources. These extractive industries are interested only in profits, as neoclassical economic theory dictates, and so the most fiscally efficient means to drive their processes is via fossil fuel combustion. As such, divestment from polluting industries at the individual level, via sustainable consumer lifestyles, and at the community, via self-sufficiency projects, will help engineer a more ecologically harmonious future for Canadians.

The province of Ontario has made great strides to take matters of sustainability into its own hand, with great success thus far. Canadians must petition every level of government to do the same in their provinces and territories and lead by example in their communities. Even in the absence of federal policy, a sustainable future may yet be achieved in Canada—through effective campaigning and constant focus on environmental issues, policy battles can be won at the upper echelons of governance.

CONCLUSION

Imagine a society in which cultural diversity leads to cooperative innovation. Narratives and views of nature from all backgrounds form the basis of social education of children, who now learn to appreciate natural ecosystems and form bonds with all species, seeing themselves as their compatriots and guardians. Imagine a community in which schools, town halls, and all other public buildings have been retrofitted with solar panels primarily through donations—skills sharing workshops have taught household owners to install their own solar roofs, for which they received a guaranteed market from the provincial feed-in tariff scheme. Rooftop and backyard farming are so common that they have become the rule rather than the exception—imagine more than three-quarters of food consumed in the city is produced within a few-mile radius of the city center. Imagine that rainwater and snowmelt are harvested and used for drip irrigation and cooling and that domestic wastewater is treated in bioreactors to produce hydrogen fuel or electricity directly. New buildings are designed to maximize natural lighting, equipped with living walls and geothermal power supply. Public transit is completely carbon neutral and carpooling has been subsidized heavily.

This is a future that is all but lost—but it is not impossible. The money saved through these practices can be reinvested to improve health care and education and to restore damaged ecosystems. The city of Tokyo projected that 50% coverage through rooftop garden initiatives would result in saved heating costs to the tune of 1.2 million USD per day (Yuen & Hien, 2005)—if true, in 2 years time this would more than surpass the current budget of Canada's SDTC biofuel initiative. Feed-in tariff contracts, which introduce increasing amounts of renewable energy capacity while diminishing the incentives to encourage low-cost technology innovations, are on track to helping many EU nations achieve complete fossil fuel independence.

Sustainable design requires local empowerment, variation in energy resources, and overall reduction in consumption and pollution. This chapter explored the initiatives that are in place and their benefits and shortcomings; the extent to which environmentally friendly technology and efficiency incentives can go with respect to achieving sustainable communities; and finally the steps and obstacles that communities will ultimately encounter in pursuing this goal. But behind all these projects, policies, initiatives, and campaigns is a collective willingness to change and a determination to seek out that sustainable, agile future that is difficult to obtain, but indeed within reach. Whether Canada will seize the opportunity to transform society and create a balance with natural ecosystems lies ultimately in the hands of its people.

REFERENCES

100% renewable electricity supply by 2050, UNCSD. Accessed: http://www.uncsd2012.org/index.php?page=view&type=99&nr=24&menu=137.
2009 energy efficiency report card, CEEA.
39th parliament, 1st session, 2006. Parliament of Canada.
About the Grange, Brooklyn Grange Farm. Accessed: http://www.brooklyngrangefarm.com/aboutthegrange/.
Alberta's oil sands, 2006. Government of Alberta.
Altman, R.C., 2009. Foreign affairs.
Atlas of Canada, NRCan. Accessed: http://atlas.nrcan.gc.ca/site/english/maps/economic/energy/oilgas/1.
Baillie, R., Renewable energy world. Accessed: http://www.renewableenergyworld.com/rea/news/article/2012/12/crunch-time-for-canadas-wind-sector.
Beringer, A., Wright, T., Malone, L., 2008. Int. J. Sustainability Higher Educ. 9 (1), 48–67.
Bioenergy success stories, CanBIO. Accessed: http://www.canbio.ca/article/bioenergy-success-stories–180.asp.
Budget, 2012. Canada won't spare a penny for clean energy. Canada won't spare a penny for clean energy, Pembina Institute. Accessed: http://www.pembina.org/blog/616.
Bus case studies, Ballard. Accessed: http://www.ballard.com/files/PDF/Bus/Bus_Case_Studies_one_pager.pdf.
California local ordinances, Plastic bag laws. Accessed: http://plasticbaglaws.org/legislation/state-laws/california-2/.
Canada boosting hydro power to 88.5 GW, Clean Technica. Accessed: http://cleantechnica.com/2011/11/16/canada-boosting-hydro-power-to-88-5-gw-to-replace-us-coal/.
Canada Customs and Revenue Agency's Administrative Monetary Penalty System, Office of Energy Efficiency, NRCan. Accessed: http://oee.nrcan.gc.ca/regulations/9874.

Canada to withdraw from Kyoto Protocol, BBC News. Accessed: http://www.bbc.co.uk/news/world-us-canada-16151310.

Canada's action on climate change, Climate change, government of Canada. Accessed: http://www.climatechange.gc.ca/default.asp?lang=En&n=72F16A84-1.

Canada's emissions trends, 2012. Environment Canada.

Canadian Energy Efficiency Alliance. Accessed: http://www.energyefficiency.org/.

Canadian wind energy industry, Industry Canada. Accessed: http://www.ic.gc.ca/eic/site/wei-iee.nsf/eng/home.

CanmetENERGY, NRCan. Accessed: http://canmetenergy.nrcan.gc.ca/renewables/solar-thermal/437.

China's first ecovillage proves a hard sell, The Age. Accessed: http://www.theage.com.au/news/world/chinas-first-ecovillage-proves-a-hard-sell/2006/08/25/1156012740582.html?page=fullpage.

Chretien signs Kyoto agreement, National Post. Accessed: http://www.nationalpost.com/national/story.html?id=%7B2E3F4E3C-BA8B-4DC2-8EEA-4D955F996A2E%7D.

Clothes washers, Office of Energy Efficiency, NRCan. Accessed: http://oee.nrcan.gc.ca/regulations/products/5648.

Cote, R., Booth, A., Louis, B., 2006. J. Cleaner Prod. 14, 542–550.

Drake solar landing community, DSLC. Accessed: http://www.dlsc.ca/.

Dusyk, N., Berkhout, T., Burch, S., Coleman, S., Robinson, J., 2009. Energy Efficiency 2, 387–400.

Earthships for sale, Earthship Biotecture. Accessed: http://earthship.com/earthships-for-sale.

ecoENERGY efficiency, Office of Energy Efficiency, NRCan. Accessed: http://oee.nrcan.gc.ca/corporate/14511.

ecoENERGY for Biofuels Program, Office of Energy Efficiency, NRCan. Accessed: http://oee.nrcan.gc.ca/transportation/alternative-fuels/programs/10163.

Enbridge completes Sarnia solar farm, CBC News. Accessed: http://www.cbc.ca/news/technology/story/2010/10/04/sarnia-enbridge-solar-farm.html.

EnerGuide home rating system, Office of Energy Efficiency NRCan. Accessed: http://oee.nrcan.gc.ca/residential/personal/16352.

Energy conservation program, Arctic Energy Alliance Northwest Territories. Accessed: http://aea.nt.ca/programs/energy-conservation-program.

Energy efficiency improving in Canada, Green living online. Accessed: http://www.greenlivingonline.com/article/energy-efficiency-improving-canada.

energy efficiency indicators US and Canada, 2012. Institute for Building Efficiency. Accessed: http://www.institutebe.com/Energy-Efficiency-Indicator/2012-EEI-US-Canada-Results.aspx?lang=en-US.

Federal budget fails to extend support for new wind energy development, CanWEA. Accessed: http://www.canwea.ca/media/release/release_e.php?newsId=76.

Feed-in tariff program, Ontario power authority. Accessed: http://fit.powerauthority.on.ca/.

FleetSmart, Office of Energy Efficiency, NRCan. Accessed: http://fleetsmart.nrcan.gc.ca/index.cfm?fuseaction=fleetsmart.home.

Folke, C., Carpenter, S., Elmqvist, T., Gunderson, L., Holling, C.S., Walker, B., 2002. Ambio. 31 (5), 437–440.

Gamtessa, S.F., 2013. Energy Buildings 57, 155–164.

Greenhouse gas emissions by province, Environment Canada. Accessed: http://www.ec.gc.ca/indicateurs-indicators/default.asp?lang=en&n=18F3BB9C-1.

Greenhouse gas emissions per person, Environment Canada. Accessed: http://www.ec.gc.ca/indicateurs-indicators/default.asp?lang=en&n=79BA5699-1.

Harper vs Kyoto, where does that leave Quebec? Behind the numbers. Accessed: http://behindthenumbers.ca/2012/02/02/harper-vs-kyoto-where-does-that-leave-quebec/.

Heller, P., Harilal, K.N., Chaudhuri, S., 2007. World Development, vol. 35 (4), 626–648.
Hopper, N., Barbose, G., Goldman, C., Schlegel, J., 2009. Energy Efficiency 2, 1–16.
Hybrid vehicle tax incentive, PEI Department of Finance, Energy and Municipal Affairs. Accessed: http://www.gov.pe.ca/finance/index.php3?number=1017738&lang=E.
Hydrogen buses pass 1.5 million mile mark, Pique News Magazine. Accessed: http://www.piquenewsmagazine.com/whistler/hydrogen-buses-pass-15-million-mile-mark/Content?oid=2327085.
Improving energy performance in Canada, 2010-2011. NRCan. report.
India blackout, CNN News. Accessed: http://www.cnn.com/2012/08/01/world/asia/india-blackout/index.html.
International Energy Statistics, US EIA. Accessed: http://www.eia.gov/cfapps/ipdbproject/iedindex3.cfm?tid=6&pid=37&aid=12&cid=regions&syid=2001&eyid=2010&unit=BKWH.
Irving, J., 2010. Hydropower in Canada. Canadian Hydropower Association.
Keystone XL pipeline project, Transcanada. Accessed: http://www.transcanada.com/keystone.html.
Kyoto Protocol, UNFCCC. Accessed: http://unfccc.int/kyoto_protocol/items/2830.php.
Liming, H., Haque, E., Barg, S., 2008. Renewable Sustainable Energy Rev. 12, 91–115.
LiveSmart BC home efficiency program extended, Government of British Columbia. Accessed: http://www2.news.gov.bc.ca/news_releases_2009-2013/2010EMPR0014-000437.htm.
Madhavan, K., The Hindu. Accessed: http://www.thehindu.com/todays-paper/tp-national/tp-tamilnadu/article3157071.ece.
National energy strategy, CanREA. Accessed: http://www.canrea.ca/site/national-strategy/.
National inventory report greenhouse gas sources and sinks in Canada 1990-2007, 2009. Environment Canada.
National survey report of PV power applications, 2010. IEA Co-Operative Programme.
Nuclear Power in Canada, World Nuclear Association. Accessed: http://www.world-nuclear.org/info/inf49a_Nuclear_Power_in_Canada.html.
Nyboer, J., Lutes, K., 2009. A review of renewable energy in Canada, NRCan.
OERD programs, NRCan. Accessed: http://www.nrcan.gc.ca/energy/science/programs-funding/1477.
Office of Energy Efficiency, NRCan. Accessed: http://oee.nrcan.gc.ca/.
Offshore wind energy blows into Canada, Earthtechling. Accessed: http://www.earthtechling.com/2011/03/offshore-wind-energy-blows-into-canada/.
PERD BEST, Canadian Biomass Innovation Network. Accessed: http://cbin.gc.ca/6.
Petz, B., 2012. Ecology global network. Accessed: http://www.ecology.com/2012/05/31/eu-2010-greenhouse-gas-emissions/.
Power Smart First Nations Program, Manitoba hydro. Accessed: http://www.hydro.mb.ca/your_home/first_nations/index.shtml.
Recuperateur douche, Gazifere. Accessed: http://www.gazifere.com/en/en-residentiel_recuperateur_douche.php.
RECYC-FRIGO, Hydro-Quebec. Accessed: http://www.hydroquebec.com/microsite/residential/geothermie/.
Reducing transport greenhouse gas emissions, 2010. International Transport Forum.
REN21 global status report 2006.
REN21 global status report 2012.
Renewable power production incentive, HRSDC. Accessed: http://www.hrsdc.gc.ca/eng/workplaceskills/sector_councils/renewable_energy/section11_2.shtml.
Renewable theology vs. economic reality, The Billings Outpost. Accessed: http://www.billingsnews.com/index.php/commentary/3991-renewable-theology-vs-economic-reality-part-2.

Report of the individual review of the annual submission of Canada, 2010. UNFCCC.
Sandler, R.L., 2013. Environmental virtue ethics, Blackwell Publishing Ltd.
Scrap-it, B.C., Government of British Columbia. Accessed: http://www.scrapit.ca/.
SDTC funds, SDTC. Accessed: http://www.sdtc.ca/index.php?page=sdtc-profile&hl=en_CA.
Solar photovoltaic map, CanmetENERGY. Accessed: http://canmetenergy.nrcan.gc.ca/renewables/solar-photovoltaic/562.
Solar Schools, 10:10 UK. Accessed: http://www.solarschools.org.uk/.
South Africa - plastic bags regulation, ELAW. Accessed: http://www.elaw.org/node/1779.
South Korea's drive for renewable energy, BBC News. Accessed: http://www.bbc.co.uk/news/business-15984399.
Stephen Harper's democracy award a sad joke on Canadians, The Star. Accessed: http://www.thestar.com/opinion/editorialopinion/article/1255830–stephen-harper-s-democracy-award-a-sad-joke-on-canadians.
Sustainable endowment fund, Environment British Columbia. Accessed: http://www.env.gov.bc.ca/epd/recycling/resources/reports/sef.htm.
TakeCHARGE!, Government of Newfoundland & Labrador. Accessed: http://takechargenl.ca/.
Taylor, A., 2012. Business insider. Accessed: http://www.businessinsider.com/china-plastic-bag-ban-2012-6.
Technology roadmap, Marine Renewables Canada. Accessed: http://www.marinerenewables.ca/technology-roadmap/.
What is ENERGY STAR?, Office of Energy Efficiency, NRCan. Accessed: http://oee.nrcan.gc.ca/equipment/energystar/11980.
What is R-2000?, Office of Energy Efficiency, NRCan. Accessed: http://oee.nrcan.gc.ca/residential/new-homes/r-2000/7334.
Whitlock, J., Canadian Nuclear FAQ. Accessed: http://www.nuclearfaq.ca/cnf_sectionG.htm.
Wind energy, CanWEA. Accessed: http://www.canwea.ca/wind-energy/index_e.php.
Wind vision, CanWEA. Accessed: http://www.canwea.ca/windvision_e.php.
World economic outlook database, 2012. International Monetary Fund. Accessed: http://www.imf.org/external/pubs/ft/weo/2012/02/index.htm.
Yuen, B., Hien, W.N., 2005. Landscape Urban Plan. 73 (4), 263–276.

CHAPTER 12

Energy Management in a Small-Island Developing Economy
The Case of Mauritius

Khalil Elahee
Chairman, Energy Efficiency Committee Member, National Energy Commission Associate Professor Mechanical and Production Engineering Department, Faculty of Engineering, The University of Mauritius, Mauritius

Contents

Introduction	293
Energy Management and Climate Change	294
Buildings: Caught in a Vicious Circle	296
Assessing Energy Efficiency	299
The Way Forward	302

INTRODUCTION

Mauritius is a tropical island in the southeast of the Indian Ocean with a population of 1.3 million and an area of 2040 km^2. It has a warm and humid climate with a mean summer temperature of 28 to 32°C and a humidity of 80 to 90%. In winter, temperatures average 20°C, but there is increasing disruption in weather patterns. Microclimatic conditions prevail in different localities, not to mention heat island effects. With a very high population density of 631 inhabitants per km^2, it ranks fifth in the world. It has a fragile ecosystem with only 1% of its indigenous forests left, surrounded by threatened coral reefs and lagoons from which derives largely its attraction toward some one million tourists annually.

Since 1960, buildings in Mauritius have been constructed in concrete to face cyclonic risks. However, this has not been with consideration of ecological and energy-efficiency aspects. Consequently, the current use of air conditioners, driven by grid electricity, causes excessive peak power demands in the summer. Imported fossil fuels are responsible for the 82% of total primary energy demand, which includes oil, coal, and liquefied petroleum gas, despite the available potential of renewable energy sources. The rise in prices of the former makes it a necessity to shift toward sustainable energy, including optimal efficient utilization of energy. Climate change risks also imply that Mauritius must lead the way in mitigating greenhouse gas emissions, particularly through energy management, given its own vulnerability as a small island. Other sister islands and islets within the Republic of Mauritius such as Rodrigues, Agalega, and St. Brandon are no less affected by climate change, particularly their fishing and agricultural communities.

ENERGY MANAGEMENT AND CLIMATE CHANGE

In 2011, an energy efficiency act was adopted in Mauritius with provisions for the setting up of an energy efficiency management office (EEMO). The main objectives of the EEMO are to (1) promote the efficient use of energy and (2) promote national awareness for the efficient use of energy as a means to reduce carbon emissions and protect the environment. Energy efficiency should ensure energy security, stimulate economic growth, and reduce pollution.

China, the United States, the European Union (EU), and Japan have adopted new energy efficiency measures, and the World Energy Outlook 2012 forecasts a reduction in global energy intensity [energy consumption per unit of gross domestic product (GDP)] of 1.8% a year through to 2035, a major improvement compared with only 0.5% per year over the last decade. According to the same source, a significant share of the economic potential of energy efficiency—four–fifths in the buildings sector and more than half in industry—remains untapped, mostly due to nontechnical barriers. In the Republic of Mauritius, from 2010 to 2011, the energy intensity dropped by 4%. This implies that had the energy intensity remained the same as in 2010, the import bill would have been at least Rs 1 billion (about USD 30 million) higher with more than 200,000 tons of CO_2 emitted. A corresponding increase in the maximum peak demand of at least 20 MW and an extra power consumption of about 200 GWh would have been recorded. Sustaining such a drop in energy intensity will become a challenge, but as the structure of the economy moves away from energy-intensive industries such as textile manufacturing and agriculture to tertiary services, a decoupling of GDP with economic growth should be possible. Consequently, this will avoid an investment in installed capacity, particularly an emergency recourse to fossil-fuel power plants. Investment in energy efficiency is also more than compensated by a reduction in fuel bills, as the case of developed and emerging economies seems to prove.

Special adviser to the Prime Minister, Joel de Rosnay has proposed the efficient, decentralized combination of not less than 12 renewable sources of energy potentially available in Mauritius (Le Mauricien 2012). These are (1) solar photovoltaics, (2) solar thermal, (3) wind power, (4) hydropower, also with the latter in storage as pumped hydro, (5) wave power, (6) sustainably grown biomass, (7) biogas, (8) advanced biofuels from algae, (9) geothermal power, (10) ocean thermal energy, (11) offshore wind or offshore sea current, and (12) hydrogen generated by electrolysis using any of the available renewable sources. He recommends taking advantage of the existing national power grid, fully reticulated throughout the island, to transfer electricity from one point to another as per real-time demand and supply requirements. For instance, the G3-PLC standard has been mentioned as implementable across the Mauritius smart grid connecting a few hundred thousand prosumers. Storage ability can be included, as well as some back-up facilities, but the problem of demand–supply matching, unpredictability of renewable sources, or the impossibility of cross-border transmission in small islands will be avoided. Moreover, de Rosnay sees the interfacing of smartphones with the smart grid as a world of endless possibilities in terms of energy management, including savings and efficiency as well as renewables. Buildings that today consume electricity from the grid supplied by fossil-fuel power plants will thus become net producers of clean energy.

To undertake such a transformation that should render meaningless the correlation between GDP and energy demand, a number of challenges have to be addressed. Institutional and regulatory reforms in the most radical manner, technology transfer, capacity building, and rigorous implementation programs will have to be initiated. This can be triggered only if there is committed leadership and shared vision.

The globalization of telecommunications and information technology is occurring at a fascinating speed with little to stand in its way. The same is not happening with renewable energy and energy efficiency, some having even called the latter case an "epic failure" given the magnitude of the potential. The paradigm shift to decentralized sustainable energy systems, even in a small island, is set to meet strong reticence and resistance. This comes not just from the oil and coal lobbies, but from those whose immediate self-interests in the dominant economic system lie in favoring the business-as-usual practices.

The World Energy Scenario 2012 proposes indeed relevant policy action in six categories, namely, (1) increasing the visibility of energy efficiency through measurement and communication; (2) integrating efficiency into the decision making; (3) creating appropriate business models and financing instruments; (4) mainstreaming the most efficient technology options and discouraging the least efficient ones; (5) making it real by implementing monitoring, verification, and enforcement activities; and (6) making it realizable by increasing governance and administrative capacity at all levels. This is probably a proper starting point toward materialization of the vision proposed by de Rosnay, assuming that it is shared by all stakeholders. Time is of the essence and action should start as soon as possible.

The urgency for such action has been highlighted by the World Bank itself in its "turn down the heat" report (World Bank 2012). Mauritius as a small-island developing state cannot be a passive spectator given the gravity of the threat on its future. Small-island developing states and least developed countries have identified global warming of 1.5°C as warming above which there would be serious threats to their own development, if not to their survival. According to the World Bank, which itself has its share of responsibility for favoring fossil fuels in so many recommendations, current trends put us on a path toward 4°C warming within this century. It warns that high-temperature extremes in the tropics will cause significantly larger impacts on agriculture and ecosystems. A sea-level rise is likely to be 15 to 20% larger in the tropics than the global mean. Increases in tropical cyclone intensity are likely to be felt disproportionately in low-latitude regions. Increasing aridity and drought are likely to increase substantially in many developing country regions located in tropical and subtropical areas. Moreover, it adds that there is a risk that the sea-level rise exceeds the capabilities of controlled, adaptive migration, "resulting in the need for complete abandonment of an island or region."

BUILDINGS: CAUGHT IN A VICIOUS CIRCLE

The last decade has experienced record high mean annual temperatures globally, but also locally in Mauritius. To counter days of heat waves, people turn on air-conditioning units at home, in industry, and in the services sector. The Building Control Act, which requires energy efficiency in buildings, has yet to be enforced. Similarly, the introduction of labels and minimum performance standards for appliances is not yet a reality in Mauritius. Hence, a vicious circle sets in as power from fossil-fuel combustion is being used for short-term cooling of buildings through air conditioning. Such a practice increases, in the long term, greenhouse gas emissions and accelerates global warming, the very cause of extreme high temperatures.

As a first measure, the government is now proposing differentiated excise duties on appliances, 0% on the most efficient and up to 25% on those with minimum performance standards (MOFED 2012). The EU labeling system is being adapted for this purpose. While this is a proper initiative, it is doubtful if its impact will be significant. It may lead to rendering appliances too expensive for lower income groups. A bonus on the return and scrapping of old inefficient appliances targeting the latter, as well as tariff structures discouraging higher consumption of electricity, may also be needed to boost energy efficiency. It is also feared that without focused education and sensitization of stakeholders, particularly households, the aimed objective of energy efficiency may not become a reality.

In the case of air conditioning, a coherent holistic approach in terms of energy management will also require the urgent enforcement of energy efficiency regulations under the building control act. However, the sheer complexity of the matter, involving regulatory and institutional changes, is leading to much delay in such enforcement. Training and monitoring mechanisms are also taking too long to be set up. This is probably largely due to the low critical mass or economies of scale related to small islands such as Mauritius. The same reason lies, to some extent, behind the slow emergence of an energy services sector. International cooperation exists but has been mostly happening in the form of desktop studies by foreign consultants, not as technology or know-how transfer to the local population.

The decentralized combination of renewable energy sources, smart grid, introduction of labels and standards, and several other means of achieving sustainable energy management in Mauritius rely much on the use of new techniques and technologies. They have administrative and management costs, are subjected to bureaucratic hurdles, and rest upon top-down implementation that has to be reckoned right from scratch, that is, from the design stage.

A complementary approach is next discussed with reference to heat recovery for air conditioning in the Mauritius context. It promotes a more bottom-up, end-of-pipe grassroot method applicable in the case of existing buildings. It relies on current, often locally available materials and knowledge. Its emphasis is not on sophisticated methodologies or technologies. It favors behavioral change, even simple traditional ways of achieving efficiency such as good housekeeping, maintenance, and waste minimization. Although this aspect is not fully discussed here, this complementary approach can even go to the extent of questioning our real needs, the economic model being used, and our philosophy of consumption, if not the purpose of life itself.

According to Pandoo (2012), air conditioning in the domestic sector in Mauritius contributes to between 12 and 18% of the total electricity consumption, mostly used in summer at night and present in about one-third of households. However, the power consumed by air conditioners can represent up to half of the maximum demand at a given instant. The waste heat potential for cooling in the domestic sector was found out to be almost null due to few sources of waste heat with corresponding temperatures of less than 50°C, low availability (not constant and present only less than 4 hours per day), and a relatively low quantity of heat to run the thermal air conditioners. For better energy management, the use of fans and air coolers and regular maintenance, as well as proper operation of air-conditioning units, should be an absolute priority. Badurally and colleagues (2009) have shown that peak power demand is predominantly a function of ambient temperature. Hence, measures such as the proper energy-efficient design of buildings, as well as the planting of trees to reach better bioclimatic conditions, should be a priority. It was also noted that most consumers do *not* know how to operate and maintain an air-conditioning unit properly, and salespersons often have no idea of the meaning of power consumption of what they sell. Education and sensitization are essential.

In the commercial sector in Mauritius, air conditioning consumes about one-third of the electricity demand, contributing up to 35% of its maximum demand (Pandoo 2012). This is due to the fact that the units run year round with at least 25% still operating in the winter where temperatures are well within the comfort zone. Hotels are included in this category, which run 24 hours to ensure quality standards for their clients. The best waste heat sources in hotels are with boilers and generators. Heat recovery of at least 10% can be achieved using economizers in boilers. The main area of an energy-saving opportunity, however, is to change the attitude of commercial use toward doing without air conditioning in the winter. Solar thermal hybrid units should also be introduced to promote sustainability.

According to Pandoo (2012), air conditioning is less than 4% of the total electricity in the industrial sector and its power demand is 3 to 4 MW. However, the potential of waste heat is very large: up to 24% recovery can be obtained from steam boilers, 60 to 80% recovery could be obtained from drying and finishing equipment, and 75 to 85% recovery can be obtained from wastewater. Wastewater can be used as a source of heat for chillers in many cases. Maintenance and good housekeeping, as well as a culture of energy management in industry, will reduce energy consumption by at least 10% with a negligible investment, including the use of air-conditioning equipment (AMM, 2012). A similar potential should exist in the power industry itself.

ASSESSING ENERGY EFFICIENCY

The two complementary approaches given earlier combine, to some extent at least, to give the following proposals toward energy management, particularly through heat recovery to cool down buildings in the Mauritian context. The following recommendations, while not being exhaustive, are holistic in nature toward breaking the vicious circle discussed earlier. A virtuous circle is proposed here where cold is generated from heat, particularly where waste heat is recycled with maximum efficiency. The recourse to renewable energy, not considered per se in Table 1, will enhance the virtuous circle.

The key issue related to energy conversion from one form to another is how to measure energy efficiency. If heat and cold are to be equated as forms of energy flowing due to temperature differences, it is important to consider not just the quantity of energy but also its quality. While it is not the place here to undertake a detailed analysis in terms of thermodynamic exergy or work availability, henceforth it will be worth considering also each source of energy in terms of its entropy or order as well. Such an analysis may indeed boost the value of renewable sources at the expense of more destructive or chaotic fossil fuels.

In the meantime, far from thermoengineering considerations or even transparent social and environmental criteria, development projects are dominated by market economics, not to say the perceived influence of lobbies and other obscure practices. This is particularly true for the energy and power sector. The building industry is also one with enormous vested interests, related also to the control of land use. In Mauritius, the case is not much different from that of other countries with equal "ease of doing business."

Measuring energy efficiency is a challenge, as noted by the World Bank and the International Energy Agency (WB/EIA) recently in the context of the development of a global tracking framework (UN-SEFA, 2012). Indeed, the GDP-related energy intensity links energy efficiency with structural economic shifts that occur as a normal part of the development process. Precise data are needed, which are not always available in developing countries. The WB/EIA proposes to explore the energy intensity of six key

Table 1 (Elahee & Pandoo 2012)

Recommendations	Measures	Estimated benefits
No air conditioners (ACs) with coefficient of performance (COP) <3	Banning of sales of ACs with COP less than 3	24 MW power saved on peak demand and 21 GWh saved on energy demand
	25% excise duty on window-type ACs and lower COP ACs	Less ACs with low efficiencies on the local market
	Sensitization campaigns about effect of efficiency on electricity bill	Part of population may shift to ACs with better COP
Promotion of inverter technology	Laws to force all ACs suppliers to sell also inverter ACs	Population will have wider range of products to choose from
	Tax deduction on inverter ACs	Peak power saved may delay installations of new power plants with high costs
Greener design	Provide engineers and architects with courses on green design of buildings	Better natural ventilation, insulation from sun, and lower cooling load for ACs
Switching demand curve of ACs to lower value in commercial sector	Promoting flexi time for wholesale and retail trade	Lower peak demand due to shifting in demand across time period
	Promoting centralized large commercial centers with central units than decentralized units	Higher COP of 4.96 compared with split units of 3.66
	Giving construction permits in higher Plaines Wilhems where comfortable zones are present	Switching off of some ACs will decrease peak load
Installation of economizers in boilers in commercial sector/industry	Sensitization campaigns in hotels on benefits of economizers	Savings in annual consumption of 2–5%
	Law forcing boiler suppliers to propose a packaged boiler–economizer assembly at sales	
	Introduce local suppliers of heat exchangers, including economizers	Availability, proximity, and consultation services of heat exchangers locally

Table 1 (Elahee & Pandoo 2012)—cont'd

Recommendations	Measures	Estimated benefits
Heat recovery in industrial and power sector	Creation of an institution responsible to perform waste heat audits and propose long-term heat recovery technologies in local industries	10 to 24% recovery in steam boilers
		60 to 80% recovery in hot air
	Green fines of 10–15% on low-efficiency equipment.	75 to 85% recovery in hot wastewater
	Setting up of cogeneration in power sector where possible	24 to 42% recovery in exhaust
		25 to 50% recovery in cooling water
	Decentralization of the power sector and setting up of partnership between them and local industries to promote flexi factory concept	More cogeneration plants with high efficiencies of 90% to provide steam or hot water or cooling for air conditioning or refrigeration to industrial processes.
	Centralization in the textile, food processing, and beverage industry to increase heat recovery potential	Higher efficiency processes with lower payback periods
Waste heat cooling in large factories	Setting up of a national framework to locate necessary conditions for propagation of waste heat cooling	To be determined
Waste heat cooling in power plants	All IPPs, Fort George, and Saint Louis should install waste heat absorption chillers	To be determined

energy-consuming sectors, including residential, transport, agriculture, industry, services, and electricity supply. However, the problem remains as prosperity and well-being, or simply progress, are still limited to the concept of economic growth.

A study is under way to integrate noneconomic dimensions in an analysis of the sustainability of energy efficiency measures (Elahee & Ramatally

2012). The case of the cane industry is considered, and multicriteria sustainability assessment is carried out to assess, for example, the switch to higher calorific value biomass. It was found that the energy intensity of the cane industry can increase while the overall energy intensity of the country tends toward sustainability. Environmental indicators, such as CO_2 emission per unit electrical energy production and carbon emission from energy production and use per capita, can show a trend away from sustainability despite an overall energy intensity improvement in the country. A ratio such as agricultural food supply land per capita also moves away from sustainability despite energy efficiency gains.

THE WAY FORWARD

The seemingly endless possibilities and potentials of renewable energy sources, energy efficiency, and smart grids combined are enough to stir hope in a more sustainable future. The case of a small-island developing economy such as Mauritius is even more interesting. Its limited size can be an asset, easing the task of transformation. However, in an era of globalization, the lack of economies of scale against the current dominant market forces, not to mention vested interests, rather plays against the case of Mauritius. Moving away from renewables can be a tedious process if there is no strong political will, courage to stand against business-as-usual aficionados, and a shared commitment to undertake a paradigm shift. The fact that Mauritius already has a well-established power grid, providing 100% access of electricity indirectly, disfavors attempts at decentralizing the system.

There is also a battle to fight on the opposite front. It is about changing people's attitudes to make them energy conscious. The popular reflex to look for inexpensive electricity at any cost must disappear. Energy must be understood in terms of quality as well, not just quantity. Efficiency has to be translated into concrete parameters reflecting the most valued outcomes in terms of sustainability. This requires a radical change in mindsets from the short-term, self-interest considerations to longer term common objectives, including revisiting our relationship with nature. Through education and sensitization, integration of environmental and social costs will have to be embedded in our politicoeconomic system, but also in the decision-making processes at all levels. Sustainability also implies acting against greed, waste, and overconsumption. This is a real challenge in a world of rampant materialism.

The way ahead is one of patient perseverance struggling to improve ourselves, believing that human nature will rise above our destructive instincts. It is also about applying our intelligence collectively through visionary leadership and innovative endeavors.

Energy management as a systemic effort through engineering and organizational techniques to optimize energy production, distribution, and usage for specific political, economic, and environmental objectives is indeed one of the precious keys to a sustainable future.

REFERENCES

AMM, 2012. Association of Mauritian Manufacturers, Newsletter 'Maitrise de l'Energie', no. 1.

Badurally, A.N., Dauhoo, M., Elahee, M., 2009. On the influence of weather and socio-economic factors on peak electricity demand in Mauritius. International Conference on Energy and Environment, Singapore.

Elahee, K., Pandoo, R., 2013. Heat from cold: the recovery of waste heat for air-conditioning (to be published).

Elahee, K., Ramatally, W., 2013. Sustainability criteria for energy for Mauritius (to be published).

Le Mauricien, 2012. http://www.lemauricien.com/article/joel-rosnay-%C2%AB%C2%A0maurice-peut-atteindre-l%E2%80%99autonomie-energetique-en-2040%C2%A0%C2%BB >. (03.12.12.).

MoFED, 2012. Ministry of Finance and Economic Development, budget speech 2013. Available from http://www.gov.mu/portal/goc/mof/files/20122013/BudgetSpeech2013.pdf. (03.12.12.).

Pandoo, R., 2012. Waste heat for ar-conditioning in Mauritius. B.Eng (Hons) dissertation, Faculty of Engineering. The University of Mauritius.

UN-SEFA, 2012. Sustainable energy for all newsletter. Available from: <http://globalproblems-globalsolutions-files.org/gpgs_files/SEFA/sefa%20july%202012%20email.html.>. (03.12.12.).

World Bank, 2012. Turning down the heat. Available from: <http://climatechange.worldbank.org/sites/default/files/Turn_Down_the_Heat_Executive_Summary_English.pdf.>. (03.12.12.).

CHAPTER 13

Energy System of the Baltic States and its Development

Jurate Sliogeriene
Associate Professor at Vilnius Gediminas Technical University, Department of Construction Economics and Property Management, Lietuva, Vilnius, Lithuania

Contents

Infrastructure of the Energy Industry and Its Economic Significance	306
Objectives of Sustainable Development of the Energy Industry	309
Energy System of the Baltic Countries	316
Estonian Energy System, Its Development Objectives	318
Latvian Energy System, Its Development Objectives	322
Lithuanian Energy System, Its Development Objectives	327
Renewable Energy Production Technologies in the Baltic States: The Case of Lithuania	333
Use of Hydropower	334
Development of Wind Energy Industry	336
Use of Biomass in the Energy Sector	339
Solar Energy Industry in Lithuania	341
Use of Geothermal Energy in Lithuania	343

As most former Soviet-bloc countries, the Baltic states—Lithuania, Latvia, and Estonia—have inherited a technologically inefficient and resource-guzzling centralized energy sector, which relies on Russia's natural resources. Although these countries are not rich in traditional energy resources (oil, gas, and coal), all their principal energy production technologies rely on these resources and thus their imports. The supply of energy resources from Russia has therefore become a means to exert economic pressure and demand for political concessions, while growing resource prices have become a source of social exclusion and poverty. Moves to achieve energy independence and integrate into international economic, energy, and other structures demand a massive overhaul of the systems and development of new infrastructure to ensure energy production from independent sources. Today the Baltic states must make a decision: to continue relying on ever-pricier resources, to choose and develop nuclear energy production technologies received controversially by the public, or to bet their future on constantly improving renewable energy technologies. This issue is now paramount in attempts to determine the future of the Baltic states' energy systems. Joint efforts and a solution acceptable to all three countries would be the most effective way to handle the issue of national energy security. Energy is "business without borders." Whether it is securing primary resources, supplying the produced energy to the market, or handling such issues as the effect of energy facilities on the environment, countries depend on each other. Keeping in mind that handling of energy issues individually is difficult or even impossible, countries must find ways to collaborate and establish joint structures that ensure economic and political support.

As countries are looking for the best approaches to the increasing energy demand and secure supply, they come to realize that technological parameters and economic efficiency are not critical in securing welfare. A higher regard for the environment and the quality of life demand solutions that would focus more on public attitudes and accepted values, public favor or hostility toward technologies, and the role of a community's self-determination in the process of important decision making. Any energy system operates to serve the needs of society as a whole; the decisions on its development, choice of technologies, and operation must therefore respect environment protection and public values. Any energy infrastructure must be developed and long-term decisions made looking for ways to link economic benefits, public attitudes, and technological solutions. A promising approach is the analysis of energy-sector development based on the choice of technologies that are the most acceptable to people rather than the most efficient. New and innovative energy production technologies—whether we speak about the energy supply to industry, their use in transport, or for individual needs—are the main tools that can help reach the objectives of the sustainable development of society.

This part reviews the Baltic states' energy sector, its principal energy production technologies, and the objectives of integration into the European energy system. It also describes the development of renewable energy production technologies in these countries and the place of these technologies in energy systems

INFRASTRUCTURE OF THE ENERGY INDUSTRY AND ITS ECONOMIC SIGNIFICANCE

The energy industry is a complex field of the economy, having huge impacts on a country's economics, dynamics of growth, rational distribution of investments, and payment balance. It covers several interrelated systems:

electrical power industry, production and supply of central heating/cooling, and supply of primary energy sources—oil and its products, natural gas, etc. These are the areas that consume most of the energy resources. The key consumers of primary energy resources are the sectors of electrical energy production and central heating/cooling.

The electrical power industry has received special attention in the past decades. Expenses for building the energy industry infrastructure, export and import of energy resources, and their processing products and energy prices for the end user have very strong influences on the development of economies, making tendencies of economic indicators for the energy sector the primary index of an economy's future development. The energy industry, with its specific technological, economic, social, ecological, and other aspects, is integrated closely into the key areas of social life: politics, education, science, economic and social spheres, and environmental protection. The economic community of today's countries is undergoing rapid integration and forming a single economic system, where the energy industry's infrastructure is becoming a component part and linking element of the economy. Due to its exceptional significance for the economic and social systems, the performance of energy systems is the object of state regulation in many countries.

Energy sector enterprises have one common feature: they perform a very specific function within a particular territory, using a complex infrastructure that has been built specially for them. Infrastructures of energy industry enterprises are created during the long-time process of their performance and make up the major part of the structure of the enterprise's capital. Due to big capital investments, it is not so simple to compete with these enterprises; in most cases, such competition would not be a well-reasoned deal. In most territories or regions, energy industry enterprises operating there serve the needs of the economy and residents of this particular territory. However, significant shifts in the European Union's (EU) energy policy, application of market tools within the energy industry, development of renewable resource technologies, and investments by private capital into the infrastructure call for fundamental changes to the energy industry. To increase performance efficiency, energy enterprises began applying multiple forms of ownership and management models, and there have been processes of privatization and segmentation of enterprises as well as separation of activities. The goals outlined for reforming the energy industry provide measures for establishing competition, ensuring transparency of activities, availability of resources, secure supply, and lesser reliance on external suppliers.

Due to an insufficiently developed infrastructure, the supply tends to be very limited, systemic failures occur, and there are delays in meeting growing industrial and consumer demands. For that reason, the focus of energy enterprises should be on building and developing purpose-specific infrastructures and more active application of multiple forms of ownership, thus increasing the concentration of capital resources and achieving higher levels of performance efficiency. The Energy Commission of the European Union holds the view that the level of investments into infrastructures of electricity transmission networks in EU member states is insufficient, which, in turn, makes the return of investments into networks low. EU promotes development of the infrastructure by financing projects for trans-European electricity networks. It is a common consideration that in order to achieve effective management of the energy industry, the main focus should be on the management of energy demand and measures promoting investments into the infrastructure, namely transborder connections, which are the sole prerequisite for successful operation of the EU's internal energy market (Bačauskas 1999). The total budget of the European Grid Infrastructure for 2014–2020 will be 50 billion EUR, from which 9.1 billion EUR will be allocated for the development of the energy infrastructure.

The most effective means to balance out fluctuations in energy production, that is, its lack or excess (particularly when renewable resource technologies are applied), is more extensive integration into international energy systems. This makes energy industry systems to associate themselves into systemic alliances—both physically and from the management point of view. Managers of 36 transmission systems operators (TSO) from 31 countries signed a declaration in Prague in 2008 concerning establishment of the European Network of Transmission System Operators for Electricity (ENTSO-E). ENTSO-E seeks to harmonize efforts of TSOs in creating a single European energy market. As far back as 2009, ENTSO-E comprised 42 transmission systems operators from 34 countries, covering five synchronic zones, including the three Baltic states. The goal is closer partnership between energy systems in creating a single European energy market. Naturally, physical laws that regulate the functioning of energy industry systems remain the same: the final integration will be achieved only after new connections have joined continental networks and started functioning synchronically within the network of continental Europe. Currently, energy systems of the Baltic countries function synchronously with the Integrated Power System (IPS)/Unified Power System (UPS) of Russia system, connecting systems of Russia, Belarus, and the Baltic region. Management of the electrical energy system of the Baltic countries within the IPS/UPS system is centralized and coordinated from Russia. Such a situation with energy systems of the Baltic countries does not meet the EU requirements for electrical energy systems of the European Union.

Today's Europe encounters increase in energy consumption and its potential lack. The more powerful the economic potential of the country, the more energy it produces and consumes. In this framework, in order to achieve economic and social goals targeted at ensuring the well-being of the population, the major focus should inevitably be on developing the energy industry. The energy industry is a fundamental part of the whole infrastructure designed for "serving the entire reproduction process and permitting this process" (Webster's Third New International Dictionary 1976). In today's knowledge-based economy, the role of the energy industry is even more increasing: systemic failures in energy systems may damage most of the technologies indispensable for successful functioning of the society. Furthermore, due to the progress of other technology fields, new consumers emerge, for example, telecommunications and electrical transport means. Almost all the multiple level scenarios for energy industry's development are forecasting increase in electrical energy consumption (Gwartney, Stroup & Soubel 1997).

The energy industry is responsible for most of the environmental pollution and carbon dioxide emission into the atmosphere. There are serious environmental protection tasks that the electrical power industry is supposed to face in the future: gradual transition to more efficient and "cleaner" technologies and setting up technical facilities for decreasing carbon dioxide emission within the transport and heating/cooling systems.

OBJECTIVES OF SUSTAINABLE DEVELOPMENT OF THE ENERGY INDUSTRY

The *Europe 2020* strategy agreed by the European Commission on March 26, 2010 puts forward five flagship initiatives, which are of primary significance for achieving sustainable development, social cohesion, and economic convergence of European nations. One of these initiatives, "20-20-20," is targeted at transforming energy industry systems. The targets to be met by 2020 are the following: reduction of greenhouse gas emissions by 20%, coming of 20% of the energy consumption from renewable sources, and improving energy efficiency by 20%. Political and regulatory measures are not sufficient for achieving these goals. There should be efforts for modernizing the energy sector, accumulating financial resources, attracting and channeling necessary investments, using the scientific potential, and entering into a useful dialogue with social partners. The following major activity areas will have to be addressed when transforming and developing energy industry's infrastructure: (1) diversification of energy production sources and ensuring security and technical safety of the energy supply; (2) promotion of the use of local resources best corresponding to the natural environment and expansion of the network; (3) implementing means for saving energy and its efficient use; (4) collaboration of the private and public capital and promotion of private investments into the energy sector; and (5) application of new and more effective management models for energy systems.

1. *Diversification of energy production technologies.* Many countries have been reconsidering their priorities for the energy industry in recent years as a result of their efforts to reduce reliance on ever-pricier imported energy resources and stock, and awareness of negative environmental impacts of fossil fuels. Tangible climate changes and diminishing natural resources, as well as the Chernobyl and Fukushima Daiichi nuclear disasters and their severe outcome, stimulated interest in renewable resource technologies. It has become popular to highlight deficiencies and threats of old fossil fuel-based technologies and emphasize advantages of alternative energy technologies. This tendency is often caused by society's attitude and views, which are not always based on factual information. Such countries as Germany, Japan, Italy, and Switzerland, under the pressure of social opinion, no longer use nuclear technologies and have started navigating toward renewable and environmentally friendly power engineering technologies. Still, such solutions do not fit some countries, and for others, they may cause an economic burden. There is no reason to entirely deny advantages of nuclear energy technologies, especially for countries that possess limited primary energy resources. Nuclear energy will have its place in the future, as it still serves as a major source of electrical power with low carbon emissions. In fact, there is no one preferred technology—all of them have certain strengths and weaknesses, and all energy sources are able to compete in the market without any additional support.

    Currently, European countries do not possess technologies allowing them to reduce significantly the use of fossil fuels within the energy industry, namely the electrical power industry and heating/cooling systems. At present, fossil fuels account for 79% of the EU's gross inland consumption (see Figure 1).

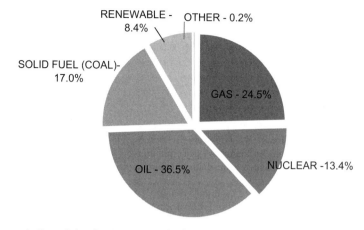

Figure 1 Gross inland consumption by fuel, EU-27, 2008. *(Source: Eurostat, 2010).*

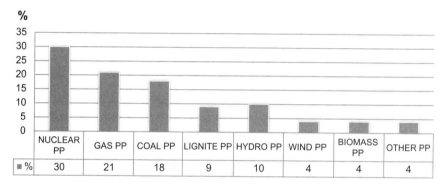

Figure 2 Mix of EU's energy sources in 2010. *(Source: Eurostat, 2011).*

All EU countries have focused their political, economic, and technological resources on gradual reduction of the use of fossil fuel technologies and advancement of renewable resource technologies. Governments are devising and implementing various mechanisms, motivating designers and manufacturers of new technologies and investors to change their approach to both consumption of energy within production processes and development of power engineering technologies—they are encouraged to create new more energetically efficient and environmentally friendly technologies and enter the market of the changing energy sector. Use of renewables in the electrical power industry does not much exceed 18%. Figure 2 presents the mix of EU's energy sources in 2010. Use of renewable energy resources has a rich potential and may presumably result in supplying comparatively clean and, what is very important, mostly local energy. Despite the fact that technologies for the use of renewables in power engineering are still more expensive than traditional production methods, use of these resources is expanding. Some facilities of wind, small hydro, and geothermal energy are starting to compete in wholesale energy markets. Other facilities, such as photodetectors, solar water heaters, and biomass, if subsidized appropriately, may provide services in regions with insufficient energy networks. Hydro and wind energy are still predominating within the renewable energy industry. Figure 3 shows the contribution of renewables to electricity production in 2009.

The Baltic states as well will have to contribute to the development of renewable resource technology by applying various promotion tools. Table 1 presents renewable energy action plans of the Baltic states.

2. *Development of local resources, integration, and centralized systems.* A major precondition for a sustainable and safe energy system is increased consumption of energy from renewable resources. Countries differ not only in their natural resources, but also possibilities for applying renewable

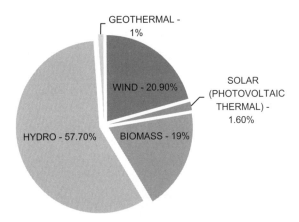

Figure 3 Contribution of renewable resources to electricity production, EU-27, 2009. *(Source: Eurostat, 2010).*

Table 1 Projected % Share of Gross Final Electricity Consumption in National Renewable Energy Action Plans

Country	2010	2015	2020
Estonia	1.7%	3.5%	4.8%
Latvia	44.7%	51.4%	59.8%
Lithuania	8.0%	17.0%	21.0%

(Source: Beurskens & Hekkenberg 2011)

resource technologies and production methods. Every country chooses a production method that is most acceptable and economically reasonable. Renewables, however, are not able to guarantee the needed amounts of energy. Technologies enabling to accumulate the produced energy are still under development and expensive. In order to be able to use electrical energy produced from renewable resources, it is necessary to modernize the infrastructure of energy distribution so that increased electricity demands arising from multiple dispersed sources (e.g., solar, photovoltaic, or wind energy) could be met. Electrical power engineering is becoming more decentralized, and innovative electricity networks are being developed. A more innovative energy transmission and distribution network will be able to balance fluctuations of wind and solar energy production by using renewable resource energy produced at other locations in Europe. This could reduce the necessity for the storage of energy, expensive reserve capacities, and basic supply.

3. *Saving energy and management of demand.* It is the major priority for tomorrow's energy industry. More efficient consumption of energy brings about a whole range of benefits—saving natural resources and decreasing the emission of carbon dioxide and other pollutants, as well as optimizing the use of financial, material, and other resources within the energy sector.

In 2000, fuel expenses of the Baltic states accounted for 9.1% of the gross national product (GNP), and in 2010, this figure reached 13%. This exceeds the expenses for fuel of other EU countries significantly. Experts predict that if the prices for fossil fuels keep rising, by 2020, the Baltic states will be forced to spend more on energy resources, thus jeopardizing the efficiency of their economies and long-term development. The reason for high prices within the economies of the Baltic region and their reliance on imported energy is an extremely low efficiency of energy consumption.

Numerous tools exist for increasing the efficiency of energy consumption. One of the most powerful is increased energy efficiency of buildings. Buildings with almost zero energy must be a norm. More efficient vehicles should be employed within the transport sector. Products and facilities should conform to the highest energy economy standards, and new smart technologies, such as for household automation, should receive broader application. Motivating people to change their consumption behavior is, nevertheless, the primary measure. Responsible behavior, positive environmental attitudes, and conscious consumption—these are the most effective measures for limiting the scope of energy demands. It is estimated that by applying saving measures, the total energy consumption in 2020 may be 17% lower than it was in 2009. Households and the transport sector have the largest propensity to save energy: total savings could amount to 65% (Steimikiene & Volochvich 2010). Figure 4 presents data on final energy consumption essential for producing one GNP unit.

As evident from Figure 4, Lithuania's energy intensity exceeds the EU average by 2.69 times. All three Baltic countries (Lithuania, Latvia, and Estonia) should implement measures for saving energy.

4. *New ways for energy management.* Because production of electricity from renewable resources is increasing, which makes a steady supply of energy a risk, one of the goals to be achieved is ensuring flexibility of resources for electrical energy production—production, storage, and management of demands should be flexible. Another goal is related to the impact that production of such energy has on wholesale market prices. Under increasing competititon, prices of energy from renewable resources may start falling and facilities for generating renewable resources may attract fewer investments. Certain measures should be applied for ensuring return of the invested capital, and existing control mechanisms should stimulate investments into the renewable energy industry and contribute to creating new capacities (Sugolov, Dodonov & Hirschhausen 2003).

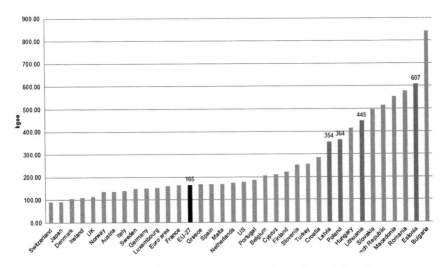

Figure 4 Energy intensity of the economy. *(Source:* http://www.lithuanian-economy.net/, *Eurostat 2010).*

Development of the renewable energy industry requires coordination: it is essential to consider solutions taken by neighboring countries and their impact on the national electricity production system. Joint effort helps lower costs and ensure reliable supply, whereas closed systems reduce security and efficiency of supply. The electricity market should be organized in such a way that makes it possible to perform all activities concerning flexible supply and management of energy demands, as well as storage and production of energy within the market.

5. *Participation of the state in the management of energy systems.* Many experts on infrastructure are of the opinion that the infrastructure is an important segment of the social life and that the principal subject of its development policy is the state, very often taking the responsibility for its management and owning the infrastructure objects or entire systems. A management model based on centralized planning and state ownership does not always justify itself due to its delayed response to shifts and insufficient economic efficiency. Still, new market-based energy industry management models have their own drawbacks. Not all infrastructure systems are able to function within the economic market system without the state's interference and regulation. To achieve economic changes, state-level initiatives for realization of reforms are needed as well as appropriate mechanisms for market regulation. This makes many countries discuss issues of infrastructure management and ownership intensively (Guthey Clark & Jackson 2008; Moreau 2004; Oss, Zeltina & Zeltins 2003).

The degree of state regulation within the energy sector is very high due to the impact on the political and economic systems. Succesful functioning and development of the energy industry require political agreements and great capital resources, and participation of the state in the sector's activities brings certain guarantees and ensures the viability of development projects. The scope of application of the model for partnerhip between public and private sectors has been expanding lately. Although the state's influence within the management of the energy sector is not likely to diminish, private sector investors will retain their significance within the concept for the market-oriented energy policy.

Implementation of sustainable development is becoming a distinct priority for the energy industry. Sustainable energy systems are the reality that is attainable, although requiring a long-time process with plenty of urgent issues that need to be solved. Growing energy costs call for increasing energy efficiency and investments; they also restrict the availability of modern energy services for the entire population and impede the state's efforts to ensure development of a competitive economy. The key long-time objective for the energy industry should be a gradual transition from today's most popular organic fuel technologies, which are wasting natural resources, toward new renewables-based technologies. Major priorities within this long-time process are education, promotion of responsible business within the sector, and energy culture.

ENERGY SYSTEM OF THE BALTIC COUNTRIES

The aim of every country is to ensure a stable and reliable supply of energy resources. It is a huge political, economic, and social task for countries not rich in organic fuel. Even small countries are subject to numerous factors in developing their energy sectors—globalization processes, growth of prices for energy resources, increasing energy demands, progressing technologies, international agreements, environmental commitments, etc. It is a common view that resource prices and environmental issues will be the deciding factors in selecting development scenarios for the energy sector in the future.

The Baltic countries (Estonia, Latvia, and Lithuania) have inherited a technologically inefficient and resource-guzzling centralized energy sector relying on Russia's natural resources. The supply of energy resources from Russia has become a means to exert economic pressure and demand political concessions. Moves to achieve energetic independence and integrate into EU's and other international energetic and economic structures require a massive overhaul of the system to ensure energy production from independent sources. Governments and societies of the Baltic states must make a decision: to continue relying on ever-pricier resources and develop nuclear energy production technologies received controversially by the public or bet their future on constantly improving renewable energy technologies. These issues are now paramount in attempts to determine the future of energy systems of the Baltic states. A lack of infrastructural connections with Scandinavia and continental Europe isolates Baltic energy users from European markets; therefore, the Baltic region is still considered an "energy island" within the context of European energy integration.

A major direction of the EU energy policy is creating a single European energy market, starting with regional markets. Conclusions of the European Council of February 4, 2011 provide for creating a single European energy market and allowing a free flow of electricity. No EU member state should remain isolated from the European gas and electricity networks after 2015 or see its energy security jeopardized by lack of the appropriate connections. The energetic safety of the Baltic states has been a serious concern since regaining their sovereignty in 1991. In fact, their energy systems rely on a single supplier of oil, electricity, and gas—Russia. Since 1996, there has been investigation into the possibility of creating an energy market of the Baltic region covering Scandinavia, Germany, Poland, the Baltic states, Belarus, and Poland. Such a market is feasible only if appropriate connections are built between the Baltic and Scandinavian countries, as well as between Lithuania and Poland. New connections alone will not be able to entirely abolish energetic isolation of the Baltics—for this purpose, energy systems of these countries must become equal players within the European electricity infrastructure, market, and system, that is, start functioning synchronically within the network of continental Europe (Energy policy strategies of the Baltic Sea region for the post-Kyoto period 2020; Menezes 2009).

Due to their smallness and some shared history, the Baltic countries are often referred to collectively. Although closely allied geographically, these countries are dissimilar in their natural resources, as well as energy systems, differing in their generating sources and management structures. There is one eloquent fact—in June 2012, Lithuania imported 66% of the needed electricity and Latvia 35%, whereas Estonia consumed only 72% of the generated electricity, exporting the surplus to Latvia and Lithuania. Despite their differences, these countries have one goal in common: building connections with European energy systems and development of new sources of energy production. Figure 5 shows on the left side the energy grid of the Baltics and on the right side the power industry system of this region within the IPS/UPS system.

Against this background, issues of the projected nuclear plant in Visaginas, the biggest power capacity within the Baltics system, are being solved. As far back as 2006, heads of states of Lithuania, Latvia, and Estonia signed a comminication on cooperation in constructing a new nuclear plant in Lithuania. This new plant should cushion the energy industry's negative impacts on the import/export balance of the three Baltic states and ensure the safety of power transmission. The new power plant will require new connections between Lithuania and neighboring EU countries, as well as a closer partnership and new agreements with neighboring energy systems concerning power reservation and other systemic services. The energy industry may also be seen as a physical link binding the Baltic countries firmly. Despite the slow down of the global economy, the three Baltic states, with a population of only 6.7 million, are aiming at greater integration into

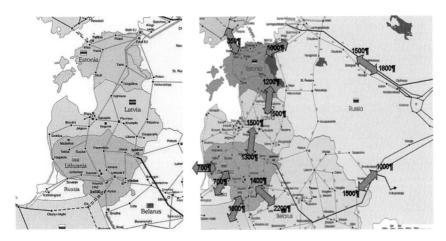

Figure 5 Transmission grid of the Baltic states, main connections. *(Source:* http://www.lrv.lt/EP/sinchronizacija.pdf*).*

the international community, and the accomplished energy projects will contribute largely to this process. The energy systems of all three Baltic states are now characterized briefly.

Estonian Energy System, its Development Objectives

The Estonian economy is highly dependent on fossil fuels. Approximately 90% of Estonia's energy is produced through the combustion of fossil fuels (Renewable energy policy review Estonia, 2009). Estonia, as well as other Baltic countries, imports the major part of its energy from Russia. Estonia's market of oil products is not regulated, open, and noncompetitive. It imports natural gas both directly from Russia and Latvia's Inčukalns underground gas repository. The fuel and energy sector is under Estonia's Ministry of Economic Affairs and Communications (Figure 6).

The share of fossil fuels in the mix of primary energy consumption is large—approximately 90%. Estonia has no local oil, natural gas, or coal. There are no oil processing enterprises either. All oil products are imported. Estonia relies on imported energy sources by about 40%.

Estonia is the only country in the Baltics that possesses oil shale, which is a significant source for electrical power production. It is the world's only country that uses oil shale as a primary energy source. Oil shale is a solid fuel characterized by a low calorific value, although its profitability is high. Oil shale is mined in the northeastern part of Estonia at a depth of 10–70 meters. Eesti Energia is the largest oil shale processing company in the

EU member since: 2004
Member of Schengen Area
Political system: Republic
Capital: Tallinn
Area: 45 000 km²
Population: 1.340 million
Estonia, the northernmost of the Baltic countries, is situated on the eastern coat of the Baltic Sea. Its climate is determined by its northern position, the Baltic Sea and the northern part of the Atlantic Ocean.
It is bordered to the south by Latvia and to the east, by the Russian Federation. Across the Baltic Sea, lies Sweden and Finland.
Estonia is one the smallest countries of Europe.

Figure 6 Main data about Estonia. (*Source:* www.estonia-eu.com, http://data.worldbank.org/country/estonia).

world, using around 15 million tons of oil shale per year for energy production. Oil shale can be used directly as a fuel for producing energy or synthetic oil. Oil shale is not well known as a fuel for energy production because it has been used much less than coal and crude oil, although the high price and diminishing reserves of crude oil and economic growth have led to interest in its possible wider use. The possibility to complement primary resources with native fuel, that is, oil shale, makes Estonia less reliant on imported resources.

Estonia's energy sector is dominated by a single entirely state-owned energy company—Eesti Energia. It is a vertically integrated joint stock company involved in producing, distributing, and transmitting electricity and providing other energy sector-related services throughout the whole of Estonia through subsidiary companies.

The Estonian electricity system comprises Estonia's power stations, network operators, and electricity consumers. This system, in turn, is part of a larger synchronized united system, BRELL, which controls AC power lines connecting Estonia to the neighboring Latvia and Russia and, through them, with their neighbors Lithuania and Belarus.

Currently, Eesti Energia comprises 23 companies, including oil shale mining enterprises. Eesti Energy produces about 98% of Estonia's electricity. Data on major power plants are given in Table 2.

The production of electricity in Estonia is based predominantly on domestic oil shale. Electricity produced from this mineral approached 95% of the total electrical energy supplied to the Estonian electrical network in 2008 (Wind Power 2010). The two largest power plants together produce

Table 2 Estonian Power Plants, 2010

Power plant	Installed capacity (MW)	Fuel
AS NarvaElektrijaamad	2000	Oil shale
Iru Power Plant (CHP)	156	Gas
Other power plants	146	Oil shale, gas
Wind turbines	140	Wind
Hydro PP	4	Water
Total in Estonia	2446	

(*Source:* Eesti Energia, AS, https://www.energia.ee/et/avaleht)

about 9 TWh of electricity. An average person consumes around 1200 kWh of electricity per year; consequently, the electricity produced in Estonia could cover the demand of 7.5 million people. The Narva Power Plant provides around 95% of the electrical energy consumed in Estonia and supplies the whole city of Narva with heat. The Eesti and Balti power plants, situated in Narva, are the world's largest power stations running on oil shale with a general capacity of about 9 TWh of electricity annually.

The generating capacity based on renewable resources includes 140 MW of wind energy and 4 MW of hydropower. Estonia's potential for renewable energy lies in heat and power generation based on wind power and bioenergy and in small-scale hydropower. One of the most important primary energy resources in Estonia is wood (Renewable energy policy review: Estonia 2009; Wind power in Estonia 2010). Forests cover more than half of Estonia's territory. Peat is also a significant source for local fuel. Figure 7 shows the mix of Estonia's renewable resource energy.

Estonia's transmission and distribution networks are also owned by Eesti Energia. The main grid of the Estonian electricity system comprises 1540 km of 330-kV lines, 184 km of 220-kV lines, and 3476 km of 110-kV lines. The Estonian electricity system is well integrated into the systems of other Baltic states and Russia. Estlink, the current direct undersea cable between Estonia and Finland, has been in operation since the end of 2006. It could serve as a symbol of convergence of power systems of the Baltic and Nordic states. Construction of the Estlink 2 undersea cable between Estonia and Finland began in 2011. Estlink 2 will be the second undersea cable between Estonia and Finland, stretching for 170 km, with about 140 km under water. New connection will start functioning in 2014. It will increase Estonia's reliability substantially and enable export of electricity produced in Estonia to the Nordic countries (Elering, AS, http://www.elering.ee/en/).

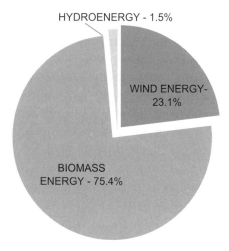

Figure 7 Estonia's renewable energy mix.

Currently, Estonia is the only exporter of electricity in the region. It exports up to 40% of the generated electricity, mostly to Latvia and Russia. Since January 1, 2013, all of Estonia will be buying electricity from the open market. Electricity in Estonia is one of the last remaining products where the price is regulated by the government, with no choice left for the customer.

Like other Baltic countries, Estonia inherited an inefficient extensive centralized heating system. Its successful operation is guaranteed by thermal power plants and boiler stations. Transformation of the heat industry involves a whole range of measures: renovation of networks, more extensive use of renewables for heating, increased energetic efficiency of buildings, and innovative solutions for management of the heating industry.

Despite the fact that the Estonian energy industry is entirely state owned and market tools are still in their embryo stage within the sector, the system operates quite effectively. Nevertheless, during preparation for entering the market of continental Europe and meeting the requirements of the EU's Third Package, the Estonian energy sector will have to be liberalized. In addition to the newly established infrastructure, integration of the electrical energy sector will involve changes in the approach to system management, splitting of the production, transmission and distribution activities, and the property into independent companies. This transformation is essential for operating synchronically with European energy networks.

Latvian Energy System, Its Development Objectives

For some historical reasons, Latvia has always relied greatly on import. At the initial period of its sovereignty, 1900–1995, Latvia was forced to import no less than 80–90% of energy resources in order to satisfy the needs of the industry and energy sector. Latvia is poor in natural resources. It possesses abundant resources of building raw material: limestone, gypsum, dolomite, clay, and gravel. Among the raw material essential for the energy sector, there are peat and wood (see Figure 8).

The model of the Latvian energy sector consists of three main blocks: resources, production, and consumption. Figure 9 presents the architecture of this model (Skribans 2010).

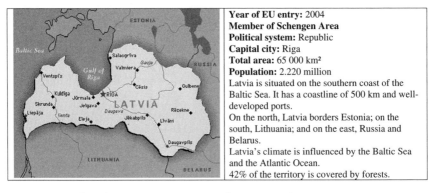

Figure 8 Main data about Latvia. *(Source:* www.latvia-eu.com*).*

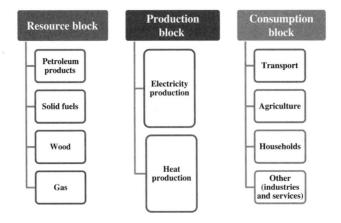

Figure 9 Main blocks of Latvia's energy sector (Skribans 2010).

Latvia is situated at a strategic point—at the crossroads of the east–west energy trade. Large amounts of crude oil and products of refined oil are transported through Latvia by pipelines and railway—the ports of Ventspils, Liepaja, and Riga. Differently than Lithuania and Estonia, Latvia's energy sector has a significant advantage —one of the largest (the third) in Europe gas storage facilities inherited from the former USSR, the Inčukalns UGS. Although having no natural energy resources, Latvia is able to completely satisfy its annual demand and re-export some of the surplus gas into neighboring countries due to its underground gas reservoirs. The Inčukalns gas repository allows the Latvian government to sell gas to Latvian consumers for a slightly lower price than to all others. The total volume of its repositories is 4.44 billion m^3, and gradual expansion of their capacity is on the agenda. The pipeline network consists of 1255 km of main gas pipelines, 47 pressure regulating stations, a border gas measurement station, and other necessary infrastructure. The capacity of the pipeline network is sufficient to cover the demand of all power plants, both small and medium, almost throughout the whole of Latvia's territory. After the pipeline from Russia to Germany was built under the Baltic Sea, the significance of the Inčukalns repository increased even more (International Energy Agency 2011). Naturally, the main natural gas supplier to Latvia is Russia. The average import of natural gas from Russia is 1.3–1.6 billion m^3.

Natural gas is one of the main fuel sources in Latvia, making some 32% of the total demand for energy. Latvia's gas consumption depends on several key consumers, the largest of which is the national electrical power company AS Latvenergo. Latvenergo consumes 34% of all gas that is sold to Latvia and supplies its two main thermal power plants with gas.

Latvia imports fuel oil and other oil products mainly from Russia and the Commonwealth of Independent States. The major drawback of such import is unpredictability of costs. Fuel oil is used mainly for generating heat in small thermal power stations and boiler stations. Fuel oil competes with gas, although its use brings about some quite serious technical and environmental problems. It is required that fuel oil should contain very small quantities of sulfur (less than 1%). In addition to fuel oil, Latvia imports small amounts of shale oil from Estonia and coal from Russia, Kazakhstan, Ukraine, and Poland. Coal is used in small boiler stations and for household needs.

Latvia has a centralized system of heat industry, which is quite efficient due to the large concentration of the population in major cities. For example, in Riga, even 75–80% of heat is supplied in a centralized way from thermal power stations. In other cities this percentage is lower—approximately 57%. In rural areas, it is only 4%. Riga's centralized heating system is one of the most innovative and efficient in Latvia.

Although Latvia inherited an energy sector that is largely reliant on import, it underwent major changes in recent years. Significant increase in the consumption of local resources, mainly wind and biomass (wood), allowed decreasing the share of imported energy resources by 20–25%. Another major change has been a decreased consumption of fuel oil and coal. These resources have been replaced by a cleaner fuel—natural gas. Figure 10 shows the mix of primary energy consumption in Latvia in 2008.

Currently, the local fuel used for heating is mainly wood and peat in Latvia. Because both fuels are powdery material of low heat value, their competing advantages are short transportation distances and production of highly efficient fuel.

Fuel wood is the main local energy source in Latvia, making 28% of the total energy consumption. Export prices of wood exceed the export-stimulating local market prices. Furthermore, Latvia has a sufficiently developed industry of processing wood waste into products suitable for fuel use, that is, wood pellet, which goes for export.

Latvia produces about 500,000 metric tons of peat each year. All peat production companies in Latvia have been privatized, although most of the peat deposits are still owned by municipal governments, which rent the deposits for extraction. There are now some 25 peat extraction companies

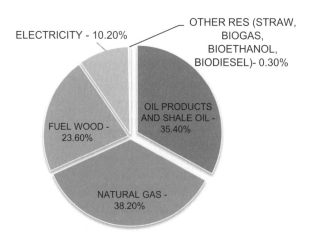

Figure 10 Mix of primary energy consumption in Latvia in 2008 (Renewable energy policy review 2009).

in Latvia. Peat covers about 10% of Latvia's territory, with the heaviest concentration in the eastern plains near Riga. Although peat is seen as an important local resource in Latvia, its consumption is rather limited. All thermal power stations have abandoned peat in favor of a more efficient fuel, that is, gas. Several investigations have been carried out and recommendations prepared concerning the use of peat in smaller plants, although not one of them has been implemented yet.

The predominant company within the electrical power industry sector is the state-owned AS Latvenergo. It owns all main generating sources and the system of power supply and distribution. The operator of transmission systems JSC Augustsprieguma is also subjected to Latvenergo. According to Latvia's law on energy, Latvenergo, together with its subsidiaries, must be state owned; hence it must not be privatized or separated into units. It makes liberalization of the Latvian electrical energy sector rather complicated.

Latvenergo produces about 65% of all electrical energy consumed in Latvia. Approximately 30% of energy is imported and 5% is produced by small generating units, basically windmills. Latvia gets almost all its electricity from hydropower (62% in 2009) and gas (36%). Wind and biomass account for 1% each (AS "Latvenergo", http://www.latvenergo.lv/)(see Figure 11). Electricity generation from coal and oil stopped in 2004.

The electrical energy generation structure of Latvenergo comprises two types of energy-generating sources (AS "Latvenergo"):
1. Two main cogeneration gas plants: Riga Thermal Power Station 1 (electricity capacity up to 144 MW and heat capacity up to 616 MW) and

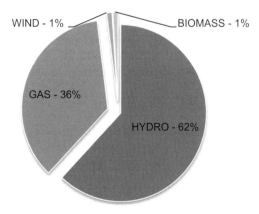

Figure 11 Structure of sources of electricity production, 2009. *(Source: Central Statistical Bureau of Latvia).*

Table 3 Latvian power plants, 2010

Power plant	Installed capacity(MW)	Fuel
Pļaviņas Hydro Power Plant	883.5	Water
Riga Thermal Power Station 2	600	Gas
Riga Thermal Power Station 1	144	Gas
Riga a Hydro Power Plant	402	Water
Kegums Hydro Power Plant	264	Water
Liepaja Thermal Power Station	12	Wind
Grobiņa Wind Farm	9.6	Wind
Liepaja Wind Farm	2.0	Wind
Ainiži Wind Farm	1.2	Wind
Aiviekste Hydro Power Plant	0.8	Water
(Approximately 150 "small" HPP)	<5 each	Water
Total in Latvia	2319.1	

(Source AS Latvenergo)

Riga Thermal Power Station 2 (electricity capacity up to 390 MW and heat capacity up to 1279 MW), as well as several smaller regional boiler stations.

2. Power plants using renewable sources (hydro, bio, solar, and wind): Plavinas HPP, Riga HPP, Kegums HPP, Aiviekste HPP, Ainazi HPP, etc.

Table 3 presents Latvia's main power-generating sources in 2010.

The role of renewables within Latvia's power industry is very significant. Hydropower is this sector's predominant energy resource. Latvia has also begun constructing a solar power plant in Daugavpils. The Latvian hydropower industry can be divided into two groups:

1. large hydropower plants on the Daugava River operated by Latvenergo
2. small units, having no impact on the installed capacity, but with drawbacks from an environmental point of view

The three largest hydropower stations of Latvia are on the Daugava River, forming a cascade of hydropower plants: Plavinas HPP, Kegums HPP, and Riga HPP. The latter generates about 70% of the total electricity of Latvia.

The Plavinas HPP is the most powerful hydrostation not only in Latvia, but also in the whole Baltic region. It generates the major part of Latvia's electricity. The distinctive feature of this plant that singles it out from most plants with such capacity is the compact nature of its structure and the fact that its water dams and generating facilities are situated in the same block. In 2001, the plant was fully modernized, thus increasing its capacity and efficiency.

The Riga HPP is the newest in the cascade and the second largest power plant in Latvia. In addition to generating electrical energy, it also serves as a synchronous compensator for regulating current and frequency within the energy system.

Small hydropower plants generate about 0.1% of Latvenergo's total electricity volume. One of them is Aiviekste HPP with a capacity of 0.8 MW.

Although reducing Latvia's reliance on foreign resource suppliers, use of hydropower, like other renewable resources, has its drawbacks. Water resources are harder to predict over the longer term. The total annual production of hydropower stations reaches about 2.7 TWh, which makes some 45% of the entire annual demand. During spring floods, which usually last a month or two, the cascade generates 40% of the total annual volume; when this brief period is over, Latvia has to use other generating capacities and start importing electrical energy.

Latvia started constructing the first and largest solar power plant in the Baltics in Daugavpils. Its construction will cost 11.5 million EUR. The area of its solar cells will be 36 480 m^2 and the capacity of its power facilities will be 4.6 MW. The forecast is that the plant will meet one-fifth of the demand of the city of Daugavpils. In 2010, construction was suspended due to financial difficulties.

Latvia's power transmission grid comprises 330- and 110-kV lines and substations. The 330-kV lines supply the main consumption units; they are also used for transmitting power to neighboring systems.

Still, the most sizeable and demanding tasks that Latvia will have to face are related to restructuring management of the power industry, notably the electricity generation sector, and liberalization of the electricity market. In order to reduce reliance on Russia's natural resources, Latvia is preparing to join the energy systems of continental Europe and build new facilities. Latvia is also involved, as a project partner, in construction of the new Visaginas nuclear power plant.

Lithuanian Energy System, Its Development Objectives

Lithuania inherited a well-developed energy system from the former Soviet Union oriented toward supplying industry throughout the country. After upheaval of the geopolitical and economic situation, the industry succumbed to competitive pressures of other countries. The energy system also lost its effectiveness and failed to answer the nation's needs. Lithuania's energy system is characterized by a major reliance on energy resources acquired from a single source, as the gas and electricity networks have no direct connections with energy systems of western Europe and Scandinavia. Following a provision of the EU entry agreement, Lithuania terminated

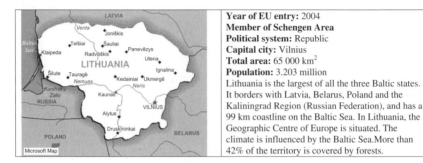

Figure 12 Main data about Lithuania. *(Source:* www.lithuania-eu.com*).*

exploitation of the Ignalina nuclear plant. Closure of this plant brought about negative social and economic consequences, making Lithuania reliant not only on primary energy resources, but also on the supply of the generated electricity. However, it stimulated reconsideration of the energy industry's priorities, built new generating facilities, developed a renewable energy industry, and prepared joining the west European energy networks. There was another innovation: active involvement of the private capital into the energy sector (Figure 12).

Lithuania's energy infrastructure is rather well developed, although there are almost no primary energy resources in its territory and limited possibilities for acquiring inexpensive resources; as a matter of fact, the potential of local renewables is not fully exploited either. Energy production, supply, and use are ineffective; the intensity of energy consumption is high, notably within the industry and household sector—almost two-thirds of Lithuania's population live in blocks of flats that are not renovated and built prior to 1990. Like in other Baltic countries, Lithuania's cities and towns have centralized heating systems; consequently, rising resource prices result in a heavy financial burden for the population. Experts are of the opinion that transformation and modernization of the centralized heating system should be set as the principal goal for achieving the well-being of the population.

Gas, oil, and coal reach Lithuania through a well-developed infrastructure connecting it with neighboring countries. Oil is supplied from Russia through the Klaipėda and Būtingė oil terminals. The Būtingė oil terminal has been built to provide the export and import of raw oil. Its capacity is, respectively, eight and six million tons of exported and imported raw oil.

The major oil supplier in Lithuania is the oil-processing plant Mažeikiųnafta, the only such plant in the Baltic region. Lithuania possesses small quantities of oil in its western part and the Baltic Sea. Oil shale gas has

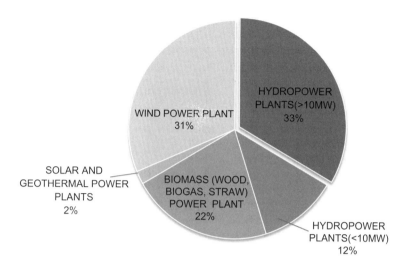

Figure 13 Lithuania's renewable energy mix, 2011. *(Source: JSC Lietuvosenergija, http://www.le.lt/en/).*

been discovered as well, although possibilities for its use and mining are still in the investigation stage. Extraction of local oil may be continued for several decades, providing the annual extraction level will not exceed 0.3–0.5 million tons.

Coal is transported by rail from Russia and Poland. Small quantities of coal are used in regional and local boiler stations and households. It makes up only about 1% of the primary energy mix.

Natural gas is imported by pipeline from Belarus, which, in turn, gets it from the oil fields of Siberia. The gas network of northern Lithuania is connected to the Latvian gas system. This provides an alternative for importing natural gas from Latvia's Inčukalns and Duobele underground repositories.

Recent years have witnessed a significant increase in the use of renewable resources. It must be noted that development of the renewable resource energy system is based entirely on initiatives of the private capital. The main renewable sources are wood, peat, and hydropower. The western part of Lithuania is rich in geothermal water; this potential, however, has not been exploited sufficiently. Development of renewables in the energy sector is related to implementation of EU and local programs and promotion schemes. Figure 13 presents Lithuania's renewable energy mix.

The Lithuanian energy system has been reorganized following EU requirements. Three principal blocks could be distinguished within this system: energy production (JSC Lietuvosenergija) comprising Lithuania's main

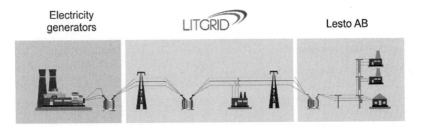

Figure 14 Lithuanian energy system. *(Source: JSC Litgrid, http://www.litgrid.eu).*

power stations; high-voltage electricity transmission lines and facilities, transmitting electricity to the biggest remote consumers, including the energy distribution company JSC Litgrid; and low and medium voltage electricity distribution network, transmitting energy to consumers (JSC Lesto). The main shareholder of these companies is the state. Figure 14 shows the architecture of Lithuania's electrical power industry.

As far back as 2009, when the Ignalina nuclear plant was still operating, Lithuania produced all the needed electricity, and some of it was exported. However, since 2010, when the nuclear plant closed, Lithuania has been importing 60% of total consumed electricity. At present, over 50% of electrical energy is imported from generating sources of EU countries situated at a distance more than 1000 km away from Lithuania. The main energy producer, Lietuvosenergija, generates only about 20% of the total demand of Lithuania (Lietuvosenergija, JSC). Table 4 presents the main energy-generating sources of Lithuania.

Commerce activities of Lietuvosenergija contribute about 27–28% to the total demand of Lithuania. Since establishing the electricity exchange market in 2010, all imported and exported electrical energy is traded exclusively in this market. The electricity market now involves more than 30 companies, which buy and sell energy through the exchange market. Nevertheless, ensuring competitiveness of the sector requires participation in the widest possible market. In this framework, the connection of Lithuanian and European markets acquires not only commercial significance—it is important from a physical point of view as well. It will be realized through building electricity connections between Lithuania and Sweden (in 2015) and Lithuania and Poland (in 2015). Figure 15 shows coverage of Lithuania's electricity demand.

Currently, the Lithuanian energy system encounters serious problems related to its inability to produce the needed amount of energy and cover the demand of the population. Generating sources operating in Lithuania are based mainly on burning imported gas or fuel oil and technologically

Table 4 Lithuanian power plants, 2010

Power plant	Installed capacity (MW)	Fuel
Lithuanian power plant (Elektrenai)	1800	Gas/oilGgasss s
Vilnius thermal power plant	360	Gas
Kaunas thermal power plant	170	Gas
Petrašiūnai power plant	8	Gas
Mažeikiai thermal power plant	160	Gas
Klaipėda thermal power plant	11	Gas
Panevėžys power plant	35	Gas
Other thermal power plants	96	Gas/oil/coal
Kauno hydropower Plant	101	Water/river dam
Small hydropower Plants	108	Water river dam
Kruonis hydro storage power plant	900	Water/accumulation tank
Wind farms	161	Wind
Biopower plants	45	Biomass (wood, straw)
Klaipėda geothermal power plant	35	Geothermal water/gas
Total Lithuania	3990	

(*Source:* Ministry of Energy of the Republic of Lithuania)

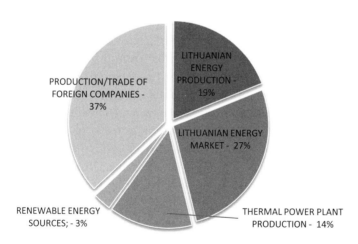

Figure 15 Coverage of Lithuania's electricity demand in 2011. *(Source: JSC Lietuvosenergija JSC, http://www.le.lt/en/).*

obsolete facilities. The net cost of electricity generated by thermal plants is higher than that of the imported electrical energy bought in the exchange market. In order to cover the ever-growing electricity demand, it is essential to systematically update the existing facilities and build new generating sources based on up-to-date technologies. One important aspect to be considered is time: realization of energy projects is very time-consuming and there should be a long-time perspective envisioned for the development of energy systems in line with the development of the entire industry. The dynamics of energy industry's development should exceed that of the entire economic system; otherwise, the energy system may become a deterrent of progress.

The bulk of the fuel consumed in Lithuania and the whole of the deficit electrical energy is imported. Furthermore, Lithuania has not yet been connected to energy systems of western countries. In fact, all the fuel and energy resources are acquired from Russia as it was prior to regaining sovereignty. In order to safeguard the possibility of choosing fuel and energy suppliers and purchasing resources under favorable prices, Lithuania has to transform its energy industry and integrate into European energy systems. Moreover, it should possess sufficient facilities to satisfy its energy demand and be able to participate and compete within common European markets. For these purposes, large-scale infrastructure projects are under implementation; the most significant of them are building electricity connections with Sweden and Poland, a terminal of liquefied gas in Klaipėda, and a nuclear power plant in Visaginas. All these infrastructure projects have been approved by the EU. There is one more precondition to energetic safety—an effective and liberal energy market, that is, implementation of the so-called Third Energy Package, which aims at destroying monopolies within the energy industry and ensuring successful operation of local competitive energy-generating facilities. New energy production facilities will have to be ensured by the projected Visaginas nuclear power plant. Accomplishment of all these projects will allow securing energetic safety. Figure 16 presents the mix of Lithuania's existing and projected energy sources.

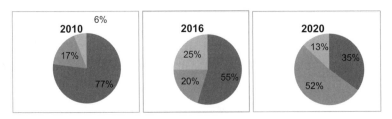

Figure 16 Mix of Lithuania's primary energy sources (Source: Ministry of Energy of the Republic of Lithuania). (Blue) Import of electrical energy and energy generation resources from a single supplier (Russian Federation); (red) local production; and (green) alternative import of energy (EU).

At present, energetic safety is a national prerogative and not that of the EU. There is no doubt that the safety of every member state adds to the total safety of the entire EU. The goal is that there will remain no such "energy islands" as today's Lithuania within the EU. Consequently, Lithuania may consider its energetic safety as a possibility to be part of the common EU energy system and freely choose primary energy resources and suppliers, and as a variety of energy production sources.

RENEWABLE ENERGY PRODUCTION TECHNOLOGIES IN THE BALTIC STATES: THE CASE OF LITHUANIA

One of the key measures for reducing the influence on climate change and quantities of emitted pollutants is gradual transition toward an energy industry based on new renewable energy production technologies. In order to promote development of these technologies, on April 23, 2009, the European Parliament and Council adopted Directive 2009/28/EB on promotion of the use of energy from renewable sources. The directive endorsed national percentage targets of the share of renewable energy in overall energy consumption for 2020 and a mandatory 10% minimum target to be achieved by all member states for the share of renewables in the transport sector. Lithuania committed itself to reach a 23% share of renewable source energy within the mix of the produced energy by 2020 (Europa: Summaries of EU legislation). Table 5 compares commitments of Lithuania, Latvia, Estonia, and several other EU countries.

Table 5 Commitments of Lithuania, Latvia, Estonia, and EU Member States

Country	Share of Renewable Source Energy in Gross Final Energy Consumption in 2005	Planned Target for the Share of Renewable Source Energy in Gross Final Energy Consumption by 2020
Ireland	3.1%	16%
Austria	23.3%	34%
Denmark	17.0%	30%
Estonia	**18.0%**	**25%**
United Kingdom	1.3%	15%
Latvia	**32.6%**	**40%**
Poland	7.2%	15%
Lithuania	**15.0%**	**23%**
France	103%	23%
Finland	28.5%	38%
Sweden	39.8%	49%
Germany	5.8%	18%

Advancement of renewable source technologies in the Baltic states has much in common, although there are priorities depending on differences in natural conditions—preference is given for technologies that can be used to the utmost efficiency in a particular region. Conditions in all the three Baltic countries are favorable for using biomass in the energy industry, as there are sufficient wood resources and land suitable for fossil energy plants. Latvia is rich in hydro resources, whereas Estonia and Lithuania have territories suitable for developing a wind energy industry.

Use of Hydropower

Flowing water is the most extensively exploited renewable energy resource. Furthermore, technologies designed for "harnessing" this type of energy are most mature, having reached the farthest limits of their technological potential. There are various ways for extracting energy from water: hydropower plants, pumped storage power plants, exploiting energy of the tides and waves, and ocean thermal energy conversion. Production of hydropower does not require the use of fossil fuel and there is no emmission of CO_2 responsible for the greenhouse effect and climate warming. In contrast to the electrical energy produced by burning fuel, hydropower production is not subjected to taxes on nitrogen oxide, sulfur oxide, and emission of particulates, as well as the tax on CO_2 gas emission. Environmentalists and green organizations, however, are of the opinion that hydropower plants do have their share in environmental pollution (Vieira da Rosa, 2009). Construction of very sizeable plants changes natural river beds, dams are being built, and large territories are being flooded. For that reason, environmentalists urge intensifying requirements for the construction of new hydropower plants.

Lithuania's is not very rich in hydropwer resources, and their efficient use is very dependent on the hydrological conditions of rivers. The hydroenergetic potential of Lithuania is about 400 MW or 3.5 TWh of electrical energy per year, although only 30–40% of hydropower facilities can be used for economic needs. Currently, Lithuania's hydropower plants generate about 1.1 TWh of electricity, that is, 1.6% of the primary energy (about 8% of Lithuania's total electricity demand).

The biggest hydropower energy sources in Lithuania are the Kaunas hydropower plant (KHPP)(Figure 17) and the Kruonis pumped storage power plant (KPSPP)(Figure 18). The KHPP (capacity 100.8 MW) has been operating since 1959. At present, the KHPP generates more than 40% of Lithuania's renewable resource energy or 5% of all electricity consumed

The KHPP capacity – 100. 8 MW, 4 units of 25. 2 MW.
The largest unit – 24.6 m, length of pressure front – approximately 1.5 km; an average perennial discharge – 259 m³/s, water permeability under normal conditions – 3030 m³/s.
4 generators are rotated by reactive turbines with flexible blades. One turbine allows a flow of 158 m³/s of water (at maximum operational capacity), total volume of turbines - 632 m³/s of water.
The hydroelectric plant generated 0.386 TWh of electricity in 2011.

Figure 17 Kaunas hydropower plant data (Lietuvosenergija, JSC).

The KPSPP capacity – 900 MW, 4 units of 225 MW.
Cycle efficient use rate: 0.74.
Unit's operational range in generator mode: 0 – 225 MW.
Fixed capacity in pump mode: 220 MW.
The KPSHP generated 0,534 TWh of electricity in 2011.
The KPSHP is intended to balance electricity supply and demand Currently the KPSHP is capable of ensuring 94% of the total necessary energy reserves for Lithuania in case of emergency.

Figure 18 Kruonis pumped storage power plant data (Lietuvosenergija, JSC).

in Lithuania. The KHPP is one of two plants in the Lithuanian energy system that can be started automatically in case of a total system blackout.

Lithuania's hydropower industry includes one object that is quite unique—the Kruonis pumped storage power plant (KPSHP). This 900-MW plant ensures the system's backup supply, balances out load fluctuations, regulates voltage and frequency, and covers about 6% of Lithunia's total electricity demand.

During periods of low demand (usually at night), the KPSHP is operated in pump mode and uses inexpensive surplus energy. It raises water from the lower Kaunas reservoir to the upper 303-ha reservoir that is 100 meters over the level of Kaunas reservoir waters. During peak periods (day time), with normal energy demands, it operates as a traditional hydroelectric plant. In order to prevent and liquidate system accidents, it is important for units to ensure a rapid reserved capacity; in less than 2 minutes, if necessary, it can be switched on at full power. The KPSHP units can be started automatically by the anti-accidental system and can thus cover any deficit in power. Other equally important functions of the KPSHP are to level the system load balance, regulate voltage and frequency, and start the system after a total system blackout.

In order to promote production of energy from renewable resources, the KPSHP began using the largest (in terms of Lithuania) system for solar energy extraction. Starting in June 2012, a battery of solar collectors began operating, and it is predicted that water heated in this way will be sufficient to cover the hot water demand for the plant's internal consumption (Lietuvosenergija, JSC).

One of the alternatives for Lithuania is a small hydropower industry. The number of small hydropower stations increased significantly during the two decades of Lithuania's independence, and in 2011, there were already 82 small HPP with a total installed capacity of about 108 MW. Nevertheless, there is still a great unused potential for the small hydropower industry, exploitation of which is compromised by environmental obstacles and strict reglamentation of hydropower resource development. Hydropower resources, if assimilated, could supply about 8% of Lithuania's electricity demand.

Against the perspective of a sustainable energy industry, expenses for the construction of both a sizeable and a small hydropower plant are very similar. Investment of about 8000–1200 EUR is needed in order to produce a 1-kW capacity. The leading position in the development of hydropower energy in the Baltics belongs to Latvia—the share of hydropower in the mix of Latvia's energy production is more than 70%.

Development of Wind Energy Industry

Within the wind energy development process in Lithuania, there was a primary assessment of wind energy resources performed in 2008 based on historical records of observations from weather stations, and methods for their calculation were designed. According to surveys, wind power energy development is feasible and justifiable from an economic point of view. In western Europe and Lithuania, it is mandatory to perform measurements of wind parameters using appropriate equipment in a partiular territory 6 to 12 months before beginning to construct wind power plants. It allows to choose appropriate facilities for wind power plants, draw up their operation schedules, predict their energy output, and define economic indicators. Building wind plants requires analyzing variation of wind parameters, formation of windblasts, profiles of wind velocity according to the earth's surface roughness and residential density of the territory, and formation of wind flows beyond natural and urbanistic barriers (Vieira da Rosa 2009).

As is seen from the Lithuanian wind atlas (see Figure 19), there is a sufficient number of territories in Lithuania where the average wind velocity exceeds 5 m/s and which are suitable for constructing wind plants in them. Unfortunately, there is a major obstacle—rather high residential density and lack of free sites for building wind plant farms.

Development of the wind power industry in Lithuania started in 2002, when the first wind station with a capacity of 160 kW was built near Skuodas. In 2004, a 630-kW plant was built in Vydmantai and connected to the electricity network. The largest wind plant farm was established in the coastline area of Lithuania, between Palanga and Klaipėda. Its capacity reaches 30 MW, and the annual production of electricity is about 64 MWh. According to data from the Lithuanian Wind Power Association, until May 2012, wind power plants operating in Lithuania generated 430 million kWh of electricity, saved 42.6 million m^3 of gas, and prevented emission of 78,900 tons of CO_2 (European Wind Energy Association 2011; Tallat-Kelpsaite, Polocka & Spizley 2011).

Development of the wind power industry has also encountered technical difficulties. The major of these are balancing energy due to variable wind

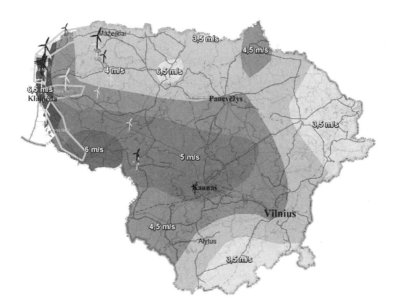

Figure 19 Lithuanian wind atlas. *(Source:* http://www.vejoekspertai.lt/lietuvos_vejo_ze melapis.html*).*

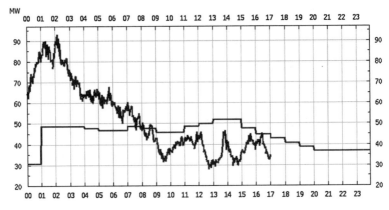

Figure 20 Wind energy balance in March 2012, Lithuania (Litgrid, JSC, http://www.litgrid.eu).

flows. Figure 20 shows wind energy produced in March 2012 against the same month's forecasts. In order to ensure a steady supply of electrical energy to users, it is essential to have backup facilities installed and a network allowing to transmit unstable flows of produced energy.

In order to ensure the balances of electrical energy and capacity, as well as safe functioning of the power industry's system, the maximum capacity of wind stations within the Lithuanian power industry system should not exceed 170 MW. Such capacity is sufficient for meeting the EU commitments. A capacity of the coastline wind plants exceeding 170 MW would cause an increased loss of energy within the transmission network, too heavy power flows, and voltage regulation problems.

Wind plants generate inexhaustible clean energy, although there are opinions that wind stations may influence health in the long term; moreover, opponents hold the view that power plants may damage ecosystems by changing customary migration routes of birds and insect populations, thus having a negative impact on local agriculture, etc. Expansion of wind plant farms is also prevented by limited land areas, large sanitary zones, and complicated provisions for connecting to transmission and distribution networks, as well as population's opposition. For that reason, the world's common practice is to build powerful offshore wind power farms.

Problems related to wind power technologies are solved by improving technologies themselves. The present level of technologies in Lithuania allows only part of the territory (preferably, the coastline area) to be used for extracting wind energy. Unfortunately, Lithuania's coastline stretch is not very large and almost the whole of it is used as a recreation zone. The Curonian Spit cuts across bird migration routes. It is considered that only a few dozen powerful wind plants could be built during the next few decades.

Considerable drawbacks of the wind power industry still could be considered their slow payoff and high net cost of produced energy. In many countries, notably EU, energy generated by wind power stations is subsidized or otherwise supported. The efficiency of wind plants has been increasing lately; the wind power industry, therefore, is considered one of the most perspective modes of electricity generation. The most rapid progress of the wind power industry in Europe is observed in such countries as Denmark, Spain, and Germany, which now may boast of the most powerful potential for wind energy production.

Use of Biomass in the Energy Sector

One of the most perspective sources for the energy sector in all of the Baltic region, including Lithuania, is biomass. Biomass has the longest usage history among all the energy types. After the emergence of more efficient fuel types and possibilities for their fast transportation, biomass gave up its position in the energy mix. In industrial countries, it takes only 10–20%. Lithuania aims to set up conditions and measures for stimulating the use of biofuels by 2020. The share of biofuel energy and the total share of biofuels should reach 10% in Lithuania by 2020.

Currently, biomass accounts for almost half of the renewable energy consumed in the EU (The Baltic Sea Region Program, 2007–2013). Biomass has numerous advantages in comparison with conventional energy sources and even with some types of renewable energy—above all due to its comparatively low expenditure, lesser reliance on brief weather changes, promotion of the development of regional economic structures, and provision of alternative revenue sources to farmers. Use of biomass creates all preconditions for being less reliant on fossil fuel, reducing emission of greenhouse gas, and revitalizing economic activities in rural areas. EU directives set a priority target to grow and use agricultural stock for biofuel production. Significantly, one of the priority goals for 2020 in biofuel production and usage is to use biodiesel in diesel engines. This makes it important to evaluate possibilities for the use of biodiesel in agriculture and public and inland water transport. Biofuel development increases the demand for biomass of various crops: oleaginous (rape, sunflower, and soy) used for producing biodiesel; starchy (wheat, sugar beet, and potatoes), for bioethanol; and green mass (corn, perennial plants, and fodder beet), for biogas. Agricultural areas used for bioenergetic crops tend to expand. The forecast is that within achieving the target of the biofuels directive (replacing 5.75% of all used fuels), energetic crops will take from 4 to 13% of the entire agricultural area of EU countries (the Baltic Sea Region Program, 2007–2013).

Lithuania possesses a sufficient amount of resources to make renewable resources a serious rival for fluid fuels and ensure energetic independence. Due to climatic conditions, Lithuania's energy mix very much differs from those of other European regions. Western Europe is generally concerned about a more intensive use of renewables for producing electrical energy and transport fuels, whereas in Lithuania the preference should be for the consumption of energy in heating. Currently, the heating of buildings actually is the most urgent problem in Lithuania.

Lithuania has considerable potential for using biofuels in energy production. Lithuania produces about 3 million tons of communal waste annually. Naturally decomposing waste makes about 0.3–0.5 million tons of the annual general waste production. It could be used for energy production after separating it from other waste and recycling in biogas reactors. Lithuania has a steady annual amount of unexploited arable land reaching about 300–500 thousand ha; there is also some infertile land where crops could be raised for energetic purposes. About 10–15% of the agricultural area could be used for raising energetic crops. It is estimated that some 0.72 TWh of energy could be produced from energetic crops annually. Figure 21 shows Lithuania's potential for the use of biomass.

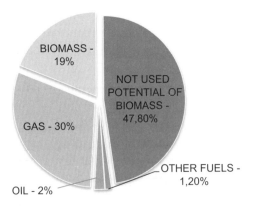

Figure 21 Lithuania's potential for the use of biomass. *(Source: Lithuanian biomass energy association,* http://www.biokuras.lt*)*.

There are six biogas plants in Lithuania: three biogas plants recycling municipal wastewater, one plant recycling food industry's waste, and two plants recycling animal manure, together with waste from food industry enterprises. The total capacity of biogas plants is about 17.7 MW (heat capacity about 15.6 MW and electrical capacity about 2.1 MW). In 2000, the largest biogas plant in Lithuania was built within the Kaunas water purifying plant (see Figure 22), which recycles municipal wastewater into biogas. It has two bioreactors (metatanks), 8800 m^3 each, and produces about 2.8 m^3 of high-quality (concentration of methane 70%) biogas. Part of the produced biogas is burned in two 1.9-MW water heating boilers. From there, the heat is supplied to the city's university, whereas electricity is transmitted to the 10-kV electricity network.

The major drawback of biofuels is its high net price. The production expense of biofuels exceeds that of conventional fuels by 2.8 times. For that reason, in order to allow biofuels entering the market, additional taxes are imposed on oil fuels and various subsidies and privileges are granted for producers and users of biofuels (Lithuanian Energy Institute 2012).

Although the bioenergy industry is not the "cleanest" and most innovative renewable energy sector, Lithuania's legislation on this sector is the most exhaustive and favorable in comparison with other industries.

Solar Energy Industry in Lithuania

It is a common view that Lithuania cannot boast of sufficient solar energy resources, making the use of solar energy not rational. Naturally, such countries as Greece, Spain, or Portugal are significantly richer in solar energy resources. Still, comparison of the number of sunny hours in Vilnius with other European cities (see Table 6) suggests that Lithuania has a sufficient amount of solar energy resources.

The total annual number of sunny hours is measured by 11 weather stations throughout Lithuania. The sunny period is longest in Lithuania's coastline region and gets shorter toward the eastern border, as the expectation

Figure 22 Biogas plant in Kaunas (Lithuanian Energy Institute 2012).

Table 6 Comparison of Solar Energy Recources

Country (city)	Number of sunny hours
Lithuania (Vilnius)	1690
Great Britain (Manchester)	1360
Germany (Hamburg)	1570
Sweden (Kiruna)	1470
Switzerland (Geneva)	1500
Italy	2500
Spain, Portugal	3000

(*Source:* Lihuanian Energy Institute).

Figure 23 The Arginta experimental solar power plant. *(Source:* http://www.arginta.lt/).

of sky cloudiness gets higher toward the east. The average annual number of sunny hours within the coastline area reaches 1840–1900. On the eastern border, it does not exceed 1700 hours per year (Tallat-Kelpsaite, Polocka & Spizley 2011).

The solar energy industry in Lithuania does not operate on an industrial level. There is an experimental solar power plant within JSC Renerga near the Benaičiai 1 wind power plant and experimental 149-kW solar plant within JSC Arginta (Figure 23). The latter plant has been designed not so much for producing electrical energy as for presenting scientific conclusions concerning which of the 35-volt photovoltaic equipment used in this solar plant and which installation methods are most suitable within the conditions of Lithuania.

In fact, solar energy in Lithuania has a more widespread use only on a household level—for heating water for residential buildings.

Lithuania has invented a technology for producing silicon solar elements, allowing to produce solar elements with 15% efficiency. Furthemore, it is able to produce the world's most popular (85% of the total use) monocrystalline silicon solar elements with the capacity up to 1–2 MW per year. This would cover the entire demand of Lithuania and create surplus for export.

Use of Geothermal Energy in Lithuania

Lithuania has significant reservoirs of geothermal water. Most of it is concentrated in western Lithuania. Geothermal resources situated near the Earth's surface are the easiest to exploit and are supplied to users with the help of heat pumps. These are the so-called shallow (up to 100 m) geothermal and mixed resources.

Within the sedimentary mantle, three regional hydrothermal horizons (complexes) are distinguished, which are suitable for extracting geothermal water with a different temperature (30–90°C) and mineralization and then returning it to the same mantles through another borehole after extracting part of the heat. Such a method for using geothermal resouces is applied in Klaipėda's geothermal power plant. Under the estimation of Lithuanian geothermal experts, western Lithuania's geothermal energy reserves suitable for heat production allow installing centralized heat supply facilities with an estimated total capacity up to 41,600 MW. Maximum heat demand of this region is 517 MW, making 1.2% of the current geothermal energy reserves, that is, this region's geothermal energy resources and reserves are actually inexhaustible (Suveizdis 2000).

The first geothermal water boreholes in Lithuania were drilled in 1989 in Vydmantai. Their depth reaches 2.5 km, and the temperature of the geothermal water is 74°C. In 1997, with the support of the Global Environmental Fund and Danish Environmental Protection Agency, construction of the Klaipėda geothermal demonsration plant was started. This project aimed at supporting technologies that reduce the emission of greenhouse gas. Since 2001, this plant has been supplying heat to the heating network of the city of Klaipėda, and this supply has been growing annually. This plant is the first such (with the closed circulation system) power plant in the Baltic region. The so-called immersion pumps raise mineralized (93 g/liter) geothermal water with the temperature of 38°C to the surface through two boreholes of about 1 km each. Its temperature is raised using heating pumps and then the heat is supplied to Klaipėda's thermal heating network. The used geothermal water is returned to the same mantle through two injectional boreholes. The total capacity of this plant is 35 MW (13.6 MW of heat from geothermal water and 21.4 MW from heated water of the boiler). When in full operation, the plant is able to decrease the burning of rather large amounts of organic fuel, thus making Klaipėda's atmosphere cleaner. Unfortunately, this project was not entirely successful either. Chemical composition of the water and unsuitable technologies, as well as economic conditions, make the plant's operation unprofitable.

In addition to the aforementioned geothermal water plant, small low-capacity (up to 55 kW) heating systems using shallow (up to 100 m) geothermal resources paired with aditional water heaters have been emerging in Lithuanian cities. In order to retain the needed temperature within a building, the entire such system is piloted by a microprocessor. The important aspect is that the use of geothermal energy does not depend on seasonal or meteorological conditions; however, in Lithuania, this resource is not used for generating electricity. Lithuania is the only country in eastern Europe having reserves of geothermal energy suitable for producing electrical energy (Lithuanian Energy Institute 2012). What is lacking is the consent of institutions responsible for the development of Lithuania's energy industry, as well as precise research analysis and engineering solutions.

Although geothermal energy is considered one of the cleanest energy types, it has not escaped environmental problems either. Drilling industrial boreholes, stimulating fractures, and extracting geothermal water may be the cause of artificial seismic activity, earth failures, gas escape (sulfur), and other unacceptable phenomena. Moreover, the development of geothermal systems depends not only on the knowledge of physical, mechanical, hydrodynamic, and other processes, but also the entirety of circumstances—economic conditions, political solutions, level of environmental protection, society's views, etc.

REFERENCES

AS "Latvenergo." Available from: https://www.latvenergo.lt.
Bačauskas, A., 1999. Investments in the Lithuanian power sector to secure adequate and reliable supplies of energy. In: International Conference Investment in Energy in the Baltic Sea Region. Conference Proceedings, Riga, pp. 122–125.
Beurskens, L.W.M., Hekkenberg, M., 2011. Renewable energy projections as published in the National Renewable Energy Action plans of the European member states, executive summary. Eur. Environ. Agency ECN-E-10–069.
Eesti Energia, AS. Available from: <https://www.energia.ee/et/avaleht>.
Elering, AS, Transmission system operator. Available from: <http://elering.ee/en/>.
Energy policy strategies of the Baltic Sea region for the post-Kyoto period, A study of possible energy strategies for the Baltic Sea Region for 2020 and after. Available from: <http://www.ea- energianalyse.dk/projects- english/1112_energy_policy_strategies_bsr_post_kyoto_period.html>.

Europa: summaries of EU legislation. Available from: <http://europa.eu/legislation_summaries/energy/renewable_energy/index_en.htm>.

European Wind Energy Association, 2011. Wind energy and EU climate policy. Available from: <http://www.ewea.org/fileadmin/ewea_documents/documents/publications/reports/20110909_ClimateReport.pdf >.

Guthey, E., Clark, T., Jackson, B., 2008. Demystifying business celebrity. Routledge, New York.

Gwartney, J.D., Stroup, R.I., Soubel, R.S., 1997. Economic. Private and public choice, eighth ed. The Dryden Press.

International Energy Agency, 2011. Gas emergency policy: where do IEA countries stand?. Available from: <http://www.iea.org/publications/freepublications/ publication/name, 3985, en.html >.

Lietuvosenergija, JSC. Available from: <http://www.le.lt/en/>.

Litgrid, JSC, Electricity transmission system operator (Lithuania). Available from: <http://www.litgrid.eu>.

Lithuanian Energy Institute, 2012. Biogas. Available from: <http://www.ena.lt/doc_atsi/Atsi_EI.pdf >. (15.04.12.).

Menezes, F.M., 2009. Consistent regulation of infrastructure business: some economics issues. Econ Papers 28 (1), 2–10.

Moreau, F., 2004. The role of state in evolutionary economics. Cambridge J. Econ. 28, 847–874.

Oss, A., Zeltina, L., Zeltins, N.V., 2003. Cooperation between developed countries and the Republic of Latvia in the field of energy globalization. Int. J. Global Energy 19 (2/3), 212–224.

Renewable energy policy review, Estonia, 2009. Available from: <http://www.erec.org/fileadmin/erec_docs/Projcet_Documents/RES20>.

Skribans, V., 2010. Development of the Latvian energy sector system dynamic model. MPRA Paper No. 25067. Available from: <http://mpra.ub.uni-muenchen.de/2506/>.

Steimikiene, D., Volochvich, A., 2010. The impact of household behavioral changes on GHG emission reduction in Lithuania. Renewable. Sustainable Energy Rev. 15, 4118–4124.

Sugolov, P., Dodonov, B., Hirschhausen, C., 2003. Infrastructure policies and economic development in east European transition countries. Inst. Econ. Res. (IER) Working Paper, WP-PSM-02, 39.

Suveizdis, P., Rastenienė, V., Zinevicius, F., 2000. Geothermal potential of Lithuania and outlook for its utilization. Proc. Geothermal Congress, 461–468.

Tallat-Kelpsaite, J., Polocka, A., Spizley, J.B., 2011. Integration of electricity from renewables to the electricity grid and to the electricity market – RES – INTEGRATION, national report: Lithuania, Ecleron GmbH. Oko-Institute. Available from: <http://www.eclareon.eu/sites/default/files/lithuania_-_res_integration_national_study_nreap.pdf>.

The Baltic Sea Region Program, 2007-2013. Baltic Sea Region Bioenergy Promotion Project. Available from: <http://eu.baltic.net/Energy.21783.html>.

Vieira da Rosa, A., 2009. Fundamentals of renewable energy processes. Elsevier Academic Press, Boston 844.

Webster's Third New International Dictionary 1976, Springfield, MA, 1, 1161.

Wind power in Estonia, 2010. An analysis of the possibilities and limitations for wind power capacity in Estonia within the next 10 years. Prepared by Energy Analyses for Elering OU, 46. Available from: <http://ea-energianalyse.dk/reports/1001_Wind_Power_in_Estonia.pdf>.

CHAPTER 14

Renewable Energy Generation: Incentives Matter
A Comparison Between Italy and Other European Countries

D. Chiaroni, V. Chiesa, F. Frattini
Dipartimento di Ingegneria Gestionale, Politecnico di Milano, Piazza Leonardo da Vinci, Milan, Italy

Contents

Introduction: Forms of Incentives for Renewable Energy Generation	347
Renewable Energy Generation in Italy: An Overall Picture	351
The Case of Photovoltaics in Italy	354
The Case of Biomasses in Italy	362
Renewable Energy Generation: Incentives Matter	366

INTRODUCTION: FORMS OF INCENTIVES FOR RENEWABLE ENERGY GENERATION

Since 1997 when the Kyoto Protocol was ratified stating a 5.2% reduction of greenhouse gas emission levels of 1990 by the period 2008–2012, the interest for renewable energy sources (RES), due to their ability to produce electricity (and heat) with no (e.g., photovoltaics) or limited emissions (e.g., biomasses) of greenhouse gas, grew significantly and became a global issue for all major countries worldwide. Also, more recently in 2011 the Durban Conference set an extension of the effects of the Kyoto Protocol to a "second phase" in the period of 2013–2017 and defined a path for drafting a new protocol to take effect no later than 2020, thus confirming this will also remain a crucial issue for the coming future.

Ever since the beginning, the interest of Europe for RES was even greater than that of other countries. The initial goal of the Kyoto Protocol was immediately increased to 8% (+53%) of overall reduction of greenhouse gas emissions. Later in 2007, the year before the practical enforcement of the Kyoto Protocol, the European Commission adopted an action plan entitled "An Energy Policy for Europe" for further promoting environmental sustainability and fighting climate change.

The action plan was detailed in 2008 into the well-known "20–20–20" Climate Energy Package: (i) reducing, by 2020, at least 20% of greenhouse gas emissions resulting from energy consumption in the EU-27 compared to 1990 levels (mandatory target); (ii) achieving a 20% share of energy from RES of total energy consumed by 2020 (mandatory target) and a minimum of 10% for biofuels in the total consumption of petrol and diesel in the EU by 2020 (mandatory target); and (iii) forcing a 20% improvement in energy efficiency (reduction of energy consumption) of the EU compared to projections for 2020.

The second mandatory pillar of the climate energy package set an ambitious goal for Europe as a whole and forced each country to adopt adequate incentive systems for ensuring the achievement of the stated objective in the due date.

The need for incentives was indeed clear since the beginning, mostly due to the combination of two factors.

- Available technologies (at that time and, in most cases, still nowadays) for producing energy (mainly electricity) from renewable sources were not cost competitive against traditional production plants from fossil fuels. The levelized energy cost[1] of RES in Europe was, at that time, on average in a range between 0.15 €/kWh (for certain types of biomass plants) and 0.50 €/kWh (for photovoltaics), against an average cost for electricity production from fossil fuels of 0.06 €/kWh.
- A RES also implies a significant change in the energy production paradigm—from a "centralized" production (typical of large plants working with fossil fuels) to a "decentralized" and more "distributed" production (with small plants localized in proximity of the energy users). As a consequence, there is a need to overcome the inertia of the final users in adopting technologies for producing RES and in assuming also the potential risks and obligations of becoming an energy producer.

In this respect, incentives should act as a sort of catalyst for the adoption of RES by making it economically viable. It is worth mentioning, and thus reinforcing the metaphor of the catalyst, that the rationale for the incentives is also in reducing (by leveraging innovation and economies of scale) the levelized energy cost of RES progressively; it worked accordingly to the

---

[1] Levelized energy cost (also known as levelized cost of energy) is the result of the economic assessment of the cost of an energy-generating system that distributes all the costs over its lifetime (initial investment, operations and maintenance, cost of fuel, cost of capital) on the overall energy produced by the system.

purpose if, for example, it is considered that currently on average the cost of producing electricity with photovoltaic plants is around 0.15–0.20 €/kWh.

The strategy followed by European countries about the form (and the amount) of incentives has been rather variegate. It is possible to recognize at least three main forms of incentives, listed accordingly to an increasing level of risk and volatility for the electricity producer as well as for level of diffusion among European countries.

- **Feed-in tariff**, where the government sets a fixed and "all inclusive" price, higher than the market price, for the energy generated from renewable sources for a given number of years. The amount depends on the type of renewable source and should be adapted progressively to the reducing distance to the economic viability. This bonus applies to producers as an incentive to invest in the development of innovative and green technologies.
- A feed-in tariff is the easiest and therefore also most common system throughout Europe and has been adopted in Germany, Italy, France, Spain, Portugal, and the Czech Republic, but also in non-EU countries such as the United States and China.
- **Feed-in premium**, where the price actually paid for renewable energy is composed of two factors: (i) the market value of electricity, subject to supply and demand fluctuations; and (ii) a premium determined by the public authority. Differently from the feed-in tariff, the energy producer is here more exposed to the risk of price volatility, whereas the government sets clearly in advance the total amount of payments due. This incentive scheme, as detailed later, has been used in Italy for photovoltaic.
- Usually, in order to reduce the aforementioned exposure of the producer to variations in the electricity price, the feed-in premium system is paired with other contractual agreements (i) either under the form of a power purchase agreement, that is, a contract stating the price (flat, escalate over time, or negotiated) for the electricity agreed upon between the producer (seller) and a given buyer (usually a local utility or another energy trader) or (ii) by introducing a governmental entity (e.g., in Italy with the case of Gestore dei Servizi Energetici) with the mandatory task of purchasing the electricity generated particularly by smaller producers at a given price, defined by law every year.
- **Quota obligations**, where the government fixes that a certain share of the electricity sold on the market by energy producers should come from RES. For each MWh generated using RES, a certain number of trading certificates is recognized to the producer. To

equalize the different RES (which have different costs of production), the number of certificates given for each MWh produced changes according to the type of source used. Certificates are therefore in the hands of RES energy producers who can sell them (for the part exceeding their obligations) to other energy producers. The assumption is that energy producers from RES represent the supply and energy producers from fossil fuels represent the demand of the market for trading certificates.

- The producer from RES therefore has two forms of revenues: from the selling of the electricity at its own price (such as the feed-in premium system) and from the selling of the trading certificates, which is also subject to price volatility due to the demand–supply dynamics. This system has been used in the United Kingdom, Italy, Denmark, Sweden, and Poland.

The aforementioned forms of incentives can be paired eventually with a **tender or auction system**. Several players can participate in an auction for the approval of incentives for a given amount of power that the public authority will assign for a certain number of years. The definition of the value of the incentive (feed-in tariff or eventually feed-in premium) is through a competitive bidding downward and will coincide with the lowest value of submitted claims in the auction in accordance with fixed values of caps and floors. Auctions will take place on a periodic basis and usually define minimum requirements in terms of technical expertise and financial strength for bidder's projects and mechanisms to ensure delivery of authorized facilities, including by setting deadlines for the entry into operation. In Brazil, Portugal, the United Kingdom, Denmark, Ireland, the United States, Morocco were defined fruitfully arrangement of this kind of incentive systems. In other countries, such as Italy, this mechanism has been adopted only very recently.

Figure 1 shows the distribution of the different forms of incentives for RES in European countries at the end of 2012. It is worth highlighting the fact that most countries select a single form of incentives for the different renewable energy sources, using the level of the tariff and the number of years of application to differentiate among RES. In a few cases (6 out of 26), two forms of incentives coexist, but only in Italy have all the available forms (including also the auction system) been used for supporting the development of the electricity production from RES.

The case of Italy therefore appears of particular interest and deserves the attention paid in this chapter.

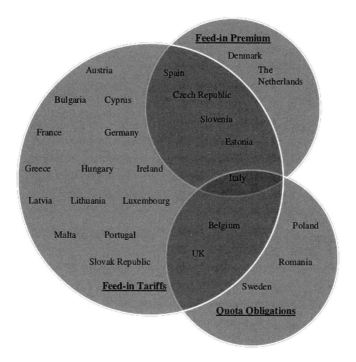

Figure 1 Forms of incentives for RES in European countries. *Source: elaboration from Energy & Strategy Group on publicly available data.*

RENEWABLE ENERGY GENERATION IN ITALY: AN OVERALL PICTURE

During the period of 2008–2012, that is, the 5 years covered by the Kyoto Protocol, the newly installed renewable energy generation power for electricity production in Italy has been 47.7 GW, more than doubling the total installed base of 23.6 GW at the end of 2008.

Table 1 lists all the RES that have been subject to incentive systems since 2008 according to the growth rate of installations in the period. Hydro power, mostly due to the topology and the history of energy production in Italy, is still the largest contributor to the RES installed base (about 37% of total), even if the compound annual growth rate(CAGR) measured since 2008 clearly shows that it is a mature and almost saturated market, similar (even if with very different absolute numbers) to that of geothermal sources for electricity production. Although incentives were also fixed for the hydro and geothermal sources, there was not a viable potential for exploitation in practice.

Table 1 Installed Base for Electricity Generation Power in Italy by RES and by Year (2008–2012)[a]

	Installed base for electricity generation power (MW)					5 years CAGR (%)
	2008	2009	2010	2011	2012	
Photovoltaics	431	1.144	3.470	12.750	16.281	106.7
Biomass	1.361	2.016	2.349	2.824	3.257	19.1
Wind	3.538	4.898	5.814	6.936	8.144	18.1
Geothermal	711	737	772	772	772	1.6
Hydro	17.623	17.721	17.876	18.092	18.080	0.5
Total	23.664	26.516	31.494	42.340	47.719	15.1

[a]*Source:* Energy & Strategy Group elaboration on data from Gestore dei Servizi Energetici.

A very different story is the one of biomass and wind energy, of which the growth rates are double digits, as well as for photovoltaics, passing from less than 0.5 GW of installed power in 2008 to more than 16 GW at the end of 2012, doubling the newly installed power almost every year.

Italy has been the largest worldwide market for photovoltaics in 2011, and still in 2012 is ranked second in Europe after Germany (by far the world leader for the total installed base with more than 32 GW).

In 2012, Italy occupied third place in Europe for electricity production from biomasses (10% of total), immediately after the United Kingdom (11%) and again far from Germany (30% of total European production); that, however, has a significantly longer tradition in supporting this industry.[2]

In wind energy, at the European level, Italy ranked fourth place in 2012 for newly installed power generation, again after Germany (31 GW of total installed base), Spain (22 GW), and the United Kingdom (9 GW).

The effect of incentives is therefore rather evident in these cases, considering that the installed base for these sources (photovoltaic, biomasses, and wind) at the end of 2008 was still marginal (5.3 GW, 22% of total), the market potential was quite relevant (due to characteristics of Italy in terms of availability of sun, of biomasses particularly from agriculture, and, even if limited to a few areas, of wind) and the presence of local players of a certain relevance in renewable industries was also rather still marginal in 2008. In other words, starting almost from a green field (particularly, as discussed later, in biomasses and photovoltaics) for what concerns the installed base and the industrial value chain, Italy was at that time an ideal setting for studying the impact of incentives on RES.

[2] The first relevant incentive system for energy production from biomasses was, indeed, set in Germany in 2000 under the Erneubar-Energien-Gesetz framework.

Table 2 Energy Production and Total Incentives Committed in 2012 for RES in Italy[a]

	Electricity production from RES 2012 (GWh)	RES production mix (%)	Incentives 2012 (mln €)	Incentives mix (%)
Photovoltaics	18.323	19	6.522	65
Biomass	14.236	15	1.484	15
Wind	12.373	13	1.001	10
Hydro	43.322	45	931	9
Geothermal	5.854	6	115	1
Total	94.108	100	10.053	100

[a]*Source:* Energy & Strategy Group elaboration on data from Gestore dei Servizi Energetici.

The result of the incentives system adopted in Italy for the three sources mentioned earlier is certainly a successful increase in the total installed power base (27.7 GW, 58% of total). However, the amount of financial resources committed by the Italian government was also extremely relevant (see Table 2).

About 65% of the total 10.2 € billion of incentives committed yearly for the year 2012 went to photovoltaics, while biomasses accounted for 15% and wind for 10%. Hydro and geothermal sources received, respectively, less than 1 billion (9%) and 100 million (1%), thus confirming their marginal role in the incentives' framework of the Italian government.

In 2008, the amount of financial resources committed yearly to RES was less than 3 € billion. The increase in the period is therefore around 240% and explains clearly the numbers given previously and listed in Table 1.

Data presented in this paragraph, however, by looking at the total energy generation power installed are only surfacing the phenomenon. The huge amount of resources committed clearly generated a market for RES in Italy, but the form of each incentive system (feed-in tariff, feed-in premium, quota obligations) and the related procedures also explain much more about the characteristics of the market, its main segments, the typology of energy producers, and of industrial players mostly involved. It is possible to argue that the incentive system in Italy has been used, intentionally or even unintentionally, to shape the RES market and industry.

To support this thesis, two stories are presented and discussed in more detail in the next sections about the energy generation from photovoltaics (where incentives are in the form of feed-in premium and feed-in tariff) and from biomasses (feed-in tariff, quota obligations, tender, and auction) in Italy.

THE CASE OF PHOTOVOLTAICS IN ITALY

The installation of photovoltaic plants has been subsidized in Italy since 2005, when the First "Conto Energia" was launched. It introduced feed-in premiums paid to the owner of the plant for each kWh of electricity generated. It should be noted that these premiums were given on top of the money earned by the owner of the plant through the sale of the electricity and on top of the savings resulting from autoconsumption. Other countries (e.g., Germany, Spain) used instead a feed-in tariff, which included the price of the electricity sold. This made the Italian incentive system much more convenient than the ones available in other countries.

The First "Conto Energia" has several problems that reduced its effectiveness as a mechanism to promote the diffusion of photovoltaic installation. In particular, the feed-in premiums were higher the larger the size of the plant. This promoted large-scale installation, without taking into account the economies of scale characterizing large plants. Moreover, only plants smaller than 1 MWp in terms of installed power had access to the incentive scheme. Furthermore, there was a maximum amount of photovoltaic installation per year, equal to 85 MWp. This maximum threshold was reached very quickly and therefore new installations were frozen.

After 2 years since its introduction, the First "Conto Energia" was replaced by the Second "Conto Energia." Table 3 gives an overview of the levels of feed-in premiums established by the Second "Conto Energia," which are different depending on the characteristics of the plant (being nonintegrated, partially integrated, and fully integrated into a building) and depending on its size. Differences in the feed-in premiums depicted in Table 3 clearly indicate the goal of the Italian government to favor the diffusion of small, residential installations consistently with a paradigm of distributed electricity generation.

The feed-in premiums were given to the owner of the plant for 20 years and they had constant values. The Second "Conto Energia" established a

Table 3 Incentive Tariffs Provided by Second "Conto Energia" (€/kWh)[a]

		Plant type		
		Not integrated	Partially integrated	Fully integrated
Plant size (kW)	1–3	0.40	0.44	0.49
	3–20	0.38	0.42	0.46
	>20	0.36	0.44	0.40

[a]*Source:* Energy & Strategy Group.

maximum level of 1.2 GW of installed photovoltaic power that can have access to the incentive system. It was already established that, after the achievement of this goal, the Italian government had the opportunity to delay the end of the incentive mechanism. Considering the levels of the feed-in premiums established by the Second "Conto Energia," Italy became one of the European countries with a more generous mechanism for subsidizing photovoltaic installations. Many foreign funds and financial players began looking at Italy as an opportunity to earn high profits through photovoltaic installations.

However, there were strong barriers related with the very complicated bureaucratic process that had to be followed to authorize and install the plant, especially for a large size installation. These included authorization for the construction of the plant, the connection to the electricity grid, and the request to have access to the incentive mechanisms. Some analysts showed that, on average, to install a photovoltaic plant with a large size (e.g., 1 MW) in 2008 in Italy, the investor had to fill in 70 documents, whereas in Germany this number was 10 times lower. This resulted in high opportunity costs that reduced the overall profitability of the investment in a large plant. Despite the length of the authorization process, it should be noted that the main issue with it was that it required very heterogeneous steps and procedures depending on the region or province in which the plant was to be installed. This created a lot of additional organizational burden on the potential investors.

The Second "Conto Energia" determined a high growth in the rate of new photovoltaic installations in Italy in 2008 and 2009, thanks to the generous feed-in premiums available for the owner of the plant. Moreover, especially during 2009 and 2010, large photovoltaic plants started to grow in importance, with many installations, especially in the southern regions characterized by high levels of direct radiation. This was due to the significant reduction of the turnkey price of large plants, which made the investment especially profitable for the potential investors.

The maximum amount of installed photovoltaic power that could access the Second "Conto Energia" was reached during 2010. This is why, with the Decreto Ministeriale 06/08/2010, the Italian government introduced a radically revised version of the incentive mechanism, known as the Third "Conto Energia." It started to be applied from January 2011, had a maximum duration of 3 years (2011–2013), and lasted until 3 GW of newly installed photovoltaic power would be installed. The application of the feed-in premium, paid on top of the money earned from the sale of electricity and from autoconsumption, was confirmed in the Third "Conto Energia." However, the level of the premiums was reduced significantly. Table 4 depicts the level of the premiums for the plants installed in 2011. These premiums were reduced by 6% in 2012 and 2013.

Introduction of the Third "Conto Energia" was preceded by very tumultuous months for the Italian photovoltaic market. In particular, Law n. 129, published on August 13, 2010, postponed the deadline by which new photovoltaic installations could access the Second "Conto Energia." This new law gave the opportunity to access the Second "Conto Energia" to plants that were built by the end of 2010, although not yet connected to the grid (whereas the rule was and is that the incentive level is that in place at the date in which the first kWh is produced). This gave rise to an impressive rush to install new plants in the last 5 months of 2010, before the feed-in premiums were reduced due to introduction of the Third "Conto Energia." Studies showed that around 4 GW of new photovoltaic power was installed in this very short period of time.

When the Third "Conto Energia" began at the beginning of 2011, the level of installed power in Italy had already reached very high levels (i.e., 3.27 GW of connected plants, 7.22 GW considering also installed power not yet operational). Due to worsening of the economic crisis, the Italian government was especially worried about the impact on the Italian financial conditions of the rapidly increasing cost of the feed-in premiums. This is why, in March 2011, the Italian government promoted a decree that determined the end of the application of the Third "Conto Energia" in May 2011, only after 5 months since its start.

On May 5, 2011, a Fourth "Conto Energia" was approved that radically changed the approach for subsidizing the installation of photovoltaic plants

Table 4 Incentive Tariffs Provided by Second "Conto Energia" (€/kWh)[a]

Plant size (kW)	Plant connection Jan–Apr 2011		Plant connection Jan–Apr 2011		Plant connection Jan–Apr 2011	
	Rooftop	Ground	Rooftop	Ground	Rooftop	Ground
1–3	0.402	0.362	0.391	0.347	0.380	0.333
3–20	0.377	0.339	0.360	0.322	0.342	0.340
20–200	0.358	0.321	0.341	0.303	0.323	0.285
200–1000	0.355	0.314	0.335	0.309	0.314	0.266
1000—5000	0.351	0.313	0.327	0.289	0.302	0.257
>5000	0.333	0.297	0.311	0.275	0.287	0.244

[a]*Source:* Energy & Strategy Group.

in Italy. In particular, it introduced a cap on the newly installed power that could access the feed-in premiums evaluated for each semester in terms of a maximum annual cost. In case the maximum amount of installed power determines a total cost higher than the established threshold, the cost available for the ensuing semester is reduced. This mechanism was inspired by the experience of Germany, where an incentive system based on these dynamic caps was present since 2009. It was introduced with the aim of controlling more closely the growth of the cost for subsidizing photovoltaic installations that skyrocketed in the second half of 2010. Table 5 represents the maximum cost for each semester of application of the Fourth "Conto Energia."

Moreover, the new incentive system further reduced the level of feed-in premiums paid to the owner of the plant, as illustrated in Table 6. There was a clear tendency toward reducing the profitability of installing large plants, designed for the sale of electricity, especially those built on the ground.

Finally, the Fourth "Conto Energia" established that, for plants larger than 1 MW for a rooftop plant and 200 kW for ground-mounted plants, access to the incentive premiums was regulated through a public procedure to which the owners of the plants had to apply. A number of criteria were

Table 5 Cost Limits and Targets for Large Photovoltaic Power Plants from June 2011 to December 2012 with Fourth "Conto Energia"[a]

	Second semester 2011	First semester 2012	Second semester 2012	Total
Cost limit (mln €)	300	150	130	580
Installation target (MW)	1.200	770	720	2.690

[a]*Source:* Energy & Strategy Group.

Table 6 Incentive Tariffs Provided by Fourth "Conto Energia" (€/kWh)[a]

Plant Size (kW)	December 2011		Rooftop 2012		Other Plants 2012	
	Rooftop	Ground	First Semester	Second Semester	First Semester	Second Semester
1–3	0.298	0.261	0.274	0.252	0.24	0.221
3–20	0.268	0.238	0.247	0.227	0.219	0.202
20–200	0.253	0.224	0.233	0.214	0.206	0.189
200–1000	0.246	0.189	0.224	0.202	0.172	0.155
1000–5000	0.212	0.181	0.182	0.164	0.156	0.14
>5000	0.199	0.172	0.171	0.154	0.148	0.133

[a]*Source:* Energy & Strategy Group.

introduced to determine which plants that applied to the procedure could have access to the incentive system. Only those plants ranked among the top could actually benefit from the feed-in premiums. This created several problems due to the uncertainty on the level of revenues that the owner of the plant could earn, which were unknown until the end of the public procedure. Banks and financial institutions started to avoid giving loans to the owners of these plants, given this uncertainty, which further hindered the rate of new installations for large plants. Taken together, these chances introduced by the Fourth "Conto Energia" signed a marked reduction in the growth of the Italian photovoltaic market, especially if compared with the tremendous levels of new installations experienced in 2010.

The Fourth "Conto Energia" established a maximum level of cumulated annual cost for the incentives to photovoltaic plants, which was 6 bln €. This threshold was reached on July 6, 2012, and a Fifth "Conto Energia" began at the end of August 2012. In comparison with the previous version of the "Conto Energia," the new one, first of all, changed the incentive mechanism from a feed-in premium on all the electricity generated to a feed-in tariff (including the incentive and the price of electricity) on the electricity sold and a premium on the electricity self-consumed. Moreover, the level of the incentives was reduced further, as suggested by Table 7.

The level of the feed-in tariffs and the premium on self-consumed electricity point to the intention of the Italian government to strongly favor the installation of small residential and industrial plants, especially those that ensure a high percentage of self-consumption and minimize the amount of electricity sold. Furthermore, access to the incentive system without passing through the public admission procedure was ensured only for the smaller

Table 7 Incentive Tariffs Provided by Fifth "Conto Energia" (€/kWh)[a]

	Rooftop		Other Plants	
Plant size (kW)	Feed-in Tariff	Self-Consumption Premium	Feed-in Tariff	Self-Consumption Premium
1–3	0.208	0.126	0.201	0.119
3–20	0.175	0.093	0.168	0.086
20–200	0.126	0.044	0.12	0.038
200–1000	0.196	0.114	0.189	0.107
1000–5000	0.142	0.06	0.135	0.053
>5000	0.119	0.037	0.113	0.031

[a] *Source:* Energy & Strategy Group.

plants, with a size lower than 12 kWp (in comparison with the threshold of 1 MW for rooftop plant or 200 kW for ground-mounted plants introduced in the Fourth "Conto Energia"). Most importantly, the Fifth "Conto Energia" defined a maximum level of 6.7 bln € in terms of total annual cost for the incentive, after which the Fifth "Conto Energia" will end up. It is likely that this threshold will be reached no later than June 2013. After that, the photovoltaic market in Italy will have to run with its own legs, as no further incentive systems will be available to investors. Based on available data, it is likely that only residential plants installed in the center or south of Italy and industrial plants (e.g., 200–300 kWp) with a high percentage of self-consumption (between 60 and 80% of the total electricity produced) will be in grid parity in 2013 and can be installed without additional incentives.

Box 1 The Comparison with Germany, Inspiration for the Italian Regulatory Scheme

The incentives to photovoltaics in Germany were introduced for the first time in 1991, with the "Electricity Feed Law," which was then replaced in 2000 by the "Renewable Energy Source Act" (EEG: Erneubar-Energien-Gesetz). EEG fixed the value of 20 years feed-in tariffs for the photovoltaic plant until 2004, when an amendment changed the value of reducing tariffs and establishing the principle for which, every year, they will be cut by 5% in value. A further revision of tariffs was then introduced in June 2008, stating that since January 2009 the rates had to be reduced further and rooftop plants larger than 1 MW and ground systems of any size will obtain a lower tariff. Similarly, it increased the rate of the annual reduction rate, differentiating on the basis of size (with greater reductions for larger plants). In this way the industry could deal with an incentive system that laid down, in the long run, clear rules on the type and value of tariffs. The trajectory of a gradual reduction of tariffs pushed the technology to an ever greater cost reduction.

The overall objective confirmed by the German system has favored from the outset small photovoltaic systems with energy self-consumption. This trend was confirmed further through a May 2013 launch of an incentive program for energy storage systems for photovoltaic applications in order to enlarge the self-consumption share of energy generated and better integrate PV plants in the national grid.

The overall effect of an incentive system based on a long-term horizon, with rules of dynamic adaptation of the scales set "a priori," has allowed Germany to be the world leader for photovoltaic installations up to 2012 (over 7 GW of new annual installed capacity between 2010 and 2012) and the point of reference for the incentive systems of other countries, such as Italy, which have adapted their incentive systems along the lines outlined by the German case.

Thanks to the different generations of the "Conto Energia" presented here, photovoltaic power installed in Italy has grown along the years, as depicted in Figure 2.

It emerges clearly that development of the Italian photovoltaic market has suffered from an uneven growth rate, with an important peak of new installation that took place when termination of the Second "Conto Energia" was postponed after December 2010. This created a rush to reduce the annual cost for subsidizing photovoltaic plants from political institutions with the introduction of the Third, Fourth, and Fifth "Conto Energia" in a very short period of time. Moreover, this caused the termination of any support mechanism (in mid-2013) earlier than what would happen if a gradual reduction of the level of incentives was planned in advance, as for instance was done in Germany.

This is clearly the result of a lack of long-term planning of the evolution of the incentive schemes in Italy and an inability to link the reduction of the feed-in premiums consistently with lowering of the turnkey price of photovoltaic plants, especially the largest ones, characterized by important economies of scale (their price dropped by 66% between 2011 and 2013). As a result, the Italian photovoltaic market has not been guided toward grid parity step by step, forcing industrial players to cope with a dramatic reduction of the size of the market as happened in 2012 and 2013.

Introduction of the Fourth and, especially, the Fifth "Conto Energia" had the positive effect of reducing the average size of the installed plants, as shown in Figure 3, limiting the speculative effects that became clear in 2010 and 2011 with the boom of installation of solar farms, which occupied large

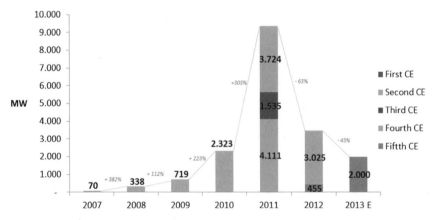

Figure 2 Annual power that has come into operation in Italy by incentive schemes. *Source: Energy & Strategy Group.*

areas of land in the search for superior returns on the investment ensured by the very generous feed-in premiums ensured to these plants by the Second "Conto Energia."

Another problem concerns the high level of bureaucracy characterizing the authorization procedures in Italy, which have had negative effects on the return on investment in new plants. With a simpler authorization and installation process, the level of feed-in premiums could be lower, ensuring the same rates on returns for the investors. As a result, more photovoltaic power could have been installed, with lower costs for the Italian people.

Figure 4 depicts the evolution of the overall revenues generated by the photovoltaic market in Italy, which clearly follows the evolution of the installed power.

Figure 4 also shows the revenues obtained by Italian firms. It emerges clearly that a very large part of the overall value generated by the photovoltaic market has gone to foreign players that have taken advantage of the generous incentive premiums. The fact that an effective incentive system was introduced only in 2007, when firms (e.g., manufacturers of photovoltaic cells and modules) in other countries (e.g., Germany and the United States) already had a strong international presence, prevented Italy from the development of a strong local photovoltaic industry. This was also due to the lack of subsidies to research and development and to the installation of manufacturing capacity provided to Italian firms, which made it impossible to close the competitive gaps with already developed photovoltaic industries. Finally, continuous changes in the incentives regimes that were discussed earlier contributed further to undermine the solidity of the Italian

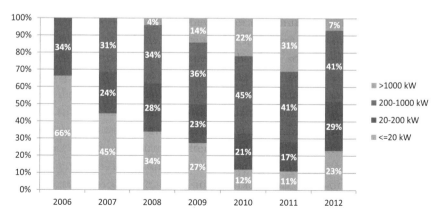

Figure 3 Segmentation of PV installed capacity in Italy by year. *Source: Energy & Strategy Group.*

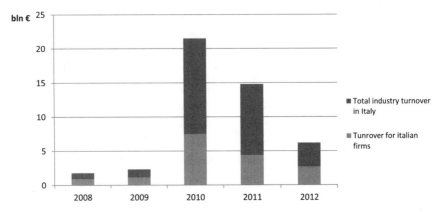

Figure 4 Revenues generated by the photovoltaic market in Italy by year. *Source: Energy & Strategy Group.*

industrial value chain, with players in the upward phases (e.g., manufacturing of cells and module) that could not plan accurately their investments in a new manufacturing capacity and firms in the downstream (e.g., system integrators and distributors) that could hardly forecast the expected demand.

Overall, the Italian photovoltaic market has grown too quickly under the effect of an incentive mechanism that was too focused on promoting the installation of new plants instead of favoring investments in new manufacturing capacity. This has created serious challenges to the sustainability of the business models of the Italian firms that have invested in this business over the years, which are actually suffering from competition in 2012 and 2013, when the growth rates in the local market have reduced dramatically.

THE CASE OF BIOMASSES IN ITALY

In the years '00, the first intervention of the Italian government regarding incentives for electricity production in the biomasses goes back to 2003 with introduction (for all renewable sources with the exception of photovoltaic) of the so-called "green certificates," a quota obligation according to which trading certificates for electricity production from RES are recognized for each MWh generated during the year. Certificates are issued for the first 15 years of life of the plant, but only for production plants with an installed power capacity greater than 1 MW.

Moreover, in the case of biomasses, the number of "green certificates" assigned for each MWh of electricity produced is incremented by a multiplication coefficient of 1.3 to consider the effects of potential burdensome, for example, in terms of shortage of raw material supply.

"Green certificates" were at the basis of the first development of electricity production from biomasses in Italy, but they presented a number of limitations.
- First of all, they were only available to medium and large plants, and indeed were used at the beginning for the installation of relatively large power production units (on average greater than 5 MW) fuelled with wood biomass mostly imported by foreign countries.
- The market for "green certificates" that was thought to be supported by the strong demand of large electricity producers and utilities willing to cover their quota obligations with the exceeding electricity production from RES by smaller producers was, in fact, rather poor in terms of actual transactions due to the fact that most of large players exploited incentives to made their own investments in renewables.

The value of "green certificates" therefore went down from more than 100 €/MWh to less than 60 €/MWh (−40%) in 2007.

As a consequence, after the initial interest and a total installed base that grew at about 1.2 GW from 2003 to 2007 (at a pace of less than 250 MW per year), the market was almost frozen in 2008 with about 170 MW of newly installed generation capacity, mostly related to plants that were already under construction (or already authorized) during the previous years.

It was clear at that time that the incentive system designed for biomasses was not effective. The government had to face at least two challenges: (i) ensuring a reasonable market liquidity and market price for "green certificates" upon which the previous plants have been based for their business plans and, more relevantly, banks provided the monetary resources for the installation[3] and (ii) further fostering the electricity production from biomasses, unlocking the potential from smaller and distributed plants, thus leveraging the abundance in Italy of local agricultural products and subproducts.

During the year 2008 and with the effect from January 2009, a relevant change occurred in the incentives framework for biomasses.
- First, a governmental entity (namely Gestore dei Servizi Energetici) was entitled with the mandatory task of purchasing "green certificates" at a given price. The price was established on the basis of the prices at which

---

[3] Please note that in the years 2006–2008, banks were used to finance RES investments in Italy with a leverage close to 90% and that the investment for a 5-MW plant was, on average, about 15–18 € million.

Table 8 Evolution of Energy Generation from Biomasses in Italy (2008–2012)[a]

	Biomasses				
	2008	2009	2010	2011	2012
Newly installed electricity generation power (MW)	163	655	333	475	433
Average plant size (MW)	4.4	2.9	2.0	1.5	1.2

[a]*Source:* Energy & Strategy Group.

green certificates have been negotiated in the previous 3 years. In other words, the government introduced a "hybrid form" of a feed-in tariff by transforming a formal quota obligations mechanism into a substantial feed-in tariff mechanism, reducing the risk for the producers significantly. This "hybrid form" was, however, well defined only until the year 2011, with limited information about its potential further evolution for the coming years.

- Second, a formal feed-in tariff system was introduced for plants with a power generation capacity lower or equal to 1 MW. The feed-in tariff was fixed at 280 €/MWh (+55% in comparison to that of larger plants) and over a period of 15 years from installation of the plant.

The impacts of this change appear clearly considering the newly installed electricity generation power during the following years (see Table 8).

The new incentives system allowed reducing the average plant size significantly, particularly starting from 2010, whereas in the year 2009 a number of larger plants went to completion that were stopped during 2008 due to the aforementioned "crisis" of the "green certificates." The new electricity producers are mostly farmers (exploiting local agricultural products or leveraging on the production of biogas) in the northern regions of Italy, which brought a significant reduction in the relative weight of import in the "fuel" supply for biomasses plants. In this respect, the Italian market became more similar to that of Germany, to which probably the government took the inspiration for drafting the new incentives system in 2008 (see Box 2).

The success of the new incentives also resulted, however, in a relevant increase in the amount of public financial resources needed to sustain the incentives system. As a result, in mid-2012, the government decided to revise its approach to electricity production from biomasses significantly.

As it happened in 2008, the change was extremely relevant and with effects from January 2013:

> **Box 2 Energy Generation from Biomasses in Germany**
> Germany is, as already mentioned, the leading country at the European level for the production of energy from biomasses, with more than 30 TWh of electricity generated in the year 2012. About 63% of the total comes from the "transformation" of biogas, which has been subsidized with a feed-in tariffs system since 2000 within the Erneubar-Energien-Gesetz framework. Like the Italian case, even if on a larger scale, the largest share of energy producers from biomasses in Germany (about 75%) is represented by farmers. The average size of plants is about 5–6 MW and is therefore significantly higher than in Italy, but they still exploit mostly local agricultural products and subproducts. Indeed, the average size of a farm in Germany is about 46 hectares against 8 hectares in Italy. It is worth mentioning that the incentive system in Germany shows specific and relevant bonuses for the coproduction (together with electricity) of heat to be used locally for the needs of the farm.

- a cap was introduced for the newly installed generation power over the 2013–2015 period. As a whole, 490 MW is available over the 3 years for small and medium size plants (less than 5 MW) and 120 MW is available for larger plants;
- a feed-in tariff system was defined for small and medium size plants by cutting, on average, 30% from the previous tariffs and differentiating them further by size (the larger the size, the lower the tariff available). A system has been introduced for ranking plants against the power for new installations available in each year, as mentioned in the previous point. Ranking ("Registro impianti") is done once a year, and the relative positioning is based mainly on the size of the plant (the lower the better) and on the date of the authorization (the older the better). A new and relevant risk has therefore been introduced for producers, that is, a risk that the plant is not accessing the incentives system;
- an auction mechanism based on a feed-in tariff system was introduced for larger plants, forcing players to compete for accessing the 120 MW of allowed incentivized power capacity.

The expected results (see also Table 9) are (i) a drop in the newly installed power due to the presence of a cap along the 3 years of 610 MW (i.e., less than half the power capacity installed every year during the last three years); and (ii) an ever more significant reduction in the average size of new plants, thus further forcing the adoption of a decentralized and distributed electricity generation system.

Table 9 First Results for Biomasses in Italy of the New Incentives System for 2013[a]

Biomasses	Available power capacity (MW)	Number of requests (n)	Requested power capacity (MW)	Average size (MW)	Cap saturation (%)
<5 MW	170	229	220.6	0.96	129.8
>5 MW	120	1	13	—	10.8

[a]*Source:* Energy & Strategy Group elaboration on data from Gestore dei Servizi Energetici.

As a whole, the new system is attempting to properly solve the trade-off between keeping the overall amount of resources to be committed to RES under control and fostering the development of a national market and industry value chain for biomasses in Italy. In this respect, exploiting the local agricultural products and subproducts more, as well as focusing on farmers as the focal market segment for producers, resulted in an increase since 2008 of the number of national players involved in the different stages of the value chain. The small size of the plants claimed for more localized engineering and construction, as well as operations and maintenance services.

Differently from the photovoltaics industry, whose exponential growth was too fast for developing a national industrial system, in biomasses the slowest pace of development (and of normative changes), as well as the characteristics of the focal market segment, represented a favorable environment for fostering the creation of industrial players well suited to serve at least the national market.

RENEWABLE ENERGY GENERATION: INCENTIVES MATTER

In introducing the analysis of photovoltaics and biomasses for electricity production, it has been argued that the presence of incentives is of paramount importance for ensuring the development of a market for RES.

The two cases investigated in detail in this chapter, however, revealed that the impact of incentive systems is more articulated and that the following issues have to be addressed carefully in their design.
- The choice of the form of incentive (feed-in tariff, feed-in premium, quota obligations) does not seem to play a crucial role in the development of the market for RES, as industrial players and investors are usually able to adapt their evaluation mechanisms to any form of incentive system. Nevertheless, feed-in tariffs, being comparatively less exposed to the volatility of energy prices, are a form of incentives that ensure maximum viability, that is, are

adopted and understood more easily by any typology of energy producers from RES, from individuals to small and large players.
- The procedures designed for accessing the incentive systems are significantly more relevant. The higher their complexity and/or the more they are exposed to exogenous factors (e.g., to a cap for the allowed power capacity), the higher the risk for energy producers from RES. The higher the risk, the lower the chance to have the investment financed by banks or, alternatively, the higher the cost of financing.
- Similarly, the process for having an electricity production plant installed and put in operation as a whole, including all the authorizations required for construction, as well as for grid connection, has a strong impact on the actual development of the market for RES. The longer the time required for having the plant ready to produce electricity, the higher the risk and the costs for energy producers.
- The incentives system, particularly through differentiating incentives by size of plants and/or characteristics of the energy producers, is a strong instrument not only for developing the market for RES, but also for shaping the industry for RES in a given country. In this respect, two factors emerge as key:
  (i) the stability of the incentive system, that is, the fact that, once designed, the system lasts for a significant portion of time, allowing industrial players, particularly those involved in the activities of the value chain far from the final market (e.g., the production of components or technologies), to define their investment plan properly and to have enough time to gain payback for their investments.
  (ii) the absolute level of the incentives, which drafts the growth curve of the market directly. Indeed, the faster the market, the lower the chance to develop a national industry for RES, particularly if, as it occurred in Italy, the decision to invest in the development of photovoltaics comes once other European countries have already developed strong industry players (the first to be ready for exploiting the market potential).

Moreover, the faster the market, the higher the chance that, without a strict control mechanism such as a cap for the installed capacity or the expenses for incentives, governmental expenditures for supporting RES go out of control, thus undermining the stability of the incentives system, which goes rapidly into a vicious cycle. A slower but more controlled growth rate, as happened in Italy for biomasses, where the system was also designed in 2008 to target the most appropriate market segment, is going to provide better results in the long run.

CHAPTER 15

Germany's Energiewende

Eric Borden[a], Joel Stonington[1, b]
[a]German Chancellor Fellow, Alexander von Humboldt Foundation, Berlin, Germany
[b]Visiting journalist, Spiegel Online International and German Chancellor Fellow, Germany

Contents

Primary Goals and History of the Energiewende	369
Background and History of the Energiewende	370
Renewable Energy Law and Feed-in Tariff Mechanism	373
Cap and Trade: Putting a Price on Carbon in Europe	374
Achievements of the Energiewende	376
Challenges and Impediments to the Energiewende	379
Conclusions and Going Forward	385

PRIMARY GOALS AND HISTORY OF THE ENERGIEWENDE

The "Energiewende," or energy transition, is the term for the change of Germany's energy system from conventional, fossil-fuel, and nuclear-based means of energy production to cleaner, sustainable production and consumption. The Energiewende's far-reaching ambition touches virtually every sector of German society. The primary goals of the Energiewende, as set out by the government, begin with a reduction in greenhouse gas emissions. This is achieved primarily through a reduction in energy consumption via increased energy efficiency, an increase in the share of renewable energy, and a shift away from oil in the transport sector toward electric vehicles and cleaner-burning fuels. The primary goals as set forth in Germany's Energy Concept of 2010 are summarized in Table 1.

Although ambitious goals were set years earlier, the March 2011 Fukushima nuclear disaster in Japan galvanized a rapid response from Germany with regards to nuclear power. After vociferous public opposition and protests, the German government decided that nuclear power should be phased out of the energy mix. Eight nuclear reactors were immediately shut down and the remaining nine are scheduled to be phased out of the electricity sector by

---

[1] Eric Borden and Joel Stonington are German Chancellor Fellows with the Alexander von Humboldt Foundation, conducting energy research in Germany. Funding for the research on this chapter has been provided by the Alexander von Humboldt Foundation in the framework of the German Chancellor Fellowship Program.

Table 1 Primary Goals of the Energiewende

	2020	2030	2040	2050
Greenhouse Gas Emissions (compared to 1990)	-40%	-55%	-70%	-80% to -95%
Energy Efficiency (compared to 2008)				
Primary Energy Consumption	-20%			-50%
Renewable Energy				
Gross Final Energy Consumption	18%	30%	45%	60%
Share in Electricity Production	35%	50%	65%	80%
Transport (compared to 2005)				
Final Energy Consumption	-10%			-40%
Number of Electric Vehicles	1,000,000	6,000,000		

Source: BMWi, BMU, 2010. *Energy Concept for an Environmentally Sound, Reliable, and Affordable Energy Supply,* September 2010. Federal Ministry of Economics and Technology and Federal Ministry for the Environment.

2022. Nuclear energy was not a minor portion of Germany's electricity generation mix, accounting for over 20% of its electricity production at the time.

These events help illustrate the vital role electricity plays in Germany's energy sector. The sector is responsible for about 43% of emissions in Germany, and, particularly for a modern industrial nation that prides itself on its manufacturing sector, the stability of supply and price of this good is fundamental to German society, business, and way of life.

This chapter devotes itself primarily to how the Energiewende movement has influenced the electricity sector and its transition to cleaner, more sustainable methods of production. The primary tools the government has used to implement the energy transition, including the Renewable Energy Law (EEG) and a market to put a price on carbon dioxide (CO_2)(Figure 1), are discussed, followed by achievements and challenges of the movement going forward. First, however, we look back to understand the deep roots from which the Energiewende continues to grow.

BACKGROUND AND HISTORY OF THE ENERGIEWENDE

The speed of the Energiewende seems, at times, dizzying. The recent rapid abandonment of nuclear power caught many by surprise, especially the industry players that lost billions from the decision. Yet Germany did not come to these goals and decisions on a whim, or even recently, but rather through the

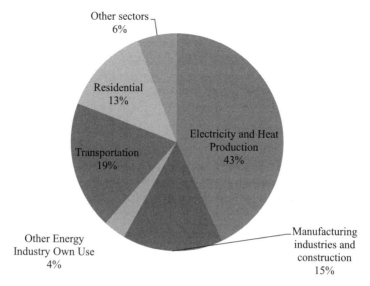

Figure 1 CO_2 emissions in Germany by sector (2010).*

* International Energy Agency 2012, *CO_2 emissions from fuel combustion highlights*, p. 69.

gradual building of institutions and collective social consciousness about energy issues over several decades. Even the term "Energiewende" was coined in the beginning of the 1980s by the Eco (Öko) Institute in Germany.

Four decades ago, Germany's energy politics centered primarily around energy independence due to the 1970 Arab oil embargo. A shift toward increasing production from coal and nuclear power was seen as prudent to move away from the Middle Eastern influence on the energy sector. However, a nuclear accident at Three-Mile Island in the United States (1979) and a catastrophe at the Chernobyl nuclear power plant in Ukraine (then a part of the Soviet Union, 1986) galvanized popular movements against nuclear power in Germany, although it was not completely taken out of the energy mix at this time. Energy efficiency and renewable energy became increasingly emphasized, although public funds for renewable energy were used primarily for the transfer of technology to developing countries, not domestic German use. The period saw lobbies, industry organizations, and interest groups, including the German Solar Energy Industry Association (1978), develop around the development of renewable technology.[2] The

---

[2] Jacobsson, S & V. Lauber, V 2006, 'The politics and policy of energy system transformation: explaining the German diffusion of renewable energy technology', *Energy Policy*, vol. 34, no. 3, 261-263.

movement also had a voice within the political establishment with the founding of the Green party in Germany.

Media reports in the mid-1980s on influential publications from the German Physical Society and other scientific organizations garnered public attention regarding the threat of climate change. Ensuing years saw increased interest in the promotion of renewable energy with limited implementation on the ground.

In 1990, Germany's feed-in tariff (FIT) scheme was introduced (predecessor to the Renewable Energy Law of 2000, discussed later). It was meant to promote small hydropower installations but came to encompass decentralized solar and wind power development. Other initiatives at the state level put locals in charge of selecting locations for wind turbines (often a controversial topic) and public awareness campaigns informed voters of the benefits of renewable energy.[3]

National expenditures on energy research over the past few decades mirror this proportional decreased focus on nuclear and coal, and a relative and nominal increase for renewable energy and energy efficiency (Figure 2).

These historical events and the development of institutional support clusters created the foundation for passage of one of the primary tools of the Energiewende, the Renewable Energy Law (EEG), which provided a stronger FIT system to promote renewable energy technology and reach emissions targets, and has remained the primary instrument of renewable policy through several iterations since the first in 2000. At the European

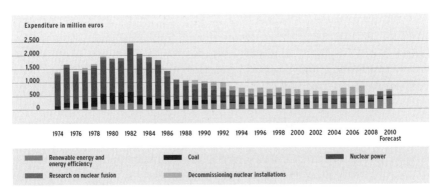

Figure 2 German federal energy expenditures by category.*

* Dollar amounts are inflation adjusted. The German Federal Energy Ministry (BMU) 2010, *Innovation through research*, p. 12.

[3] Mendonça, *Feed-in tariffs: accelerating the deployment of renewable energy*, p. 29.

level, Germany and other European Union (EU) member states created a one-of-a-kind cap-and-trade market to place a price on carbon emissions. These two central efforts of the Energiewende are described in further detail in the following sections and illustrate the domestic and international (in particular, European) components of the Energiewende.

RENEWABLE ENERGY LAW AND FEED-IN TARIFF MECHANISM

Germany's first public support of renewable energy followed almost a decade of discussion with little tangible results. The first FIT mechanism was created through a cross-party initiative to support small hydroelectric stations in the south of Germany. First suggested by Forderverein Solarenergie, Eurosolar, and an association of hydroelectric power plant owners,[4] the law served to decentralize the production of wind and solar energy as well. Utilities did not foresee damaging competition from this limited piece of legislation and did not exercise their influence at the time. The law mandated large utilities to connect decentralized renewable energy operators to the grid and pay them 65% to 90% of the "average tariff for final customers."[5] Under this incentive structure, the most efficient locations for wind and solar energy became worthy investments. However, sufficient incentive did not exist for alternative locations.

In 2000, Germany moved to update and enhance these renewable incentives through the Renewable Energy Law (EEG), which modified and vastly expanded incentives for renewable energy based on technology, location, and project scope. The new law also switched payment incentives to a fixed rate for renewable energy production. Economic studies calculate payment rates for renewable electricity that decrease over time (attempting to account for increased efficiency and technological improvements) and guarantee payments over a 20-year period for wind, solar, and other renewable energy technologies. Significantly, the law gives priority to renewable sources and obligates grid operators to purchase renewable electricity.

The costs of feed-in payments are passed down to consumers of electricity and industry participants as part of their electricity bill, although charges are not distributed equally between the two groups (discussed later). These payments have created an environment of investment and regulatory security that has increased the development of renewable energy sources rapidly

---

[4] Jacobsson & Lauber, *The politics and policy of energy system transformation: explaining the German diffusion of renewable energy technology*, p. 264.
[5] Mendonça, *Feed-in tariffs: accelerating the deployment of renewable energy*, p. 28.

in Germany. After the first law in 2000, there have been multiple iterations that have modified and adjusted rates of compensation for renewable energy technologies.

The Renewable Energy Law of 2008 provides an example of how these FIT payments work. For instance, roof-mounted solar photovoltaic (PV) panels were provided, varying payments depending on the size of the installation, displayed in Table 2 when electricity production fed into the German electrical grid.[6]

Depending on the size of the installation, the owner is paid a specified amount for electricity fed into the grid. These amounts decrease over time according to a set digression schedule and are adjusted further through iterations of the Renewable Energy Law.

This model of renewable energy development belongs to a beginning phase of the FIT policy "that provided transparency, longevity, and certainty to investors."[7] Recently, however, the structure has started to shake, induced by increasing costs, caused in part (and perhaps ironically) by a rapid decrease in the cost of solar PV panels with a dramatic increase in solar PV deployment in Germany—at a large overall cost passed down to consumers. A recent PV amendment in June 2012 went so far as to set a future cap on PV installations of 52 GW, above which solar installations will not receive additional compensation. A further look at the achievements of the Renewable Energy Law and current challenges is discussed later in this chapter.

CAP AND TRADE: PUTTING A PRICE ON CARBON IN EUROPE

Germany is not an island and is connected both politically and geographically with its European neighbors. Faced between the two main economic tools used to limit pollution, Europe chose a cap-and-trade system instead

Table 2 Feed-in Tariff for Roof-Mounted PV in Renewable Energy Law of 2008*

Size	Payment (€ cents per kWh)
< 30 KW	43.01
30 KW to 100 KW	40.91
100 KW to 1 MW	39.58
> 1 MW	33.00

* KW, kilowatt; kWh, kilowatt hour. See the Renewable Energy Sources Act of October 25, 2008, p.25.

[6] Feed-in tariff payments for solar and other technologies have since been modified further. Table 2 provides an example of how these payments work.

[7] Fulton, M, Capalino, R & Auer, J 2012, *The German feed-in tariff: recent policy changes*, Deutsche Bank, p. 1.

of a tax on carbon. According to the EU, the system "is based on the recognition that creating a price for carbon offers the most cost-effective way to achieve the deep reductions in global greenhouse gas emissions that are needed to prevent climate change from reaching dangerous levels."[8]

Under cap and trade, a limit is set on the total amount of carbon that can be released, and certificates are issued for each ton of carbon under that cap. The total amount of carbon allowed to be released each year is reduced at a set rate and companies must pay for polluting beyond the limit. The certificates can then be traded on the open market so companies that reduce emissions successfully can profit, whereas high-polluting companies must pay for the "right" to pollute. Overall, a carbon market is intended to place a price on a negative by-product (or externality) of a process—in this case, pollution—creating an economic incentive to produce less of this by-product. Trading of certificates allows for economic efficiency, so overall targets can be achieved while lessening the burden for less-adaptive industries and manufacturers.

Germany signed on to the European Emissions Trading System (ETS) along with 30 other countries in a system that now encompasses 11,000 power plants and industrial factories, as well as airline travel. The cap on carbon emissions was put in place to incentivize a 20% reduction below 1990 levels by 2020 as part of the "20–20–20" targets passed and ratified by EU member states.[9] The ETS also expands beyond Europe, accepting credits generated under the Clean Development Mechanism of the Kyoto Protocol, an international climate treaty intended to curb global emissions. A new phase of the ETS began in 2013.

The price for emitting a ton of carbon peaked in 2008, at around €30 per ton. However, the financial crisis in Europe hurt the functioning and value of the ETS, as carbon emissions dropped due to the economic downturn and consequent drop in emissions. The ETS has no mechanism to lower the output of credits and certain types of companies were over awarded free credits. The market was flooded and the price on carbon has crashed to a point where entities have little incentive to reduce their carbon output (Figure 3).

The current failures of the ETS are of major concern for Germany (more on this later). As one member of the German member of the European Parliament, Jo Leinen, wrote in *European Voice*, "The success of Germany's

---

[8] See European Commission, *EU action against climate change*. Available from: <http://ec.europa.eu/clima/publications/docs/ets_en.pdf>.
[9] Refers to European targets of a 20% emissions reduction, a 20% increase in renewable energy, and a 20% increase in efficiency by 2020.

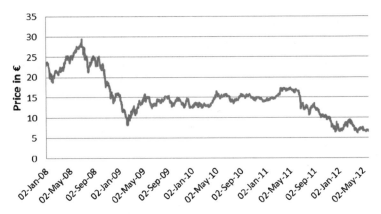

Figure 3 Carbon price since 2008 (© Europäische Union, http://eur-lex.europa.eu/).*

* European Commission 2012, *The state of the European carbon market in 2012*, p. 5.

Energiewende and the EU's low-carbon transition depends on the effectiveness of tools and policy frameworks to drive low-carbon technologies. That is why the functioning of the ETS must be reviewed."

ACHIEVEMENTS OF THE ENERGIEWENDE

The Energiewende is sometimes characterized as a high-stakes "gamble" or, alternatively, a worthy initial investment that will pay dividends in economic and sustainable outcomes. This section discusses some of the primary achievements of the Energiewende, including decreased fossil fuel emissions, increased deployment of renewable energy, establishing a price on carbon at the European level and a way to track emissions, and a social and political acceptance unparalleled in the world. It has also spurred jobs and innovation in the renewable industry. Challenges and impediments to the Energiewende are discussed in the following section.

Germany is already seeing less of a link between emissions and growth. Between 1990 and 2010, Germany's CO_2 emissions have decreased by around 20%. In comparison, emissions in the United States have increased by 10% over the same period.[10] Parallel to this, production of electricity from renewable sources in Germany has increased between 1990 and 2012, driven by the first FIT mechanism and the Renewable Energy Law and its subsequent iterations (Figure 4).

[10] International Energy Agency 2012, *CO₂ Emissions from fuel combustion highlights*, p. 48.

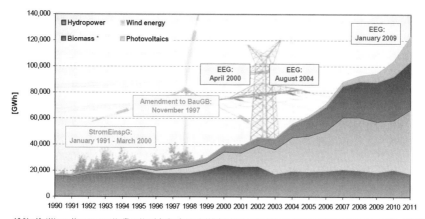

Figure 4 Renewable energy deployment in Germany from 1990 to 2011.*

* EEG= Renewable Energy Law. The German Federal Energy Ministry (BMU) 2012, *Development of renewable energy sources in Germany 2011*, graphics and tables, p. 14.

These policies have driven renewable energy development in the country—it currently makes up over 20% of annual electricity production in Germany (Figure 5).

Europe's carbon market has also achieved some early success since its inception in 2005. According to a study by the Environmental Defense Fund,[11] long-term prices between €15 and €30 per ton resulted in reductions in greenhouse gases and increased low-carbon innovation at lower-than-expected costs, just 0.01% of the gross domestic product. Perhaps most importantly, successful auctions were held of carbon credits, data were collected on emissions throughout Europe, and the carbon market began functioning, creating a financial market for a commodity that, beforehand, was emitted with little to no cost into the atmosphere in Europe, with no systematic way of holding those responsible to account.

The massive growth of the renewable energy industry in Germany has fueled lower prices for PV panels and wind turbines. For example, the system price for someone installing solar panels on their barn in Germany's

---

[11] Environmental Defense Fund 2012, *The EU Emissions Trading System*. Available from: <http://www.edf.org/sites/default/files/EU_ETS_Lessons_Learned_Executive_Summary_EDF.pdf.>

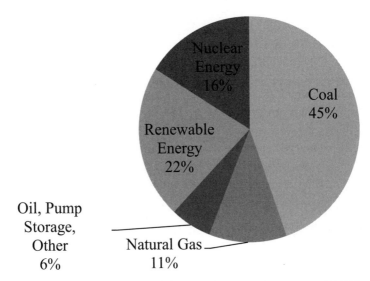

Figure 5 Gross production of electricity by source in Germany (2012).*

* Coal consists of 25.6% brown coal and 19.1% hard coal. German Association of Energy and Water Industries (BDEW) 2013, *Erneuerbare energien und das EEG: Zahlen Fakten, Grafiken*, p. 15.

Bavarian region was 5.1 euros per peak kilowatt installed back in 2006. By the end of last year, that same number had dropped to 1.75 euros.[12]

One of the most important achievements of the Energiewende is its acceptance into the mainstream of German society—at individual, social, and political levels. The question in Germany is not *should* the Energiewende and its ambitious goals be achieved, but *how* should these goals be accomplished? This is not a semantic difference—as other developed and economically advanced countries debate the merit of overhauling the way they produce energy and manufacture products, Germany forges ahead. Although the national election in 2013 will certainly affect the country's energy policy, the outcome of the election is unlikely to change the country's support of renewables or the goals of the Energiewende. It must be noted here that the perceived costs of the Energiewende and renewables in Germany have been met with calls for a fundamental overhaul of how the Energiewende is implemented (for a further discussion, see the next section).

One of the ways the Energiewende has woven itself into the fabric of German life is demonstrated by the fact that solar panels in Germany are

---

[12] German Solar Industry Association (BSW-Solar) 2013, *Statistic data on the German solar power (photovoltaic) industry*. Available from: <http://www.solarwirtschaft.de/fileadmin/media/pdf/2013_2_BSW-Solar_fact_sheet_solar_power.pdf.>

often owned by individuals and cooperatives, effectively democratizing energy production (while cutting into the profits of large fossil fuel-dependent energy producers). The result is that more than 80,000 Germans are invested in cooperative power plant projects that amount to more than 800 million euros in investments. This broad base of citizen investment has made the Energiewende far more approachable for average Germans. It has also put profits into the hands of co-ops and individuals.

Despite the immense achievements of Germany's energy transition, challenges have been encountered, and remaining hurdles to the ultimate success of the transition continue to loom large.

CHALLENGES AND IMPEDIMENTS TO THE ENERGIEWENDE

Most envision energy little further than a light switch or a plug in the wall. But while turning on a new computer is just a matter of plugging it in, a new energy source requires significant financial backing, long-term investment, interconnection, and, in the case of increasing amounts of fluctuating sources of renewable energy, a modernization of how electricity markets work. Challenges and impediments to the Energiewende include rising costs, disproportionate allocation of resources to some renewable technologies, loss of public confidence in how the Energiewende is being managed, integration of increasing amounts of renewable energy through more flexible operation of the electricity system and interconnection, and low spot market prices on Germany's energy-only market—all within the context of an unstable economic situation in Europe that may threaten Germany's export-oriented economy.

The up-front costs of the shift to renewable energy are one of the largest challenges the movement will continue to face. Figure 6 shows a summary of recent costs, primarily via the FIT mechanism, that have been invested in renewable energy, followed by how these costs are paid for.

Over the last few years, particularly in 2011 (as well as 2012), investment in solar has significantly outpaced investment in other renewable technologies, which has also been a significant contribution to rising electricity prices in recent years (Figure 7).

How these costs are *perceived* may be even more important than their nominal total. With the Renewable Energy Law, consumers are charged an additional amount, or surcharge, to a "base cost" to pay for the FIT mechanism, discussed earlier. These charges are listed clearly on consumer bills. Household bills do not, however, enumerate the financial support enjoyed

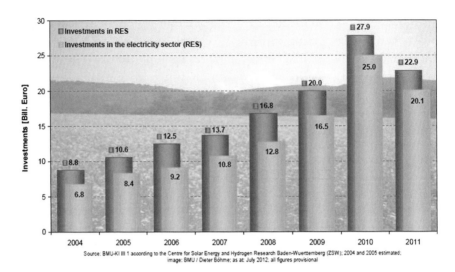

Figure 6 Investments in construction of renewable energy sources to 2011. RES, renewable energy sources.*

* The German Federal Energy Ministry (BMU) 2012, *Development of renewable energy sources in Germany 2011*, graphics and tables, p. 47.

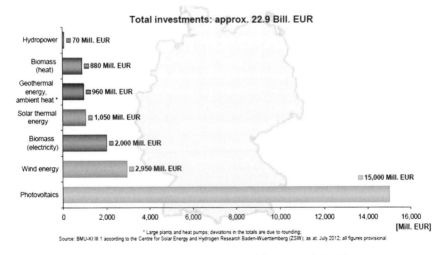

Figure 7 Investments in renewable energy (2011).*

* The German Federal Energy Ministry (BMU) 2012, *Development of renewable energy sources in Germany 2011*, graphics and tables, p. 46.

by fossil fuels and nuclear power over the last several decades, which far exceed support for renewable energy.[13] This may create a false impression that renewables receive excessive support or are more expensive than fossil fuel-driven energy production.

Figure 8 demonstrates that overall prices for an average three-person household have increased by 51% since 1998, with base prices (which make up the majority of electricity prices) increasing by 10% over this period, but taxes, charges, and levies, a portion of which is the renewable energy apportionment (or EEG surcharge), increasing 179%. In 2012, out of the (on average) 26 cents per kWh paid by a three-person household, 3.6 cents was allocated to the Renewable Energy Law apportionment.[14] In 2003, this apportionment was just 0.4 cents. The Eco (Öko) Institute calculates that between 2003 and 2013, 44% of the electricity price increase can be attributed to the increasing surcharge to promote renewable electricity. The charge in 2013 is expected to rise to around 5.3 cents per kWh, or an extra

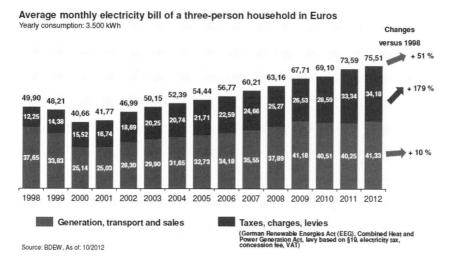

Figure 8 Increasing electricity prices in Germany.*

* Kuhlmann, German Association of Energy and Water Industries (BDEW) 2012, *The German "Energiewende": German gamble or modern vision?* p. 20.

---

[13] Küchler and Meyer calculate that between 1970 and 2012, around 54 billion was spent on renewable energy research and deployment in Germany compared to 177 billion for coal and 187 billion for nuclear power. See Küchler & Meyer 2012, Greenpeace and Wind Energy Industry Group, *was Strom wirklick kostet*, p. 15.

[14] German Association of Energy and Water Industries (BDEW) 2012, *BDEW-Strompreisanalyse Oktober 2012*, p. 6.

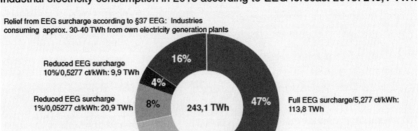

Figure 9 Industry payment for the renewable energy surcharge (2013).*

* Kuhlmann, German Association of Energy and Water Industries (BDEW) 2012, *The German "Energiewende": German gamble or modern vision?* p. 23.

50 euros per year for an average household, partially due to a rapid increase in solar panel deployment, discussed later.

A portion of the increase in the renewable energy apportionment for households is also due to exemptions from the charge allocated to German industry participants. The argument, generally, is that German industry and manufacturing must remain competitive with internationally. In a climate of climbing prices, these industry exemptions will become increasingly controversial and likely untenable in their current form.

Figure 9 demonstrates that for expected 2013 consumption levels, 47% of industry's electricity consumption will be subject to the full Renewable Energy Law surcharge. Without this exemption, the energy ministry calculates the surcharge for 2013 would be a full 1 cent lower for consumers.[15]

These rapidly increasing costs have begun to shake the pillars of Energiewende's development. Dangerous proposals have been floated in the latest election cycle, such as the idea to take away a portion of past profits from renewable energy generators (the idea has since been taken off the table). Nevertheless, investor confidence has been shaken. Further, German support in *how* the Energiewende is progressing is seeing a decline (see Figure 10). This progression comes amidst a European economic crisis

---

[15] It is important to remember that all costs incurred by renewable energy have, in addition, associated benefits via reduced emissions, revenues to renewable energy companies, and employment (almost 400,000 jobs in the renewable energy sector in 2011). The exact cost–benefit analysis, including foregone emissions, is not presented here.

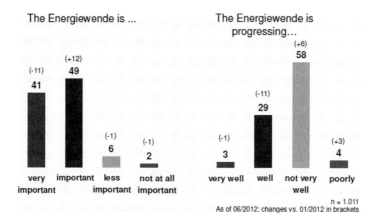

Figure 10 Public perception of the Energiewende.*

* Kuhlmann, German Association of Energy and Water Industries (BDEW) 2012, *The German "Energiewende": German gamble or modern vision?* p. 33.

that is affecting Germany, although not yet to the extent of many of its neighbors. Germany's central bank forecasted just 0.4% growth in 2013.

In addition, although renewable energy has been integrated into Germany's system without significant stability issues to this point, the technical and economic challenges of integrating fluctuating sources of energy, namely, wind and solar power, are likely to play an increasing role in coming years. Unlike a base load plant such as coal or gas that is turned on and can provide regularly scheduled amounts of power, wind and solar go through cycles between zero output and spikes of energy. The integration of large amounts of this energy is necessary for Germany to achieve its goals and to maximize investment returns in these technologies.

The primary tools that Germany (as well as other developed electrical systems) will utilize to integrate larger and larger amounts of fluctuating renewable energy are balancing from thermal power plants, interconnection via transmission lines both within Germany and throughout the European area, curtailment of renewable energy, demand-side management, and energy storage.

In the short term, fossil-fueled power plants are expected to continue to provide much of the balancing required to stabilize the system, for example, to ensure that the demand of electricity and supply always match. However, as more and more fluctuating sources of renewable energy come online, other technologies and solutions will have to pay a bigger role. Interconnection via the build-out of transmission lines is likely the most critical of these solutions in the short to medium term. Transmission between production of renewable electricity and areas of largest demand must be, to a greater extent, connected physically so that the system can operate most effectively. For instance, the majority of Germany's wind production occurs in the north of Germany, whereas a large amount of Germany's demand is in the south. At the moment, insufficient interconnection exists to link the two, and progress has been slower than expected.

Storage of electricity will also begin to play an important role to optimize and "smooth out" production from both wind and solar, although technical and cost parameters currently impede large-scale implementation of electricity storage. Pumped-hydro storage, however, is a mature technology that will also play a role in balancing electrical load, particularly with storage facilities in the Alps and Scandinavian countries.

On the demand side of the equation, demand-side management allows, with relative ease compared to large infrastructure projects, the ability of industry and consumers to shift portions of their demand to nonpeak times of electricity consumption, when plenty of excess capacity exists. One can envision a system where electricity demand more closely tracks production of renewable energy. All of these measures require some degree of technological innovation and modernization of how the electricity system currently operates.

Presently, the integration of increasing amounts of renewable energy has caused a slowdown in investment for traditional power plants in Germany, which currently relies on an "energy-only" market for generators to make profits. One such instance occurs with solar power. The sun produces the most energy at the height of the day, causing solar energy to cut into the profits of fossil fuel plants—this creates financial disincentives for energy producers to invest in highly flexible power plants, necessary in the short to medium term to balance the system. As more standby plants may be needed to fill gaps in renewable energy output, it is unclear how these plants could be profitable. Although there is very little danger of brownouts or blackouts during the next few years, energy is a long game, with investments that are meant to pay out over decades.

There is further danger of grid disturbances due to behind-schedule upgrades to the high-voltage grid used to transport energy throughout the country rapidly and, even more complexly, throughout Europe. If interconnection continues at a turtle's pace, even more base load plants may be needed. But current spot prices, meaning the wholesale price of electricity, have been depressed largely due to the feed-in of renewables, which is prioritized in Germany over other forms of electricity and has a near zero marginal cost of production. This phenomenon has created the effect that energy from many gas plants is often sold at a loss. In January, one large producer, Düsseldorf-based E.ON, announced a number of closures of gas-fired plants that had been losing money due to low electricity prices on the market. Although natural gas emits about half the amount of carbon as a coal power plant when burned, Germany is once again investing in coal as the price has dropped. The country currently has 11 new gigawatts of coal plants in production, as if signifying the ongoing difficulty of phasing out these traditional forms of energy production. Further complicating the issue is the institutional influence coal power and its related industry partners continue to have in Germany's energy landscape.

Due to low market prices on Germany's energy exchange, many worry that the current pace of investments in power plants is insufficient to replace older plants and integrate larger amounts of renewable energy. Thus, it seems likely the country will adopt some kind of capacity mechanism, or additional payment for investment in additional power plant capacity, in the next years.

CONCLUSIONS AND GOING FORWARD

The Energiewende in Germany transforms the way energy is produced and consumed, incorporating sustainability to a degree not before attempted by a large, developed, industrial society. Germany's Renewable Energy Law has produced a wave of investment and deployment in renewable technologies, while a carbon market has attempted to put a price in Europe on the primary negative externality of energy production that contributes to climate change. Achievements have been seen from this course in the form of emission reductions, renewable energy deployment, job creation, and social acceptance of the movement. Challenges have been and will continue to be encountered, primarily through economic, technical, and political barriers and impediments that will alter how the Energiewende is implemented going forward.

The primary measures discussed in this chapter, the Renewable Energy Law, the FIT mechanism, and the carbon market, represent the domestic and international, namely, European, elements of the Energiewende. Yet, as has been seen, the primary successes of the Energiewende have been achieved domestically. Although Germany would like to see more success at the European level, resistance from its neighbors may be stiff. To the east, Poland produces more than 90% of its electricity from coal and, in the context of an economic crisis, this is not likely to change. To the southwest, France has built a system based around nuclear power, whose interests politically and economically will continue to dominate debate. Cooperation will likely occur in the near future in a build-out of existing transmission between neighboring countries—this may also meet resistance as fluctuating renewable sources from Germany can be difficult to integrate.

Higher energy prices have become a major issue in the 2013 national election and reflect how energy issues can be politicized quickly. Peter Altmaier, the head of Germany's energy ministry, and others in his party have called for a "brake" or slowdown of energy prices, almost always in the context of the rising surcharge for renewable energy. Other charges and fees, of which there are several, are not discussed with the same intensity as the renewable energy surcharge. Additionally, there has been no mention of creating something like a price floor for carbon to compensate for the unusually low carbon prices in the ETS. He has also not mentioned reducing subsidies to the nuclear and fossil fuel sectors. Finally, while Germany's environmental ministry has supported structural changes to the ETS that would help counteract low prices, the economy ministry has been against such changes and Angela Merkel has been unwilling to step in and resolve the dispute during an election year. Angela Merkel's challenger, Peer Steinbrück of the Social Democratic Party, has been gaining political ground from the growing discontent with the Energiewende, demonstrating the centrality of the energy issue in German politics.

Thus, there are a number of ongoing political debates about the high costs of renewables and how those costs are passed on to the consumer. All of this is in light of the main question, which is how to make renewables more competitive in a market that is far from free? Offshore wind desperately needs government support to get to competitive cost levels. The grid needs major investments to handle these new types of energy. Storage technologies range in development, but require fundamental research, development, and demonstration to help integrate renewable energy further down the road. All of this is within the context of a major, ongoing economic crisis in Europe.

Nevertheless, what the Energiewende has achieved and stands to accomplish in coming years is significant in the scope of the undertaking and the depth of change that has already occurred. The hurdles are not small, but the gains, for Germany and the on-looking world, are enormous.

CHAPTER 16

Educational Programs for Sustainable Societies Using Cross-Cultural Management Method

A Case Study from Serbia

Jane Paunković
Professor, Faculty of Management Zajecar, Megatrend University, Serbia

Contents

Introduction	387
Educational Program "Strategic Leadership for Sustainable Societies" in Serbia	390
Sustainable Development Methodology	392
Program Activities	393
Activity "Strategic Visioning Conference"	393
Activity "Training of the Trainers"	394
Activity "Workshop for Professions"	394
Activity "Development of Pilot Projects for Implementation of Sustainable Development in Organizations"	395
Activity "Implementation of Pilot Projects for Sustainable Development in Organizations"	395
Activity "Final Conference"	395
Cross-Cultural Management Approach	395
Organizational and National Culture	397
Research Objective and Study Design	397
Results	398
Discussion	398

INTRODUCTION

Sustainable development became a globally recognized concept through the Brundtland Commission Report "Our Common Future," instigated by the World Commission on Environment and Development (WCED), an entity of the United Nations. The WCED related sustainability to environmental integrity and social equity, but also to corporations and economic prosperity as "development that meets the needs of the present without compromising generations to meet their own needs" (WCED 1987).

The Earth Summit in Rio de Janeiro in 1992 initiated general acceptance of this definition by business leaders, politicians, and nongovernmental organizations(NGOs)(Dyllick & Hockerts 2002). For organizations, it was a challenge to simultaneously improve social and human welfare while reducing ecological impact and ensuring the effective achievement of organizational objectives (Sharma & Ruud 2003).

In the European Union (EU), environment action programs (EAP) have directed the development of EU environment policy since the early 1970s. The 5th Environment Action Program (1992–1999), "Towards Sustainability," emphasized the integration of environmental concerns into wider economic and social policies and included recommendation that sustainable development should be the core concern of EU policies. The Sixth Environmental Action Program "Environment 2010: Our Future, Our Choice," which was in action from 2002 to 2012, focused on four priority areas: climate change, biodiversity, environment and health, and sustainable management of resources and wastes. The Sixth EAP accentuated that high environmental standards are also an engine for innovation and business opportunities and must work to decouple environmental impacts and degradation from economic growth: "Business must operate in a more eco-efficient way, in other words producing the same or more products with less input the ability of future and less waste, and consumption patterns have to become more sustainable."

For the "NEW EAP," which was presented in 2012, to be effective until 2020, the European Commission has a proposal entitled "Living Well, within the Limits of Our Planet" that aims to enhance Europe's ecological resilience and transform the EU into an inclusive and sustainable green economy. The commission set out the vision of where it wants the union to be "green" by 2050:

> In 2050, we live well, within the planet's ecological limits. Our prosperity and healthy environment stem from an innovative, circular economy where nothing is wasted and where natural resources are managed in ways that enhance our society's resilience. Our low carbon growth has long been decoupled from resource use, setting the pace for a global sustainable economy.

The new EAP added two priority objectives to focus on: enhancing the sustainability of EU cities and improving the EU's effectiveness in addressing regional and global challenges related to the environment and climate change.

The United Nations Conference on Environment and Development organized the "Rio+20 Conference" in June 2012 in Rio de Janeiro to mark the 20th anniversary of the 1992 and the 10th anniversary of the 2002 World Summit on Sustainable Development in Johannesburg. The outcome of the conference, "The Future We Want," states,

> *We, the Heads of State and Government and high-level representatives, having met at Rio de Janeiro, Brazil, from 20 to 22 June 2012, with the full participation of civil society, renew our commitment to sustainable development and to ensuring the promotion of an economically, socially and environmentally sustainable future for our planet and for present and future generations.*

Nevertheless, the supporting documents for the conference ("A Pocket Guide to Sustainable Development Governance") once again state clearly that despite the growing number of institutions, instruments, and processes addressing sustainable development, environmental problems have intensified globally. It also outlines that continued degradation of the global environment has not been caused solely by governance weaknesses, but rather by a multitude of drivers, including prevailing economic models and patterns of consumption and production.

For many years, governments and civil society jointly put emphasis on education as a key policy instrument for bringing about transition toward sustainable behavior: education in terms of the lifelong process of learning, action, and critical reflection involving all citizens. The United Nations Educational, Scientific and Cultural Organization (UNESCO) declares that "Education for Sustainable Development (ESD) allows every human being to acquire the knowledge, skills, attitudes, and values necessary to shape a sustainable future." It defines ESD as not a particular program or project, but rather an umbrella for many forms of education that already exist, and new ones that remain to be created. Because these essential characteristics of ESD can be implemented in myriad ways, ESD programs reflect the unique environmental, social, cultural, and economic conditions of each locality. Furthermore, ESD increases civil capacity by enhancing and improving society through a combination of formal, nonformal, and informal education. In December 2002, the United Nations General Assembly proclaimed the UN Decade of Education for Sustainable Development, 2005– 2014 (DESD), emphasizing that education is an indispensable element for achieving sustainable development. Since the launch of the DESD in 2005, many sustainability issues have been included in education: peace education, global education, development education, HIV and AIDS education, citizenship education, intercultural education, and holistic education, along with established environmental education and health education.

A report of the United Nations Conference on Sustainable Development held in Rio de Janeiro in June 2012 reaffirms commitments to the right to education and commitment to strengthen international cooperation to achieve universal access to primary education, particularly for developing countries. It reaffirms that full access to a quality education at all levels is an essential condition for achieving sustainable development, poverty eradication, gender equality, and women's empowerment, as well as human development. It emphasizes the importance of supporting educational institutions, especially higher educational institutions in developing countries, to carry out research and innovation for sustainable development, including in the field of education, to develop quality and innovative programs, including entrepreneurship and business skills training, professional, technical, and vocational training, and lifelong learning, geared to bridging skills gaps for advancing national sustainable development objectives. One of the recommendations of the report is to put education in the core of the sustainable development goals.

EDUCATIONAL PROGRAM "STRATEGIC LEADERSHIP FOR SUSTAINABLE SOCIETIES" IN SERBIA

The National Strategy Sustainable Development of Serbia (NSOR) was presented and adopted at the government level in 2008. NSOR defined major national priorities for achieving sustainable development in Serbia for the next 10 years, including education and lifelong learning, as an important component that will provide a qualified and skilled workforce for a knowledge-based economy in accordance with the EU Lisbon strategy.

In that context, NSOR discusses the current situation in the Serbian educational system and emphasizes that a knowledge-based economy could not be founded on "factographic," rigid, scholarly, or textbook knowledge, but on an array of skills, competencies, and interests that create innovation, problem solving, and cooperation for social advancement. Learning processes should not be based on answers to questions on "what" and "who," but also on "why" and "how." The national strategy concludes that the educational system in Serbia is generally unsustainable, is not adequately efficient or effective, does not include all categories of children, and does not have a quality outcome at any level. Consequently, education for sustainability in Serbia does not imply simple program adjustment, but a new educational approach is essential for a knowledge-based economy. NSOR also recognizes that education about sustainable development principles is fundamental for good governance, decision making, and capacity building at all levels.

Regrettably, 5 years after the adoption of NSOR, there is still a general lack of awareness about sustainability at all levels in Serbia. Some research data (Miladinovic & Paunković 2012) revealed that less than 5% of the employees in local municipalities actually involved in sustainable development programs and actions were able to explain the concept of sustainability in common terms. The general public and students (high school and university) are even less informed, despite proclamations about the UN "Decade of education for sustainability."

Although strategies of sustainable development have been presented at the government level, there is a lack of professional and organizational knowledge of internationally recognized campaigns for the sustainable development of cities and towns in local communities. There is also inadequate coordination among possible implementing subjects, including the general public, municipal officers, NGOs, professional associations, and, more importantly, among legislators and key decision makers in local communities.

A few members of the Faculty for Management Zajecar (FMZ) have designed the program "Strategic Leadership for Sustainable Societies" with the goal to assist local communities in east Serbia in their attempt to adopt sustainability. The FMZ, located in east Serbia, a part of Megatrend University, Belgrade, introduced courses on sustainable development at the undergraduate and graduate levels, including international research doctorate studies, in 2007. FMZ was the first business school in the region to include sustainability in the accredited curriculum. As of now, several hundred students with comprehension and skills in management and sustainability have graduated from this faculty. It is a considerable contribution from the FMZ and a significant foundation for the advancement of this less developed region in Serbia.

The goal of the program "Strategic Leadership for Sustainable Societies" was to educate policymakers, municipal officers, and elected officials, as well as NGO staff who are, or should be, directly responsible for local sustainable development planning and implementation. Furthermore, the number of organizations in industry, tourism, agrobusiness, education, health, communications, press, entrepreneurs, and so on is also expected to benefit from the implementation of this highly structured and practical management system, with defined methodology for sustainable development interventions.

The desired outcome of the program for each participant is to have an understanding of the framework for sustainable development and be able to apply the framework in a range of situations for an analysis of problems as well as a creation of solutions. For a selected group of participants of this course ("trainers"), there is additional education in organizational and personal learning, as well as improved presentation, facilitation, and coaching skills.

Sustainable Development Methodology

There is a wide range of existing codes of conduct and best practice, business principles, and guidelines for sustainable development. In the process of planning of our activities we have used the framework derived from a number of key sources, including UNESCO ESD-Sustainable Development Action Plan for Education and Skills, the Rio Declaration, World Business Council on Sustainable Development, SIDA, National Strategy for Sustainable development of Serbia, and so on. For a guiding methodology for an introduction of sustainable development in organizations, we decided on a combination of the SIGMA (Sustainability Integrated Guidelines for Management) project (SIGMA Guiding Principles and SIGMA Management Framework) and the Natural Step Framework (TNS). The SIGMA project was launched in 1999 by the British Standards Institution, Forum for the Future, and Account Ability, with support of the U.K. Department of Trade and Industry (DTI). SIGMA recognizes that a key issue for organizations that want to respond to the challenge posed by sustainable development is how to take effective action. SIGMA provides a clear, practical, integrated framework for organizations. It allows an organization to build on what it has, to take a flexible approach according to its circumstances, and to reduce duplication and waste by seeing how different elements can fit together.

The SIGMA Management Framework is a cycle of four flexible implementation phases: leadership and vision, planning, delivery, and review, feedback, and reporting. Organizations may enter and move through the phases at different speeds and give different phases and different emphasis depending on their individual circumstances, the availability of resources, and the level of maturity of their sustainable development policies, strategies, and programs. The SIGMA Management Framework may be used to integrate existing management systems, build on existing approaches, and establish a stand-alone management system as guidance to deepen and broaden existing management practice without the formal structure of a management system. In order to ensure compatibility with the existing practice, the SIGMA Management Framework is modeled on approaches widespread in management systems. The "Plan, Do, Check, Act" model that underpins the SIGMA Management Framework is familiar to many organizations and has the benefit of being both practical and effective in delivering improved organizational performance.

The TNS framework is a methodology developed by The Natural Step, an international organization that helps organizations move strategically toward sustainability. It enables organizations to create optimal strategies for dealing with the present-day situation by incorporating a perspective of a sustainable future. Today's perception of what can be achieved never determines the direction of change, solely pace. This results in investments and activities that not only move the organization toward sustainability, but also maximize short-term profitability and long-term flexibility. The Natural Step framework is used by a number of organizations, including many global corporations in Europe and the United States, to provide strategic direction for their sustainability initiatives. The framework does not prescribe detailed actions. Once an organization understands the framework, it identifies and specifies the detailed means by which to achieve the strategy, because it knows its business best. Steps in the planning process are understanding and discussing the system conditions for sustainability, describing and discussing how the company relates to the system conditions in today's situation, creating a vision of how the company will fulfill its customers' needs in the future while complying with the system conditions, and specifying a program of actions that will take the company from today's situation to the future vision (Checkland 1999; Martin & Hall 2002).

Program Activities

The main activities of the educational program "Strategic Leadership for Sustainable Societies" have been developed on the principles described earlier. Authors have reported results of their investigation in detail previously (Paunković et al., 2007). An overview of the main activities of the program is presented in Table 1.

Activity "Strategic Visioning Conference"

The initial step in this program, a part of a strategic visioning process, is organization of the top-level "strategic visioning conference." The goals of this activity are to present the main concepts of sustainable development to policymakers and opinion leaders in municipalities and business communities and to raise awareness of the general public. Presentations at the conference are based on interrelated themes:
- Principles of sustainability
- Business benefits of sustainable development
- Action planning

Table 1 Brief Overview of the Main Activities of the Program

1. Strategic visioning conference
   - Top-level meeting for municipality officials
   - Top-level business meeting
2. Training of trainers
3. Workshops for professions
4. Development of pilot projects for implementation of sustainable development in organizations
5. Implementation of pilot projects for sustainable development in selected organizations
6. Final conference
   - Top-level meeting for municipality officials
   - Top-level business meeting
   - Exhibitions

The expected outcomes of the conference are to disseminate effective messages about sustainable development and to create shared strategic vision and compelling aspirations within the framework of a sustainable society in municipalities.

Activity "Training of the Trainers"

"Training of the trainers" is organized on practices of The Natural Step, SIGMA, and Professional Practice for Sustainable Development (Baines 2001). The training structure is based on a framework developed by Martin and Hall (2002). We have also developed basic organizational management training based on Hofstede's methodology (Hofstede 2001).

Activity "Workshop for Professions"

Trainers educated in the previous phase of the program are expected to design an original educational platform for particular business enterprises based on the same shared principles. Their requirement is to deliver 1- or 2-day workshops for professionals adapted to meet the needs of their specific audience, including organizational management principles. Workshops for professions are organized for members of management teams from industry, tourism, agrobusiness, education, health, communications, entrepreneurs, municipal officers, and NGOs. Participants of this course are responsible for delivering the concept of sustainability and the development of projects for the implementation of sustainable development in their own organizations.

Activity "Development of Pilot Projects for Implementation of Sustainable Development in Organizations"

An expert team is established to support organizations involved in adopting a sustainable development practice into their management framework. This activity is based on the SIGMA Management Framework, and additional education is organized for participants from these organizations. At the same time, an expert team supports the development of particular projects for the implementation of sustainable development measures in organizations.

Activity "Implementation of Pilot Projects for Sustainable Development in Organizations"

Few small-scale pilot programs developed in the previous phase of the project are implemented in organizations. The following activities in organizations are supported: measures to promote energy and water saving, efficient waste management, rationalization of consumption and encouragement of usage of renewable resources in the facilities, and measures to promote healthy lifestyles.

Activity "Final Conference"

The final conference is organized on the same principles developed for the "strategic visioning conference." This kind of event has an important purpose in raising public awareness for sustainability. The main activity of the conference includes media presentation of the project activities. Exhibitions of sustainable development implementation projects in organizations contribute to audiovisual effects of presentations. Communication has a crucial role in providing understanding and support of key actors in the process of education for sustainable development. It is also important to attain the attention of existing mass media and availability of other channels of communication at the community level. The awareness and practices of people with regard to the culture of sustainability also receive special attention.

Cross-Cultural Management Approach

There are opinions in the literature that a widespread resistance to adopting sustainable habits, despite the apparent environmental crisis, can be ascribed to two main factors: insufficient efforts to finding viable and visible alternatives and the failure to re-examine dominant cultural paradigms thoroughly (Gambini 2006).

One of the recognized definitions of culture is that culture is learned programming of the mind, which differentiates one group from another. It could be identified by observing the external manifestations of culture, values, perceptions, behaviors, and attitudes of the individuals who make up that group. Cross-cultural analyses are important in showing that what may work in one culture may not be appropriate in another (Hofstede 2001). Because people from different cultures may have different values, perceive situations differently, act differently in the same situation, and approach life in different ways, attempts to transport Western practices to other nations where the culture is incompatible with the practices are likely to fail (Gomez-Mejia & Palich 1997). As such, culture is conceptualized and measured through different value dimensions identified and measured by numerous scholars (Hofstede 1980).

Although many different cultural dimensions have been identified over the years, the most replicated and with a high practical value are Hofstede's cultural dimensions. Based on surveying attitudes of 116,000 employees within subsidiaries of IBM in 40 countries and three regions, in 1980 Hofstede described four basic cultural dimensions, largely independent of each other: (1) individualism vs collectivism, (2) power distance, (3) uncertainty avoidance, and (4) masculinity vs femininity. Hofstede identified the degree that a society accepts inequality and distribution of power within that society in the dimension power distance (PD); the degree to which a culture feels comfortable in unstructured or ambiguous situations, uncertainty avoidance (UA); the degree to which individuals in a culture define themselves as individuals or according to their place in groups, individualism/collectivism (IDV); and masculinity/feminism, the degree to which a culture demonstrates certain characteristics considered to be masculine (e.g., valuing achievement) or feminine (such as valuing relationships)(Hofstede 1983). Collectivism is measured by the individualism index (IDV), ranging from 0 (low individualism, high collectivism) to 100 (high individualism). Power distance is measured by the power distance index (PDI), ranging from 0 (small PD) to 100 (large PD). Uncertainty avoidance is measured by the uncertainty avoidance index (UAI), ranging from 8 (lowest UA country) to 112 (highest UA country). Masculinity vs femininity is measured by the masculinity index (MAS), ranging from 0 (low masculinity) to 100 (high masculinity).

In 1988, one additional dimension was described by Bond and was named Confucian dynamism, to be renamed later, to long-term orientation versus short-term orientation (Hofstede & Bond 1988). A number of newer and older findings by Asian and European researchers suggested the need for expanding the dominant five-factor model of personality traits, known as the "Big Five," with a sixth factor, "dependence on others," to keep the model culturally universal (Hofstede 2007).

The original research conducted by Hofstede included former Yugoslavia as the only socialist county. After the dissolution of Yugoslavia in 1991, Hofstede revisited the original Yugoslav samples in order to obtain cultural dimension scores for three former Yugoslav republics: Slovenia, Croatia, and Serbia. In the second edition of *Culture's Consequences*, Serbian national culture is characterized by high power distance, 86; high uncertainty avoidance, 92; collectivism–low individualism, 25; and high-to-medium femininity–low-to-medium masculinity, 43.

Organizational and National Culture

Because cultures have an important impact on management approaches, cultural differences call for differences in management practices (Newman & Nollen 1996). The appropriate design of an organization depends on many factors, but Hofstede (1983) has argued that organizational systems work best when their design is consistent with the underlying values and culture of the society in which they function. For an organizational culture to function effectively as a part of a managerial mechanism, the organizational culture and the formal organizational structure must be interrelated harmoniously (Worley, Hitchin & Ross 1996). Thus, the structure and culture of an organization must be aligned with the demands and predispositions of the national culture in which the organization is embedded (Trompenaars & Hampden-Turner 2004).

RESEARCH OBJECTIVE AND STUDY DESIGN

A significant part of the educational program "Strategic Leadership for Sustainable Societies" is the development and implementation of projects for sustainable development in the participants' organizations. To improve the outcome of our educational program, our objective was to investigate the problems our participants are confronted with in their own organizations. We arranged a study with the main goal to explore the factors important for the facilitation of sustainable development programs in organizations.

Our hypothesis was that problems in the implementation of the projects are mostly organizational in their origin and correlate with the dominant national culture. To understand these problems, we made an attempt to analyze a number of organizational characteristics and to correlate them with certain cultural dimensions, with the goal of exploring the optimal organizational design for projects, aligned with the predominant national culture.

Consistent with Hofstede's cultural dimensions findings for Serbia—high PDI (86), high UAI (92), and collectivism–low individualism (25)—we made a subhypothesis that a successful organizational design of sustainable development projects has to be strongly supported by leadership, but with a dominant collectivistic character.

Our extended investigation involved over 200 persons, but only 67 participants were included in the investigation conducted by both unstructured interviews and questionnaires to assess participant's views on optimal organizational design in reference to the implementation of sustainable development projects. Thirty-four of the participants were employed at the local institutions involved in development (18 females, 16 males) and the other 33 came from different professions: health, education, and business. Participants had mostly higher education (44 with higher education) and work experience (as a rule over 5 years). Participants in the survey were asked to grade (1, not important; 5, very important) particular organizational characteristics. These organizational characteristics were found in the literature to correlate with organizational culture and structure (Doktor, Bangert & Valdez 2005).

RESULTS

Results of the investigation of participants' views on certain organizational characteristics are presented in Table 2 and Figure 1. Average marks and standard deviations for each investigated characteristic for local communities employees group (34 participants)(X1) and the entire group (X2)(67 participants) are presented numerically in Table 2 and Figure 1.

A presentation of pooled values for organizational characteristics associated with PDI (1,3), COL (6,7), and IND (8,9) and PDI 1 (1,3), COL 1 (6,7), and IND 1 (8,9) is shown in Figure 2.

DISCUSSION

Good governance in municipalities with an efficient and effective use of resources is very important for the sustainable advancement of Serbia. Unfortunately, there is still a general lack of awareness about sustainability at all levels in local municipalities. Previous research revealed that less than 5% of employees in local municipalities were able to explain the concept of

Table 2 Average Mark (X1, X2) and Standard Deviation (SD)

	Organizational Characteristic	Local Communities Employees Group (n = 34) X1	SD	Entire Group (n = 67) X2	SD
1	Support from superiors	4.6	0.9623	4.4	1.0882
2	Involvement of superiors	3.4	1.1997	3.5	1.2129
3	Clear instructions from superiors	4.4	0.7918	4.4	0.8729
4	Independence in choosing own work style	4.3	0.5401	4.3	0.5682
5	Decision making in own line of work	4.5	0.6170	4.4	0.6766
6	Good working relations with colleagues	4.8	0.4647	4.4	0.9780
7	Good communication with superiors	4.4	0.6629	4.1	0.9980
8	Acknowledge of individual performance through salary	4.2	0.9984	4.1	1.0423
9	Career advancement through individual pperperformance	4.3	0.9180	4.3	0.9988
10	Support for continuing education	4.6	0.7513	4.4	0.7912

sustainability in common terms. General public and high school and university students are even less informed, despite global efforts such as "Decade of education for sustainability." Local communities still suffer from a deficiency in professional and organizational knowledge of internationally recognized campaigns for sustainable development.

The educational program "Strategic Leadership for Sustainable Societies" has been created with the goal to educate local policy makers, municipal officers, and elected officials directly responsible for local sustainable development planning and implementation. We have also included a number of organizations in the fields of industry, tourism, agrobusiness, education, health, communications, etc. In addition to more profound knowledge of sustainability in organizations, this program comprises a basic training in organizational behavior and culture, founded on our own research results.

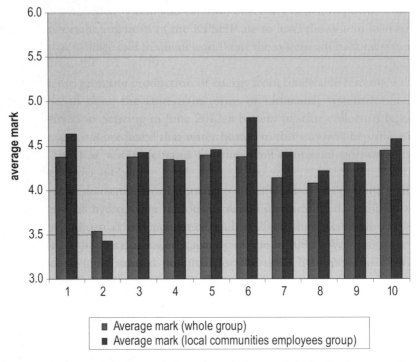

Figure 1 Average marks for each investigated characteristic for local communities employees group and the entire group.

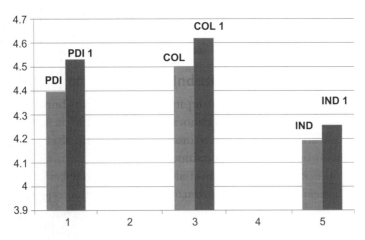

Figure 2 Pooled values for PDI (1,3), COL (6,7), and IND (8,9) and PDI 1 (1,3), COL 1 (6,7), and IND 1 (8,9) for the entire group and local communities employees group.

As part of the educational program "Strategic Leadership for Sustainable Societies," we have made an attempt to analyze opinions and views on optimal organizational design for sustainable development projects in organizations. Participants in our study are employees engaged in development projects in local communities in Serbia, as well as representatives from other professions (health, education, business).

Data reveal that participants in this investigation have recognized communication between colleagues and superiors, support from superiors, and clear instructions from superiors as the most important. Acknowledgment of individual performance and independence in choosing their way of working were found less important. The least important for the participants was actual involvement of their superiors. Results correlated with our hypothesis that organization of the projects has to be strongly supported by leadership (high PDI)(but participants emphasized that they should not be involved directly), with dominant collectivistic conduct (very important working relations with colleagues and good communication with superiors, clear instructions from superiors), and minor individualistic performance (acknowledge individual performance through salary, career advancement through individual performance). Nevertheless, some of the characteristics that could be associated with individualism (independence in choosing own work style and decision making in own line of work) were ranked high.

Studies on implementation of information technology projects in health care have already been published (Paunković et al. 2010). Results of these investigations delineated interdependence and team work, along with acknowledgment of individual performance and clear instructions from superiors and independence in choosing their way of working. The least important for the participants, as a rule, was actual involvement of their superiors.

Exploring similar organizational categories with U.S. information technology(IT) professionals, Bangert and Doktor found involvement and strong leadership and acknowledgment of individual performance as the most important, while support for continuing education and new things were ranked as least important. In the same study, IT professionals from South Korea ranked having clear rules to follow and loyal fellowship as the most important.

Once again, we have found Serbian organizational culture in investigated organizations to be "somewhere in between" ("west–east orientation"). If we are to make propositions about the design of sustainable development project organization in Serbia depicted from this study, we could delineate that there should be compelling leadership support for the project (without actual involvement of superiors, but with good communication and clear instructions from them), with a strong emphasis on harmonious team work and achievement (but with the opportunity to make decisions about the line of work), and support for continuing education.

Hofstede's work has often been criticized for a simplistic view of the multifaceted, complex dimensions, but the usefulness of the categories he developed remains very popular and is utilized by scholars in a variety of fields (Gerhart & Fang 2005; Sondergaard 1994). Our research has been performed in a country where this context has not been studied frequently, with the exception of the pioneering work of Jovanovic (2004), Jovanovic and Langovic (2006), and Milikić (2009). Our primary aim was to facilitate the process of the adoption of sustainability practices by understanding organizational and cultural factors. Sharing our research findings with the participants of our educational program has been a rewarding experience. Understanding the basic concepts of dominant organizational and cultural factors contributed to the capacity of our participants to analyze and improve current situation in their organizations and to explore inventive solutions further.

In cooperation with local NGO-Community Development Association Bor, over 250 people have been included in basic activities of the educational program "Strategic Leadership for Sustainable Societies." We have already trained 16 trainers and began project planning in 11 organizations, with pending funding for the projects. In addition to standard educational activities of the Faculty of Management Zajecar, the educational program "Strategic Leadership for Sustainable Societies" will make a substantial contribution to a network of practitioners in sustainable development who will be able to implement sustainable development projects in their own organizations in east Serbia. Additionally, it will raise public awareness about sustainability in local communities, information and experience exchanges between participants of the program, and development of sustainable business partnership networks, as well as development of joint projects and presentation in the future.

REFERENCES

Baines, J., Brannigan, J., Martin, S., 2001. Professional partnerships for sustainable development. Institution of Environmental Sciences, London.

Bogićević Milikić, B., 2009. The influence of culture on human resource management processes and practices: the propositions for Serbia. Econ. Ann. LIV (181).

Doktor, R., Bangert, D., Valdez, M., 2005. Organizational learning and culture in the managerial implementation of clinical e-health systems: an international perspective. Proceedings of the 38th Hawaii International Conference on System Sciences - (HICSS'05) –, Track 6.
Dyllick, T., Hockerts, K., 2002. Beyond the business case for corporate sustainability. Business Strategy Environ. 11, 130–141.
Gambini, B., 2006. Cultural assumptions against sustainability: an international survey. J. Geography Higher Educ. 30, 263–279.
Gerhart, B., Fang, M., 2005. National culture and human resource management: assumptions and evidence. Int. J. Human Resource Manag., 971–986.
Gomez-Mejia, P., 1997. Cultural diversity and the performance of multinational firms. J. Int. Business Stud. 28 (2), 309–335.
Hofstede, G., 1980. Motivation, leadership and organization: do American theories apply abroad? Organizational Dynamics, Summer, 42–46.
Hofstede, G., 1983. The cultural relativity of organizational practices and theories. J. Int. Business Stud. Fall, 75–89.
Hofstede, G., 2001. Culture's consequence. Sage Publications, Thousand Oaks, CA.
Hofstede, G., 2007. Asian management in the 21st century. Asia Pacific J. Manage vol. 24, 411–420.
Hofstede, G., Bond, M.H., 1988. The Confucius connection: from cultural roots to economic growth. Organizational Dynamics 16, 4–21.
Jovanovic, M., 2004. Interkulturni menadzment. Megatrend, Beograd.
Jovanovic, M., Langovic-Milicevic, A., 2006. Interkulturni izazovi globalizacije. Megatrend University, Belgrade.
Martin, S., Hall, A., 2002. Sustainable development and the professions. Planet 3, 17–18.
Miladinovic, V., Paunković, D.Z., 2012. Odrzivi razvoj u lokalnim samoupravama. Ecologica. br. 66.
Nacionalna strategija održivog razvoja, Vlada Republike Srbije, Beograd, 2008. ("Službeni glasnik RS", br. 57/08).
Newman, K.L., Nollen, S.D., 1996. Culture and congruence: the fit between management practices and national culture. J. Int. Business Stud. 27 (4), 753–779.
Paunković, J., Jovanović, R., Stojković, Z., Stojković, I., 2010. Sustainable implementation of information and communication technology in health care: case study of organizational and cultural factors. Sibiu Alma Mater University Journals. Series A , Econ. Sci. 3 (3), 1–8.
Paunkovic, J., Paunkovic, N., Milutinovic, S., Zikic, S., 2007. Education for sustainable development, October Annual of the University of Mining and Geology. St. Ivan Rilski 50.
Sharma, S., Ruud, A., 2003. On the path to sustainability: integrating social dimensions into the research and practice of environmental management. Business Strategy Environ. 12, 205–214.
Sida, 1998. Care of the environment: Sida's programme for sustainable development. Sida, Stockholm.
SIGMA, 2003. The Sigma guidelines. Putting sustainable development into practice: a guide for organizations.
Sondergaard, M., 1994. Research note: Hofstede's consequences: a study of reviews, citations and replications. Organ. Stud. 15 (3), 447–456.
Trompenaars, F., Hampden-Turner, C., 1998. Riding the waves of culture: understanding cultural diversity in global business, second ed. McGraw-Hill, New York.
Trompenaars, F., Hampden-Turner, C., 2004. Managing people across cultures. Capstone Publishing.
Worley, C.G., Hitchin, D.E., Ross, W.L., 1996. Integrated strategic change: how OD builds competitive advantage. Addison-Wesley Publishing Company, New York.

WEB SITES

http://ec.europa.eu/environment/newprg/index.htm.
http://ec.europa.eu/environment/newprg/pdf/7EAP_Proposal/en.pdfref.
http://eur-lex.europa.eu/LexUriServ/LexUriServ.do?uri=CELEX:32002D1600:en:NOT.
http://www.desd.org/.
http://www.dfes.gov.uk/sd/action.shtml.
http://www.earthsummit2012.org/resources/useful-resources/1157-the-future-we-want-rio20-outcome-document.
http://www.europarl.europa.eu/summits/lis1_en.htm.
http://www.naturalstep.org/.
http://www.projectsigma.co.uk/.
http://www.srbija.gov.rs/vesti/dokumenti_sekcija.php?id=45678.
http://www.thecommonwealth.org/files/247497/FileName/PocketGuidetoSDGovernance (second ed.).pdf).
http://www.uncsd2012.org/content/documents/814UNCSD%20REPORT%20final%20revs.pdf.
http://www.unesco.org/new/en/education/themes/leading-the-international-agenda/education-for-sustainable-development/.
http://www.un-documents.net/wced-ocf.htm.
http://www.dfes.gov.uk/sd/action.shtml.

CHAPTER 17

Business Ventures and Financial Sector in the United Arab Emirates

Robert Ruminski
Chair of Banking and Comparative Finance, Faculty of Management and Economics of Services, University of Szczecin, Szczecin, Poland

Contents

Introduction	405
Small and Medium Enterprises and Government Financial Support	409
Ease of Doing Business in the United Arab Emirates	411
Investment Bodies: Key Players and Contributors	413
Legal Aspects of Conducting Business Activity	415
Banking Sector: Structure and Recent Developments	421
Other Key Players of the Financial Market	434
Business Cooperation between the UAE and Poland	439
Prospects for Polish Enterprises in Persian Gulf States	440
Persian Gulf Markets	441
Entry Barriers for Polish Entrepreneurs on the Example of Polish Companies of Inglot and Can-Pack/Arab Can	442
Factors Determining Expansion of Polish Enterprises in the UAE	444
Market Chances and Threats	445
Arabic Investments in Poland	446
Market Strengths, Weaknesses, Opportunities, and Threats (SWOT) Analysis of the UAE Market: Business Perspective	446
Conclusions and Summary	448

INTRODUCTION

Sustainable development is widely defined as "development that meets the needs of the present without compromising the ability of future generations to meet their own needs."[1] The United Nations (UN) has supplemented this definition with the following statement: "At the heart of operationalising sustainable development is the challenge of evaluating and managing the

---

[1] *Brundtland Commission, 1987, Our Common Future, Chapter 2: Towards Sustainable Development from A/42/427.* Report of the World Commission on Environment and Development available at http://www.un-documents.net/ocf-02.htm#I.

complex interrelationships between economic, social and environmental objectives."[2] Meeting the challenges of sustainability is vital both globally and locally.

According to the UN (2002), "... critical objectives for environment and development policies that follow from the concept of sustainable development include:
- reviving growth
- changing the quality of growth
- meeting essential needs for jobs, food, energy, water, and sanitation
- ensuring a sustainable level of population
- conserving and enhancing the resource base
- reorienting technology and managing risk
- merging environment and economics in decision making."

This chapter is devoted to the economic dimension of sustainability—some of the macro-and microeconomic processes and interrelationships between different financial and nonfinancial entities executing well-balanced public policies leading to sustainable growth of the United Arab Emirates (UAE). It refers to the effects of the UAE policies rather than the policy itself.

Sustainable growth is supported by the following initiatives:
- issuing effective laws and legislations to encourage the business environment and develop national industries and exports
- strengthening investments and promoting small and medium business sectors
- protecting consumer rights and intellectual property rights
- diversifying trade activities under the leadership of qualified national resources while adhering to international standards of excellence and the tenets of knowledge economy, thus ushering in balanced and sustainable growth for the UAE[3]

The United Arab Emirates has a superb location for international trade, which makes it a natural gateway into the other Gulf Cooperation Council (GCC) countries of Saudi Arabia, Qatar, Bahrain, Kuwait, and Oman. The investment environment built a global reputation in an unprecedented length of time and demonstrates how global business thrives with visionary leadership and commercial cooperation thanks to sustainable growth policy. The UAE remains the region's most attractive destination for foreign investment. Recently, the country ranked 27 among 142 countries in the Global

---

[2] The United Nations Department of Economic and Social Affairs (2002).
[3] Ministry of Economy, United Arab Emirates: http://www.economy.ae/English/Pages/VissionAndMission.aspx.

Competitiveness Index (2011–2012), where it was the only Arab economy categorized as innovation driven, and 33 among 183 countries in the ease of doing business (Doing Business 2011)[4]; 35 among 183 countries in the Index of Economic Freedom (2012)[5]; and 28 among 178 countries in transparency and accountability (Corruption Perceptions Index 2011).[6] In addition, the UAE has been a contracting party to the General Agreement on Tariffs and Trade (GATT) since 1994 and a member of the World Trade Organization (WTO) since 1996. It is also a member of the Greater Arab Free-Trade Area in which all GCC states participate.[7]

The country is committed to maintaining a policy of economic openness, actively seeking to develop economic/business projects that are in harmony with the changes taking place in the world. The improved credit and loan provision, as well as major public investment project expansion measures, contributes to building the country's positive economic growth, despite the current unfavorable economic environment.

In order to attract strategic foreign investments and to add value to the existing business community, as well as to provide a stable and attractive business environment, the UAE government continues to launch initiatives and incentives to boost its investment environment. Nevertheless, the volume of investment might contract as a result of the restructuring of global businesses.

The UAE economy has been recovering slowly from the 2009 crisis, and its balance sheets have improved. Authorities strengthened the banking sector through liquidity support, recapitalization, and deposit guarantees, and the emirate of Abu Dhabi provided financial support to the emirate of Dubai. The Dubai Financial Support Fund was called to support troubled entities in the emirate and has now almost exhausted its funding of 20 billion USD. Several factors might undermine economic recovery. They are the following:

- *Massive property projects and uncertainty regarding their size*—the excess supply of property in Dubai, which will increase further (as unfinished projects come to completion), will continue to weigh on property prices and growth prospects. Abu Dhabi's strategy to increase its housing supply may also pose risks by placing additional pressure on the property market.

[4] http://www.ukiet:doingbusiness.org/rankings
[5] http://www.heritage.org/index/ranking
[6] http://cpi.transparency.org/cpi2011/results/
[7] http://www.polishbusinessgroupuae.com/page.aspx?l=1&pg=8&md=pagedetail

- *GREs[8] debt rescheduling*—according to the International Monetary Fund (IMF), GREs have an estimated 32 billion USD of debt due in 2011–2012 (of which at least 5 billion USD is in the real estate sector).[9] For that reason, Dubai may continue to face significant rollover risks in the short term, which may raise the cost of borrowing. The Dubai World (DW) debt restructuring was completed (March 2011—final DW agreement to restructure 14.7 billion USD of its debt with all of its creditors; DW will divide its liabilities in two tranches—with 4.4 billion USD to be repaid in 5 years and the remaining 10.3 billion USD in 8 years), but several other troubled GREs are still in the process of restructuring.
- *International sanctions on Iran*—one of the UAE's largest trading partners, as a result, sanctions on Iran could weaken the UAE's recovery.
- *Political unrest in the region poses downside risks to the outlook*—it may result in more difficult market conditions, as evidenced by the sharp drop in equity markets during Q1-2011. However, there are also indications that the UAE may benefit from increased tourism and investments. Moreover, higher oil prices are also benefiting the UAE as it exports oil and hydrocarbons.[10]

In 2011, the economic recovery continued. While the construction and real estate sectors still remained subdued in the aftermath of the crisis, real gross domestic product (GDP) growth reached an estimated 4.9%, supported by high oil prices and production in response to disruptions in Libya. Nonhydrocarbon growth strengthened to around 2.7%, backed by strong trade and a buoyant sector of tourism. The current account surplus increased significantly, to around 9% of GDP. Average inflation remained at 0.9%, largely due to declining rents. The UAE economic recovery was supported by an expansionary fiscal policy, and the nonhydrocarbon primary deficit rose to nearly 42% of nonhydrocarbon GDP in 2011 (36% in 2010) due to Abu Dhabi's increased current and development expenditures and its support for the weakened real estate sector. Nevertheless, the overall fiscal balance, backed by high oil prices, improved significantly. The UAE economy is proving to be extremely resilient in a difficult global economic climate.

---

[8] Government-related entities.

[9] A renewed worsening of global financing conditions could make it more difficult to roll over some of the GREs' maturing external debt and would raise the overall cost of their borrowing from international markets.

[10] *KAM CO Research, United Arab Emirates (UAE), Economic Brief and Outlook 2011*, April 2011 pp. 5–6.

SMALL AND MEDIUM ENTERPRISES AND GOVERNMENT FINANCIAL SUPPORT

Small and medium enterprises (SMEs) are recognized as an engine of economic growth and a source of sustainable development. Within this sector, micro and small enterprises are of special importance because they are considered the cradle of entrepreneurship, particularly in environments facing high unemployment. Entities of that size play a vital role in both developing and well-developed economies and are perceived as those creating new jobs and enhancing competition. SMEs control 99% of the business sector worldwide, account for 50% of the global GDP, and employ 85% of the world's labor force.

The UAE private sector has been the major vehicle driving the UAE economy, providing the major part of investments and the biggest contribution to the growth rate. This would have not been possible without the efforts and support of the government resulting from the policy. The government has been focusing on identifying a number of ways to support SMEs in order to boost economic growth and employment in line with country Vision 2021.[11] The government defines a small business as a company that employs less than 50 employees and a medium business as a firm employing between 50 and 100 employees.

Small and medium enterprises in the UAE are considered the backbone of the economy, representing 85–95% of the total business sector. The SMEs' contribution is a critical element of the UAE economy, driving growth and prosperity and providing diversification, as well as employing large numbers of people in the country. The Ministry of Economy is currently drafting a new SMEs law to regulate investment in this sector.

In order to facilitate development of the SME sector, the government launched a loan scheme. The Abu Dhabi Emirate established the Sheikh Khalifa Fund[12] for SMEs, which offered 246 million AED in funds (2008) for 154 new enterprises (industry and services). The fund also conducted training programs for young men and women on ways to design and choose projects. It aims to create a new generation of Emirati entrepreneurs by enriching the culture of investment among young people, as well as supporting and developing small- to medium-sized investments in the Emirate.[13]

---

[11] http://www.uaeinteract.com/docs/Cabinet_releases_UAE_Vision_2021_(full_text)/39555.htm
[12] http://www.khalifafund.gov.ae/En/AboutUs/Pages/MessagefromtheChairman.aspx
[13] The Khalifa fund was launched in 2007 with a total capital investment of 2 billion AED.

In 2009, His Highness Sheikh Mohammed bin Rashid issued Law No. 23 of 2009 concerning the Mohammed Bin Rashid Establishment for SME's development, one of the Department of Economic Development establishments, recently renamed the Mohammed Bin Rashid Establishment for Young Business Leaders. One of the establishment's key objectives is to help address the challenges of the funding gap for SMEs and make capital available to them. This comprehensive financing program is to ease the UAE entrepreneurs' financial challenges and facilitate the startup.

The establishment has set up a dedicated 700 million UAE fund based on Islamic banking principles in conjunction with the Dubai Islamic Bank to provide access to capital at preferential terms for new entrepreneurs. The application procedure is simple, and repayment terms are highly favorable. For existing small and medium business owners, the establishment provides loans at preferential terms via a network of affiliated banks. Moreover, the establishment performs as an advisor to entrepreneurs seeking financing. Key benefits of the program are the following:

- dedication to small and medium enterprises
- simple and streamlined application process
- competitive interest rates
- highly favorable financing terms (repayment)
- financing both existing and new entrepreneurs[14]

The new law (Law No. 23 of 2009) confirms the commitment of the Dubai government to support the development of SMEs. This sector represents 98.5% of registered businesses in Dubai and 61% of the workforce employed. The law aims to strengthen Dubai's position as a center for entrepreneurship and enterprise development based on innovation and intellectual property.

The emirate of Ras Al Khaimah is focusing on SMEs by encouraging them to invest in untapped areas. It increases their management skills and ensures high sales of their products. The Saud bin Saqr program for SMEs supports business ventures in the field of information technology where young investors can succeed. There is high demand for these services from the market.

HSBC introduced the SME international fund[15] in line with the government efforts in enhancing the role of SMEs in the country. A 100 million USD fund is dedicated to support UAE-based international SME business. The fund is a part of HSBC's global strategy to support internationally focused SMEs, helping them to grow and conduct business internationally.[16]

[14] http://www.sme.ae/english/index.html
[15] http://www.ameinfo.com/hsbc-launches-dhs1bn-international-trade-sme-300518
[16] *Investor's Guide to the UAE 2010–2011, Your one-stop information resource*, Ministry of Economy, 2011.

Apart from the aforementioned programs supporting SMEs, there are significant government policies in place[17]:
- Small and medium enterprise development support policies, e.g., access to finance, IP, exports and internationalization, bankruptcy, and company closures
- Industry-specific policies, e.g., information and communications technology, creative business sectors, emerging businesses
- Small and medium enterprise professionalization and capability development program covering skills and leadership development, corporate governance, accounting, financial and investment management, business continuity, operational and productivity excellence, etc.

EASE OF DOING BUSINESS IN THE UNITED ARAB EMIRATES

The Doing Business[18] report is one of the key instruments providing quantitative measures of regulations for
- starting a business
- dealing with construction permits
- getting electricity
- registering property
- getting credit
- protecting investors
- paying taxes
- trading across borders
- enforcing contracts
- resolving insolvency

These issues are concerned regarding domestic small- and medium-sized enterprises. This internationally recognized report assesses countries on how easy it is for small- and medium-sized enterprises to conduct business (Table 1).

In 2010, the UAE climbed 14 places in the Doing Business report compiled by the World Bank and its international finance corporation. The UAE rose to the 33 position in global rankings for regulatory reform, partly as a result of the government's decision to abolish a 150,000 AED minimum capital requirement for some startups.

---

[17] *PKF–Doing business in the UAE–Financial Reporting and Auditing*, PKF International Limited, UAE 04.2012.
[18] http://www.doingbusiness.org

Table 1 Doing Business 2010 Report[a]

	Doing Business 2010 Ranking	Doing Business 2009 Ranking	Doing Business Change in Rank
Doing Business	33	47	+14
Starting a Business	44	118	+74
Dealing with Construction Permits	27	54	+27
Employing Workers	50	45	−5
Registering Property	7	7	0
Getting Credit	71	68	−3
Protecting Investors	119	114	−5
Paying Taxes	4	4	0
Trading Across Borders	5	13	+8
Enforcing Contracts	134	135	+1
Closing a Business	143	143	0

[a] *Source*: http://www.doingbusiness.org/~/media/giawb/doing%20business/documents/profiles/country/ARE.pdf

Two other key reasons for the country's rise was a streamlining of the process involved in obtaining construction permits and improving the capacity at Dubai ports.[19]

The following important improvements were carried out in 2010:
- business startup was eased by simplifying the documents needed for registration
- abolishing the minimum capital requirement
- removing the requirement that proof of deposit of capital be shown for registration
- time for delivering building permits was shortened by improving its online application processing system
- greater capacity at the container terminal
- eliminating the terminal handling receipt as a required document
- increasing trade finance products

In 2012, the following reforms were introduced (according to the Doing Business report).
- The UAE made starting a business easier by merging the requirements to file company documents with the Department for Economic Development, to obtain a trade license, and to register with the Dubai Chamber of Commerce and Industry.

[19] Vine, P (ed.) 2010, *UAE 2010 yearbook*, Trident Press Ltd. and the National Media Council, Mayfair, London.

- The country improved its credit information system through a new law allowing establishment of a federal credit bureau under supervision of the central bank.
- Access to credit was enhanced by setting up a legal framework for the operation of a private credit bureau and requiring that financial institution to share credit information.
- The UAE streamlined document preparation and reduced the time to trade with the launch of Dubai customs' comprehensive new customs' system (Mirsal 2).[20]

INVESTMENT BODIES: KEY PLAYERS AND CONTRIBUTORS

According to the UN Conference on Trade and Development, between 2003 and 2008 the UAE was the third largest recipient of foreign direct investment in west Asia, behind Saudi Arabia and Turkey.

Investments in overseas markets have been integral to the country's strategic plan to create a security net for future generations facing the prospect of a depletion of hydrocarbon reserves. The major international investment bodies in the Emirates contribute significantly to the sustainable growth of the country. They are the following:

- **Abu Dhabi Investment Authority (ADIA)**[21]—its mission is to secure and maintain the prosperity of the emirate through management of its investment assets. ADIA is a leading international investor and, since the late 1970s, has established itself as a trustworthy and strong investor (supplier of capital). The ADIA supervises a substantial global-diversified portfolio of assets across varying sectors, regions, and asset classes (private equity, property, public-listed equities). ADIA does not seek active management of the entities it invests in, only long-term sustainable financial returns.
- **Abu Dhabi Investment Council (ADIC)**[22]—responsible for investing part of Abu Dhabi's surplus financial resources. It executes a globally diversified investment strategy focused on gaining positive capital returns across a range of asset classes.
- **Invest AD**[23]—a subsidiary of ADIC; a government investment vehicle similar to ADIA, established in 1977. Invest AD allows outside investors

---

[20] http://www.doingbusiness.org/reforms/overview/economy/united-arab-emirates#
[21] http://www.adia.ae/En/home.aspx
[22] http://www.adcouncil.ae/
[23] http://www.investad.com/

to put their money in it (alongside). Invest AD invests on behalf of the government and attracts capital from external investors.
- **The Investment Corporation of Dubai (ICD)**[24]—investing to create stability and foster diversification. It owns 60% of Borse Dubai,[25] a holding company performing as a holding company for Dubai Financial Market and NASDAQ Dubai.
- **Dubai Holding**[26]—one of Dubai's major holding companies, divided between the Dubai Holding Commercial Operations Group (DHCOG) and the Dubai Holding Investment Group (DHIG). It was formed in 2009 when Dubai Group and Dubai International Capital (DIC) were combined. Dubai Holding has reorganized its companies into property, business park, hospitality, and investment units.
- **Dubai Holding Commercial Operations Group**[27]—composed of property developers Dubai Properties Group, Sama Dubai and Tatweer fall under DHCOG. Moreover, DHCOG holds the hotel operator Jumeirah Group and the business park operator TECOM Investments.
- **Dubai Holding Investment Group**[28]—formed after combining previously separate entities of Dubai Group and Dubai International Capital. It controls six financial companies under the responsibility of the Dubai Group, including Dubai Capital Group, Dubai Financial Group, Dubai Investment Group, Dubai Banking Group, Dubai Insurance Group, and Noor Investment Group. Focused on the private equity asset class, DIC operates through global buyouts specializing in secondary leveraged buyouts in Europe, North America, and Asia. DIC owns stakes in the Travelodge hotel chain, the Middle Eastern operations of the property consultancy CB Richard Ellis, and the U.K. engineering company Doncasters.
- **Dubai World**[29]—a holding company that has been at the forefront of Dubai's rapid growth. It operates in different industrial segments and invests in four main sectors:
  - transport and logistics
  - dry docks and maritime
  - urban development
  - investment and financial services

[24] http://www.icd.gov.ae/
[25] http://www.borsedubai.ae/
[26] http://dubaiholding.com/
[27] http://dubaiholding.com/media-centre/news/2012/Dubai-Holding-Commercial-Operations-MTN-Limited-Bond-Coupon-Payments/
[28] http://www.dubaigroup.com/aboutus/dubaiholding_en_gb.aspx
[29] http://www.dubaiworld.ae/

- Dubai World's portfolio includes some of the world's best-known companies:
  - DP World (maritime terminal operator)
  - Drydocks World and Dubai Maritime City
  - Economic Zones World (operates several free zones around the world)
  - Nakheel—the property developer behind The Palm Islands and The World
  - Limitless (international real estate master planner)
  - Leisurecorp—a sports and investment group
  - Dubai World Africa
  - Istithmar World

LEGAL ASPECTS OF CONDUCTING BUSINESS ACTIVITY

The UAE legal system is essentially a civil law jurisdiction influenced heavily by French, Roman, and Islamic laws. The increasing presence of international law firms from common law jurisdictions has demonstrated the application of common law principles in commercial contracts. This indirectly has further influenced the UAE legal system.[30] Establishing a business in the UAE is subject to licensing requirements as well as foreign investment restrictions. Businesses can be set up in the following two investment locations:

a. mainland UAE
b. free trade zones (FTZ)

In order to regulate matters such as commercial transactions, commercial agencies, civil transactions, labor relations, maritime affairs, intellectual property, and commercial companies, a number of codified federal laws have been passed. In addition, a number of local laws have also been passed in various areas by individual emirates. There are two main types of laws in the UAE:

- federal—applicable to the UAE as a whole and issued either by the legislative body or by the ministers of each ministry (ministerial order)
- local—decrees and orders apply only to a particular emirate (passed by the ruler or crown prince of a particular emirate and issued by a member of the royal family of that emirate)

All emirates have brought their judicial systems into the UAE Federal Judicial Authority except Dubai and Ras Al Khaimah. Dubai has retained its

---

[30] *Doing business in the UAE, a business and tax profile*, PKF accountants and business advisers, April 2012.

own independent courts and judges, which are not a part of the UAE Federal Judicial Authority. Dubai's courts first apply federal laws (e.g., companies law or the civil code), as well as the laws and decrees enacted by the ruler of Dubai, whereas federal law is absent.

In 2004, Dubai International Financial Center (DIFC) courts were founded and are an independent common law judiciary based in the Dubai International Financial Center with jurisdiction governing civil and commercial disputes.

Dubai International Financial Center courts are composed of international judges from a common law jurisdiction, such as England, Malaysia, and New Zealand, and their procedural rules are largely modeled on English civil procedure rules. The official language of DIFC courts is English. All proceedings are also conducted in English.

The Dubai government has expanded the jurisdiction of DIFC courts, which allow any parties (even not incorporated within the DIFC free zone) to use the DIFC courts to resolve commercial disputes.[31] Currently, parties in the region and internationally can agree to use the DIFC courts in the event of a dispute (the parties should then agree to incorporate the jurisdiction of the DIFC courts into their contracts prior to taking the dispute to the DIFC courts).

The expansion of DIFC courts' jurisdiction represents an important policy shift and will give the business community an unprecedented access to DIFC courts. The move is likely to be welcomed by both legal and business communities, as international parties may be more likely to wish to resolve their disputes in a more familiar forum using the common law English model.[32]

One of the crucial decisions that entrepreneurs need to make in order to set up their business entity is to choose the appropriate legal form of conducting business activity. The way the business is structured and operated will most likely influence the access to future financing opportunities.[33] It is essential for the potential business owner to understand how each legal structure works and then pick the one that best meets the entrepreneur's needs. Because each business form has its own unique legal and financial ramifications, it is advisable that the choice is made in conjunction with a

---

[31] Previously, only companies based in the DIFC or those that had an issue related to the DIFC could use the DIFC courts.
[32] *Doing business in the UAE, a business and tax profile*, PKF accountants and business advisers, April 2012
[33] Sitarz, D 2001, *The complete book of small business legal forms*, 3rd edn., Nova Publishing Company, Carbondale, IL, p. 21.

lawyer and an accountant. The choice of the form of business organization can have a great impact on the success of the business. It may influence:
- how easy it is to obtain financing
- how taxes are paid
- how accounting records are kept
- whether personal assets are at risk in the venture
- the amount of control the owner has over the business, etc.

Foreign investors can choose between several types of cooperation and partnerships for conducting business in the UAE. Entrepreneurs can also conduct business activity through the UAE branch office. Limited liability companies (LLCs) are used more commonly by foreign investors.

The UAE company law determines a total local equity of not less than 51% in any business entity. Moreover, it defines the following seven categories of business organizations allowed to be established in the UAE.

1. **General partnership**—formed by two or more partners jointly liable to the extent of all their assets for the company liabilities.
2. **Simple limited partnership**—formed by one or more general partners liable for the company liabilities to the extent of all their assets, and one or more limited partners liable for the company liabilities to the extent of their respective shares in the capital only.
3. **Joint venture**—a company concluded between two or more partners sharing profits or losses of one or more businesses being performed by one of the partners. Local equity participation must be at least 51%.
4. **Public joint stock**—any company whose capital is divided into equal value negotiable shares and a partner therein is only liable to the extent of his share in the capital.
5. **Private joint stock**—not less than three founder members may incorporate. Shares are not offered for public subscription. Founder members will fully subscribe to the capital (not less than two million AED).
6. **Limited liability company**—An LLC can be formed by a minimum of 2 and a maximum of 50 persons whose liability is limited to their shares. Most companies with expatriate partners have opted for this LLC due to the fact that this is the only option that will give maximum legal ownership, that is, 49% to expatriates for a trading license. Fifty-one percent UAE nationals participation is the general requirement (normal share holding pattern is local sponsor, 51%; foreign shareholder, 49%). There is no minimum capital requirement for establishing a company. The foreign equity capital cannot exceed 49%, but the profit and loss distribution can be mutually agreed. Management of an LLC may be

performed by foreign and national partners or third parties. Formation of a company takes approximately 1–2 weeks from the date of receipt of all the documents.
7. **Share partnerships**—a company formed by general partners jointly liable to the extent of all their assets for the company liabilities and participating partners liable only to the extent of their shares in the capital.

Foreign direct investments are encouraged through branches and representative offices of foreign companies, and 100% foreign ownership is permitted in the FTZ. Partnerships are generally open only to UAE nationals.

As mentioned before, LLCs are used more commonly by foreign investors, and the following documents are required for a LLC:
- certificate of capital contribution from a bank
- auditor's certificate for shares of all kinds
- all other items requested in the application form

There are five steps to set up a LLC.
1. Approval of company name and activity from the relevant office of economic development, municipality, and chamber of commerce
2. Articles of association must be notarized according to the requirements of each emirate(s)
3. Application package must be delivered to the department of economic development or the municipality as appropriate
4. Following approval, the new company will be included in the commercial register and the articles of association published in the Bulletin of the Ministry of Economy
5. A license will then be issued by the Department of Economic Development (Dubai and Sharjah) or the municipality or the chamber of commerce of the other emirates.

The Commercial Companies Law allows for setting up branches and representative offices of foreign companies. They may be 100% foreign owned, provided a local service agent[34] is appointed.

A **branch office** (regarded as part of its parent company) is a full-fledged business, permitted to realize contracts or conduct other activities similar to those of its parent company as specified in its license. Activities are approved on a case-by-case basis.[35]

---

[34] Only UAE nationals or companies 100% owned by UAE nationals may be appointed as local service agents.

[35] *Expanding your horizons? A guide to setting up business across the Middle East and North Africa region*, KPMG International Cooperative, 2010.

A **representative office** is limited to promoting its parent company's activities (e.g., gathering information and soliciting orders and projects to be performed by the company's head office). Representative offices are also limited in the number of employees that they may sponsor.

Local service agents (also referred to as sponsors) are not involved in operations of the company but assist in obtaining visas, labor cards, etc. The time required for setting up a branch of a foreign company is approximately 3 to 4 weeks from the date of receipt of all documents. In order to prove the credibility of the company, a business plan and current profile, as well as the last 2-year financial statements, should be submitted.

In the case of **professional firms**, 100% foreign ownership, sole proprietorships, or civil companies are permitted. They may engage in professional or artisan activities. A UAE national must be appointed as a local service agent.

The basic requirement for all types of business activities in the UAE is one of the following three categories of **licenses**:

a. *commercial license*—covering all kinds of trading activity
b. *professional license*—covering professions, services, craftsmen and artisans
c. *industrial license*—establishing industrial or manufacturing activity

In order to set up a company to engage in certain activities (including financial institutions), official approval is required from the appropriate government ministry or department. After receiving the license, companies are also required to register with the local chamber of commerce. These institutions are very important and investors should regard them as effective resources of information (databases, business literature) to establish their projects. Moreover, chambers provide an extensive range of basic and more sophisticated services for entrepreneurs.

Another opportunity for the foreign investor is to set up a company in one of the free trade zones. They are special economic areas established to promote foreign investment and economic activities within the UAE. FTZs are governed by an independent free zone authority responsible for issuing FTZ operating licenses and assisting entrepreneurs with establishing their businesses. The procedures for establishing a business are relatively simple. Entrepreneurs (investors) can either register a new company in the form of a free zone establishment or simply establish a branch or representative office of their existing or parent company based within the UAE or abroad. An FZE is a limited liability company governed by the rules and regulations of FTZ in which it is established. There are currently 36 FTZs with many more in the pipeline[36]:

---

[36] http://www.uaefreezones.com

1. Masdar City
2. Abu Dhabi Ports Company
3. Abu Dhabi Airport Free Zone
4. Khalifa Industrial Zone
5. ZonesCorp
6. twofour54
7. Dubai Airport Freezone
8. Dubai Silicon Oasis
9. Jebel Ali Free Zone
10. Dubai Multi Commodities Center
11. Dubai Internet City
12. Dubai Media City
13. Dubai Studio City
14. Dubai Academic City
15. Dubai Knowledge Village
16. Dubai Outsource Zone
17. Enpark
18. Intl Media Production Zone
19. Dubai Biotech Research Park
20. Dubai Auto Zone
21. Gold and Diamond Park
22. Dubai Healthcare City
23. Dubai Intl Financial Centre
24. Dubai Logistics City
25. Dubai Maritime City
26. Dubai Flower Centre
27. Intl Humanitarian City
28. Sharjah Airport Free Zone
29. Hamriyah Free Zone
30. Ahmed Bin Rashid FZ
31. Ajman Free Zone Authority
32. RAK Investment Authority
33. RAK Free Zone
34. RAK Maritime City
35. Fujairah Free Zone
36. Fujairah Creative City

Companies operating in the FTZs are treated as being offshore or outside the UAE for legal purposes. The major advantages in setting up in a free zone are:

100% foreign ownership of the enterprise
100% import and export tax exemptions
100% repatriation of capital and profits
exemption from all import and export duties
corporate tax exemptions for up to 50 years
no personal income taxes
assistance with labor recruitment
inexpensive energy and workforce
additional support services (e.g., administration, sponsorship, and housing)
companies at FTZ can operate 24 hours a day

The FTZ also has some limitations:

a company set up in FTZ is not allowed to trade directly with the UAE market
a company can undertake the local business only through the locally appointed distributors
custom duty of 5% is applicable for the local business[37]

BANKING SECTOR: STRUCTURE AND RECENT DEVELOPMENTS

Islamic finance is a US $400 billion dollar industry, growing at a rate of over 15% per annum with an expected high growth rate for the next 15 to 20 years. Each Islamic market has developed relatively independently, setting its own regulations and standards and developing a wide variety of products with different benchmarks and pricing techniques.[38]

The UAE financial services sector has served as an important element of growth toward diversification of the UAE's strategy. Abu Dhabi and Dubai financial sectors constitute the majority of the UAE's financial system. There is a sound, modern, and competitive banking industry. In 2011, 23 locally incorporated banks and 28 branches of foreign banks were operating in the country. The international financial crisis has had a relatively mild influence on the banking sector, thanks to government interventions (the improvement of banks' liquidity), which gave a boost to economic activity (the UAE Central Bank AED 50 billion facility to support local lenders and the UAE Ministry of Finance AED 70 billion liquidity support scheme).

---

[37] http://www.polishbusinessgroupuae.com/page.aspx?l=1&pg=8&md=pagedetail
[38] http://www.difc.ae/

The crisis witnessed in Dubai in 2008 led its neighbor Abu Dhabi to intervene and provide financial aid.

In consequence, capitalization of the UAE's banking system remains sound as the capital adequacy ratio increased from 13.3% at the end of 2008 to 21% in 2010. Moreover, the banking sector has a strong deposit base (it increased from 923 billion AED in 2008 to 967 billion AED in 2010). The strong capital base has allowed banks to increase lending in the UAE, even during the crisis.[39] The sector remains resilient to shocks, backed by a solid capital base, including money injected by the government, and strong earnings, despite the doubling of nonperforming loans since the global financial crisis struck.

A renewed worsening of global financing conditions could make it more difficult to roll over some of the GREs' maturing external debt and would raise the overall cost of their borrowing from international markets. About 32 billion USD of sovereign and GRE debt is estimated to mature in 2012, of which $15 billion in Dubai.[40]

The UAE banking sector's capitalization remains sound and is sufficient to absorb debt problems faced by Dubai government-related entities. Banks' credit portfolios in the UAE have extensive exposures to trade, real estate, and construction sectors. Bank lending is reviving, but credit growth remains sluggish. Lending was up 1.4% in 2010, reaching 972.1 billion AED[41] (264.6 billion USD), indicative of the bank's general cautiousness in lending to the economy. Lending to the government maintained its upward trend, which is in line with the government's continued expansionary policies (100 billion AED in 2010; 27.2 billion USD). Lending for construction decreased 2.6% in 2010, whereas credit extended for real estate mortgage loans has raised 15%. Moreover, stress tests on aggregate banking data indicate resilience to shocks, despite nonperforming loans doubling since the crisis.

Effective bank governance is fundamental to support financial sector soundness. In light of the government's control of banks and the banks' high exposure to GREs, a clear governance framework is needed to safeguard the banks' financial integrity (Table 2).

The Central Bank of the UAE has also made important progress in strengthening its financial stability approach, reorganizing the regulatory framework and developing macro prudential policies.[42]

---

[39] *Investor's guide to the UAE 2010–2011.*
[40] International Monetary Fund, Middle East and Central Asia Department, United Arab Emirates, 2012 article IV consultation concluding statement, March 19, 2012.
[41] 2010 data from the Central Bank of the UAE.
[42] *Doing business in the UAE, a business and tax profile*, PKF accountants and business advisers, April 2012.

Table 2 Selected Monetary and Banking Indicators (in AED)[a]

	Dec 2009	Mar 2010
Total bank assets (net of provisions)[b]	1519.1	1533.1
Certificate of deposits held by banks	71.9	63.3
Bank deposits	982.6	967.0
Loans and advances (net provisions)[b]	1017.7	1022.0
Personal loans	209.8	212.2
Letters of credit	102.8	104.2
Total private funds (capital + reserves)[c]	231.4	252.8
Specific provisions for nonperforming loans	32.6	34.4
General provisions	10.7	13.7
Total investments by banks		
Capital adequacy ratio: Banking system	19.2%	20.3%
Banking institutions (total numbers)		
UAE incorporated banks		
Head offices	24	23
Branches	674	687
Electronic banking service units	25	25
Pay offices	71	73
Foreign banks		
Head offices	28	28
Branches	82	82
Electronic banking service units	43	45
ATMs	3599	

[a]*Source*: *United Arab Emirates: selected issues and statistical appendix*, IMF Country Report No. 12/136, June 2012.
[b]Net of interest in suspense, specific provisions, and general provisions.
[c]Excluding current year profit.

As mentioned previously, the number of locally incorporated commercial banks stood at 23 during 2011, while the number of their branches increased from 732 at the end of December 2010 to 768 at the end of December 2011, the number of their electronic/customer service units remained at 26. The ratio of nonperforming loans of national banks stood at 6.2% in 2011, a sharp increase from 2008 precrisis levels, while that of Dubai banks was higher at 10.6%. Stress tests showed that the domestic banking system could absorb a significant increase in nonperforming loans.

The following national banks are currently registered in the UAE (Table 3):[43]

---

[43] http://www.centralbank.ae/en/index.php

Table 3 UAE National Banks and Distribution of their Branches (2011)[a]

No	Name of the bank	Head Office	Abu Dhabi	Dubai	Sharjah	Ras Al Khaimah	Ajman	Umm-Al Qaiwain	Fujairah	Al Ain	Total Number of Branches	Pay offices	Electronic Banking Service Units
1	National Bank of Abu Dhabi	Abu Dhabi	39	18	10	2	1	1	3	12	86	42	0
2	Abu Dhabi Commercial Bank	Abu Dhabi	20	11	3	1	1	0	2	7	45	5	1
3	ARBIFT	Abu Dhabi	3	4	1	0	0	0	0	1	9	0	0
4	Union National Bank	Abu Dhabi	19	14	8	2	2	1	1	7	54	10	0
5	Commercial Bank of Dubai	Dubai	3	17	1	1	1	0	1	1	25	5	0
6	Dubai Islamic Bank PJSC	Dubai	9	32	12	4	2	1	2	6	68	0	5
7	Emirates NBD Bank	Dubai	15	83	7	3	1	1	2	3	115	18	0

8	Emirates Islamic Bank	Dubai	4	17	5	1	1	1	1	3	33	1	0
9	Mashreq Bank PSC	Dubai	13	33	9	2	3	1	2	3	66	0	7
10	Sharjah Islamic Bank	Sharjah	1	3	20	0	0	1	1	1	26	1	0
11	Bank of Sharjah PSC	Sharjah	1	1	1	0	0	0	0	1	4	0	0
12	United Arab Bank PJSC	Sharjah	2	3	3	2	1	0	1	1	13	0	0
13	InvestBank PLC	Sharjah	2	2	4	1	1	0	1	1	12	0	0
14	The National Bank of R.A.K	RAK	5	12	4	7	1	0	0	1	30	1	4
15	Commercial Bank International	Dubai	3	5	2	3	1	1	1	1	17	1	0

Continued

Table 3 UAE National Banks and Distribution of their Branches (2011)[a]—cont'd

No	Name of the bank	Head Office	Abu Dhabi	Dubai	Sharjah	Ras Al Khaimah	Ajman	Umm-Al Qaiwain	Fujairah	Al Ain	Total Number of Branches	Pay offices	Electronic Banking Service Units
16	National Bank of Fujairah PSC	Fujairah	2	4	2	0	1	0	5	1	15	0	0
17	National Bank of U.A.Q PSC	U.A.Q	2	6	2	1	2	2	1	1	17	1	7
18	First Gulf Bank	Abu Dhabi	7	3	2	1	2	0	1	2	18	0	0
19	Abu Dhabi Islamic Bank	Abu Dhabi	28	11	8	3	2	1	2	11	66	0	0
20	Dubai Bank	Dubai	4	13	3	1	1	0	1	1	24	0	0
21	Noor Islamic Bank	Dubai	3	9	2	0	0	0	0	1	15	0	2
22	Al Hilal Bank	Abu Dhabi	10	8	1	1	0	0	0	2	22	0	0
23	Ajman Bank	Ajman	3	2	1	0	4	0	0	1	11	2	0
	Total		196	311	111	36	28	10	28	69	791	87	26

[a] Source: www.centralbank.ae.

1. Abu Dhabi Commercial Bank[44]
2. Abu Dhabi Islamic Bank[45]
3. Al Hilal Bank[46]
4. Arab Bank for Investment and Foreign Trade (Al Masraf)[47]
5. Commercial Bank of Dubai[48]
6. Commercial Bank International[49]
7. Dubai Bank[50]
8. Dubai Islamic Bank PJSC[51]
9. Emirates NBD Bank[52]
10. Emirates Islamic Bank[53]
11. First Gulf Bank[54]
12. Mashreq Bank[55]
13. Noor Islamic Bank[56]
14. RAK Bank[57]
15. Sharjah Islamic Bank[58]
16. United Arab Bank PJSC[59]
17. Union National Bank[60]

The complete list of commercial banks and representative offices in the UAE is the following (Table 4):

1. ABN Amro Bank NV
2. Abu Dhabi Commercial Bank
3. Abu Dhabi Islamic Bank
4. Al Ahli Bank of Kuwait
5. Al Rafidain Bank
6. Al Hilal Bank

---

[44] http://www.adcb.com/general/chargesandfees/chargesandfees.asp
[45] http://www.adib.ae/savings-account
[46] http://www.alhilalbank.ae/
[47] http://www.al-masraf.ae/
[48] http://www.cbd.ae/cbd/index.aspx
[49] http://www.cbiuae.com/
[50] https://www.dubaibank.ae/?item=%2fcontent%2fdefault&user=extranet%5cAnonymous&site=website
[51] http://www.dib.ae/support/schedule-of-charges
[52] http://www.emiratesnbd.com/en/
[53] http://www.emiratesislamicbank.ae/default.aspx
[54] http://www.fgb.ae/en/
[55] http://www.mashreqbank.com/personal/service-charges/overview.asp
[56] http://www.noorbank.com/english/help/contact-us
[57] http://rakbank.ae/rakbank/personalbanking/personalbanking.jsp
[58] http://www.sib.ae/en/retail-banking/products-services/fees-and-charges.html
[59] http://www.uab.ae/cms/index.php?option=com_content&view=article&id=60&Itemid=92
[60] http://www.unb.ae/english/inner.aspx?p=1&mid=403

Table 4 Commercial Banks Operating in the UAE (2010 and 2011)[a]

	2010	2011			
	December	March	June	September	December
National Banks					
Head Offices	23	23	23	23	23
Branches	732	736	745	757	768
Electronic/Customer Service Units	26	29	27	27	26
Cash Offices	86	86	86	87	87
GCC Banks					
Main Branches	6	6	6	6	6
Aditional Branches	1	1	1	1	1
Other Foreign Banks					
Main Branches	22	22	22	22	22
Aditional Branches	82	82	82	82	82
Electronic/Customer Service Units	50	48	47	47	50
Cash Offices	1	1	1	1	1
Number of ATMs	3,758	3,846	3,963	4,053	4,172

[a]*Source*: www.centralbank.ae.

7. Arab African International Bank
8. Arab Bank for Investment and Foreign Trade
9. Arab Bank PLC
10. Bank Melli Iran
11. Bank of Baroda
12. Bank of Sharjah
13. Bank Saderat Iran
14. Blom Bank France SA
15. Banque Du Caire
16. Calyon Bank
17. Al Khaliji France
18. BNP Paribas Bank
19. HSBC Bank Middle East
20. CitiBank NA
21. Commercial Bank International
22. Commercial Bank of Dubai
23. Dubai Islamic Bank
24. Dubai Bank PJSC
25. El Nilein Bank

26. Emirates Bank International
27. Emirates Islamic Bank
28. First Gulf Bank
29. Habib Bank Limited
30. Habib Bank AG Zurich
31. Invest Bank
32. Janata Bank
33. Lloyds Bank TSB
34. Mashreq Bank
35. National Bank of Abu Dhabi
36. National Bank of Bahrain
37. National Bank of Dubai
38. National Bank of Fujairah
39. National Bank of Oman
40. National Bank of RAK
41. National Bank of UAQ
42. Noor Islamic Bank PJSC
43. Sharjah Islamic Bank
44. Standard Chartered Bank
45. Union National Bank
46. United Arab Bank
47. United Bank Limited

Representative offices are as follow:
1. Bank of Bahrain and Kuwait
2. Barclays Bank
3. Doha Bank
4. ED & F Investment Products Ltd.
5. ICICI Bank
6. Korea Exchange Bank
7. Standard Bank London Ltd.
8. Union Bancaire prive'e (CBI-TDB)
9. Westdeutche Landesbank

In 2011, two licenses were granted to wholesale banks, namely Deutsche Bank AG and Industrial & Commercial Bank of China. Moreover, two investment banks commenced operation in the country—Arab Emirates Invest Bank and HSBC Financial Services (Middle East) Limited.

The number of GCC banks in 2011 remained unchanged at 7, while the number of other foreign banks remained unchanged at 22, the number of their branches at 82, and the number of their electronic/customer service units remained at 50.

The number of automated teller machines (ATMs) in the UAE increased from 3758 ATMs at the end of 2010 to 4172 at the end of 2011.

There are 11 foreign banks operating in the UAE. They are as follows:
1. Bank of Baroda[61]
2. Barclays Bank[62]
3. Bank Saderat Iran[63]
4. Citi Bank[64]
5. Doha Bank[65]
6. Habib Bank A.G. Zurich[66]
7. HSBC Middle East Limited[67]
8. Lloyds TSB Bank PLC[68]
9. National Bank of Bahrain[69]
10. National Bank of Oman[70]
11. Standard Chartered Bank[71]

The number of licensed foreign banks' representative offices operating in the UAE reached 110 at the end of 2011(Table 5). The new representative offices licensed during 2011 are the following:
1. AXIS Bank Ltd.
2. Falcon Private Bank Ltd.
3. Doha Bank
4. Bank of Montreal
5. SBI Funds Management Private Ltd.
6. Bank of the Philippine Islands
7. Liechtensteinische Landesbank (Liechtenstein) Ltd.
8. ABN Amro Bank N.V.
9. Banque Privee Edmond De Rothschild SA
10. Fairbairn Private Bank

The list of investment banks, wholesale banking, and companies that conduct finance and investment activities is relatively short:
- Arab Emirates Invest Bank–P.J.S.C.—Dubai

---

[61] http://www.bankofbaroda.com/
[62] http://www.barclays.ae/accounts/savings-account/
[63] http://www.banksaderat.ae/OpenAcc.html
[64] http://www.citibank.com/uae/homepage/index.htm
[65] http://www.dohabank.ae/en/Personal/Accounts/SavingsAccount.aspx
[66] http://www.habibbank.com/
[67] https://www.hsbc.ae/1/2/
[68] http://www.lloydstsb.ae/personal-banking/savings-investment/instant-access.html
[69] http://www.nbbonline.com/default.asp?action=category&ID=48
[70] http://www.nbo.co.om/
[71] http://www.standardchartered.ae/personal/Important-Information/en/service-charges.html

Table 5 Foreign Banks in the UAE (2011)[a]

SLNO	Name of the bank	Head Office	Abu Dhabi	Dubai	Sharjah	RasAl Khaimah	Ajman	Umm-Al Qaiwain	Fujairah	Al Ain	Total Number of Branches	Pay Offices
1	National Bank of Bahrain	Abu Dhabi	1	0	0	0	0	0	0	0	1	0
2	Rafidain Bank	Abu Dhabi	1	0	0	0	0	0	0	0	1	0
3	Arab Bank PLC	Abu Dhabi	1	2	1	1	1	0	1	1	8	0
4	Banque Misr	Abu Dhabi	1	1	1	1	0	0	0	1	5	0
5	El Nillein Bank	Abu Dhabi	1	0	0	0	0	0	0	0	1	0
6	National Bank of Oman	Abu Dhabi	1	0	0	0	0	0	0	0	1	0
7	Credit Agricole – Corporate and Investment Bank	Dubai	1	1	0	0	0	0	0	0	2	0
8	Bank of Baroda	Dubai	1	2	1	1	0	0	0	1	6	7
9	BNP Paribas	Abu Dhabi	1	1	0	0	0	0	0	0	2	2
10	Janata Bank	Abu Dhabi	1	1	1	0	0	0	0	1	4	0

Continued

Table 5 Foreign Banks in the UAE (2011)[a]—cont'd

SLNO	Name of the bank	Head Office	Abu Dhabi	Dubai	Sharjah	RasAl Khaimah	Ajman	Umm-Al Qaiwain	Fujairah	Al Ain	Total Number of Branches	Pay Offices
11	HSBC Bank Middle East Limited	Dubai	1	3	1	1	0	0	1	1	8	16
12	Arab African International Bank	Dubai	1	1	0	0	0	0	0	0	2	0
13	Al Khaliji (France) S.A.	Dubai	1	1	1	1	0	0	0	0	4	0
14	Al Ahili Bank of Kuwait	Dubai	1	1	0	0	0	0	0	0	2	0
15	Barclays Bank PLC	Dubai	1	1	0	0	0	0	0	0	2	3
16	Habib Bank Ltd.	Dubai	1	4	1	0	0	0	0	1	7	0
17	Habib Bank A.G Zurich	Dubai	2	5	1	0	0	0	0	0	8	1
18	Standard Chartered Bank	Dubai	2	7	1	0	0	0	0	1	11	3

19	Citi Bank N.A.	Dubai	1	2	1	0	0	0	0	1	5	7
20	Bank Saderat Iran	Dubai	1	3	1	0	1	0	0	1	7	0
21	Bank Mell Iran	Dubai	1	2	1	1	0	0	1	1	7	1
22	Blom Bank France	Dubai	0	1	1	0	0	0	0	0	2	1
23	Lloyds TSB Bank PLC	Dubai	0	1	0	0	0	0	0	0	1	5
24	The Royal Bank of Scotland N.V.	Dubai	1	1	1	0	0	0	0	0	3	3
25	United Bank Ltd.	Dubai	3	3	1	0	0	0	0	1	8	2
26	Doha Bank	Dubai	0	1	0	0	0	0	0	0	1	0
27	Samba Financial Group	Dubai	0	1	0	0	0	0	0	0	1	0
28	National Bank of Kuwait	Dubai	0	1	0	0	0	0	0	0	1	0
	Total		27	47	15	6	2	0	3	11	111	51

[a]Source: *United Arab Emirates: selected issues and statistical appendix*, IMF Country Report No. 12/136, June 2012.

- HSBC Financial Services (Middle East) Limited—Dubai
- Deutche Bank AG—Abu Dhabi
- Industrial & Commercial Bank of China—Abu Dhabi
- Mubadala GE Capital–P.J.S.C.—Abu Dhabi

The UAE banking sector has served as an important element of growth toward diversification of the UAE's growth strategy. It appears to be sound, modern, and very competitive. Nevertheless, "Dubai SME"—part of the Department of Economic Development—revealed that 86% of SMEs have not sought bank finance. It reflects how difficult it is for this sector to get a bank loan. This shows the big gap between the government directives and the banks' policies. Bank financing is usually the only option and is the predominant source of external financing for most SMEs. However, banks consider SMEs to be relatively high risk as most of their businesses are service activities, which lead to charging higher interest rates.

Most commercial banks are keen on funding the working capital needs of businesses, but less on funding startups. Thus, there is a need for dedicated banks, working on commercial principles but devoted to financing of the SMEs startups. Currently, not only the UAE but the entire GCC lack institutions that specialize in funding SMEs. Challenges of the banks in serving the SME sector include:

- quality of financial reporting (usually poor, especially in case of microenterprises)
- lack of credit history
- inadequacy of collateral (especially in case of startups)
- informal management (especially in micro- and small enterprises)
- short-term planning horizon
- weak cash flow management

Despite structural challenges, innovation is widespread in the banking community.

OTHER KEY PLAYERS OF THE FINANCIAL MARKET

The **Dubai International Financial Center**[72] is a financial center that offers a convenient platform for financial institutions and service providers (financial intermediaries). The center was established as part of the vision to position Dubai as an international hub for financial services and as the regional gateway for capital and investment. The DIFC's ambition is to become the global hub for Islamic finance.

[72] http://www.difc.ae/

The city of Dubai has an established track record of realizing projects in a safe environment and is perceived as one of the fastest growing cities in the world. It has a well-diversified economy based on international trade, banking and finance, information and communication technology, tourism, and real estate. In 2005, oil contributed less than 6% of Dubai's GDP and in 2010, it was less than 1%. This economic diversification is continuing with the establishment of new industries, private sector growth, and increased regional economic integration.

The DIFC seems to fill the gap between the financial centers of Europe and southeast Asia—the region comprising over 42 countries with a combined population of approximately 2.2 billion people. This region, stretching from the western tip of North Africa to the eastern part of south Asia had (until 2004) been without a world class financial center.

The DIFC aims to meet the growing financial needs and requirements of the region while strengthening links among the financial markets of Europe, the Far East, and the Americas. The DIFC's mission is to be a catalyst to facilitate the mobilization of capital for regional economic growth, development, and diversification by performing as a globally recognized financial center. The center attracts regional liquidity back into investment opportunities. It has attracted international firms such as Merrill Lynch, Morgan Stanley, Goldman Sachs, Mellon Global Investments, Barclays Capital, Credit Suisse, Deutsche Bank, and many other leading international financial institutions. The DIFC intends to be the regional gateway for investment banks, as well as other financial institutions wishing to establish underwriting, M&A advisory, venture capital, private equity, foreign exchange, trade finance, and capital markets operations to service this large and relatively untapped market.

The objectives of the DIFC are the following:
- to attract regional liquidity back into investment opportunities within the region and contribute to its overall economic growth
- to facilitate planned privatizations in the region and enable initial public offerings by privately owned companies
- to give impetus to the program of deregulation and market liberalization throughout the region
- to create added insurance and reinsurance capacity 65% of annual premiums are reinsured outside the region
- to develop a global center for Islamic finance (serving large Islamic communities stretching from Malaysia and Indonesia to the United States)

The DIFC focuses on the following main financial services sectors:
- banking and brokerage
- capital markets
- wealth management
- reinsurance and captives
- Islamic finance
- ancillary services

Benefits of establishing an institution in the DIFC include:
- 100% foreign ownership
- 0% tax rate on income and profits
- modern office accommodation and sophisticated infrastructure
- double taxation treaties available to UAE-incorporated entities
- no restriction on foreign exchange
- world-class English language court system based on the common law
- freedom to repatriate capital and profits (no restrictions)
- high standard laws, rules, and regulations
- high standard operational support

The DIFC offers a wide range of investment opportunities, such as
- mutual funds
- exchange traded funds
- open- and closed-ended investment companies
- index funds
- hedge funds
- consultant wrap accounts
- Islamic-compliant funds

Moreover, the DIFC supports the operational needs of financial institutions and provides an ideal environment and a highly skilled workforce to asset management firms and private banks. These services include accounting and legal practices, actuaries, management consultants, recruitment firms, and market information providers, among others.

An independent regulator, the Dubai Financial Services Authority (DFSA), supervises the DIFC, and the independent status of the center is enhanced further by the DIFC courts, providing the highest international standards of the legal procedure.

The DFSA's primary functions include[73]:
- policy development
- enforcement of legislation and authorization
- supervision of DIFC licensees

---

[73] www.dfsa.ae

It manages companies offering asset management, banking, securities trading, Islamic finance, and reinsurance and regulates the Nasdaq Dubai exchange.

The **Dubai Financial Market**[74] (DFM) was established as a public institution with its own independent corporate body[75] and started operations in March 2000. The DFM operates as a secondary market for trading of
- securities issued by public joint stock companies
- bonds issued by the federal government or any of the local governments and public institutions in the country
- units of investment funds
- other financial instruments (local or foreign) accepted by the market

As decided by the executive council decree in December 2005, the DFM is set up as a public joint stock company in the UAE.[76] It operates on an automated trading system (screen based), which offers a major advantage over traditional floor trading in terms of transparency, efficiency, liquidity, and trading of prices. The market has collaborated with renowned international experts in trading systems. Safeguarding market efficiency and integrity of trading requires the DFM to conduct regular monitoring and controlling. The market control section monitors compliance with the trading rules and regulations of ESCA.[77] The brokers' licensing and inspection section monitors the brokers' conduct to ensure integrity of the brokers' activities and that the best service is given to investors.

In 2010, the DFM launched a global service with the iVESTOR card—a revolutionary and innovative solution enabling retail investors to instantly receive DFM dividends directly into their investor card. Investors no longer have to wait for dividend checks or have the hassle of depositing checks into their bank accounts. The card will enable the DFM to credit any future dividends directly into the cardholders' balance. Additionally, the cardholder can withdraw cash easily from ATMs.

Finance companies[78] undertake one or more of the following major financing activities:
- extend advances and/or personal loans for personal consumption purposes
- finance trade and business, opening credit and issuing guarantees in favor of customers

---

[74] http://www.dfm.ae/Default.aspx
[75] Resolution from the Ministry of Economy in 2000
[76] http://www.dfm.ae/pages/default.aspx?c=801
[77] UAE Securities and Commodities Authority
[78] Central Bank of the United Arab Emirates, report 2011 at www.centralbank.ae.

Table 6 Finance Companies in the UAE (2011)[a]

1.	Osool "A Finance Company" L.L.C. – Dubai	
2.	Gulf Finance Corporation – Dubai	
3.	HSBC Middle East Finance Co. Ltd – Dubai	
4.	Max Orix Finance P.P.C. – Dubai	
5.	Finance House P.J.S.C. – Abu Dhabi	
6.	Dubai First P.P.C. – Dubai	
7.	Reem Finance P.J.S.C. – Abu Dhabi	
8.	Majid Al Futtaim JCB Finance L.L.C. – Dubai	
9.	Al Futtaim GE Finance P.P.C. – Dubai	
10.	Dunia Finance L.L.C. – Abu Dhabi	
11.	Abu Dhabi Finance P.P.C. – Abu Dhabi	
12.	Amlak Finance P.J.S.C. – Dubai	
13.	Tamweel P.J.S.C. – Dubai	
14.	Al Wifaq Finance Company P.P.C. – Abu Dhabi	
15.	Mashreq Al Islami Finance Co. P.P.C. – Dubai	
16.	Islamic Finance Co. P.P.C. – Dubai	
17.	Aseel Finance "Aseel" P.P.C. – Abu Dhabi	
18.	Mawarid Finance Co. P.P.C. – Dubai	
19.	Abu Dhabi National Islamic Finance P.J.S.C. – Abu Dhabi	
20.	Islamic Finance House – Abu Dhabi	
21.	Emirates Money Consumer Finance – Dubai	
22.	Abu Dhabi Commercial Islamic Finance Company P.P.C. - Abu Dhabi	
23.	Siraj finance Company P.P.C. - Abu Dhabi	
24.	Amex (Middle East) B.S.C. - Dubai	

[a]*Source*: www.centralbank.ae.

- subscribe to the capital of projects and/or issues of stocks, bonds, and/or certificates of deposit

Contribution of the financing company to the capital of projects, issues of stocks and/or bonds, or certificates of deposit should not exceed 7% of its own capital. The paid-up capital of a finance company should not be less than 35 million AED, and national shareholding should not be less than 60% of total paid-up capital, without prejudice to provisions of Federal Law No. 8 of 1984 and any subsequent amendments.

The number of finance companies licensed to operate in the UAE increased from 23 in 2010 to 24 in 2011 (Table 6) due to a license given to AMEX (Middle East) BSC.

One new investment company was issued a license in 2011, namely Masdar Investment,[79] thereby increasing the number of licensed investment companies from 21 in 2010 to 22 in 2011 (Table 7).

[79] http://www.masdar.ae/en/home/index.aspx

Table 7 Financial Investment Companies (2011)*a*

1.	Oman & Emirates Investment Holding Co. - Abu Dhabi
2.	Merill Lynch International & Co. C.V. – Dubai
3	Emirates Financial Services – Dubai
4.	Shuaa Capital P.S.C. – Dubai
5.	The National Investor - Abu Dhabi
6.	Islamic Investment Co. P.J.S.C. – Dubai
7.	Abu Dhabi Investment House P.J.S. - Abu Dhabi
8.	Al Mal Capital P.S.C.
9.	Injaz Mena Investment Company P.S.C. - Abu Dhabi
10.	National Bonds Corporation P.S.C. – Dubai
11.	Noor Capial P.S.C. – Abu Dhabi
12.	Unifund Capital Financial Investment P.S.C. - Abu Dhabi
13.	Daman Investment P.S.C. – Dubai
14.	Allied Investment Partners P.J.S.C. - Abu Dhabi
15.	Gulf Capital P.S.C - Abu Dhabi
16.	CAP M Investment P.S.C. - Abu Dhabi
17.	Royal Capital P.P.C. - Abu Dhabi
18.	Al Bashayer Investment Company L.L.C. - Abu Dhabi
19.	Dubai Commodity Asset Management P.P.C. - Abu Dhabi
20.	ADIC Investment Management P.P.C. - Abu Dhabi
21.	ADS Securities – L.L.C. – Abu Dhabi
22.	Masdar Investment – L.L.C. - Abu Dhabi

a Source: www.centralbank.ae.

BUSINESS COOPERATION BETWEEN THE UAE AND POLAND

Legal and treaty foundations of cooperation between the UAE and Poland arise from the following bilateral agreements:[80]

a. Double Taxation Avoidance Agreement between Poland and the UAE of 1993.

b. Agreement between the government of the United Arab Emirates and the government of the Republic of Poland for the Promotion and Protection of Investments—1993r.[81]

c. Civil Aviation Agreement of 1994.

---

[80] *The market guide for entrepreneurs: the United Arab Emirates*, Polish Agency for Enterprise Development, Warsaw 2010.

[81] (1) Desiring to create favorable conditions for greater economic cooperation between and particularly for investments by investors of one contracting state in the territory of the other contracting state. (2) Recognizing that the encouragement and reciprocal protection under international agreements of such investments will be conductive to the stimulation of business initiative and will increase prosperity in both contracting states.

Table 8 Polish Exports in 2010 (mln PLN)[a]

Saudi Arabia	580.3
Bahrain	43.8
Iraq	151.7
Iran	370.1
Oman	58.1
UAE	742.8
Qatar	48.8
Kuwait	84.1

[a]*Source*: Central Statistical Office, http://www.stat.gov.pl/gus/index_ENG_HTML.htm.

The aforementioned agreements, along with liberalization of the regulations concerning foreign investments in Poland, create favorable conditions for capital cooperation and capital influx from the UAE to Poland. In 1994, the Polish party submitted a draft trade agreement to the UAE; however, its negotiation was unnecessary due to the fact that both Poland and the UAE are signatories of a multilateral GATT 94/WTO agreement, which regulates the rules of international trade in a comprehensive and obligatory manner.[82]

Prospects for Polish Enterprises in Persian Gulf States

The Persian Gulf states feature a high level of money reserves. The last financial crisis forced some Polish enterprises to undertake a more intensive expansion abroad, which for some of them may contribute to establishing cooperation with the countries in that part of the world. In some of the states, where a government change had taken place, certain shifts in politics started, yet pundits' opinions are divided on whether the revolutions in a few countries of North Africa can actually affect the economic relations with those countries (Table 8).

Currently, it is chiefly construction companies that seek to win foreign contracts. A significant drop in the number of public tenders in Poland gave the impetus for undertaking actions in that direction. In the course of 2009 and 2010, the value of tender offers realized for GDDKiA[83] exceeded 40 billion PLN; however, in 2011, orders stood at 4 billion PLN. However, a large part of the Polish construction companies market has foreign shareholders who vie for contracts in the Near East on their own behalf. They do not welcome any competitors from Poland—hence, they may oppose the expansion of affiliated enterprises into the Arab region.

[82] http://www.obserwatorfinansowy.pl/2012/02/20/przedsiebiorcom-z-polski-marza-sie-kontrakty-z-1001-nocy-jest-potencjal/
[83] http://www.gddkia.gov.pl/en/1618/News

Table 9 Budget Income and Spending in the GCC (bln USD)[a]

Country	Budget income	Budget spending
Saudi Arabia	293.1	210.6
Bahrain	7.93	8.297
Qatar	68.79	37.88
Kuwait	92.23	61.16
Iraq	69.2	82.6
Iran	130.6	92.22
Oman	23.75	23.21
UAE	120.8	102.9

[a]*Source: CIA the world factbook* at https://www.cia.gov/library/publications/the-world-factbook.

Persian Gulf Markets

Saudi Arabia, Qatar, and Kuwait register significant budget surpluses (amounting to 20.5% GDP; second highest in the world), as do Iran, Oman, and the United Arab Emirates. Only Bahrain and Iraq follow the example of the Western world and spend more than their budget income (Table 9).

Buying something from Arab states is much easier (chiefly oil and oil products) than selling anything in those markets. All the Persian Gulf states show a positive—and distinctly so—balance of trade. For instance, Saudi Arabia ranks as the 15th global exporter, but only the 33rd importer. The largest global economies, such as the United States, Japan, China, South Korea, Germany, Italy, Great Britain, and India, are the trade partners of the Persian Gulf states. Thailand is another major importer of goods from Oman, Singapore – from Saudi Arabia, while Turkey plays an important role in trade with Iran and Iraq. Iraq is the fifth most important export market for Turkey (5.3%), while Iran is the seventh in respect of the value of import volume (Table 10).

Poland's trade with the Persian Gulf states has been small so far. It amounts to hundreds of millions of PLN in the case of Saudi Arabia, while it is counted in the tens of billions of PLN with any of Poland's top 10 trade partners (with Germany; even in hundreds of billions of PLN). The Polish export to Saudi Arabia is chiefly furniture, along with medical equipment and lamps. Enterprises of certain industries managed to establish their position in those markets. Annual trade with the United Arab Emirates amounts to nearly 320 million USD.[84] The UAE is one of Poland's biggest trade partners in the area of North Africa and the Near East. A higher total turnover in trade between Poland and the UAE is only recorded in the exchange

[84] Central Statistical Office, http://www.stat.gov.pl/gus/index_ENG_HTML.htm

Table 10 Imports and Exports of GCC Members (bln USD)[a]

Country	Import	Export
Saudi Arabia	106.5	350.7
Bahrain	16.8	20.23
Qatar	25.33	104.3
Kuwait	22.41	94.47
Iraq	53.93	78.38
Iran	76.1	131.8
Oman	21.47	45.53
UAE	185.6	265.3

[a]*Source*: *CIA the world factbook* at https://www.cia.gov/library/publications/the-world-factbook.

of trade with Israel and Saudi Arabia, but the UAE ranks first in respect of the volume of Polish exports in this area.

The number of foreign investments in the UAE is rising month by month, mostly in the largest metropolises—Dubai and Abu Dhabi. The amount of Polish capital is also on the increase. It is allocated on highly favorable conditions, inter alia into real estate, purchasing of which guarantees a relatively quick and certain profit (Table 11).

Polish entrepreneurs do not limit their expansion exclusively to the largest countries of the Persian Gulf. "Inglot"[85]—a producer of cosmetics for women—has already opened 21 shops in the Near East.[86] It sells its cosmetics in the UAE (six outlets in Dubai alone, additional shops in Abu Dhabi and Sharjah), Saudi Arabia, Qatar, and Bahrain, with two outlets in each of these countries, and in Kuwait and Oman with one outlet in each. The company is also present outside of the Gulf states—in Lebanon.

Mr. Wojciech Inglot—founder and president of the cosmetics company—is very satisfied with the company's operations in the Near East:

> The region is inhabited not only by Arabs, but also by people coming from all over the world, mostly from the south-east Asia. In some countries, e.g., in the UAE, there are more foreigners than there are native residents. We are planning further expansion in the region: in Saudi Arabia and Kuwait (Wojciech Inglot[87])

Entry Barriers for Polish Entrepreneurs on the Example of Polish Companies of Inglot and Can-Pack/Arab Can

Poles find it hard to establish their footing in the Arab markets, and practitioners offer a number of theories explaining why it is so. Stanisław Waśko,

---

[85] http://inglotcosmetics.com/
[86] http://biznes.newsweek.pl/wojciech-inglot--wladca-kolorow,79347,1,1.html
[87] http://www.ekonomia24.pl/artykul/864257.html#

Table 11 Polish Imports from GCC Members in 2010 (bln PLN)[a]

Saudi Arabia	636.9
Bahrain	150.3
Qatar	8.1
Kuwait	5.3
Iraq	0.0175
Iran	205.9
Oman	24.7
UAE	214.8

[a] *Source*: Central Statistical Office, http://www.stat.gov.pl/gus/index_ENG_HTML.htm.

vice-president of Can-Pack[88]— a manufacturer of packaging—claims that the largest problems encountered in Near East countries are that it is impossible to hold a majority package in the share capital of local companies and the need to have a local partner. It is a significant discomfort, particularly in case of production companies, which by their very nature need to invest a lot. Inglot solved that problem with the help of a franchise partner.

In 1999, Can-Pack founded Arab Can in the UAE, in 2004 it started up a second plant in Dubai, in 2009 it established a joint venture company of Can-Pack Linco[89] in Cairo, and in 2010 it opened a beverage can production facility in Casablanca in Morocco.

In turn, it is more difficult for construction companies to succeed in markets where cheap local companies, along with well-recognized international brands, already operate. Polish construction companies are unable to compete against local ones and have not yet achieved the position in the market that would enable them to compete against international concerns. Finding qualified workers in Arab states is also a problem. In practice, workers are recruited from among immigrants coming from India, Pakistan, the Philippines, and so on, and assembling an efficient production facility team requires a lot of effort.

Can-Pack is one of the few Polish companies that has found out what it is like to operate—and not only sell—in the Near East. The fear of the situation in Iraq is too great for most companies. Companies in Poland demonstrate great interest at the stage of talks. However, when it comes to concrete actions, only a small percentage of companies decide to go there. Fear and a misconception of the situation in Iraq constitute the largest problem. Many entities in Poland are interested in cooperation, but news of explosions and other types of dangerous events refrain them from action.

[88] http://www.canpack.eu/index.php?lang=en
[89] http://www.canpack.eu/index.php?lang=en&action=1

If an entrepreneur is determined to take subsequent steps aimed at starting cooperation with partners in the Persian Gulf region, then the next—highly significant stage—is going there. Arab states do not conduct business at a distance. Some companies tried to start operations remotely, but without a physical presence locally, not much could be achieved. Some experts claim that the time of Polish enterprises in the Near East is about to come. Companies without references that want to implement contracts in the Near East must start as subsuppliers or subcontractors. That is the strategy adopted by Spanish companies, which, since the early 1980s, have achieved unprecedented global expansion. Nevertheless, in order to enter new foreign markets, Polish companies need to take a greater advantage of the financing of national institutions that support and promote such types of undertakings (banks or insurance corporations). For instance, a Swedish equivalent of KUKE S.A.[90] (Export Credit Insurance Corporation Joint Stock Company, which establishes conditions for the safe and stable functioning of Polish enterprises by securing export and domestic transactions and by facilitating access to external financing) signed contracts to the tune of 25 billion EUR over a 1-year period, whereas KUKE – only to the amount of several hundred million PLN[91].

Factors Determining Expansion of Polish Enterprises in the UAE

The most important factors contributing to the cooperation and expansion of Polish enterprises in the territory of the UAE include

a. political stability of the UAE—lack of significant influence of fundamentalist movements on the internal policy (a consistently implemented vision of a modern state, in which old Arab traditions and the fundamental principles of Islam combine with modernity and the progress of civilization)
b. good economic situation (high GDP growth, significant budget surpluses, surpluses in foreign trade balance)
c. no restrictions on profit transfer or capital repatriation
d. low import duties (less than 5% for virtually all goods), nonexistent in the case of items imported for use in free zones
e. competitive labor costs; corporate and personal taxes are nil, numerous double taxation agreements and bilateral investment treaties are in place

---

[90] http://www.kuke.com.pl/home.php
[91] http://www.obserwatorfinansowy.pl/2012/02/20/przedsiebiorco-z-polski-marza-sie-kontrakty-z-1001-nocy-jest-potencjal

f. excellent infrastructure as well as a stable and safe working environment
g. development of new branches of industry in the UAE (chemical, metallurgical, machine industries, trade, and construction)
h. rich deposits of natural resources (approximately 10% of global oil deposits and fifth largest gas deposits)
i. growth recorded in certain areas outside of oil sector (apart from industry—in construction, trade, services, and banking)
j. substantial financial reserves (estimated at over 250 billion USD), invested in foreign markets (western Europe, the United States, the Near East)
k. dynamic development of financial services sector (e.g., establishment of the DFIC), constituting a bridge that links the financial markets of the East and the West
l. increase in trade between Poland and the UAE
m. most advanced telecommunication technologies
n. access to numerous trade fairs and international conferences
o. high-quality office and residential space
p. security of energy supplies
q. presence of Polish citizens in the UAE (chiefly engineering personnel, architects, medical personnel)
r. successful development of cultural and scientific contacts

Factors limiting cooperation with the UAE include high competition of local and international companies, as well as moderate interest demonstrated by local enterprises in conducting business with Poland.

Market Chances and Threats

As already mentioned, the UAE is without a doubt a highly attractive trade partner, especially for exporting of goods. There are no financial or administrative barriers restricting access to that market. Moreover, Poland and the UAE share a rich tradition of trade, and the UAE has an extensive, modern infrastructure at its disposal. A varied group of consumers in the Emirates alone creates favorable conditions for export expansion, but above all, it provides wide-ranging possibilities of re-export of goods from other states of the Persian Gulf region.

Apart from the oil and gas sector, crucial for the UAE, the following branches of industry can be seen as promising from the point of Polish investment export: chemical industry, machine-making sector, power engineering, and industrial and residential construction, as well as food processing. Numerous duty-free zones operating in the UAE are a chance for Polish (production and trade) enterprises (the UAE ranks third—after Hong Kong and Singapore—in respect of the size of a re-export center size in the world). The fact that in general the country uses a zero rate corporate tax and personal income tax seems to be a significant investment incentive.

Arabic Investments in Poland

Apart from conducting business with the Near East, it is worthwhile drawing capital from there. The Persian Gulf states are classified as the strongest players in respect of their capital power. Saudi Arabia invested 18 billion USD abroad in 2010, Kuwait 39 billion USD, and Qatar 23.5 billion USD. By way of comparison, on average, Poland receives 10 billion EUR of foreign direct investments in a year, and its accumulated value stands at 150.4 billion EUR.

A bond issue by the ministry of finance or a large city addressed to Islamic investors could be a good step. Several years ago the German land of Sachsen-Anhalt issued bonds worth 100 billion USD. Thanks to presentations of the region during a road show, contracts for sale of agro-food products were concluded that were worth three times more than the bond issue itself. In order to enable the operation of Islamic banks in their markets, some countries of western Europe introduced a provision in their regulations permitting application of the Quran law (the Sharia). Above all, it concerns a ban on calculating interest on loans and credits. The economic principle of Muslim banking instruments is largely based on market patterns of traditional banking. Interest is replaced with other equivalent financial constructions.

MARKET STRENGTHS, WEAKNESSES, OPPORTUNITIES, AND THREATS (SWOT) ANALYSIS OF THE UAE MARKET: BUSINESS PERSPECTIVE

Taking into consideration all of the aforementioned recent and future UAE market developments and country characteristics, including market data, as well as the policies realized within the last decade, it is a good reason to draw appropriate conclusions by making the SWOT analysis from the potential foreign investor's perspective. Market strengths, weaknesses, opportunities, and threats are as follow.[92]

---

[92] Hong Ju Lee and Dipak Jain, *Dubai's brand assessment success and failure in brand management*, pp. 234–246 (August 2009).

Strengths:
- political neutrality and stability of the UAE—lack of significant influence of fundamentalist movements on the internal policy (a consistently implemented vision of a modern state, in which old Arab traditions and the fundamental principles of Islam combine with modernity and the progress of civilization)
- booming economy—good economic situation (high GDP growth, significant budget surpluses, surpluses in foreign trade balance)
- no restrictions on profit transfer or capital repatriation
- low import duties (less than 5% for virtually all goods), nonexistent in the case of items imported for use in free zones
- competitive labor costs; corporate and personal taxes are nil, numerous double taxation agreements and bilateral investment treaties are in place
- excellent infrastructure, as well as stable and safe working environment
- development of new branches of industry in the UAE (chemical, metallurgical, machine industries, trade, and construction)
- rich deposits of natural resources (approximately 10% of global oil deposits and fifth largest gas deposits)
- growth recorded in certain areas outside of oil sector (apart from industry—in construction, trade, services, and banking),
- substantial financial reserves (estimated at over 250 billion USD), invested in foreign markets (western Europe, the United States, the Near East),
- dynamic development of financial services sector (e.g., establishment of the DFIC), constituting a bridge that links the financial markets of the East and the West
- increase in trade between Poland and the UAE
- most advanced telecommunication technologies
- access to numerous trade fairs and international conferences
- high-quality office and residential space
- security of energy supplies
- numerous and well-balanced policies leading to sustainable growth of the country

Weaknesses:
- negative image of the Middle East
- barren desert, lack of natural resources (other than oil and gas)
- only 20% of UAE nationals
- lack of fundamental infrastructure (e.g., transportation, water)
- luxuries may appeal to too small a segment

Opportunities:
- increasing oil price
- increasing job opportunities for immigrants and natives
- growing luxury market
- increase in foreign investment
- proactive attitude
- well-developed meetings, incentives, conventions, and exhibitions environment

Threats:
- strong competitors—within the region: Abu Dhabi, Qatar; outside of the region: Singapore, Hong Kong
- oil running out in 30 years
- terrorism and war could further negative image of Middle East
- limited media coverage

CONCLUSIONS AND SUMMARY

The UAE financial services sectors, as well as SMEs, have served as important elements of growth toward diversification of the UAE's strategy. There is a sound, modern, and competitive banking industry, and innovation is widespread in the banking community. Abu Dhabi and Dubai financial sectors constitute the majority of the system.

The country's sustainable growth is supported by issuing effective laws and legislations (encouraging the business environment and developing national industries) and strengthening investments and promoting the SME sector (numerous incentives for foreign investors, e.g., FTZs), as well as protecting consumer and intellectual property rights and diversifying trade activities.

All of the aforementioned government initiatives and incentives adhere to international standards of excellence and the tenets of knowledge economy. They aim to attract strategic foreign investments and add value to the existing business community, providing a stable and attractive business environment. UAE market strengths (according to the presented SWOT analysis) are numerous. The country is committed to maintaining the policy of economic openness, seeking actively to develop projects that are in harmony with the changes taking place in the world.

Despite the numerous policies in place, there are some deficiencies regarding SMEs' debt financing. "Dubai SME"—part of the Department of Economic Development—revealed that 86% of SMEs have not sought bank finance. It reflects how difficult it is for this sector to get a bank loan.

This shows the gap between the government directives and the banks' policies. Bank financing is the predominant source of external financing for most SMEs. However, banks consider SMEs to be relatively high risk, as most of their businesses are service activities. Most commercial banks are keen on funding the working capital needs of businesses, but less on funding startups. There is a need for dedicated banks, working on commercial principles but devoted to financing SMEs, especially startups. Currently, not only the UAE but the entire GCC lack institutions that specialize in funding SMEs.

CHAPTER 18
Development of a Sustainable Disabled Population in Countries of the Cooperation Council for the Arab States of the Gulf

Simon Hayhoe
Faculty (Educational Technology), Sharjah Women's College, UAE; Research Associate, Centre for the Philosophy of Natural & Social Sciences, London School of Economics, UK

Contents

Introduction	451
The Context of the Study	453
Open Coding Phase	455
Axial Coding Phase	455
Selective Coding Phase	456
Findings from Available Literature	456
First Analysis: Open Coding Phase	456
Second Analysis: Axial Coding Phase	457
Family Structures	457
Gender	458
Economic Extravagance	459
Islamic Religion	460
Third Analysis: Selective Coding Phase	461
Conclusion	464

INTRODUCTION

This chapter presents results of a grounded theory literature search on sustainability of the disabled population in countries of the Cooperation Council for the Arab States of the Gulf (GCC), which are composed of the countries surrounding the southern coast of the Persian Gulf and Oman. Research focused on the adaptability of the GCC's Arabic culture to accommodate a growing number of disabled people in its population, the potential areas of an increase or a decrease in its disabled population, and the advantages and potential problems in the societies of the GCC's constituent countries.

The analysis of the literature focused on the cultural factors and gaps in research and our understanding of disability in the GCC, particularly on the influences of external factors on the evolution of its epistemology, using the epistemological model of disability (Hayhoe 2012) as a framework, after its first phase of analysis. The study is necessary at this time as the countries of the GCC are currently undergoing a shift in the structure of their economies from being largely dependent on the mining of oil to those of *postoil*, knowledge, service sector, and industrial-based economies—the result of fluctuating oil prices and a growing population.

Furthermore, governments of the GCC have targeted sustainability as one of its goals and aims to countenance the most severe effects of its changing social environment (GCC Secretariat General 2011a). This makes it necessary to reevaluate the institutions, the distribution of resources, and the education of its indigenous population in order to protect the quality of life and rights of its disabled citizens, as set out by the United Nations' Millennium Goals (United Nations 2000) and the UN Convention on the Rights of Persons with Disabilities (United Nations, downloaded).

It is thus the primary aim of this research to start a debate and to define foci of future research needed in the GCC in order to sustain its disabled population. Gharaibeh (2009) finds that there has been scant investigation of such issues or their relevance to a general understanding of the role of Arabic culture on disability in general, and it appears that there is no evident strategy in the GCC on such issues. The secondary aim of this research is to provide the reader with an introduction to the broader debate on the nature and role of extrinsic historical and cultural factors on the attitudes toward and treatment of disability in the societies of this part of the Middle East and to provide an understanding of their epistemology of disability.

This chapter investigates these issues through three stages of analysis of secondary source literature, including research reports, surveys of statistics, reports on relevant laws, and philosophical essays on the nature of disability in GCC countries. It has three main findings: (1) that four main concepts (gender, Islamic religion, economic extravagance, and family structures) have affected the structure of disability in GCC countries; (2) that the development of a disability agenda is in its infancy in the constituent countries of the GCC, and thus structures such as statistical information on the nature of disability from their governments are lacking; and (3) despite cultural, economic, and social ties among countries of the GCC, there has been no evident coherent strategization of resources, institutions, policies, and education for disabled people. Before presenting the results of this study, however, this chapter addresses the context of the work, including the notion of a sustainable population, the foundation and objectives of the GCC, and the methodology used to conduct this study.

THE CONTEXT OF THE STUDY

The Cooperation Council for the Arab States of the Gulf was founded in May 1981 (21st Rajab 1401 AH in the Islamic calendar) in Abu Dhabi, United Arab Emirates, by the governments of the United Arab Emirates (UAE), Bahrain, Saudi Arabia, Oman, Qatar, and Kuwait. The council was established on a number of objectives—the most important of which were to coordinate, integrate, and interconnect all six countries' populations, recognizing their shared cultural and familial heritage; develop homogeneous economic, business, financial, legal, and administrative regulations; maintain cross-border scientific and technical research and education in industry, mining, agriculture, water, and animal resources; and create research centers, joint ventures, and cooperation within private industry. As its charter states, the founding of the block was created to develop an

> *institutional embodiment of a historical, social and cultural reality. Deep [Islamic] religious and cultural ties link the six states, and strong kin relations prevail among their citizens. All these factors, enhanced by one geographical entity extending from sea to desert, have facilitated contacts and interaction among them, and created homogeneous values and characteristics... It is also a fulfillment of the aspirations of its citizens towards some sort of Arab regional unity (GCC Secretariat General, downloaded).*

In the new millennium, new objectives were mooted. In particular, it was proposed that by 2010 the GCC should launch a single currency throughout its member states in a manner similar to that of the single European currency (the euro)(Khan 2009; Sturm & Siegfried, 2005), as it was felt that this would improve the economic prospects of its constituent countries, as well as ease partnerships and cooperation between institutions and industrial and business projects. However, following problems with the euro after a number of local and global economic problems, these steps toward changing the objectives and constitution of the GCC have been abandoned in the short to medium term (Vine et al., 2010).

Despite instabilities in the European economic union and political and social unrest in neighboring regions, there has generally been continued mutual economic and social development in the GCC states (Vine et al. 2010). This economic development in particular has produced a significant growth in population and quality of life among the indigenous and expatriate populations of its constituent states due in part to falling infant mortality rates and increased adult life expectancy (GCC Secretariat General 2011b). Consequently, these changes have the potential to make a significant impact on its demography, the number of disabled people, and the nature of services to sustain this population's quality of life (United Nations 2006).

Simultaneously, in 2011, the GCC published a number of revised future plans up until the year 2025 (GCC Secretariat General 2011a), part of a strategy that began at the start of the new millennium. Central to these plans were four revised main goals and strategic objectives, the first and foremost of which was to move toward a wholly sustainable population across its member states. This was defined thus

> The main strategic and integrated goal of the GCC states will be achieved within the framework of the comprehensive concept of sustainable development. Therefore, the first strategic objective is embodied in the following:
>
> The comprehensive concept of sustainable development should be promoted over the time period in which this strategy will be implemented. This is because the concept of sustainable development stresses the fact that development is a continuous process transcending generations and that it is the outcome of human interaction with the existing resources as well the prevailing conditions that cause constant advancement of society and increases the efficient use of human, material and technological resources.
>
> That requires adopting the following approaches:
> 1. Optimal utilization of the available resources and allocation of human and material resources in an appropriate manner.
> 2. Deriving maximum benefit from the technical capabilities and adapting their use for promoting growth and enhancing human capacities.
> 3. Enhancing understanding of the modern functions of government, which ensure sustainable development and adopting policies that ensure economic and social stability and performance in terms of development.
> 4. Participation of all community institutions in the development process and seriously handling the options and priorities.
> 5. Developing the institutional capacities and creating a good environment for the general economic and social policies.
> 6. Emphasizing correlations between productive work, consumption patterns and development of human resources.
> 7. Participation of the work force in productive economic activities and guaranteeing their rights and constantly rehabilitating and training them for the job market...
>
> (GCC Secretariat General 2011a, pp. 18–19)

This leads us to the following research questions: If we use the United Nations' definition that "sustainable development is development that meets the needs of the present without compromising the ability of future generations to meet their own needs" (United Nations General Assembly 1987) and take the existing effects of this changing population into consideration, (1) what are the potential challenges to these goals and objectives of a sustainable disabled population in the GCC and (2) how does this changing disabled population affect the institutions and cultures of the countries of the GCC?

To investigate these questions, a survey of secondary source statistics and research studies was planned in order to evaluate an existing understanding of disability in the GCC and to find areas of research that showed strength or needed improvement or further consideration. The methodology used was a grounded theory framework of data collection and analysis (Glaser & Strauss 1967). This methodology included adapting elements from a basic structure of the three phases of data collection described by Glaser and Strauss (1967) as *coding phases*. These are laid out here in their chronological order of implementation.

Open Coding Phase

This was defined by Glaser and Strauss (1967) as the stage of in which all initial data are analyzed without preconceptions, or as few preconceptions as possible, in order to determine whether there are initial connections between data, the initial nature of these connections, and whether they can be classified and recorded in a meaningful way. Hayhoe (2012) finds that this stage of research is analogous to starting a theatrical work, during which characters, relationships, and personalities are defined. In the open coding phase of this research, literature was read without any prior hypothesis being assumed as closely as possible, allowing factors surrounding the disabled population, such as population dynamics and cultural attitudes, to arise and connections between different literatures to develop. The literature selected for this section of the research consisted largely of smaller scale research studies and reviews of previous research in order to identify the need for and sources of more detailed statistical reports for later analysis.

Axial Coding Phase

Glaser and Strauss (1967) defined this as the stage during which further links between data are made and refined and initial crude hypotheses are drafted and compared to existing data. During this stage, data that do not appear to be initially relevant or do not induce a pattern are put aside—although not rejected as it may contain some later significance. Hayhoe (2012) argued that this analysis was equivalent to developing the plot of a play, working out interconnected story lines and planning the beginning, middle, and end of the narrative—the hypothesis. In this research, the refined literature and statistics were again compared, and further memos were drafted and analyzed. From these, a more targeted literature search using these newly found categories was created and a tentative first hypothesis was developed, with any potentially false assumptions refuted.

Selective Coding Phase

Glaser and Strauss (1967) described this phase as being a testing of new tentative hypotheses with fresh data, which are then refined, refuted, or used to support it in its entirety. Hayhoe (2012) finds that this phase is analogous to the final refining of a script after its initial reading, where characters and stories are reformulated and the production made ready for its initial performance—publication. In this study, the tentative hypothesis was to be compared to secondary source statistical data selected for its similarity to the chosen categories, although this was not largely possible. As a result, an additional hypothesis based on the need for reliable data collection and analysis was formulated and further secondary source literature used.

What now follows is a presentation of the analysis of the literature gathered in all three phases of the investigation. This analysis is presented in the three chronological data coding stages discussed previously.

FINDINGS FROM AVAILABLE LITERATURE

First Analysis: Open Coding Phase

Literature from this phase immediately presented a problem. No general discussion of disability across the GCC countries appeared to exist; no definitions of how the GCC defined disability and no discussion of the concept of disability as a whole appeared to have been debated by the GCC Secretariat General, either in academic literature or statistical reports of umbrella international organizations relating to the GCC. There also appeared to be little discussion of the politics of disability in the Middle East as a whole or the formation of various social or cultural models of disability, such as those discussed by many western disability theorists (Barnes & Mercer 2003; Oliver, 2001), although a small number of standard political works on disability mentioned Arab or Middle Eastern cultures in general within their discussions of regional differences (see, e.g., Albrecht, Seelman, & Bury 2001; Aruri & Shuraydi 2001; Charlton 1998).

Furthermore, in the small amount of literature surveying the state of disability in the constituent countries of the GCC, the majority referred to Arabic culture as a whole (see, e.g., Al Thani 2006; Gharaibeh 2009) or could be subdivided into literature referring to the Gulf Region or the Middle East as a whole; although there is further work on migrants from this region who have settled in the United States, and the effects of their culture on their treatment and any perceived differences in their behavior (see, e.g., Brodsky 1983; Budman, Lipson & Meleis 1992; Lamorey 2002; Stone 2004).

The largest body of research on this region appeared to be on medical factors relating to a specific disability causing illnesses within specific countries or subregions of the GCC, either singly or in conjunction with other Arab countries [see, e.g., literature on the epidemiology of mental health problems (El-Islam 2008; Hamdi, Amin & Abou-Saleh 1997) or multiple sclerosis within a Middle East context (Yaqub & Daif, 1988)]. These, however, saw disability not as a social phenomenon, but as a traditional medical problem, requiring only therapy and institutional care away from mainstream society, referred to in disability theory as the medical model (Shakespeare 2006).

However, four particular epistemological trends emerged in the analysis of disability within the countries of the GCC and with Arab culture in general. These trends appeared similar in their structure to those identified by Hayhoe (2008) in the development of attitudes toward the blind community in Western societies, although there were social differences among the Arab communities in Europe based on their distinct cultural characteristics. In particular, it was observed that gender, Islamic religion, results of economic extravagance, and family structures created foci for the literature on Arab and Middle East societies (Al Thani 2006; Balcazar et al. 2010; ElHesser 2006; Gharaibeh 2009). These trends thus became the categories of analysis in the axial coding phase.

Second Analysis: Axial Coding Phase

During the axial coding phase, the four categories identified in the open coding phase were addressed individually, with any interconnections noted within the final hypothesis. These are outlined here.

Family Structures

In this literature, a primary topic of discussion was consanguineous marriage, that is, the practice of marriages being arranged between members of the same family. This was found to be most frequent between cousins and was often the result of social–cultural factors such as a desire to maintain family bonds, attempting to secure a positive social compatibility between married couples, lessening of a need for a dowry, securing of land rights, and the result of geographical isolation, such as that found in desert communities (Abdulkareem & Seifeddin 1998; Bener, Abdullah & Murdoch 1993; Bener et al. 1996; Saggar & Bittles 2008).

Numerous articles have observed that surveys of populations that include GCC Arab communities have shown that the results of consanguineous marriage can cause various physical and learning disabilities, as well as an increased risk of infant mortality (Modell & Dar 2002; Yunis, El Rafei, & Mumtaz 2008; Saggar & Bittles 2008). However, more targeted studies of small Arab communities in Kuwait find that such inherited disorders are more likely to manifest themselves through learning and developmental disabilities rather than physical disabilities (Al-Kandari & Crews 2011). In particular, Abu-Rabia and Maroun (2005) found that the more closely related parents are, the more likely that their children are to have a reading disability. Furthermore, Swadi and Eapen (2000) found that Arab children within the UAE with similar forms of learning difficulties were also more susceptible to depressive conditions, although it was unknown whether this was due to genetic disorders; the stigmatization of their disability; or other social, economic, or cultural factors.

Gender
A further significant issue affecting the disabled Arab population within the GCC is that of gender. ElHessen (2006) and Nagata (2003) argue that disabled Arab women are doubly oppressed, having less life chances in education and employment in many Arab societies because they are women, and also from a social stigma and perceived ineffectiveness because of their disability. Similarly, Frank (1989) found that as women are allowed to attend to men in an Arab family but the opposite is not true, disabled Arab women are at particular risk if they do not have other female family members to attend to any needs they cannot fulfill themselves. Likewise, Crabtree (2007) found that in a study of mothers in the United Arab Emirates, these women bore the greatest responsibility for their disabled children's welfare and care.

Permanent injury from violence within the family is also a potential cause for impairment and disability, with Al Gharaibeh (2011) finding that wives are at the greatest risk of such violence, including impairments caused by such ritualized practices as tongue lashing; with young boys being the second most vulnerable group in Arab states. Similarly, Lightfoot-Klein (1994) found that a number of women immigrating into the United States from Arab countries had disabling injuries caused by ritualized female genital mutilation.

However, it was also observed that it is not only biological impairments that can cause disabilities in the Middle East. Despite large investments in education since the early 1970s, Al Gharaibeh (2011) found that illiteracy rates among Arab families are still high in Middle Eastern states compared to those in the West, with women's literacy rates being significantly lower than men's; he also finds that literacy is at its highest in the Ghaza Strip, which is one of the poorest parts of the Middle East. However, Nagata (2003) observed that in the small number of Arab countries that collect such statistics, the number of illiterate biologically disabled people was significantly higher than their able-bodied counterparts and that disability appears to be a more important factor than gender for predicting literacy. She also argued that women from higher class families in the Persian Gulf are considered to be more privileged than men as they find it relatively easy to find well-paid work, yet are not expected to provide for their families as many men in the region are.

Economic Extravagance

Although there is no literature related directly to economic extravagance and disability, two particular effects of the rapid increase in living standards and incomes in countries of the GCC have led to unforeseen contemporary causes of disability. The first of these causes is the exponential rise in traffic accidents caused by increasingly large and fast cars on the road in the region south of the Persian Gulf (Bener 2005; Bener & Crundall 2005; McIlvenny 2006). Bener and colleagues (2002) found that despite stringent road laws, many drivers in the UAE still have dangerous behavior and habits when driving; although such dangerous habits tend to be different for men and women, the authors of this study recommend more stringent traffic laws and education as a remedy for such attitudes.

Bener and co-workers (1992, 2007a) found that the introduction of stringently applied seatbelt laws reduced the rate of permanent injury and death significantly in the Middle East. In relation to the outcomes of such injuries, Eid (2009) and Al-Naamani and Al-Adawi (2007) observed that adults are susceptible to head injuries in the Gulf region, leading to paralysis and neuropsychiatric syndromes, whereas Bener and colleagues (2007b) found that the Gulf's children are susceptible to severe bone fractures, which can be particularly dangerous to their long-term physical health and welfare.

The second factor occurring as a result of the new and sudden wealth in GCC countries is the threat of diabetes, caused by a massive increase in the consumption of sugary and high calorific foods; decreased amounts of exercise, partially as a result of increased car use; and the epidemic of morbid obesity. This can result in conditions such as kidney and heart failure, strokes, blindness, and amputation of limbs. Novo Nordisk A/S (2010) found that diabetes is nearing critical levels in the Middle East and North Africa region of Arab countries. In particular, their survey found that 14% of all health care spending in this region (approximating to just under US $3000 per head of population in Qatar, for instance) will be on diabetes or conditions resulting from diabetes. Furthermore, Novo Nordisk A/S estimated that around 40% of all people in this region will contract diabetes at some point in their lives.

Islamic Religion

ElHessen (2006) found that few contemporary works focus on Islam and disability, although there is a significant amount of scholarship on Islam by disabled scholars in the Middle East from Antiquity. She argued that the modern dearth of work on disability as a social issue in particular was difficult to understand, as religion is such an important tenant of modern Arabic society. "For example, the materials from the twelfth and thirteenth centuries compiled by al-Marghinani and Ibn Khallikan suggest the wide range of disability experiences, of social responses, and of debate regarding disability among Muslim scholars and jurists" (ElHessen 2006, p. 98). She also observed that the Quran itself is broadly positive about disability and that if a person is unable to conduct an act of worship or a common act because of their impairment, then, as long as they have good intentions, they have moral equality with an unimpaired person. Similarly, Al-Mousa (2010) argues that disability equality is a fundamental tenant of Islam, as the religion does not judge people on physical, social, or economic strength or superiority, but on their piety to the laws of Islam.

Furthermore, ElHessen (2006) observed that according to Islamic scripture, disabled children can be seen as a gift from God, with the spirit of Ibsan asserting that such children are sent as a test of their parent's compassion and charity. Similarly, Morad and colleagues (2001) observe that the Quran asserts that Muslims with wealth and lesser physical dependency are obliged to conduct positive acts for those without such assets. However, they also found that there is a difference in moral attitude in Islam to those with a perceived intellectual difficulty—who may be regarded as relatively benign simpletons—and those with a mental illness, including illnesses such as schizophrenia, although both are found to be morally "incompetent" by the Islamic judiciary if they commit a crime and are not treated as harshly as an unimpaired person would be. Similarly, Yong (2007) found a comparable mixture of the special status of children with Down's syndrome and the shame of the family of such children in Arab Islamic cultures.

Further moral ambiguities toward disabilities have been observed by Gaventa and Coulter (2001) in more traditional parts of the Middle East. In these communities they find that there is a belief that intellectual disabilities are the work of an evil spirit (termed Jinn) and that having a child with such an impairment can bring shame on their families. Such beliefs, they argue, have resulted in many children in traditional communities being doubly handicapped by their disability and social and economic rejection. Similarly, Rispler-Chaim (2007) found that the Islamic law of Hijazi sees that there is a link between parents' misdeeds and their children's disabilities, with surveys discovering that such connections have contributed to feelings of guilt.

Third Analysis: Selective Coding Phase

During the axial coding stage, the unrefined hypothesis that arose was as follows: The culture of the GCC has been slower to evolve to the needs of its disabled population than the speed at which its economy has grown, leaving the development of a sustainable disabled community lagging behind. However, the Islamic tradition of giving of financial and physical help to those people in greater need and the potential to provide resourced a positive foundation on which to develop a sustainable disabled community.

During the selective coding phase it was observed initially that the most significant source of literature came from research on special educational needs, whose findings appeared to support the hypothesis. In particular, national surveys of this form of education by countries of the GCC indicated that (1) the needs of students with significant impairments have been recognized by the governments of the GCC countries, with the latter countries taking steps to educate and, if possible, include these students, with the last of the countries undertaking national initiatives in the early years of the 21st century; and (2) significant amounts of funding had been spent on these initiatives and resourcing schools for students with impairments (Alpen Capital 2012).

For example, after the introduction of inclusive education in1984, the first of its kind in the Arab world (Alpen Capital 2012), the state of Saudi Arabia observed an exponential growth of inclusive lessons for students with special needs from the early1990s (Al-Mousa 2010). Furthermore, in line with the Islamic traditions of their culture, this inclusion was conducted in cooperation with existing charitable bodies, many with royal and government connections. Similarly, the International Bureau of Education (IBE-UNESCO 2006) observes that Bahrain has had a central special needs committee since 1986 and, since this time, has promoted inclusive classes within mainstream institutions alongside special education in separate schools. Like Saudi Arabia, they find that these forms of education have also been conducted with the close cooperation of large charitable bodies that have Islamic foundations at their core.

Similarly, Gaad (2011) observed that far-reaching moves toward inclusion have been made by UAE, Qatar, Oman, and Kuwait, with Kuwait providing a particularly strong educational model of educational legislation and implementation for similar Arab countries. However, Gaad (2011) also observed that the take up of such measures and initiatives is variable across these countries, with little monitoring or checks, particularly in private educational establishments.

In terms of its cultural foundation in Islamic law, Saudi Arabia now also has antidiscrimination legislation that makes it an offense to victimize people according to language group, social status, or disability (Bureau of Democracy, Human Rights, and Labor 2010a). The UAE goes further, legislating against discrimination according to race, nationality, social status, and disability (Bureau of Democracy, Human Rights, and Labor 2010b). Like Saudi Arabia, Qatar has a general law against discrimination, which was enacted in 2004, but it goes further than the Saudi Arabian legislation and bears greater similarity to the UAE's broader standing on human rights (Bureau of Democracy, Human Rights, and Labor 2010c).

Bahrain, Kuwait, and Oman have similar legislation and, like other GCC countries, make access to education, training, and public spaces a legal right, with Kuwait also having a nongovernmental organization entitled the Higher Council for Handicapped Affairs (Bureau of Democracy, Human Rights, and Labor 2010d–f). Such legislation has largely led to specific state benefits for people with disabilities in the GCC. For example, the Japan International Cooperation Agency (2002) observed that disabled people in Saudi Arabia are entitled to a great many financial and physical benefits, such as the provision of airfares for careers, adaptation of cars, designated parking places, provision of educational materials and teaching support, purchasing of artificial limbs, and provision of ramps and other technologies that aid access to parks, roads, and public gardens.

However, although this development of resources and legislation, the implementation of access to education, and the progress of social and cultural institutions appears revolutionary within the context of what are relatively young countries, information, data gathering, and checking on education, training, employment, and implementation of access in the greater society seem either to have lagged behind this development or remain unconsidered. For example, Gaad (2011) found that there is no monitoring of educational attainment or widespread checking of the implementation of inclusive education. Similarly, ElHessen (2006) argues that there has historically been a lack of study and rhetorical information about the lives, development, and achievements of disabled people in Arab societies in general.

This lack of data is also particularly marked in the comparative health and social surveys produced by agencies of the United Nations. For example, a review of surveys on mortality over the course of four decades up to 2002 (United Nations 2004a) found only data from a single census for Saudi Arabia and the UAE for almost all categories of analysis, although there were full sets of data for Kuwait, Bahrain, Qatar, and Oman, which, being smaller countries, appeared to have had closer cooperation with the United Nation's agencies.

Similarly, in a study of economic activity, the United Nations (2004b) conducted a review of data published from national censuses since the 1960s until the time of publication, focusing on economic activity and housing, which had consequences for information on universal access to education, training, and employment. In its tables it observed that neither Oman nor Saudi Arabia had published any data from censuses on this topic since the 1960s. In the same review, it was found that the UAE had published two (in 1968, while it was a British colony; and 1974, 3 years after it had gained independence) but had failed to publish data from its following three censuses in 1980, 1985, and 1990, while Qatar had published data from one census (1986) but not from an earlier census (1970) or its later census (1997); however, Bahrain and Kuwait, which had worked with international organizations in order to improve health and education provision, had published data from all but one of its censuses, which was in line with many European and North American countries.

Although data gathering appears to have improved more recently, there are still gaps in statistical data and analysis in the most recent United Nations' surveys. For example, the United Nations (2011) 2009–2010 Demographic Yearbook showed that although all countries in the GCC appear to have crude population estimates, Saudi Arabia, Qatar, and UAE generally do not have analyses of birth rates, life expectancy at birth, gender differences, differences between urban and rural residency, and rate of infant mortality—information that could provide a sophisticated demography of their disabled populations. Furthermore, although Oman appeared to have improved its data collection and had a similar rate of recording and analysis to Bahrain and Kuwait, there were still significant gaps in their analysis of birth and death rates, alongside life expectancy and gender differences.

In addition, the Japan International Cooperation Agency (2002) found that although Saudi Arabia had made efforts to develop its information and data-gathering process on disability-related issues in the early years of the millennium, it was unknown whether legislation and initiatives still had the impact that they were intended to have originally. They argued, however, that this was not solely due to problems with a lack of checks and balances, but was largely due to broader social attitudes toward disability and patronage not keeping pace with those of development in other areas of their society, alongside a lack of broadcast information about disabled people designed to change social attitudes. As they argued in their findings,

> At a glance, services provided by the government and charitable organizations seem to be plentiful and of high quality, but it could also be said that this provision of institutionalized services or "institutionalization" is a form of "social segregation" rather than "social integration" of persons with disability. Although the lack of public awareness does not come out clearly in this report, the underlying public attitude towards persons with disabilities may be sympathy and social pity that in turn may be preventing their acceptance into mainstream society. Public awareness campaigns on the feelings and barriers faced by persons with disabilities may be necessary to lay foundations for transition from institutionalization to community-based services and rehabilitation according to international trends" (Japan International Cooperation Agency 2002, p. 19).

CONCLUSION

There is little doubt that the constituent countries of the GCC have generally shown substantial progress in developing sustainable services and legal equality for people with disabilities, particularly in the fields of education and access to physically impaired people in mainstream institutions, public buildings, parks, and gardens. There is also evidence that people with physical and intellectual disabilities have received essential training and are now finding gainful employment, an important function of this region's economic sustainability. However, although there is a positive culture of charity and assistance toward people with disabilities, there is also little encouragement of independent disability rights organizations, which will not only work to gain access to institutions, education, and employment, but also strive to raise public awareness of disabilities, work toward equality with unimpaired people, and develop an understanding that disabled people can achieve as much as their ablebodied counterparts in many circumstances. In the long term, this could have a more detrimental effect on the development of a sustainable disabled population than simple access issues.

There also appears to be resistance to the notion of complete equality outside of a number of institutions, mainly because of the social stigma of learning impairments and mental illness, but also because it conflicts with a number of important issues that are deeply embedded in more traditional areas of GCC countries, such as the difficulties that can be associated with consanguineous marriage and insecurity in the relative newness of their systems of education and government. Perhaps the most significant problem faced by countries of the GCC, however, is the lack of research literature relating specifically to GCC countries and statistical analysis conducted on factors that could provide targeted information on the changing demography of the disabled population. This is vital if the quality of life of this population is to remain sustainable as the broader population increases. It is not that raw data do not exist—it is the culture of analysis and understanding of what needs to be analyzed that needs to evolve.

Of primary importance, however, is a coordinated system of open discussion, strategization, and policy making agreed to by all of the governments of the GCC and implemented by its Secretariat General. There are many important and pioneering initiatives and models of good practice in the countries of the GCC; however, there is also little recognition or understanding of them. As it coordinates higher education, economic, and business activities to great effect, the GCC must now understand that it is as important to harmonize social development initiatives to minimize the impact of disability where possible and support people with disabilities when needed. Examples of these could include development of a Gulf-wide road safety and healthy diet campaign, the raising of awareness of disability through the press, raising social and cultural profiles of disabled people in public positions and in the media, or screening people at risk of disability in order to provide help and support at an early stage. It is only through actions such as these can the GCC develop a truly sustainable disabled population.

REFERENCES

Abdulkareem, A., Seifeddin, B., 1998. Consanguineous marriage in an urban area of Saudi Arabia: Rates and adverse health effects on the offspring. J. Community Health 23, 75–83.
Abu-Rabia, S., Maroun, L., 2005. The effect of consanguineous marriage on reading disability. Dyslexia 11, 1–21.
Albrecht, G.L., Seelman, K.D., Bury, M. (Eds.), 2001. Handbook of disability studies. Sage, Thousand Oaks, CA.
Al Gharaibeh, F., 2011. Women's empowerment in Bahrain. J. Int. Women's Stud. 12, 99–113.
Al-Kandari, Y.Y., Crews, D.E., 2011. The effect of consanguinity on congenital disabilities in the Kuwaiti population. J. Biosoc. Sci. 43, 65–73.
Al-Mousa, N.A., 2010. The experience of the Kingdom of Saudi Arabia in mainstreaming students with special needs in public schools. Arab Bureau of Education for the Gulf States. Saudi Arabia, Riyadh.
Al-Naamani, A., Al-Adawi, S., 2007. "Flying coffins" and neglected neuropsychiatric syndromes in Oman. Sultan Qaboos Univ. Med. J. 7, 5–11.
Alpen Capital, 2012. GCC education sector. Alpen Capital (ME) Limited, Dubai, UAE.
Al Thani, H., 2006. Disability in the Arab region: current situation and prospects. J. Disabil. Int. Dev. 3, 4–9.
Aruri, N.H., Shuraydi, M.A. (Eds.), 2001. Revising culture, reinventing peace: the influence of Edward W. Said. Olive Branch Press, Brooklyn, NY.
Balcazar, F.E., Suarez-Balcazar, Y., Taylor-Ritzler, T., Keys, C.B., 2010. Race, culture and disability: rehabilitation science and practice. Jones & Bartlett Publishing, Sudbury, MA.
Barnes, C., Mercer, G., 2003. Disability. Polity Press, Cambridge, England.
Bener, A., Crundall, D., 2005. Road traffic accidents in the United Arab Emirates compared to Western countries. Adv. Transportation Stud. Int. J. 6, 5–12.
Bener, A., 2005. The neglected epidemic: road traffic accidents in a developing country, state of Qatar. Inj. Control Safety Promot. 12, 45–47.
Bener, A., Abdullah, S., Murdoch, J.C., 1993. Primary health care in the United Arab Emirates. J. Family Pract. 10, 444–448.
Bener, A., Abdulrazzaq, Y.M., Al-Gazali, L.I., Micallef, R., Al-Khayat, A.I., Gaber, T., 1996. Consanguinity and associated socio-demographic factors in the United Arab Emirates. Human Heredity 46, 256–264.
Bener, A., Al Humoud, S.M.Q., Price, P., Azhar, A., Khalid, M.K., Rysavyx, M., Crundall, D., 2007a. The effect of seatbelt legislation on hospital admissions with road traffic injuries in an oil-rich, fast-developing country. Int. J. Inj. ControlSafety Promot. 14, 103–107.
Bener, A., Crundall, D., Haigney, D., Bensiali, A.K., Al-Falasi, A.S., 2002. Driving behaviour, lapses, errors, and violations and their relation to road accident involvement in a new developed Arabian country. Available from: (21.06.12.) http://www.salimandsalimah.org/Bener-Driverbehavioursurvey-UAE.doc.
Bener, A., Justham, D., Azhar, A., Rysavy, M., Al-Mulla, F.A., 2007b. Femoral fractures in children related to motor vehicle injuries. J. Orthop. Nurs. 11, 146–150.
Brodsky, C.M., 1983. Culture and disability behavior. West. J. Med. 139, 892–899.
Budman, C.L., Lipson, J.G., Meleis, A.I., 1992. The cultural consultant in mental health care: the case of an Arab adolescent. Am. J. Orthop. 62, 359–370.
Bureau of Democracy, Human Rights, and Labor, 2010a. 2010 human rights report: Saudi Arabia. U.S. Department of State, Washington, DC.
Bureau of Democracy, Human Rights, and Labor, 2010b. 2010 human rights report: United Arab Emirates. U.S. Department of State, Washington, DC.
Bureau of Democracy, Human Rights, and Labor, 2010c. 2010 human rights report: Qatar. U.S. Department of State, Washington, DC.
Bureau of Democracy, Human Rights, and Labor, 2010d. 2010 human rights report: Kuwait. U.S. Department of State, Washington, DC.

Bureau of Democracy, Human Rights, and Labor, 2010e. 2010 human rights report: Bahrain. U.S. Department of State, Washington, DC.

Bureau of Democracy, Human Rights, and Labor, 2010f. 2010 human rights report: Oman. U.S. Department of State, Washington, DC.

Charlton, J.I., 1998. Nothing about us without us: disability oppression and empowerment. California University Press, Berkeley, CA.

Crabtree, S.A., 2007. Maternal perceptions of care-giving of children with developmental disabilities in the United Arab Emirates. J. Appl. Res. Intellect. Disabil. 20, 247–255.

Eid, H.O., Barss, P., Adam, S.H., Torab, F.C., Lunsjo, K., Grivna, M., Abu-Zidan, F.M., 2009. Factors affecting anatomical region of injury, severity, and mortality for road trauma in a high-income developing country: lessons for prevention. Injury 40, 703–707.

ElHessen, S.S., 2006. Disabilities: Arab states. In: Joseph, S. (Ed.), Encyclopedia of women and Islamic cultures: family, body, sexuality & health. Koninklijke Publishing, Leiden, The Netherlands, pp. 98–99.

El-Islam, M.F., 2008. Arab culture and mental health care. Transcult. Psychiatry 45, 671–682.

Frank, A.O., 1989. The family and disability - some reflections on culture. J. Royal Soc. Med. 82, 666–668.

Gaad, E., 2011. Inclusive education in the Middle East. Taylor & Francis, London, England.

Gaventa, W.C., Coulter, D., 2001. Spirituality and intellectual disability: international perspectives on the effect of culture and religion on healing body, mind and spirit. Haworth Pastoral Press, Binghamton, NY.

GCC Secretariat General, 2011a. The revised long-term comprehensive development strategy for the GCC states (2010-2025). GCC Secretariat General, Riyadh, Saudi Arabia.

GCC Secretariat General, 2011b. Statistical bulletin (volume 19). GCC Secretariat General, Riyadh, Saudi Arabia.

Gharaibeh, N., 2009. Disability in Arab Societies. In: Marshall, C.A., Kendall, E., Banks, M.E., Gover, R.M.S. (Eds.), Disabilities: insights from across fields and around the world. Praeger Publishers, Westport, CT.

Glaser, B.G., Strauss, A.L., 1967. The discovery of grounded theory: strategies for qualitative research. Aldine Publishing Company, Chicago, IL.

Hamdi, E., Amin, Y., Abou-Saleh, M.T., 1997. Performance of the Hamilton Depression Rating Scale in depressed patients in the United Arab Emirates. Acta Psychiatrica Scandinavica 96, 416–423.

Hayhoe, S., 2008. God, money & politics: English attitudes to blindness and touch, from enlightenment to integration. Information Age Publishing, Charlotte, NC.

Hayhoe, S., 2012. Grounded theory and disability studies: an investigation into legacies of blindness. Cambria Press, Amherst, NY.

IBE-UNESCO, 2006. World data on education, sixth ed. IBE-UNESCO, Geneva, Switzerland.

Japan International Cooperation Agency, 2002. Country profile on disability: kingdom of Saudi Arabia. Japan International Cooperation Agency, Planning and Evaluation Department, Tokyo, Japan.

Khan, M.S., 2009. The GCC monetary union: Choice of exchange rate regime. Peterson Institute for International Economics Working Paper Series, Washington, DC.

Lamorey, S., 2002. The effects of culture on special education services: evil eyes, prayer meetings, and IEPs. Teaching Exceptional Child. 34, 67–71.

Lightfoot-Klein, H., 1994. Disability in female immigrants with ritually inflicted genital mutation. Women Therapy 14, 187–194.

McIlvenny, S., 2006. Road traffic accidents - a challenging epidemic. Sultan Qaboos Univ. Med. J. 6, 3–5.

Modell, B., Dar, A., 2002. Genetic counselling and customary consanguineous marriage. Nat. Rev. 3, 225–229.

Morad, M., Nasri, Y., Merrick, J., 2001. Islam and the person with intellectual disability. J. Religion Disabil. Health 5, 65–71.

Nagata, K.K., 2003. Gender and disability in the Arab region: the challenges in the new millennium. Asia Pacific Disabil. Rehabil. J. 14, 10–17.
Novo Nordisk, A./S., 2010. Diabetes in the Middle East and Northern Africa: Novo Nordisk's diabetes awareness survey in the MENA region. Novo Nordisk A/S, Bagsværd, Denmark.
Oliver, M., 2001. Disability issues in the postmodern world. In: Barton, L. (Ed.), Disability, politics & the struggle for change. David Fulton Publishers, London, England.
Rispler-Chaim, V., 2007. Disability in Islamic law. Springer. Dordrecht, The Netherlands.
Saggar, A.K., Bittles, A.H., 2008. Consanguinity and child health. Pediatrics Child Health 18, 244–249.
Stone, J.H., 2004. Culture and disability: providing culturally competent services. Sage, Thousand Oaks, CA.
Sturm, M., Siegfried, N., 2005. Regional monetary integration in the member states of the Gulf Cooperation Council. European Central Bank Occasional Working Papers, no. 31. Germany, Frankfurt.
Swadi, H., Eapen, V., 2000. A controlled study of psychiatric morbidity among developmentally disabled children in the United Arab Emirates. J. Tropical Pediatrics 46, 278–281.
United Nations, 2000. Resolution adopted by the General Assembly: 55/2, United Nations Millennium Declaration. United Nations General Assembly, New York.
United Nations, 2004a. Demographic yearbook review: national reporting of mortality data. United Nations, Department of Economic and Social Affairs Statistics Division, Demographic and Social Statistics Branch, New York.
United Nations, 2004b. Demographic yearbook review: national reporting of economic characteristics data from population and housing censuses. United Nations, Department of Economic and Social Affairs Statistics Division, Demographic and Social Statistics Branch, New York.
United Nations, 2006. Enable! Some facts about persons with disabilities. Paper presented to the International Convention on the Rights of Persons with Disabilities, United Nations, New York.
United Nations, 2011. 2009–2010 demographic yearbook: sixty-first issue. United Nations Publications, Department of Economic and Social Affairs, New York.
United Nations, General Assembly, 1987. Report of the World Commission on Environment and Development: our common future. Transmitted to the General Assembly as an Annex to document A/42/427-Development and International Co-operation: Environment; Our Common Future, Chapter 2: Towards Sustainable Development; Paragraph 1. Geneva. United Nations General Assembly, Switzerland.
Vine, P., Al Abed, I., Hellyer, P., Vine, P. (Eds.), 2010. UAE yearbook 2010. Trident Press, London, England.
Yaqub, B.A., Daif, A.K., 1988. Multiple sclerosis in Saudi Arabia. Neurology 38, 621–623.
Yong, A., 2007. Theology and Downs's syndrome: reimaging disability in late modernity. Baylor University Press, Waco, TX.
Yunis, K., El Rafei, R., Mumtaz, G., 2008. Consanguinity: perinatal outcomes and prevention – a view from the Middle East. NeoReviews 9, 59–65.

CHAPTER 19

Political–Economic Governance of Renewable Energy Systems: The Key to Creating Sustainable Communities

Woodrow W. Clark II[1], Xing Li[2]
[1]Qualitative Economist, Managing Director, Clark Strategic Partners, Beverly Hills, CA, USA
[2]Professor and Director, Research Center on Development and International Relations, Department of Culture and Global Studies, Aalborg University Aalborg, Denmark

Contents

Corporate and Business Influences and Power	469
International Cases	470
China Leapfrogs Ahead	470
China has "Leapfrogged" into the Green Industrial Revolution	471
The Western Economic Paradigm Must Change	472
Introduction and Background	473
European Union Policies	478
Japan and South Korea are Leaders in the Green Industrial Revolution	482
Distributed Renewable Energy Generation for Sustainable Communities	486
Developing World Leaders in Energy Development and Sustainable Technologies	487
Costs, Finances, and Return on Investment	489
Conclusions and Future Research Recommendations	492

CORPORATE AND BUSINESS INFLUENCES AND POWER

Corporate interests and impacts on public policy are extremely significant around the world. In the United States, the U.S. Supreme Court ruled in 2012, about 6 months before the U.S. national election for president and members of Congress, that anyone (defined now as even companies and corporations) could contribute any amount of money to people running for election to any office. This put the U.S. national election in the hands of the wealthy and corporate interests. In total, over $1 billion was put into the presidential and congressional elections. About 70–80% of the money went to Republican candidates. The results were different than what was predicted, as the public voted for Democrats, re-electing President Obama, increasing the Democratic majority in the U.S. Senate, and increasing its minority position in the U.S. House.

While the antilarge funds from individuals and corporations did not do well in the last U.S. national elections, the future is uncertain with specific needs to finance and support programs such as renewable energy. One case in point for the U.S. national election concerned the Obama administration support of a solar company for over $500 million in debt loans. In the year before the national election and less than 2 years after receiving U.S. debt funds, the company declared bankruptcy. There was no investigation (as of this chapter) into the solar company or its supporters, some of whom were associated with the Obama administration through the U.S. Department of Energy.

The U.S. national government is not alone. Such cases arise and are common in developing countries, but also in other Western-developed countries. For example, even in California on the local level, where the public votes for funds to rebuild and modernize its schools, a superintendent and his facilities manager in a wealthy southern California school district were both convicted and jailed on fraud for taking public money for themselves. Furthermore, a large national American corporation settled out of court for over $6.5million in payments back to the school district.

INTERNATIONAL CASES

For example, in the Ukraine, the former prime minister accepted 10% of natural gas funds from Russia that were piped through the Ukraine. She was voted out of office; went on trial and convicted; and then put in jail. Similar situations occur all over the world. Consider China, whose last central government administration had a high-level official whose wife was convicted of supporting a Western business person and then killing him. That resulted in a dramatic change in the central government leadership that took office in early 2013 after the Chinese New Year in February.

CHINA LEAPFROGS AHEAD

China, however, has developed its own form of non-Western government and economics (Clark 2012). Clark and Li (2004, 2010, 2013) made this point for over a decade calling China's government "social capitalism," which means that governments must be responsible for social issues, including the environment and renewable energy, as well as health and retirement issues for its people. Western economics looks at what is best for the corporation, shareholders, and employees first and foremost. This economic ideology lacks both ethics and honest use of economics, calling into question that economics is not a science but simply a way to use numbers to justify political positions and decisions. In agile energy systems (Clark & Bradshaw 2004), the point was made with a review of what happened in the California energy crisis from 2000 to 2003 and how companies manipulated "markets" to their own economic advantages while allowing brown- and blackouts throughout the state.

The Chinese governments have led in the role of government providing direction with eleven 5-year plans and funding to support these plans. Renewable energy is a key component of the 12th and then the 13th 5-year plan that began in March 2011. Each plan provides clear and formulated policies, and their intended budgets, to address environmental issues and their solutions. Denmark (and other Nordic nations, including Germany) also has national plans including renewable energy. Their current one (2012) calls for Denmark to be 100% energy independent through renewable energy by 2025. Denmark is already 45% there today.

Meanwhile, the United States has no such plans, especially in energy—and renewable energy in particular. Instead, the United States leaves the decisions on energy to "market forces," which are focused primarily on their past business models that are usually rooted in fossil fuels from the Second Industrial Revolution (2IR). It is no wonder that the United States remains behind in the Green Industrial Revolution (GIR). Clark and Cooke (2011) document the problem with the United States in comparison with the rest of the developed and developing world.

CHINA HAS "LEAPFROGGED" INTO THE GREEN INDUSTRIAL REVOLUTION

In order to avoid the mistakes of the Western-developed nations, China has moved ahead in a variety of infrastructure areas (Clark & Isherwood 2007, 2010). Also, the United States must look comprehensively into the corporate and political reactions to the 2011 Japanese tsunami and ensuing nuclear power plant explosions, as well as the 2010 BP oil spill in the Gulf of Mexico off Louisiana. The United States and other countries cannot ignore the environmental consequences and economic costs of the 2IR that have handicapped it moving into the GIR. The end result is not good for the American people, let alone the rest of the world.

The deregulation of industries starting in the Reagan and Thatcher eras was a mistake and completely naïve view of reality from neoclassical economics from Adam Smith. There has never been a society or area in the world in which the principles of capitalism have been proven to work in reality. Instead just the opposite has been the reality. Chomsky (2012) looks at the history of economics in far more concrete manner.

THE WESTERN ECONOMIC PARADIGM MUST CHANGE

Even the *Economist* in two special issues labels modern economics as "State Capitalism" (January 23, 2012) and another soon after that as The Third Industrial Revolution (April 2012), a theme from Jeremy Rifkin (2004) and his book with that title in 2012. Clark (2008, 2009, 2010, 2011) has published several articles and given numerous talks about the Third Industrial Revolution, but prefers to think of it as the Green Industrial Revolution (Clark & Cooke 2011). Basically, the GIR concerns renewable energy, smart green communities, and advanced technologies that produce, store, and transmit energy for infrastructures while saving the environment.

The point is that development of the United States into a powerful world leader had much to do with its military strength but also its economic development of fossil fuels for over a century in the 2IR, including the technologies that support them, such as combustion engines, and related technologies, including the atom bomb and nuclear power (Chomsky 2012). The growth of the United States started with businesses and their owners who control the economy today. There is little or no competition. But even more significant is that the basis for this wealth is in fossil fuels and continues to be there. Hence, the environment is continuing to be damaged in order to produce more oil and natural gas, causing a climate change. But this 2IR retards and places the United States back decades when compared to emerging economics and even other Western-developed nations.

As historians have documented, development of the 2IR in the United States was based primarily on "state capitalism," as oil companies got land grants, funding, and even trains or pipelines for transporting their fossil fuels. That governmental support continues today. Consider the issue of the United States getting shale oil from Alberta, Canada, and the massive pipelines installed through the United States to get the oil to the United States. Furthermore, these same companies get tax breaks and credits such that their economic responsibility to the United States is minimal. The argument that America will be "energy independent" with these fossil fuels is false. The United States needs to stop getting its energy from fossil fuels anywhere in the world, including domestically or from its neighbors.

Hence the argument is that China will buy the oil from Canada. Basically, Canada (and the United States) should not even extract oil from the ground, which destroys thousands of acres of land permanently, making them impossible to repair or restore. There are far more and better resources from renewable energy such as the sun, wind, geothermal, run of the river, and ocean or wave power.

INTRODUCTION AND BACKGROUND

A Green Industrial Revolution emerged at the end of the 20th century due in large part to the end of the Cold War that dominated the globe since the end of World War II. The Second Industrial Revolution had dominated the 20th century because it was based primarily on fossil fuels and technologies that used mechanical and combustion technologies primarily. However, the GIR is one of renewable energy power and fuel systems and smart "green" sustainable communities that use more wireless, virtual communications, and advanced storage devices such as fuel cells (Clark & Cooke 2011). The GIR is a major philosophical paradigm change in both thinking and implementating environmentally sound technologies that require a new and different approach to economics (Clark 2011).

The United States lived in denial during the 1970s and then again since the early 1990s, which became apparent for both Democrat and Republican presidential administrations in their lack of proactive polices globally through the Kyoto Accords and most recently the U.N. Intergovernmental Panel on Climate Change conference in Kopenhaven (December 2009) and Cancun (2010). However, in the early 1990s, economic changes in Europe and Asia were made due to the end of the Cold War to meet the new global economy. The Asian and European Union (EU) conversions from military and defense programs to peacetime business activities were much smoother than that of the United States. Environmental economist Jeremy Rifkin recognized this change and developed the concept of a "Third Industrial Revolution" in his book, *The European Dream* (2004). According to Rifkin, the GIR took place a decade earlier in some EU countries. He did not recognize that Japan and South Korea had been in a GIR even decades before that (Clark & Li 2004).

At the same time, Clark and Rifkin colleagues (2006) published a paper on the "Green Hydrogen Economy" that made the distinction between "clean" and "green" technologies when related to hydrogen and other energy sources. The former was often used to describe fossil fuels in an environmentally friendly manner, such as "natural gas" and "clean coal." Green, however, means specifically renewable sources such as the sun, wind, water, wave, and ocean power. In short, the paper drew a dividing line between what technologies were part of the 2IR (i.e., clean technologies such as clean coal and natural gas) and the GIR (solar, wind, ocean, and wave power, as well as geothermal). The GIR focused on climate change and changing the technologies and fuels that caused it or could at least mitigate and change the negative pollution and emission problems that impacted the earth.

In founding the science of "qualitative economics," Clark and Fast (2008) made the point about economics as it needed to define ideas, numbers, words, symbols, and even sentences due to the misuse of "clean" to mean fossil fuels and technologies that were not good for the environment. The documentary film *Fuel* (Tickell 2009) made these points too as it told the history about how "clean" was used to describe fossil fuels such as natural gas in order to placate and actually deceive the public, politicians, and decision makers. For example, Henry Ford was a farmer and used biofuels in his cars until the early 1920s when the oil and gas industries forced him to change to fossil fuels.

Hawkins and colleagues (1999) refer to environmental changes as the beginning of the "Next Industrial Revolution." This observation only touched the surface of what the world is facing in the context of climate change. The irony is that China has already "leapfrogged" and moved ahead of the United States into the GIR (Clark & Isherwood 2008, 2010). While China now leads the United States in energy demand and CO_2 emissions, it also is one of the leading nations with new environmental programs, money to pay for them, and installation of advanced infrastructures from water to high-speed rail systems.

These economic changes came first from Japan, South Korea, and the EU. Rebuilding after World War II from the total destruction of both Asia and Europe meant an opportunity to develop and recreate businesses and industries and to commercialize new technologies. The historical key in Japan, and then later in the EU, was the dependency on fossil fuels for industrial development, production, and transportation. For Japan, as an island nation, this was a critical transformation for them in the mid-19th century with the American "Black Ships" demanding that Japan open itself to international, especially American, trade. However, as recent events testify, Japan made the mistake of bending to the political and corporate pressures of the United States to install nuclear power plants, despite the atomic bombings of two of its major cities in World War II. The final results of tragedies from the 9.0 earthquake are not final yet in terms of the nuclear power plants in Fukushima and its global impact on the environment, let alone in Japan and the immediate region of northern Asia.

Soon after the end of the Cold War in the early 1990s, the GIR become dominant in Japan and spread rapidly to South Korea as well as Taiwan and somewhat to India. China came later when it leapfrogged into the 21st century through the GIR. Germany, Japan, and South Korea took the lead in producing vehicles that required less amounts of fossil fuels and were more environmentally "friendly," often again called "clean tech." Hence, their industrial development of cars, high-tech appliances, and consumer goods dominated global markets.

America ignored the fledging technological and economic efforts in the EU, South Korea, and Japan as the nation tilted into a long period of self-absorption, bubble-driven economic vitality. The nation had a history of cheap fossil fuels primarily from inside the United States and given high tax breaks and incentives (Tickell 2009). The 2IR also had survived World War II successfully. Furthermore, what the end of the Cold War meant to Americans was that they dominated the 2IR and they were in control of global economic markets. The Soviet Union had failed to challenge them. Then came 9/11 and its aftermath, along with the longest continuous war in American history, as well as the continuing battle with fundamental Islamic terrorists. With its own unique and fractured political debate and power struggles, America labored to make sense of a post-Cold War era where special interests replaced reason and any movement toward a sound domestic economic policy.

Instead, the American ideological belief in a "market economy," entrenched in the late 1960s to mid-1970s, replaced the reality of how government and industry must collaborate and work together. Evidence of the problems and hardships from "market forces" came initially from a convergence of events in the early part of the 21st century, including a global energy crisis, the dot.com collapse, and terrorist attacks. Spending and leveraging money into the market caused the global economic collapse almost a decade later in October 2008.

The *Economist* characterized the basic economic problem the best when in mid-2009, a special issue was published under the title of "Modern Economic Theory," superimposed on the Bible melting (*Economist*, July 18–24, 2009). Basically, the case was made that economics is "not a science" in large part because its theories and resultant data did not predict the global economic recession that began in the fall of 2008. From that special issue of the *Economist* in the summer of 2009, an international debate about conventional modern economics began and continues today.

The Green Industrial Revolution impacts America at the local level in a completely different perspective and rational than at regional, state, or national levels. Infrastructures of energy, water, waste, transportation, and information technology, among others, and how they are integrated are the core to the GIR (op.cit. Clark & Cooke 2011). These infrastructure systems need to be compatible yet integrated with one another. For example, renewable energy power generation must be used for homes, businesses, hospitals, and nonprofit organizations (government, education, and others) that are metered and monitored as "smart on-site grids" and also used for the energy in vehicles, mass train, and buses, among other transportation infrastructures (Knakmuhs 2011). Such "agile energy" or "flexible systems" (Clark & Bradshaw 2004) allow people to generate their own power while also being connected to a central power grid. However, both local power and central power in the GIR need to be generated from renewable energy sources, with stand-by and back-up storage capacity.

There are five key basic elements for the Green Industrial Revolution: (1) energy efficiency and conservation; (2) renewable power generation systems; (3) smart grid-connected sustainable communities; (4) advanced technologies such as fuel cells, flywheels, and high-speed rail; and (5) education, training, and certification of professionals and programs. First, communities and individuals all need to conserve and be efficient in our use of energy as well as other natural resources such as land, water, oceans, and the atmosphere. Second, renewable energy generated from wind, sun, ocean waves, geothermal, water, and biowaste must be top priority for power on-site and for central plants.

The third element is the need for smart girds on local and regional levels in which both monitoring and controlling of energy can be done in real time. Meters need to establish base load use so that conservation can be done (systems put on hold or turned off if not used) and then generate renewable energy power when demand is needed. The fourth element needs to be advanced storage technologies such as fuel cells, batteries, regenerative brakes, and ultracapacitors. These devices can store energy from renewable sources, such as wind and solar, which produce electricity intermittently, unlike the constant supply of carbon-based fuel sources. Finally, the fifth element is education and training for a workforce, entrepreneurial, and business sector that is growing and providing employment opportunities in the GIR.

In general, the GIR must provide support and systems for smart and "green" communities so that homes, businesses, government, and large office and shopping areas can all monitor their use of natural resources such as energy and water. For example, communities need devices that capture unused water and that can transform waste into energy so that they can send any excess power generated to other homes or neighbors. Best cases from around the world of sustainable communities that follow these elements of the GIR exist today (Clark 2009).

Essentially the GIR was started by governments concerned about the current and near-future societal impact of businesses and industries in their countries. The EU and Asian nations in particular have had long cultural and historical concerns over environmental issues. The Nordic nations, for example, have started programs on ecocities, as well as reuse of waste for more than three decades. Sweden, Denmark, and Norway have all either eliminated dependency on fossil fuels now for power generation or will be in the near future. All but Finland have shut down nuclear power plants and their supply of energy as well. The same has been true since the 1980s in most other EU nations except France.

However, the key factor in the EU and Asia has been their respective government leadership in terms of public policy and economics. Consumer costs for oil and gas consumption are at least four times that of the United States due to higher taxes (or elimination of tax benefits) to oil and gas companies in these other nations. The EU has been implementing such a policy for two decades, which has also encouraged people to use trains and mass transit instead of individual cars. The United States, however, continues to subsidize fossil fuels and nuclear power through tax incentives and government grants. Not so in the EU and Asia. The impact of fossil fuels on climate change was the basis for changing these policies and financial structures over two decades ago. Today the impact on the environment has become severe and thus even more significant for future generations around the world.

The historical difference has been the American contemporary economic ideology of market forces as simply a balance of supply and demand. This neoclassical economic model has failed for many reasons, but especially due to one of the two key issues presented in the *Economist* special issue (July 2009): economics is not a science. This is important for a number of reasons, but the basic one, which pertains to the GIR, is that contemporary economics does not apply to major industrial changes, such as the GIR, let alone the beginning of the 2IR. For most economists to be confronted with a challenge to their field being a science or not is disturbing. The "dismal science" may be boring with its statistics, but to not be a science brings questions to the entire contemporary field of economics and its future.

The debate is over how does a community or nation change? Economics is one of the key factors. The issue is, are "market forces" the key economic change factor? The 2IR discovered that market forces or businesses by themselves could not get fossil fuels and other sources of energy into the economy at reasonable costs. It took time, government support, and policies that provided the market with capital and incentives. Additionally, GIR economics includes economic externalities, such as environmental and health costs.

In short, the "market force" neoclassical paradigm represented American economic policies (and also the United Kingdom) for over the last four decades. Prime Minister Thatcher and then President Reagan were the embodiment and champions of this economic paradigm derived from Adam Smith (Clark & Fast 2008). Market force economics had some influence on the EU and Asia, but then demonstrated its failure in October 2008 with the global economic collapse that began in the United States on Wall Street. That failure meant some of the government programs in the EU and Asia, which had succeeded, yet now needed to be given more economic attention because they basically differed greatly from the United States and United Kingdom economic models.

These other nations have been in the GIR now themselves for several decades, which succeeded and continued to do so with a different economic model. Northern EU, Japan, South Korea, and China are clear documented examples of a different economic model. For example, a key economic government program representing the GIR in the EU is the feed-in tariff (FiT), which started in Germany during the early 1990s and was taking route successfully in Italy, Spain, and Canada, as well as nations in the EU and Asia. While there are problems in Spain and Germany has decided to cut it back, the United States has not started a FiT in any significant, long-term planned policy programs on a national, let alone a state, level. Some American communities and states have started very restrictive and modest FiT programs.

EUROPEAN UNION POLICIES

Germany jumped out in the lead of the GIR in the EU with its FiT legislation in 1990. Basically, the FiT is an incentive economic and financial structure used to encourage the adoption of renewable energy through government legislation. The FiT policy obligates regional or national electricity utilities to buy renewable electricity at above-market rates. Successful models like that exist, such as the EU tax on fuels and the California cigarette tax, both of which cut smoking dramatically in California and for people to use mass transit and trains rather than drive their cars as much in the EU, but also provide incentives and metering mechanisms to sell excess power generated back to the power grid. Other EU nations, especially Spain, followed and the policy is being developed slowly in Canada and some U.S. states and cities. Figure 1 shows the economic impact of the FiTs. Over 250,000 "green" jobs were created in Germany alone. The graphs in Figure 1 also show the growth of the solar and wind industries in Germany and how this expansion is becoming global.

Germany is now the world's leading producer of solar systems because it has more solar systems installed than any other nation based on the creation of world-leading solar manufacturing companies, solar units sold and installed are measured by sales, amount of kilowatts per site, and records kept by the local and national governments (Gipe 2011). China will take the lead at the end of 2011. The extensive use of solar by Germany is impressive, as the nation has many cloudy and rainy days, along

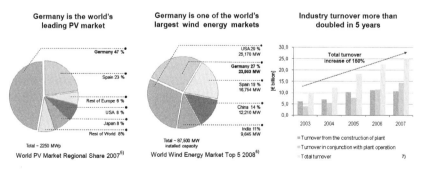

Figure 1 Germany feed-in-tariff policy and results (1990–2010).

with significant snow in the winter, which is common to northern Europe. In 2010, Japan implemented a similar aggressive FiT system in order to stimulate its renewable energy sector and regain its renewable energy technological (solar and system companies and installations) leadership that it held in the early part of the 21st century. Measurements were kept by the solar companies, as well as by local and national governments. MITI, the Japanese national research organization, measures the use of renewable energy systems on a quarterly basis. However, the aftermath of the Japanese earthquake and destruction of the nuclear power plants in April 2011 could actually expedite renewable energy growth and installation through a number of government programs and incentives that are being proposed.

Other European countries have similar GIR programs as well. Denmark, for example, will be generating 100% of its energy from renewable power sources by 2050. While trying to meet that goal, the country has created new industries, educational programs, and therefore careers. One good example of where this policy has accomplished dramatic results is in the city of Frederikshavn in the northern Jutland region of Denmark. The city has 45% renewable energy power now and by 2015 will have 100% power from renewable energy sources (Lund 2009). In terms of corporate development in the renewable energy sector, for example, one Danish company, Vestas, is now the world's leading wind power turbine manufacturer with partner companies all over the world because of its partnership and joint ventures in China. Vestas continues to introduce improved third-generation turbines that are lighter, stronger, and more efficient and reliable. They also continue to design new systems, such as those that can be installed offshore away from impacted urban areas.

Germany, Spain, Finland, France, the United Kingdom, Luxembourg, Norway, Denmark, and Sweden are on track to achieve their renewable energy generation goals. Italy is fast approaching the same goals when, in 2010, it took the distinction as having the most megawatts of solar installed from Germany. However, Denmark is one of the most aggressive countries due to its seeking 100% renewable energy power generation by 2050. Already Denmark has a goal of 50% renewable energy generation by 2015 (Clark 2009). Other EU countries are lagging behind, especially in central and eastern Europe. The EU has required all its member nations to implement programs such as those in western EU in order to be energy independent from oil and gas, especially now since most of these supplies come from North Africa, the Middle East, and Russia.

Various EU nations have widely different starting positions in terms of resource availability and energy policy stipulations. France, for example, is a stronger supporter of nuclear energy. Finland has recently installed a nuclear power plant due its desire to be less dependent on natural gas from Russia. However, Sweden is shutting its nuclear power plants down. The United Kingdom and The Netherlands have offshore gas deposits, although with reduced output predictions. In Germany, lignite offers a competitive foundation for base load power generation, although hard coal from German deposits is not competitive internationally. In Austria, hydropower is the dominating energy source for generating power, although expansion is limited.

Other EU directives toward energy efficiency improvement and greenhouse gas emission reductions also impact electricity generation demand. Many EU members have taken additional measures to limit greenhouse gas (GHG) emissions at the national level. Since the EU-15 is likely to miss its pledged reduction target without the inclusion of additional tools, the European Parliament and the council enacted a system for trading GHG emission allowances in the community under the terms of Directive 2003/87/EC dated October 13, 2003. CO_2 emissions trading began in January 2005 but has not produced the desired results due to limitations of "cap and trade" economic measures and the use of auctions over credits given for climate reduction.

After being established for 3 years, by 2007 the results were not good, however, as the economics and "markets" were not performing as predicted. Basically, the carbon exchanges have performed poorly and are not as promised to either buyer or seller of carbon credits (or other exchange mechanisms). The initial issues are emission caps not being tight enough and lack of significant EU or local government oversight (EU 2009). By 2010, many of the exchanges have closed or combined with others. The problem is often cited as a lack of supporting governmental (EU or by nation) policies, but the real issue is that the economics do not work as well as the control over carbon emissions. Furthermore, the trading and auction mechanisms do not provide direct and measurable solutions to the problem of emissions and its impact on climate change. A far more direct finance and economic mechanism, as proposed by several EU nations and China, would be to have a "carbon tax."

An important lesson from the FiT policies in Germany came from the two decades of the policies from 1990 through 2007. As Figure 2 shows, Germany learned that a moderate or small FiT was not sufficient enough

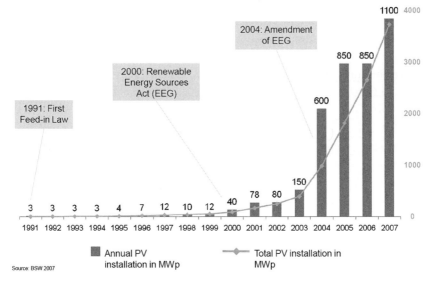

Figure 2 Germany feed-in-tariff policy economic results (1990–2007).

to push renewable energy systems such as solar into the mainstream of its economy. In short, a far more aggressive use of the FiT type of financing and/or direct carbon taxes needs to be made. On its own, the solar industry would not move fast enough into the GIR. In many ways, this is the lesson for other nations. In fact, the reality of the 2IR historically has been to have strong and continuous government incentives from the late 19th century to the present day. The definition and model of economics of a market remain critical in understanding how the United States can move into the GIR. Consider now how Japan and South Korea did just that: moved into the GIR with strong government leadership and financial support.

JAPAN AND SOUTH KOREA ARE LEADERS IN THE GREEN INDUSTRIAL REVOLUTION

While it took an extraordinary political transition to prompt Europe to open the door to the Green Industrial Revolution, Japan and South Korea in particular have taken a completely different path. And now China is moving aggressively ahead in the GIR. Most of the information and data that follow focus on China (Clark 2009). For example, China led the United States and the other G-20 nations in 2009 for annual "clean energy investments and finance, according to a new study by "The Pew Charitable Trusts" (Lillian 2010, p. 4):

> Living in a country with limited natural resources and high population density, the people of Japan had to work on sustainability throughout their history as a matter of necessity. With arable land scarce—some 70–80% of the land is mountainous or forested and thus unsuitable for agricultural or residential use—people clustered in the habitable areas, and farmers had to make each acre as productive as possible. The concept of "no waste" was developed early on; as a particularly telling, literal example, the lack of large livestock meant each bit of human waste in a village had to be recycled for use as fertilizer.
>
> Along with creating this general need for conservation, living in close proximity to others inspired a culture in which individuals take special care in the effect their actions have on both the surrounding people and environment. As such, a desire for harmony with others went hand in hand with a traditional desire for harmony with nature. Nature came to be thought of as sacred and to come into contact with nature was to experience the divine. Centuries-old customs of cherry blossom or moon viewing attest to the special place nature has traditionally held in the Japanese heart.

However, in April 2011, China became the world leader of financial investment in "clean tech" with $54 billion invested, which was over $10 billion from second-place Germany and almost double the third-place United States (*San Jose Business Journal*, 2011). Wind was the favorite sector of renewable energy with $79 billion invested globally. This article noted in particular a comment by a senior partner in a venture capital firm,

> a lot of the clean technologies are dependent on policy and government support to scale up. In some other parts of the world (not USA), you have more consistency in the way these types of funds are appropriated" (San Jose Business Journal *2011, p. 8*).

The Japanese have had a long cultural and business history in commercializing environmental technologies. The 2011 earthquake made Japan focus on that historical tradition. The future has yet to become clear and will not be defined for some months and years ahead. However, in Japan, the environment took a backseat to industrial development during the drive toward modernization and economic development that began in the latter half of the 19th century. After nearly 300 years of self-imposed isolation from the world, Japan was determined to catch up to the industrialized West in a fraction of the time it took Europe and the United States to make their transitions, eventually emerging as a great power in the beginning of the 20th century.

Economic development continued unabated until World War II, when its capacity was destroyed by American bombings. Economic growth restarted again in the postwar period at a rapid pace but with a distinctive orientation and concern for the limited nature resources of the island nation. By the 1970s, on the strength of its industry and manufacturing capabilities, Japan attained its present status as an economic powerhouse. Companies such as TOTO (concerned with water and waste conservation and technologies), along with automobile makers concerned with atmospheric pollution, emerged as global leaders. A large part of that success was the need for the government to invest in research and development organizations (e.g., METI) to support companies and business growth, what would now be called the GIR. For example, high-speed rail was started in Japan in the mid-1980s and has expanded. The transportation systems were economically efficient, as well as being environmentally sound, and provided for the public at reasonable rates.

While this incredibly successful period of development left many parts of the country wealthy, it also resulted in serious environmental problems. In addition, the oil crisis had hit Japan particularly hard because of its lack of natural resources, making it difficult for the industrial and manufacturing sectors to keep working at full capacity. To respond to the effects of pollution, municipalities began working in earnest on ways to reduce emissions and clean up the environment, while Japanese industry responded to the oil crisis by pushing for an increase in energy efficiency.

At the same time, Japan's economy was evolving more toward information processing and high technology, which held the promise of further increases in energy efficiency. Japan created new innovative management "team" systems that were copied in the United States and the EU. Many manufacturing firms saw value in establishing plants in other developed countries in part to create a market for their products, employ local workers, and establish firm and solid roots. For example, Toyota and Honda established their Western Hemisphere headquarters in Torrance, California. Other high-tech companies established large operations throughout the United States. In this way, the Japanese have worked collaboratively with local and regional communities in government, industry, and academia to reincorporate traditional Japanese ideas about conservation and respect for the environment in order to create sustainable lifestyles compatible with modern living.

Community-level government efforts in Japan, supported by national government initiatives, have led to unique advancements in energy efficiency and sustainable lifestyles, including novel ways of preventing and eliminating pollution. As it stands, Japan is responsible for some 4% of global CO_2 emissions from fuel combustion, and although this is the lowest percentage among major industrialized nations, it is still something the country intends to reduce, with a long-term goal of reducing emissions by 60–80% by 2050. With the majority of energy still coming from coal, Japan is also attempting a large shift toward renewable energy.

As of November 2008, residential-use solar power generation systems have been put in place in around 380,000 homes in Japan. A close examination of data on shipments domestically in Japan shows that 80–90% is intended for residential use and that such shipments are likely to increase, as the government aims to have solar panel equipment installed in more than 70% of newly built houses by 2020 to meet its long-term goals for reductions in emissions. Current goals for solar power generation in Japan are to increase its use 10-fold by 2020 and 40-fold by 2030, and large proposed subsidies for the installation of solar—9 billion yen or $99.6 million total in the first quarter 2009—along with tax breaks for consumers, will continue the acceleration of solar adoption by Japanese households.

In recent years, places such as Europe, China, southeast Asia, and Taiwan saw tremendous growth in energy generation almost entirely from solar power installations. However, these have mostly involved large-scale solar concentrated power facilities not fit for individual households. In Japan, however, as solar power generation systems for residential use become increasingly commonplace, they have become concentrated by creating sustainable communities through use on roofs of local homes and businesses.

The same is true with light-emitting diode (LED) light bulbs and now solar panels. Today, LED bulbs may cost a few pennies more but they last far longer than a regular light bulb and can be recycled without issues of mercury and other waste contamination. The result is better lighting for homes and offices with significantly less costs in terms of the systems and the environment. Some LED bulbs are guaranteed to last from 6 to 8 years (Nularis 2011). While energy demands in homes and offices continue to rise due to the Internet, computers, and video systems, the installation of energy-efficient and now cost-saving systems is very much in demand. Some states are even requiring by law a change over from older light bulbs to the newer LED ones.

DISTRIBUTED RENEWABLE ENERGY GENERATION FOR SUSTAINABLE COMMUNITIES

Adding more complications to the EU, Japan, and South Korea's policy decisions is the reality of an aging, undercapacity grid. The EU must crank up investment in new generation. Estimates are coming in that indicate that to meet demand in the next 25 years, they will need to generate half again as much electricity as they are now generating. According to International Energy Outlook 2010, conducted by the U.S. Energy Information Administration (U.S. Energy Information Administration 2011), the world's total consumption of energy will increase 49% from 2007 to 2035. This could result in a profound change in the EU's power generation portfolio, with options under consideration for new plants, including nuclear energy, coal, natural gas, and renewables.

Originally, when nations electrified their cities and built large-scale electrical grids, the systems were designed to transmit from a few large-scale power plants. However, these systems are inefficient for smaller scale distributed power from renewable sources (Clark 2006). Although some systems allow for individual households to either buy power or sell power back to the grid, the redistribution of power from numerous small-scale sources is not yet managed well economically (Sullivan & Schellenberg 2011). As Isherwood and colleagues (1998) documented in studies of remote villages, renewable energy for central power can meet and even exceed the entire demand for a village, hence making it energy independent and not needing to import any fossil or other kinds of fuels. This model and program have worked in remote villages, but can also be applied to island nations and even larger urban communities or their subsets.

The grid of the future has to be "smart" and flexible and based on the principles of sustainable development (Clark 2009). As the Brundtland report said in 1987, "as a minimum, sustainable development must not endanger the natural systems that support life on Earth: the atmosphere, the waters, the soils and the living beings." With that definition in mind, a number of communities sought to become sustainable over the last three decades.

Integrated "agile" (flexible) strategies applied to infrastructures are needed for creating and implementing "on-site" power systems in all urban areas that often contain systems in common with small rural systems (Clark & Bradshaw 2004). The difference in scale and size of central power plants (the utility size for thousands of customers) with on-site or distributed power can be seen in the economic costs to produce and sell energy.

Historically, the larger systems could produce power and sell it for far less than the local power generated locally for buildings. Those economic factors have changed in the last decade (Xing & Clark 2009). Now, on-site power, particularly from renewable energy power (e.g., solar, wind, geothermal, and biomass), has become far more competitive and is often better for the environment. Large-scale wind farms and solar-concentrated systems are costly and lose efficiency due to the transmission of power over long distances.

DEVELOPING WORLD LEADERS IN ENERGY DEVELOPMENT AND SUSTAINABLE TECHNOLOGIES

Some of the major benefits of the Green Industrial Revolution are job creation, entrepreneurship, and new business ventures (Clark & Cooke 2011). Considerable evidence of these benefits (Next 10 2011) can be seen in the EU, especially Germany and Spain (Rifkin 2004). Many studies in the United States have documented how the shift to renewable energy requires basic labor skills and also a more educated workforce, but one that is also based locally and where businesses stay for the long term. This is a typical business model for almost any kind of business and is what has motivated EU universities to create "science parks," which take the intellectual capital from a local university and build new businesses nearby the campus (Clark 2003a,b).

Asia's shift to renewable energy will require extensive retraining. Consider the case of wind power generation in China. In the early 1990s, Vestas saw Asia and China as the new emerging big market. Vestas agreed to China's "social capitalist" business model (Clark & Jensen 2002; Clark & Li 2004), where the central government sets a national plan, provides financing, and gives companies direction for business projects over 5-year time frames, which are then repeated and updated. Business plans are critical to any company, especially when set and followed by national governments.

A major part of the Chinese economic model required that foreign businesses be colocated in China with at least a 50% Chinese ownership. This meant that in the late 20th and early 21st centuries, the Chinese government owned companies or were the majority owners of new spin-off government-owned ventures, established international companies, or businesses started in China. Additionally, China required that the "profits" or money made by the new ventures be kept in China for reinvestments.

Additionally, the results, such as with renewable energy companies like wind and solar industries, were that all the ancillary supporting businesses also needed to support the companies from mechanics, software, plumbing, and electricity to installation, repair, maintenance, and other areas. Supporting industries were also needed such as law, economics, accounting, and planning, especially since the Chinese government began to create sustainable communities that required all these skill sets (Clark 2009, 2010). Hence, these businesses grew and became located in China.

However, the Chinese social capitalism model is not rigid with the government owning controlling percentage (over 50%) of a company. Many businesses were started by the Chinese government with its holding of 25–33% of shares, while other firms were owned by former government employees, until the companies went public (Li & Clark 2009). Yet in almost all cases, the companies are competitive globally and are performing remarkably well as demonstrated again in the renewable energy sector, where in early 2011, SunTech, a Chinese-based publically traded company, became the world's largest manufacturer and seller of solar panels (Chan 2011). According to a press release by the company in February 2011, it has delivered more than 13 million photovoltaicpanels to customers in more than 80 countries.

Today, China is a (if not the) world leader in wind energy production and manufacturing, with over 3000 MW installed in China alone (Vestas 2011). The Chinese are now following a similar business model in the solar industry (Martinot & Droege 2007–2010). As such, China and Inner Mongolia (IMAR) have contracted Vestas to install 50 MW in IMAR (Vestas 2011), according to a report from the Asian Development Bank (Clark & Isherwood 2008, 2010), which argues for targeted needs to

- create international collaborations between universities and industry
- conduct research and development of renewable energy technologies
- build and operate science parks to commercialize new technologies into businesses
- provide and promote international exchanges and partnerships in public education, government, and private sector businesses

The end results for the EU are smart homes and communities. The Green Industrial Revolution starts in the home so that energy efficiency and conservation are a significant part of everyone's daily life. The home is the place to start, but it is also the place to start with the other elements of the GIR: renewable energy generation, storage devices, smart green communities, new fuel sources for the home, and transportation.

COSTS, FINANCES, AND RETURN ON INVESTMENT

Government policy(s) and finance are critical for economic growth, especially concerning the environment and climate change. The basis of the GIR in the EU, South Korea, and Japan can be seen in their articulation of a vision and financial programs. Most of these countries also had established government energy plans. China, in fact, has had national plans since the People's Republic of China was established in 1949. Having a plan is, in fact, the basic program and purpose of most business educational programs. Governments need to have plans, as most businesses do. Business plans are for themselves and their clients. Yet the United States continues without any national energy or environment plans. Most American states do not have them either, while an increasing number of cities and communities are developing them in order to plan for becoming sustainable.

This lack of planning has both long- and short-term impacts. The finance of new energy technologies and systems (like any new technology) is often dependent on government leadership through programs in public policy and finance (Clark & Lund 2001). Fossil fuel energy systems in the 2IR have been funded and supported by the governments of Western nations through tax reductions and rebates that continue today. For the GIR, it is only logical and equitable that such economic and financial support continues, which means that the American national government should provide competitive long-term tax incentives, grants, and purchase orders for renewable energy sources instead of just fossil fuels.

Meanwhile, the EU, South Korea, and Japan took leadership in the planning, finance, and creation of renewable energy companies, while other nations, including the United States, did not (Li & Clark 2009). For example, because of the national policy on energy demand and use, Japan has one of the lowest energy consumption measurements in the developed world. This has been made possible by its continued investment in long-term energy conservation while developing renewable sources of energy and companies that make these products. Japan's per capita energy consumption is 172.2 million Btu versus 341.8 million Btu in the United States.

One critical long-term economic plan is the need for life cycle analysis (LCA) versus cost–benefit analysis (CBA). While not discussed much in this chapter about these two very different accounting processes, Clark and Sowell (2002) cover the topic in-depth elsewhere as the systems apply to government spending. Each approach is critical in how businesses learn what their cash flow is and their return on investment (ROI). The CBA model only provides for a 2- to 3-year ROI, as that is what most companies (public or government) require for quarterly and annual reports. However, for new technologies (such as renewable energy, but also even wireless and WIFI technologies), more than a few years are needed in the ROI. The same was true in the 2IR when oil and gas were first discovered and sold. Now in the GIR, longer economic and financial ROIs are needed.

Life cycle analysis covers longer time periods, such as 3–6 years, and within renewable energy systems some as long as 10–20 years, depending on the product and/or service. Furthermore, LCA includes externalities such as environment, health, and climate change factors, all of which have financial and economic information associated with them. The point is that cost–benefit analyses are limited. The basic concept is that the LCA consists of one long-term finance model in the United States today for solar systems, called a power purchase agreement (PPA) that contracts with the solar installer or manufacturer for 20–30 years. PPA is a financial arrangement between the user "host customer" of solar energy and a third-party developer, owner, and operator of the photovoltaic system (Clark 2010).

The customer purchases the solar energy generated by the contractor's system at or below the retail electric rate from the owner, who in turn, along with the investor, receives federal and state tax benefits for which the system is eligible on an annual basis. These LCA financial agreements can range from 6 months to 25 years and hence allow for a longer ROI. However, there are other ways to finance new technologies, especially if they are installed on homes, offices, and apartment buildings. Today, financial institutions and investors can see a ROI that is attractive when the solar system on a home, for example, is financed as a lease, part of tax on the home, or included in the mortgage itself such as plumbing, lighting, and air-conditioning are today.

What is interesting are some newer economic ideas on how to finance technologies that reduce "global climate change." One way to describe the GIR financial mechanisms is by looking at the analytical economic models that financed the 2IR. For example, the 2IR was based on the theory of abundance. The earth had abundant water and ability to treat waste. Hence, buildings, businesses, homes, and shopping complexes all had plumbing for fresh water and drainage for waste. The same scenario occurred in electrical systems that took power from a central grid for use in local community buildings. Locally and globally, people have found that systems work but now with the climate change need to conserve resources and be more efficient.

When these economic considerations are factored into a CBA rather than a LCA financial methodology, the numbers do not work (Sullivan & Schellenberg 2011). The financial considerations for energy transmission and then monitored by smart systems are needed, but costly. Long distances make them even more costly because then the impact of the climate (storms, tornadoes, floods, etc.) with required operation and maintenance is added today with security factors. The actual "smart" grid is at the local level where these and other uncontrolled costs can be eliminated and monitored.

The financing of water, waste, electrical, and other systems for buildings was incorporated into the basic mortgage over time for that building. In short, modern 2IR infrastructure systems were no longer outside (e.g., the outhouse or water faucet) but inside the building. What this 2IR financial model does is set the stage for the GIR financial model. Much of the 2IR financing for fossil fuels and their technologies came about as leases or building mortgages. A variation of the 2IR model, which is a bridge to the GIR, is the Property Assessed Clean Energy (PACE) program started in 2008 in Berkeley, California, whereby home owners can install solar systems on their buildings, for example, but pay for them from a long-term supplemental tax that is transferred with the sale of the property assessment on their property taxes. The financing is secured with a lien on the property taxes, which acquires a priority lien over existing mortgages. The program was put on hold in July 2010 when the Federal Housing Finance Agency expressed concerns about the regulatory challenge and risk posed by the priority lien established by PACE loans. Nevertheless, the U.S. Department of Energy continues to support PACE.

The dramatic change to the GIR, however, moves past that financial barrier. Mortgages are part of the long-term cost for owning a property. Therefore, in the GIR, the conservation and efficiency for the 2IR technologies in buildings can be enhanced with renewable energy power, smart green grids, storage devices, and other technologies through mortgages that can be financed from one owner to another over decades (20–30 years or more). This sustainable finance mortgage model is long term or a LCA framework and provides for technologies and installation costs to the consumer that make the GIR attainable with a short time. Changes, updated, and new technologies can easily be substituted and replace earlier ones. What needs to happen is that the banking and lending industries try this GIR finance model on selected areas. After some case studies the model can be replicated or changed as needed.

CONCLUSIONS AND FUTURE RESEARCH RECOMMENDATIONS

The basic point of this chapter was to highlight the need for economics to be more scientific in its hypothesis and data collection. Furthermore, the economics of the 2IR and the GIR are very similar, if not parallel, for example, the role of government, as it must often take the first steps in directing, creating, and financing technologies. As the 2IR needed the government to help drill for oil and gas, as well as mine for coal, the government needed to build rail and road transportation systems to transport the fuels from one place to another.

The GIR is very much in the same economic situation. The evidence can be seen in Asia and the EU. And especially now in China, the central government plans for environment and related technologies help a nation move into the GIR. Moreover, there is a strong need for financial support that is not tax breaks or incentives, but investments, grants, and purchasing for GIR technologies, such as renewable energy. This can be seen in the United States today with the debate over smart grids. What are they? And who pays for them? When the smart grid is defined as a utility, then the government must pay for them since they are part of the transmission of energy, for example, over long distances that must be secure and dependable.

But as the GIR moves much more into local on-site power, the costs of the smart grid are at the home, office buildings, schools and colleges, shopping mall, and entertainment centers. Local governments are also involved, as they are often one of the largest consumers of energy in any region and hence emitters of carbon and pollution. Within any building, a smart grid must know when to regulate and control meters and measure power usage and conservation. The consumer needs the new advanced technologies, but the government must support these additional costs and their use of energy as they impact the local community and larger regions' residential and business needs.

Economics has changed in the GIR. And yet, economics has a basis of success in the 2IR. Historically, 2IR economics was successful because the government was needed to support its technologies along with goods and services. Evolution into the neoclassical form of economics was far more a political strategy backed by companies that wanted control of infrastructure sectors. But the reality was that "greed" took over and has now forced a rethinking of economics as nations now move into the GIR.

ACKNOWLEDGMENTS

The authors thank Jon McCarthy for both economic details and edits, as well as Lucas Adams, Kentaro Funaki, Claus Habermeier, and Russell Vare for international perspectives and data. Namrita, Singh, and Jerry Ji checked on data contributions.

REFERENCES

Chan, S., 2011. Global solar industry prospects in 2011. SolarTech Conference, Santa Clara, CA.

Chomsky, N., 2012. Plutonomy and the precariat: on the history of the US economy in decline, Tompatch.com, pp. 1–5. Available from: < http://truth-out.org/news/item/8986-plutonomy-and-the-precariat-on-the-history-of-the-us-economy-in-decline76 >.

Clark II, W.W., 2003a. Science parks (1): the theory. Int. J. Technol. Transfer Commercialization 2 (2), 179–206.

Clark II, W.W., 2003b. Science parks (2): the practice. Int J Technol Transfer Commercialization 2 (2), 179–206.

Clark II, W.W., 2008. The green hydrogen paradigm shift: energy generation for stations to vehicles. Utility Policy J.

Clark II, W.W., 2009. Sustainable communities. Springer Press.

Clark II, W.W., 2010. Sustainable communities design handbook. Elsevier Press, New York.

Clark II, W.W., 2012. The next economics. Springer Press.

Clark II, W.W., Bradshaw, T., 2004. Agile energy systems: global lessons from the California Energy Crisis. Elsevier Press, London.

Clark II, W.W., Cooke, G., 2011. Global energy innovations. Praeger Press.

Clark II, W.W., Cooke, G., Jin, J., Lin, C.F., 2013. The green industrial revolution. [Published in Chinese].

Clark II, W.W., Fast, M., 2008. Qualitative economics: toward a science of economics. Coxmoor Press, London.

Clark II, W.W., Isherwood, W., 2010. Inner Mongolia autonomous (IMAR) region report. Asian Development Bank, 2008 and reprinted and expanded on in Utility Policy Journal, special issue. China: environmental and energy sustainable development.

Clark II, W.W., Li, X., 2004. Social capitalism: transfer of technology for developing nations. Int. J. Technol. Transfer 3 (1).

Clark II, W.W., Lund, H., 2001. Civic markets in the California energy crisis. Int. J. Global Energy Issues 16 (4), 328–344.

Clark II, W.W., Rifkin, J., et al., 2006. 'A green hydrogen economy', special issue on hydrogen. Energy Policy 34 (34), 2630–2639.

Clark II, W.W., Sowell, A., 2002. 'Standard economic practices manual: life cycle analysis for project/program finance', International Journal of Revenue Management. Interscience Press, London.

EU, CCC, 2009. Meeting carbon budgets – the need for a step change, progress report to Parliament Committee on Climate Change, presented to Parliament pursuant to section 36(1) of the Climate Change Act 2008, the stationery office (TSO). Available from: < http://www.official-documents.gov.uk/document/other/9789999100076/9789999100076.pdf >. (01.05.10.).

Gipe, P., 2011. Feed-in-tariff monthly reports.

Hawkins, P., Lovins, A., Lovins, L.H., 1999. Natural capitalism: creating the next industrial revolution. Little, Brown and Company, Boston.

Isherwood, W., Smith, J.R., Aceves, S., Berry, G., Clark II, W.W., Johnson, R., Das, D., Goering, D., Seifert, R., 1998. Remote power systems with advanced storage technologies for Alaskan Village. Energy Policy 24, 1005–1020.

Knakmuhs, H., 2011. Smart transmission: making the pieces fit. RenGrid 2 (3), p. 1+.

Li, X., Clark, W.W., 2009. Crises, opportunities and alternatives globalization and the next economy: a theoretical and critical review. Chapter 4, In: Xing & G Winther, L. (Ed.), Globalization and Transnational Capitalism. Aalborg University Press, Denmark.

Lillian, J., 2010. New and noteworthy, Sun Dial. Solar Industry, 3–4.

Lund, H., 2009. Sustainable towns: the case of Frederikshavn. 100 Percent Renewable Energy. Chapter 10. Sustainable Communities. Springer Press.

Martinot, E. & Droege, P. (series of reports from 2007-2010), Renewable energy for cities: opportunities, policies, and visions. Available from: <http://www.martinot.info/Martinot_Otago_Apr01_cities_excerpt.pdf>.

Next 10, 2011. Many shades of green, Silicon Valley and Sacramento, CA. Available from: < www.next10.org >.

Nularis, 2011. Data and information. www.nularis.com.

Rifkin, J., 2004. The European dream. Penguin Putnam, New York.

San Jose Business Journal, 2011. Clean energy financing jumps to record $243B, 8.

Sullivan, M., Schellenberg, J., 2011. Smart grid economics: the cost-benefit analysis. RenGrid 2 (3), pp. 12–13+.

Tickell, J., 2009. Director and producer, Fuel, independent documentary film. La Cinema Libra, Los Angeles, CA.

U.S. Energy Information Administration. See annual reports at: <http://www.eia.doe.gov/oiaf/ieo/world.html>.

Vestas 2011. Available from: <www.vestas.com>.

CHAPTER 20

Sustainable Agriculture: The Food Chain

Attilio Coletta
Department. DAFNE – Università degli Studi della Tuscia – Viterbo, Italy

Contents

Introduction	495
Social Implications	496
Economic Implications	497
Environmental Implications	500
Developing New Solutions	504

INTRODUCTION

Sustainability is a polysemic and ubiquitous term: a sector no longer exists where sustainability is not sought, certified, and too often boasted. This is why a correct approach requires, prior to any further analysis, a statement to which to refer while discussing the sustainability concept proposed.

This is even more important in the agricultural sector that deals with the production of food (both from vegetables and animals) by means of environmental factors (water, soil, landscape, etc.) mainly behaving like nonprivate goods and other producing factors traditionally considered private goods (labor, equipments, fertilizers, seeds, fuels, pesticides, etc.).

The shift from a traditional production approach, related to the production of commodities, toward a more comprehensive role of agriculture implied a shift of attention from outputs to inputs. This is particularly true for those producing factors that induce externalities.

Basically, sustainability could be defined as that level of resource's consumption that allows a natural regeneration of the resource itself (Godfray et al. 2010). This approach is very close to the point of view of those who focus on the resilience of the ecosystems in which human activities take place and where environmental implications of the food chain are among the priorities.

A different point of view, focusing on human beings' survival, refers to sustainability in terms of food security: sufficient food for a growing population, as pointed out at the international level by FAO (2010) whose concern regards recent estimates that indicate a world population expected to grow to nine billion in 2050 (Godfray et al. 2010).

Obviously, these two approaches overlap when considering the consequences of a lack of sustainability and focusing on the causes.

For better organization of this chapter, sustainability is analyzed in relation to social, economic, and environmental frameworks. The first and the second focus on human beings needs, while the last takes into consideration a more comprehensive approach that refers generally to ecosystems.

SOCIAL IMPLICATIONS

From a social point of view, sustainability deals with food security and food safety, as mentioned previously. Summarizing the concern, there is still the problem of how to feed the entire world population with enough and safe food (see FAO 2010; Godfray et al. 2010).

Despite concern on the expected increase in number focuses the attention of many actors, nowadays the problem of humans feeding is already present: 15% of the population does not have access to sufficient protein and energy resource in terms of available food (with respect to their purchase power). Even more people seem to be currently affected from malnourishment. Therefore, sustainability in terms of food security (at global level) deals with how it is possible to sufficiently feed the whole Earth's population today and tomorrow.

In most of the richer countries, despite a lower growth rate of population, the higher income available pushes up demand for high-value food such as processed food, meat, and fish. Such a demand pattern (like, among others, the one that China and India are showing since the recent past) is characterized by strong consumers' preference toward food requiring a greater use of natural resources and a lower efficiency in land use (animal husbandry has a conversion of around 10% of grassland food into meat). Such products seem to give consumers a higher level of satisfaction (utility achieved from the consumption of goods) and, accordingly, their consumption grows as income rises.

This is why emerging countries show a development path characterized by a strict positive correlation between per capita income and demand for a high protein diet, as Western countries show, while a healthy diet is adopted only by the richest and fewest part of the population, able to valuate health implications of an unbalanced diet. The result is a frightening dynamic: hunger (among the poorest), obesity, and protein excess due to the consumption of inexpensive food with a high energy content (sugar and alcohol mostly) in the largest part of the population that show low wage levels, high-value food (i.e., preprocessed food such as frozen food) for high-income people, and a well-balanced healthy diet only for a few most fortunate, well educated, and rich. In such situations there is a clear "health gap[1]" among countries with a different average income level per capita

Food prices are part of the problem, making poor people able to buy only unhealthy food and allowing them to get rid of hunger often with the sole alternative of obesity due to consumption of a higher energy content only via food available at a convenient price. Obesity is definitely coming back in countries affected deeply by the economic crisis (Lock et al. 2010) in which the food choice is driven strongly by a lower available income with respect to the past.

In such a critical context, food security is still a goal to be reached, mostly in those countries that show an unsatisfactory level in enforcing population rights with respect to accessing resources, mainly land and food, as happens in less developed areas of the world (i.e., land grabbing, grazing rights on common land, access to public water resource; SCAR 2011).

ECONOMIC IMPLICATIONS

By tradition, sustainability in the food sector has been referred in terms of a yield gap[2] and, accordingly, scarcity has been related to natural resources: soil water, energy, phosphorous, and nitrogen. These are the basic producing factors involved in food production. However, recent dynamics show new emerging scarcities that stem from the development path of modern society (Freibauer et al. 2011)(Figure 1).

In addition to traditional scarcities, climate change sheds a light on the future availability of rain water for crops (Iglesias et al. 2011) and on progressive desertification of cultivated land claimed by many competing uses (food production, bioenergy crops, fodder, and urbanization; Beddington et al. 2011) and therefore is becoming scarce. Referring to natural resources involved in food production, biodiversity is becoming scarce as well. Intensive agriculture implies an increasing biodiversity loss with respect to both animal and vegetable genetic resources, reducing the opportunities for future plant and animal breeding dramatically. An economic response mechanism typical of free markets is unable to alleviate the scarcity effects of those resources solely by means of prices, such as

---

[1] See Joffe and Robertson (2001) and Friel and colleagues (2008).
[2] Difference between actual productivity and best level achievable using current technologies available.

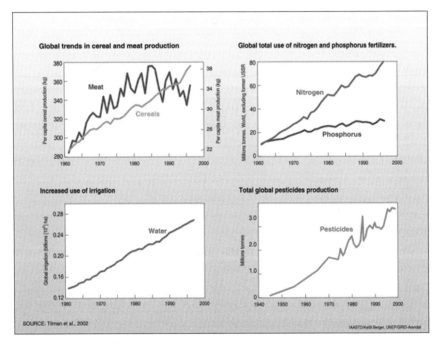

Figure 1 Trends in production and inputs use in agriculture. *Source: Tilman et al. (2002) in UNEP* http://www.grida.no/graphicslib/detail/global-trends-in-cereal-and-meat-production-total-use-of-nitrogen-and-phosphorus-fertilizers-increased-use-of-irrigation-total-global-pesticides-pr_ef80.

they usually do for private goods, showing increasing negative externalities that end up as the so-called "tragedy of the commons" (Hardin 1968).

Urbanization destroys most fertile areas, and an economic framework offers room for oligopoly, monopoly, and distribution asymmetries all reducing the development and investment opportunities in the whole economy in general and in the food sector specifically. Therefore, the challenge is a global approach that requires sustainability in the way food is produced, stored, processed, distributed, and assessed (Godfray et al. 2010).

As discussed earlier, a higher income drives a higher demand for better food, which hence encourages new investment on the distribution side. This process leads to a more industrial concentration in the retail sector, inducing in turn a higher concentration on the whole supply chain.[3]

---

[3] Concentration backward along the supply chain could be due jointly to two different reasons. It might be either a countervailing reaction of producers in order to balance increased retailers' market power or a solely business evolution to satisfy growing logistic, organizational, and supplying needs of a concentrated retail sector.

Also, bigger enterprises usually adopt less sustainable industrial processes due to the intensive use of producing factors.

In Organisation for Economic Co-operation and Development countries in which sustainability might be threatened by such a physiological industrial evolution of the food chain, measures to mitigate negative effects are often adopted (among others EU 2010, 2012). For instance, the European Union (EU) and the United States have introduced and financed agricultural policy measures devoted to strengthening environmental-friendly production (i.e., organic agriculture and low impact agriculture, both characterized by lower input necessities and lower output level). However, the debate on the environmental effects of those agricultural practices displays different opinions regarding cost/benefit results. Nevertheless, there is no doubt that those policies have generated negative externalities on less developed countries: the curtailment of total production (the EU and the United States play a key role in the global context of most commodity production) obviously causes a rise in prices hard to cope with, especially for developing countries.

Hence, a small increase in environmental sustainability in a specific country may induce a loss of global sustainability.

Economic growth in Western countries allows private companies to adopt marketing strategies oriented to stimulate and increase demand.[4] With respect to the food sector, an increase in consumption often does not imply a growth in welfare, thus causing a double cost for the consumer: cost for a higher demand for food and recover cost for the sanitary sector (due to food-related pathologies) as a taxpayer. However, the food processing industry, retail sector, and media unanimously play key roles in changing consumer habits toward a high demand consumption style.

Indeed, the consumer is encouraged by offers, price discounts, super-sized portions, and a strategic use of an anticipated "best before" date. The consumer is induced to buy an unnecessary amount of goods that inevitably become waste to be managed with additional resource consumption.[5]

However, in developing countries, a similar amount of food is wasted due to an unfit storage and distribution sector. Insects and spoilage cause a loss of one-third of the southeastern harvest, and 35–40% of Indian harvested food is lost due to the lack of cold storage systems in the retail sector (Godfrey et al. 2010). Overall, estimates indicate that a third (equal to

---

[4] The search for a profit maximization strategy requires companies to adopt behavior to increase revenues via an increase of the sold quantity, an increase of prices, or even both. Therefore, a company's survival depends strictly on an increase in consumers' consumption.
[5] Around 30% of food is wasted in developed countries (Segre & Gaiani 2011).

1.3 billion of tons) of food produced for human consumption is ether lost or wasted along the global food system (Gustavsson et al. 2011).

ENVIRONMENTAL IMPLICATIONS

Because of rapid population growth, the availability of fresh water per person has decreased from 15,900 m^3 in 1950, 10,800 m^3 in 1970, 8000 m^3 in 1990 to 6500 m^3 in 2000. In 2010 the availability of fresh water has continued to decrease, to 5800 m^3, and it is expected to decrease in the future to 4400 m^3 per person in the year 2050 (UNEP 2008).[6]

This quantity (overestimated because not all the resources of groundwater are technically/economically accessible) could be sufficient to satisfy the needs of the entire world population if it could be provided equally, but many countries of Africa, Middle East, east Asia, and some eastern European countries show an availability of fresh water much lower than the average and often under the levels of subsistence. It is estimated that in 2025 about 3.5 billion people will be part of the category of "water scarcity" with an average yearly availability of 1700 m^3 per capita (UNEP 2008).

It is important to note that freshwater source distribution is not homogeneous on the whole planet, with 60% of accessible water concentrated in only seven countries: Brazil, Russia, China, Canada, Indonesia, United States, and India. Despite being a renewable good, fresh water is not inexhaustible and is even subject to the risk of becoming scarce.

As regarding water of lower quality, agriculture is the most important consumer, using on average up to 70% of water withdrawn at the global level (FAO 2002). For irrigation systems to be sustainable, they require proper management (to avoid salinization) and must not use more water from their source than is replenished naturally. Otherwise, the water source effectively becomes a nonrenewable resource.

Irrigated agriculture plays a relevant and increasing role (particularly in the last decade) in producing food. The crop output level obtained due to irrigation is more than double the level achievable by means of solely rain water contribution (Baldos & Hertel 2012). Minimal irrigation in critical times makes it possible to obtain much better results than those achievable merely by rain water, even if with a strong rain amount. This is the result of

---

[6] These data refer to the share of renewable resources (potentially usable rainfall amounted to 40,000 km^3; FAO State of Food and Agriculture 1993); however, because rains are not all used because of their distribution, it has been estimated that the share actually usable is equal to 12,500–14,000 km^3 (UNEP 2008).

the advantage to control the quantity of water absorbed by the plants roots when they need it.

This explains why irrigation contributes to the increase of crop productivity mainly in areas characterized by arid, semiarid, and subhumid climates. Irrigation is often associated with excessive water use of both surface and underground resources, therefore causing an overexploitation associated not only with quantitative depletion, but with qualitative degradation (e.g., chemical and biological contamination of underground pure water). In addition, the creation of such negative externalities is not internalized in current irrigation costs (often set on the simple need to cover the service and delivery costs).

Moreover, evidence of the negative consequences of high intensive irrigation is the increase of marshlands and soil salinization. It is estimated that about 30% of irrigated land is affected by these two problems in a more or less extensive way; progressive salinization of irrigated areas is provoking the reduction of usable land with an increase of 1–2% per year (FAO 2002).

The effect of water on the soil is influenced strictly by water management actions. Increasing erosion effects are evident in many areas, and erosion is one of the major factors causing a loss of both organic matter and fertile soil. Estimates indicate a yearly loss of 12 million hectares of agricultural land due to soil degradation, which equals a loss of 20 million tons of grain (United Nations Convention to Combat Desertification 2011).

Erosion is still one of the most critical concerns with respect to climate change due to the intensification of heavy rain phenomena. Some of the world's most important agricultural districts are those located in mega deltas where salt water intrusion and the rising level of seas will become severe threats for food production (Beddington et al. 2011).

Climate change is moreover influenced by human agricultural activities, with animal husbandry being the most critical. Agriculture as a whole generates approximately 20–35% of greenhouse gas (GHG) emissions (IPPC et al. 2001) and 40–50% of total anthropogenic emissions of CH_4 and N_2O (up to more than 80% according to some authors; Smith et al. 2008).

Putting aside the solutions that rely on an abatement of GHG emissions by means of a reduction of livestock and those related to research effects in the field of animal diet and breeding that are far from reaching consistent outcomes, the most effective managing option to be adopted is the concentration of husbandries. This may allow managing housing and storage activities in a closed environment and applying "end of pipes strategies" to treat wastes and mitigate the emissions.

This is the most reliable choice if considering the consequences of a change in diet. Given the current genetic value of animals, an alteration of feeding protocols would determine a lower production per capita; accordingly, a higher amount of animals would serve to satisfy the market demand and, eventually, the production of GHG would increase.

Different strategies can be chosen to limit the production of GHG, however each option displays strong implications on other relevant aspect of agriculture. One of the most invoked solutions with respect to food safety and environmental-friendly cultivation practices is organic agriculture. However, organic agriculture requires a higher use of manure and slurry, which increases the soil emissions of CH_4 and N_2O. Extensive agriculture, opposite to intensive agriculture, can contribute to lowering the emission of CH_4 (thanks to the grassland diet), but at the same time leads to an increase of N_2O leaching in soil and water bodies caused by grazing livestock excreta. Extensive grazing, in addiction, lowers soil productivity (yield gap) and requires a higher amount of land to reach the same output level.

Intensive husbandry allows the efficient treatment of manure at a farmyard level (e.g., biogas production), as suggested by Monteny et al. (2006); however, it implies intensive agriculture as well on the surrounding fields. This may lead to an increase of N_2O leaching in the soil due to an increase in fertilizer use and pesticide pollution (which, conversely, are not used in grazing systems).

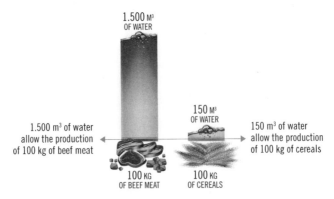

Figure 2 Meat consumption estimates. *Adapted from FAO (2006).*

Greenhouse gas reduction may be combined easily only with intensive farming contributing to a renewable energy production and a lower emission caused by the digested mass spread on the soil.

Despite a human demand for protein, foods may also be satisfied with vegetable protein (using legumes not requiring fertilization, as they are self-sufficient with respect to nitrogen) instead of animal products (meat, eggs, and milk products); the demand for meat is expected to grow rapidly due to the increase in consumption caused by the expected rise of per capita income level generally already observed all over the world in past years (Figures 2 and 3).

Energy production from manure and slurry is important as well as the potential energy produced by food chain wastes. Companies' packaging strategies drive consumers to dispose a growing amount of wastes both in terms of food unused and package not eatable. In Western countries nowadays, food packaging often has the same importance of the content in terms of energy consumption and economic value. A reuse of packaging, and eventually energy production, is a keystone for the sustainability of food consumption. Per capita food waste by consumers in Europe and North America is estimated at 95–115 kg/year, while this figure in sub-Saharan Africa and south/southeast Asia is only 6–11 kg/year (FAO 2009).

A 50% reduction of food losses and waste at the global level would save 1350 km^3 of water (for comparison, the mean annual rainfall in Spain is 350 km^3; water passing Bonn in Die Rhine is around 60 km^3 per year; the storage capacity of Lake Nasser is nearly 85 km^3)(Lundqvist 2011).

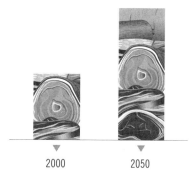

2000   2050

Estimated increase in meat consumption due to population growth and diet change

Figure 3 Water consumption required for meat and cereals production. *Adapted from UNESCO (2012).*

DEVELOPING NEW SOLUTIONS

The search for solutions follows two different approaches. The first is devoted to finding mitigation practices to manage the actual negative effects of developing the food chain. The second focuses on prevention measures by means of research and developing long-term solutions.

Furthermore, all the considerations mentioned earlier can be analyzed at different levels concerning the geographical scale: does sustainability have to be assessed referring to a specific region or to the ecosystem in which specific activities take place or even to the whole planet? Development paths usually show a growing concern about sustainability only after certain thresholds in terms of well-being have been trespassed, but measures devoted to increasing sustainability are mostly restricted to a local dimension. Induced externalities in neighboring countries or foreign markets are too often not dealt with, as decision-making processes are built at the government scale and international cooperation on such topics is difficult to achieve due to a divergence of interests between the actors involved in the decision (as international agreements on environmental matters often show). However, environmental and human diversity forbids the identification of a unique solution for the whole planet.

With respect to food scarcity, research aims to reduce the yield gap by increasing the productivity of factors and/or reducing the use of scarce resources and among these those influenced more by climate change.

The Earth's carrying capacity may be increased via plant breeding, which can help in the use of marginal soils by reaching salt, dry, and pest tolerance on the side of vegetables. With respect to animal husbandry research, improvements might combine new diets (with lower GHG production) with efficient productivity levels. Production increases, however, should not be the sole direction to follow, as agrofood system resilience is essential in ensuring food security.

Investments devoted to the reduction of food production impacts can help, mostly in mitigating the effects of by-products. The production of energy from renewable resources (as production wastes are with respect to agriculture) in the short term can lower the consumption of fossil fuels and support the reach of CO_2 the emissions level agreed at the international level.

Investments at whatever level considered (agricultural research, biotechnology, recycling technologies, etc.), however, are constrained by the uncertainty related to future scenarios and the right's ownership framework. One of the most debated fields nowadays concerns property rights on biotechnology research results. Because the legal framework is still not clear, a delay in investments and an adoption of research results may happen.

As a last remark, sustainable development implications usually imply different effects at different levels, often with opposite results—both quantitative and qualitative—that require the identification of optimal thresholds between different goals. This is why a coherent theoretical framework that takes into account different sources of information (such as multicriteria models allow) seems to be preferred.

REFERENCES

Baldos, U.L.C., Hertel, T., 2012. Economics of global yield gaps: A spatial analysis. Paper presented at the 2012 AAEA annual meeting, August 12–14, Seattle, WA.

Beddington, J., Asaduzzaman, M., Fernandez, A., Clark, M., Guillou, M., Jahn, M., Erda, L., Mamo, T., Van Bo, N., Nobre, C.A., Scholes, R., Sharma, R., Wakhungu, J., 2011. Achieving food security in the face of climate change: summary for policy makers from the Commission on Sustainable Agriculture and Climate Change. CGIAR research program on Climate Change. Agriculture and Food Security (CCAFS), Copenhagen, Denmark.

European Commission, 2010a. The CAP towards 2020: meeting the food, natural resources and territorial challenges of the future. European Commission, Brussels.

European Commission, 2012. Innovating for sustainable growth: a bioeconomy for Europe, COM(2012)60 final, Brussels.

FAO, 2006. World agriculture toward 2030/2050 Prospects for food, nutrition, agriculture, and major commodities group. Interim report. Global Perspective Unit UNESCO, 2012, UN World Water Development Report, available at http://www.unesco.org/new/en/natural-sciences/environment/water/wwap/wwdr/wwdr4-2012/

FAO, 2009. The state of agricultural commodities markets. High food prices and the food crises – experiences and lessons learned, Rome.

FAO, 2010. A conceptual framework for progressing towards sustainability in the agricultural and food sector. discussion paper.

FAO, 2002. Acqua per le colture, Corporate Document Repository.

Freibauer, A., Mathijs, E., Brunori, G., Damianova, Z., Faroult, E., Gomis, J.G., O'Brien, L., Treyer, S., 2011. Sustainable food consumption and production in a resource-constrained world, 3rd SCAR Foresight Exercise, Standing Committee on Agricultural Research (SCAR). European Commission, Brussels.

Godfray, H.C.J., Beddington, J.R., Crute, I.R., Haddad, L., Lawrence, D., Muir, J.F., Pretty, J., Robinson, S., Thomas, S.M., Toulmin, C., 2010. Food security: the challenge of feeding 9 billion people. Science.

Gustavsson, J., et al., 2011. Global food losses and food waste, FAO. Italy, Rome.

Hardin, G., 1968. The tragedy of the commons. Science 162 (3859), 1243–1248.

Iglesias, A., Quiroga, S., Diz, A., 2011. Looking into the future of agriculture in a changing climate. Eur. Rev. Agric. Econ. 38 (3), 427–447.

IPCC, 2001. Climate change 2001: the scientific background. In: Houghton, J.T. (Ed.), Cambridge University Press.

Joffe, M., Robertson, A., 2001. The potential contribution of increased vegetable and fruit consumption to health gain in the European Union. Public Health Nutr. num. 4, 893–901.

Lock, K., Smith, R.D., Dangour, A.D., Keogh-Brown, M., Pigatto, G., Hawkes, C., Mara Fisberg, R., Chalabi, Z., 2010. Health, agricultural, and economic effects of adoption of healthy diet recommendations. Lancet 376, 1699–1709.

Lundqvist, J. (Ed.), 2011. On the water front: selections from the 2011 world water week in Stockholm. Stockholm International Water Institute (SIWI), Stockholm.

Monteny, G.J., Bannink, A., Chadwick, D., 2006. Greenhouse gas abatement strategies for animal husbandry, agriculture ecosystems and environment n. 112. Elsevier, pp. 163–170.

Segre, A., Gaiani, S., 2011. Transforming food waste into a resource. Royal Society of Chemistry (RSC), UK.

Smith, P., et al., 2008. Greenhouse gas mitigation in agriculture. Philos. Trans. Royal Soc. (363), 789–813.

Tilman, D., Cassman, K.G., Matson, P.A., Naylor, R., Polansky, S., 2002. Agricultural sustainability and intensive production practices. Nature 418 (6898), 671–677.

UNEP, 2008. An overview of the state of the world's fresh and marine waters, second ed.

United Nations Convention to Combat Desertification, 2011. Desertification: a visual synthesis. Germany, UNCCD Secretariat, Bonn.

CHAPTER 21

Development Partnership of Renewable Energies Technology and Smart Grid in China

A.J. Jin[1], Wenbo Peng[2]

[1]Chief Scientist, China Huaneng Clean Energy Research Institute China Huaneng Group, Beijing, P.R China; Managing Director Haetl Ltd, a Clean Solar Thermal Electricity Company, Lafayette, CA USA;
[2]China Huaneng Group, Huaneng Clean Energy Research Institute, Haidian District, Beijing, P.R. China

Contents

Introduction	507
Solar Electricity Systems and Their Relationship with the Grid	509
Wind Power	513
Data Response and Power Transmission Lines: Examples in the United States	516
Smart Grid and Market Solution	518
China Rebuilds a Power System and Smart Grid	521
Merits of Chinese-Style Smart Grid	522
Discussion on Chinese Cases, Investment, and Forecast	523
Historical Review and Attributes of Third-Generation Grid	524
Lighting-Emitting Diode and Energy Efficiency Case Discussion	525

INTRODUCTION

The increase of greenhouse gas emission is creating numerous problems for both human health and a stable global climate. Growing energy consumption and raising oil prices are also causes of increased national security concerns. The scope of challenges in both energy and climate sectors is far reaching and relates directly to our dependence on traditional carbon-based fossil fuel. This is the heart of our global energy crisis.

There is no shortage of energy flowing to the Earth, as the sun radiates an enormous amount of power (170,000 TW) onto the Earth's surface. Although most solar renewable energy is not available to us, acquiring only about 0.01% solar energy is sufficient to meet the world's need today. As discussed here, the target of using a portfolio of renewable energies is gaining important governmental support and attracting significant private investment. Renewable energy technology is currently developing fast and is becoming economically competitive.

President Barack Obama has championed for renewable energies, as well as the smart grid, and has announced several billion dollars in U.S. government support along with private investment partnership toward that end (Associated Press news reports, 2009). The U.S. president is encouraging the new grid system to be smarter, stronger, and more secure in the future. The following sections discuss how the aforementioned targets can be achieved and what the world has achieved in terms of renewable energy generation and smart grid power transmission. To date, advanced energy technologies have shown us that the development of these technologies could potentially become the linchpin of a new system of the modern energy infrastructure. The aforementioned smart grid project would install thousands of new digital transformers and grid sensors in homes and utility substations to enable a grid-smart data system. This chapter shows examples of solar electricity and of wind power that are sustainable, abundant, and affordable energies based on the excellent uses of natural resources. It also shows that the goal of the data response system is to strengthen the grid system. The new system is smarter due to its ability to monitor and control energy consumption comprehensively in real time. It is also beneficial in terms of demand response, energy efficiency, and compatibility with a large supply of renewable energy sources. This chapter presents several cases and explores key attributes and huge merits of a Chinese-style smart grid.

The resolution of the energy and climate challenges has profound business impacts, as well as a great societal effect; the effort to meet these challenges is sometimes referred to as the third industrial revolution. Furthermore, most developed nations are facing a major challenge of upgrading their current electricity grid and energy management infrastructure. Lack of a new grid infrastructure forms a bottleneck in investing in new renewable generation and in tapping the full power of renewable energy, such as the utility-scale solar electricity system.

The strong and rapidly growing consumer demand for clean energy promotes a new resolve in meeting global needs with inexpensive electricity from clean energy sources. The advanced energy technology requires a development partnership among the elements of renewable energy, optimized energy efficiency, and smart grid. The following sections illustrate the challenges and share our knowledge of employing several renewable energies and an optimal grid infrastructure in order to reduce greenhouse gas emissions more than 25% by 2020 and more than 80% by 2050.

SOLAR ELECTRICITY SYSTEMS AND THEIR RELATIONSHIP WITH THE GRID

Solar electricity systems are anticipated to be the most likely to become successful commercially without a government rebate in a few years. This anticipation is based on the current best knowledge of cost and adoption risks of the solar electricity that offers a superior future technology trajectory. Today as the solar power business grows, the cost of solar photovoltaic (PV) panels has declined rapidly.

Under Obama's administration in the United States, the president's Solar America Initiative in collaboration with the Department of Energy (DOE) has targeted grid parity where the electricity cost based on renewable energy production is the same as the coal-fired traditional power cost.

More progress is needed in the balance-of-plant aspect of energy production, which is defined as the solar cost per installed system ready for use. Active research to accelerate the progress of the cost reduction in this area is underway. In many parts of the world the solar PV system is appropriate for cost-effective distributed generation. All the world's current and expected major electricity load centers are within practical transmission range of excellent solar radiation locations.

As shown in this section, the solar electricity system can utilize the sun's natural energy to generate electricity. This solar electricity generation can be consumed or fed back into the utility grid. This results in less energy that has to be purchased from the utility company so the consumer's monthly bill decreases (the monthly bill is then just for financing payment). During the solar system warrantee period, the solar PV system is free for electricity generation and free from maintenance. Consumers pay only for the energy used from the grid (and amortization of the system/installation costs). Solar PV systems connected to the grid can be very attractive in reducing the more expensive peak-hour costs.

The term clean tech refers to technologies that produce and use energy and other raw materials more efficiently and hence produce significantly less waste or toxicity than prior commercial products. The clean tech energy production industry has developed clean alternative energy that utilizes current benchmark technologies such as solar and wind power. Examples of alternative clean energy sources include (1) wind power, (2) solar PV cell, (3) solar thermal electric power, and (4) solar heating. Scientists and engineers continue to search for viable clean energy alternatives to our current traditional power production methods.

Even though some renewable energy sources are variable in nature, several renewable energy sources can be integrated into the grid system quite well. For example, studies show that the compatibility of sun and wind energy is complementary and may be quite manageable for integration into a grid system (Abbess 2009; Clark 2012; Jin 2010)(Figure 1).

What is a solar electricity system? A solar electricity system is a system that utilizes the abundant sun's energy in order to produce electricity or to provide heat for consumers. For example, a solar PV cell is a physical device that converts light into electricity. Figure 2 illustrates the physical mechanism of a solar photovoltaic cell in producing solar electricity. The solar electricity industry is composed of many types of competing technologies with various cost structure, efficiency, and scalability factors important to the renewable energy industry sector.

The availability and the future prospects are very promising at this time for the following three solar technologies: (1) solar thermal power, (2) solar PV panels, and (3) solar heaters. For example, utility-scale solar thermal power plants have been constructed rapidly in the last two decades. Moreover, solar PV panels offer scalable power that has been installed on thousands of rooftops in California.

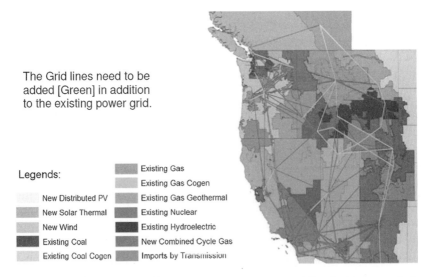

Figure 1 Map of west energy illustrates energy resources and the electric transmission grid. The diversity of energy sources and the smart grid should enable the "utility grade" power generation that is on par with current coal-based power plants.

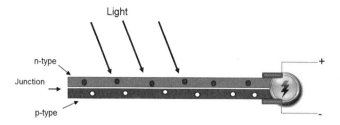

Figure 2 A schematic of a solar PV cell that converts sunlight to electricity; see text for details.

Solar PV systems, which are made up of individual solar cells, are becoming more and more affordable and reliable all the time. Solar PV panels are modular, scalable, and suitable for distributed generation. Moreover, scalable solar panels can be utilized for utility-scale power plants.

Several types of devices may be required to connect solar PV systems so they are suitable for individual consumer energy use and/or for supplying power to the electric grid. The most important unit is the inverter. The inverter unit is an electronic device that turns direct current (DC) from the solar electricity into an alternating current (AC) that is matched to the incoming main electric utilities standard and that is used by almost all home appliances and electrical devices. The concern for safety also requires the solar electric system to be enabled by circuit breakers for safe maintenance, etc. Circuit breakers are typically connected in both DC and AC sides of the circuitry path.

A solar thermal power plant (STPP) employs utility-scale steam turbine technology. As shown in Figure 3, STPP collects solar energy in a large real estate footprint for thermal energy in order to produce electricity. A circular array of solar light reflectors is used to concentrate the light on a receiver located on top of a tower. The light is absorbed as heat energy, which heats up the steam gas or air to very high temperatures that produce pressurized hot gas or air to drive the turbine. STPP has a typical footprint equivalent to the scale of a coal-fired utility power plant.

A STPP employs direct sunlight and hence requires its plant site to be in regions of high solar radiation. The thermal energy storage may be typically achieved through liquid or solid media to extend the hours of the electric cycle. Figure 3 shows a portion of a typical solar thermal power design.

The United States is the world leader in installed concentrated solar power capacity with 429 MW currently in commission. Three gigawatts of power is operational today. Seven gigawatts of total power was in development in 2010.

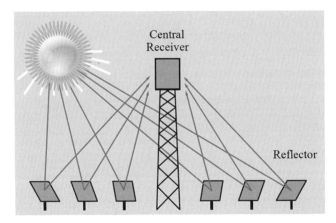

Figure 3 The solar thermal generator receives a highly concentrated sunlight beam, heats up pressurized gas in the receiver very hot, and drives hot gas through a turbine that makes use of the excellent cycle to produce electricity with a combined hot gas and turbine operation.

The United States alone anticipates powering two million homes by solar thermal power in 2020 (Environmental and Energy Study Institute, http://www.eesi.org/files/csp_factsheet_083109.pdf).

All solar power, including solar PV and solar thermal, generated about 0.1% or less of the total U.S. energy supply in 2008, but the installation of solar power is growing quickly. U.S. DOE's 2009 preliminary forecasts anticipate an annual growth rate in U.S. domestic solar PV generation of 21.3% through 2030 (and some analysts have even higher predictions for the growth rates)(U.S. Energy Information Administration, 2009). The solar PV technology is scalable. Increasing demand will bring down the cost of PV modules and solar electricity systems when they are in volume production.

Variable but forecastable renewable energies (wind turbines and solar PV power sources) are becoming more reliable in net output when integrated with each other than with one source alone. The aforementioned net output is suitable to meet the demand. Risks of security against terrorist attacks or natural disasters can be mitigated by planning geographically disperse energy sources such as microgrids. One plan is to employ a wide distribution of solar power from many sunny areas—smart grid power systems, discussed later, and backup fuel generators available from natural gases, etc.

Finally, the distributed generation of solar PV electricity can be connected to the grid. Germany is noteworthy for the nation's high-profile feed-in tariff in promoting solar electricity. Several governments have successfully offered incentive packages to promote renewable energies and energy efficiency in the world. For example, China has significant solar power investment with both a major development plan and an affluent stimulus package detailed in the Chinese national 12th 5-year plan. The plan is to construct large-scale renewable energy bases.

WIND POWER

As power is generated by new utility-scale renewable power plants such as wind power plants, the power transmission grid needs to have the capacity to fully deliver the power to consumers without a distribution block. The goals of America's clean energy market can be empowered by addressing the advanced technology platform of a nation-wide electric power system. The year 2009 ARPA-E grants of the United States have invested heavily in projects with the capability to allow intermittent energy sources such as wind and solar in order to provide a steady power flow to consumers.

An interstate power-transmission superhighway is needed for the electric power system. Immense solar power farms in America's deserts are facing a transmission challenge for moving through the power grid to the consumers. Congestion of the grid can create significant limitations that can reduce the potential advantage for a large renewable power generation to pump power into the electric grid.

What limits the renewable energies (as a commercial bottleneck) is the outdated power transmission grid mentioned in the last section. For example, the current system cannot accommodate the present and future needs of delivering hundreds of megawatts of wind power to users. In one scenario, a total of 200,000 miles of power transmission needed to be rebuilt may provoke fights among 500 divided owners and numerous property owners. In another scenario, a large power generation usually requires an extra storage system. For today's market, there is no commercially advantageous solution yet. Active development in the storage system area is underway.

The layout of the current power grid should be eventually accessible to a flexible change of total power (e.g., gigawatts) for power interconnection and transmission lines. The current transmission lines cannot increase their transmission capability by the hundreds of megawatts needed to meet the challenges. Power pumping is a challenge at times even over a distance of a few hundred miles. The commercial pain is the severe congestion today for long-distance power transmission.

We have to achieve a clean tech or renewable energy vision. The Kyoto Protocol has set a target that the world needs to reduce greenhouse gas emissions by more than 25% by 2020 and by more than 80% by 2050 (http://www.americanprogress.org/issues/2009/01/pdf/romm_emissions_paper.pdf). One of the solutions comes from an advanced energy technology, that is, wind power generation, that is very cost effective for today's commercial use.

Wind turbine manufacturing is cleaner than the volume production of solar PV cells. Today, America utilizes barely 1% of the power produced by wind energy. America's goal is to achieve 20% from wind power by 2030 (DOE news, 2008). Wind power turbines of the Maple Ridge Wind Farm near Lowville, New York, are capable of producing a total maximum of 320 MW. It has been shutting down at times due to limitations on the pumping capacity of the electric power system. Wind farms too are having power transmission challenges. One cannot easily pump a large amount of power to the grid due to various reasons, which are discussed later. In order for users to utilize the full potential of wind power or other environmentally friendly energy, it is imperative for the nation to significantly improve or to rebuild a system of populated and optimized transmission lines.

Wind power is a type of solar-induced energy. Wind is always present on our planet due to uneven heating of the Earth's surface by the sun and due to the so-called Coriolis effect, which relates to the wind being dragged by the constant rotation of the Earth on its axis.

The conversion of wind to electrical power is generated by a wind turbine. Modern wind power technology (such as the wind turbine) has been perfected over the last decade. A wind turbine power plant typically generates electricity in much the same way (through electromagnetic induction) as the alternator in a car does. As shown in Figure 4, a wind power station is usually positioned such that its rotor always faces the wind. The power engine has a drive train system that often includes a gearbox. There is a wealth of information about wind power. Interested readers are referred to http://en.wikipedia.org/wiki/Wind_power and http://www.reuk.co.uk/Calculation-of-Wind-Power.htm.

Wind power depends on three variables: (1) wind velocity, (2) radius of generator, and (3) temperature, which determines air density. The following is a simplified summary of the aforementioned relationship about the operational state of the wind turbine.

1. The power increases with the cube of velocity (e.g., a twofold increase of velocity leads to an eightfold increase of power output).

Figure 4 (top) Power is generated by a wind turbine. Various turbines have a higher capacity rating from left to right. The dashed line has a 50-meter height. (Bottom) Power is related to the following factors: sweep area (in the rotor radius squared), wind velocity (v^3), and temperature variation, which affects air density.

2. The power increases with the square of the radius (e.g., a twofold increase of velocity leads to a fourfold increase of power output).
3. The power increases with decreasing temperature (with about 3.3% of power for the change of every 10°C in air temperature).

Not only does wind power production make economic sense, there are also greater social benefits of clean energy and a sense of personal freedom (by moving toward a zero net energy residence, a type of energy independence). The current transmission lines cannot meet the goals of the advanced energy technologies. With a new grid system, the smart grid can be designed to meet challenges of and to suit ideally the demands of electricity production, distribution, and utilization.

Wind power makes good sense environmentally and economically. Turbine components are generally either recyclable or inert in the environment. The price of the wind turbine is a critical parameter for the return of investment. A typical payback period for the energy cost is about half a year. Residential wind turbine can be employed in homes or routed to storage such as battery banks. Some farmers have utilized their land for wind farms where the wind power generation does not affect how they farm, produce crops, etc.

DATA RESPONSE AND POWER TRANSMISSION LINES: EXAMPLES IN THE UNITED STATES

Power system management and optimization are really about data management bank, response, and efficiency. When the demand reaches a significantly high level or an energy reduction is needed, the smart demand response should help customers in energy conservation and reduction and thus in enhancing system reliability. The energy security is consistent with our nation's security concern so that the advanced energy technology supports the renewable power standard.

To address the transformation issue of the critical electrical power infrastructure, a major challenge is transmission lines that the government can support adequately in terms of policy and coordination. The current grid protocol of the power infrastructure has employed a century-old technology from about the turn of the 20th century (Energy Information Administration, 2009). Initially, 4000 individual electric utilities owned local grids and operated in isolation. Later, voluntary standards emerged through the electric utility industry to ensure coordination for linked interconnection operations. These voluntary standards were instituted after a major blackout in 1965 that impacted New York, a large portion of the East Coast, and parts of Canada.

Due to the limitation of transmission lines across states, thousands of megawatts of wind projects are stalled or slowed down, while many solar power deployments are experiencing similar challenges. In the United States, for example, long-distance power transmission has been the major barrier to the success of renewable power standard implementation in certain regions. Moreover, the power grid is considering limiting electricity transmission lines. For example, California had rolling blackout times dated back in Y2001 due to the transmission limitation.

To address the need of a grid transformation, the vision of the energy industry is to employ an Internet web model as follows. As shown in Figure 5, the Internet web of a smart grid takes the active system of a nerve network that determines, responds to, and controls the power needed for consumers. The network control system operates under a global scale to dispatch energy, to manage the energy flow protocol, but to distribute control around the system. For example, data response management by the network control system recovers from a power block by circumventing it. This recovery is an attribute of a self-healing power network and has attracted intense research interest.

The information exchange around the Internet web uses the concept of distributed control where a web host computer or a designated computer server acts autonomously under a global protocol. Due to the information process capability in modern Internet web technology, consumers will benefit in reduced cost by utilizing the Internet to manage the power grid effectively.

The energy efficiency comes from the consumer choosing more efficient energy options over other more costly ones. A smart grid can help utilities identify losses and support energy efficiency. The smart grid can manage its effective response to consumers. For energy consumers, power generation owners, buyers, and sellers, the nerve network in the Internet web of a smart grid will be both flexible and economical to extend the services of a power purchase transaction. An electricity system would provide supply–demand coordination and would be interconnected in the grid to dispatch power.

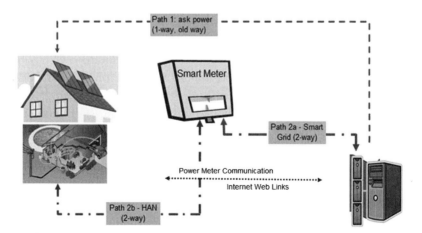

Figure 5 A smart grid system. A node with a smart meter enables a home automation network (HAN). This control system operates under a global scale to dispatch energy that determines, responds to, and controls the power needed for consumers.

Today, transmission and distribution lines have 500 owners and numerous property owners (Energy Information Administration). The coordination is mostly among three regional interconnections (western interconnection, eastern interconnection, and Texas interconnection) that their grid systems are in turn coordinated by the Federal Energy Regulatory Commission.

The diversity of energy sources and the smart grid have a designed-in specification that matches the current standard of the existing coal-based power plants. The existing power grid needs to add additional transmission lines shown in Figure 1 as green-colored lines. Additional transmission lines are required for the power transmission within each interconnection region and among the three power interconnections. By getting renewable energy sources connected to the grid and adding additional transmission lines as required, this will produce sufficient power flowing to consumers. A new power grid can adopt cleaner and more efficient power plants than just the current coal-fired power plant.

SMART GRID AND MARKET SOLUTION

In 2009, smart grid companies exhibited significant and fast-growing spots alongside the clean tech market need in the United States. The need for a smart grid is fundamental in developing modern energy networks[1] in the United States, the European Union, China, and every grid-connected nation worldwide.

What is a smart grid? A smart grid is a collection of energy control and monitoring devices, software, networking, and communications infrastructure installed in homes, businesses, and throughout the electricity distribution grid. This collective system generates a nerve system for the grid and for customers that provides the ability to monitor and control energy consumption comprehensively in real time.

Many tech giants such as Cisco and Google work to bring their products to the smart grid market. Cisco, in its May 2009 announcement about its smart grid roadmap, the expanding smart grid market is one of its "new market priorities." The advances of these products are most likely to address

---

[1] Major interests are dedicated to smart grid, e.g., March 30–31, 2010, Smart Grids Europe 2010. Please refer to the following works as well. a. Refer to Smart grid of European platform in 2006, EUR 22040. http://ec.europa.eu/research/energy/pdf/smartgrids_en.pdf; b. Refer to Smart Grid: Interop. of Energy Tech and Info Tech Operation with the grid: by IEEE, P2030/Draft 1.0 Skeletal Outline, 2009.

the challenges of our times in market demand, clean energy need, and green house gas emission reduction.

For example, Cisco expects (http://www.rechargenews.com/business_area/innovation/article296051.ece) the smart grid market to be bigger than the Internet in reach. The company has identified $15 to $20 billion in opportunities globally in the next 5 to 7 years, but this market pales in comparison to the market opportunity in China (more details are described later). At the Reuters Global Climate and Alternative Energy Summit in April 2012, Laura Ipsen, senior vice president of Cisco's smart grid unit, said:

> A lot of us looking at the China market see $60 billion by 2030 just for China alone. A lot of the big companies—the traditional GEs, IBMs, Siemens and others—are over there exploring that market.

For example, the introduction of technology of smart appliances could turn on and off themselves as provided by both the energy management and the smart grid. This technology could help the grid support a fleet of electric cars. The smart grid would improve the efficiency of transmission power.

Figure 6, which is an expanded version of Figure 5, shows a smart integrated energy system that merges Internet and grid features. This system has a smart grid with a power source(s), a data response, and a load center such as a residential home. Figure 6 illustrates several concepts, including data collection, communication, control, and smart grid system. A smart meter collects power usage data for the utilities and consumers and has Internet communication capability as mentioned previously.

Real-time data are fed back to a large distribution and transmission power grid. Moreover, energy storage is extremely important that assists in load leveling for transmitting any major power activities comparable to a typical locally rated load center. For example, a major power activity could be some solar electricity generation or a major charger for an electric car battery. Real-time data are useful for utilities to predict and to hedge power usage.

The smart grid is ideally suited to meet the challenging demands in the production, distribution, and utilization of electricity. The innovation here is to take a century-old power grid infrastructure, turn it upside down, mange it as mentioned earlier, and connect it to numerous renewable energy sources. Customers are interested in going green today. The cost of fossil fuel-based energy is rising due to both depleting resources and the cap-and-trade rules on greenhouse gas fuels.

Although solar PV and wind powers may have significant power output, they must be managed with load leveling suited to their output demand

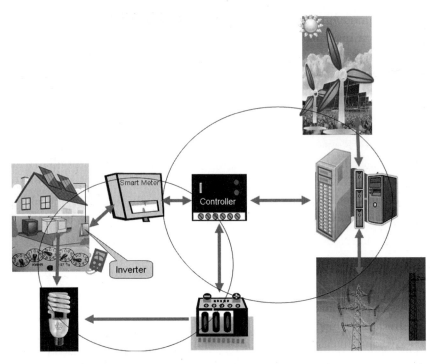

Figure 6 An integrated energy use system that has a grid to provide power (upper right circle shows distribution path), a data response system, and a load center (shown as lower left circle), such as the residential home.

profiles. Moreover, much power may be wasted because the power plants, such as nuclear power or coal fired, do not shut down when the consumer is asleep. Electricity analysis and management have to be directed toward the energy management of the existing infrastructure.

As a result, energy efficiency, modeling, and data analysis are needed and can then be fed into the smart grid. Data are for private use, and a computerized electricity management has robust integrity with regards to network security, reliability, and consumer participation. A smart grid may respond to diverse conditions that are indiscriminate to storage, resources, and electricity reliability.

There has been recent intense investor interest and corporate investment in the field of energy efficiency and carbon-free clean energy production. The smart grid is a bright and fast-growing technology sector in recent years. A sustained and deep commitment by regulators, state lawmakers, utilities, and other stakeholders is needed to achieve the cost-effective energy efficiency targets. For example, California utilities are recognized by customers as energy efficiency and demand response experts.

Transformation of the mainstream energy market requires an advanced grid infrastructure with superior energy efficiency and green technology. A smart grid is becoming increasingly important for the wide use of solar, wind, and other renewable energy. This grid reflects excellent criteria for market growth. A smart grid is an investment field in a large market sector, and many recognize the opportunity to make a substantial social contribution while providing a good return on the investment. A careful upgrade to the century-old power grids should lead to rebuilding the backbone with a smart grid system.

CHINA REBUILDS A POWER SYSTEM AND SMART GRID

China has sustained an "economic boom" for over three decades. The Chinese 12th 5-year energy plan enacted in March 2011 to develop national renewable energy systems and a smart grid in order to build its power system has support from an affluent government stimulus package. Rapid development of Chinese smart grid systems has shown and will continue to reap significant social and economic benefits.

Even though this chapter is not intended for a full-set/ thorough projection/exploration, it investigates the merits and attributes of the third-generation grid in China and the great smart grid market success within this section. In general, there is steady and astonishing growth and immense opportunities in China for its smart grid products. The Chinese-style power grid is to achieve integration of power, information, and business flow and to form a strong and secure grid.

To convey excitement in the Chinese smart grid development, let us review interesting cases in major cities. For example, China has installed smart meters (over 58 million), planned with smart charging stations and a network to serve electric vehicles (with thousands of charging piles completed), and built the world largest wind, solar energy, and storage station.

As a case in point regarding the smart grid, the Zhangbei power station (http://www.greenbang.com/china-claims-worlds-largest-battery-storage-station_21041.html) employed the world's largest hybrid green power station built until 2011. It is a demonstration project and was put into operation on December 25, 2011, in Zhangbei, the Hebei province of China. Its grid project includes clean energy sources and energy storage for demonstration.

This demonstration project was constructed by BYD and the State Grid Corporation of China, which is China's largest utilities company. This project is part of a Chinese ambitious smart grid plan. It combines 140 MW of renewable energy generation, efficient energy storage, and smart power transmission. With an initial investment of $500 million, the project combines 140 MW of renewable energy generation (both wind and solar), 36 MWh of battery power storage, and smart power transmission technologies.

The Zhangbei project is reported as a great success. Due to its complementation of wind and PV power, the utilization rate of wind turbines has been enhanced by 5–10%, and its whole renewable energy efficiency has been improved by 5–10% using its battery storage system. During its first 100 days of safe and stable operation, the power station generated over 100 GWh of electricity. There may be overflow or excess energy generation, and the excess energy can be used to feed back to the utility grid after storage has been filled. The Zhangbei project is an excellent success story, and this project provides a great perspective for the renewable energy solution for China and elsewhere.

The other case is about experimental setup of a microgrid (Jin et al. 2012). A smart microgrid PV experimental system of 50 kW has been set up by the China Huaneng Group in the Future Science and Technology City, Beijing. The company has taken up the microgrid project as a natural innovation for constructing a scalable smart grid in the near future. This project is the first smart microgrid power system and indicates that the company has begun to enter the field of distributed microgrid power generation. Based on a microgrid controller, the system has integrated 50 kW PV power, 300 VAh energy storage, grid power, and a 30-kW load. Under normal circumstances, the load is powered completely by the PV modules.

When PV power decreases, the controller can deploy battery energy to the load. Under extreme cases when the DC energy is too small to meet the load, the controller can switch the electricity supply to the grid within 8 ms to ensure a stable power supply of the load. Further reviews about the Chinese smart grids have led to the following successes.

Merits of Chinese-Style Smart Grid

The merits of transmission include factors such as safety, reliability, and stability of the smart grid while accommodating production from renewable energy sources. A highly uneven Chinese geographical distribution of electricity production and electricity demand requires China to pay more attention to a smart transmission grid on its merits. The backbone of the

Chinese smart grid has currently hosted the world's largest wind, solar, and energy storage integrated demonstration project (http://www.popsci.com/science/article/2012-01/china-builds-worlds-largest-battery-36-megawatt-hour-behemoth).

The key element to achieve for smart grid construction, according to the 12th 5-year development plan of the smart grid of the Chinese state grid, has the following goals.

- *Generation link.* The grid may meet the demand of 60 GW wind generation and 5 GW PV generation in 2015 and 100 GW wind generation and 20,000 PV generation in 2020. The capability of resource-optimized allocation is over 400 GW.
- *Transmission link.* The goal of the next 5 years of construction is to connect Chinese large-scale energy bases and major load centers in building a "three vertical three horizontal" backbone extra high voltage (EHV) grid, of which makes a high-level transmission smart grid and a transmission line availability factor of 99.6%. A "five vertical six horizontal" backbone EHV grid will be built in 2020. Meanwhile, a smart grid is fully composed.
- *Transformer link function at high voltages.* Over 6000 smart substations above 110(66) kV should be completed in 2015, accounting for about 38% of the total substations. Smart substations 110(66) kV or above will account for about 65% of the total substations in 2020.

Discussion on Chinese Cases, Investment, and Forecast

This decade set the stage for China's smart grid full-scale construction and improvement. In accordance with the Chinese development plan, from year 2011 to 2015, the smart grid construction investment amount will be over $300 billion and the total investment will reach $600 billion in 2020.

Meanwhile, national smart grid investment funds will be multiple 10-folds. According to the national smart grid plan, the smart grid radiation range is extensive. New industries such as smart city and smart transportation will also be spawned so that the market size is extremely attractive.

Moreover, smart grid construction provides a huge benefit. Specifically, by 2020, the benefits are listed here.
- The power generation benefits will be around $5.5 billion, saving the system effective capacity investment and reducing power generation costs by RMB 1–1.5 cents/kWh.
- The grid link benefit will be about $3.2 billion, grid loss will be reduced by 7 billion kWh, and the maximum peak load will be decreased by 3.8%.

- The user benefit will be about $5.1 billion, by offering a variety of services, saving 44.5 billion kWh of electricity.
- Environmental benefits will be about $7 billion, conservation of land about 2000 acres/year, emission reductions of SO_2 about 1 million tons, and CO_2 emission reductions of approximately 250 million tons.
- Other social benefits will be about $9.2 billion, increasing employment opportunities for 145,000/year, saving the cost of electricity, and promoting balanced regional development.

Historical Review and Attributes of Third-Generation Grid

Currently, China aims to become the world's largest smart grid user in the power industry. It is imperative to have advantageous elements that contain a robust and low-cost smart grid that accommodates renewable clean energy.

China's power industry began in 1882 with the birth of the Shanghai Electric Power Company, hence producing the first-generation grid of China. Until 1949, the Chinese power generation equipment-installed capacity reached 1.85 GW with a generating capacity of 4.31 billion kWh. The second-generation grid constructed in the 1970s aimed to interconnect the national grid. The Northwest Power Grid 750-kV transmission line was put into operation in 2005, and China's first 1000-kV UHV transmission lines were built in 2009. As of July 2010, China's 220-kV or above transmission lines are over 375,000 km in length, of which exceed the United States with ranking first in the world.

In fact, China's total installation reached 1 TW by the end of 2011 and the annual total electricity consumption was 4.7 trillion kWh. The grid-connected new energy power generation capacity reached 51.6 GW, of which 45.1 GW is wind power, accounting for 4.27% of total installed capacity; the grid-connected solar PV capacity is 2.1 GW, accounting for 0.2% total installed capacity; and the biomass installed power capacity is 4.4 GW, accounting for 0.4% total installed capacity, a geothermal power generation capacity of 24 MW, and an ocean energy power generation capacity of 6 MW.

The attributes of a Chinese smart grid or third-generation grid are substantially beneficial. Here is a brief list of the important attributes of a smart grid:
- Strong: robust and flexible are the bases for a future smart grid.
- Clean/green: the smart grid makes the large-scale use of clean energy possible.
- Transparent: the power grid openly shares information and is user transparent, and the grid is nondiscriminatory to users.

- Efficient: improve transmission efficiency, reduce operating costs, and promote the efficient use of energy resources and electricity assets.
- Good interface: compatible with various types of power and user; promote generation companies and users to participate actively in the grid regulation.

Lighting-Emitting Diode and Energy Efficiency Case Discussion

China offers fast growth business and market opportunities and is currently the second largest economy in the world. Chinese investment in clean technology is very positive by its government and has significant efforts in all related areas.

For example, the Chinese government's impact in the energy efficiency field is noteworthy. Beijing, the Chinese capital city, has pushed strongly on energy-efficient lighting, especially on lighting-emitting diode (LED) technology. LED lighting is a solid-state lighting product. One of Beijing city's goals is to eliminate incandescent lights. Currently, a consumer can buy a 60-watt-equivalent LED light bulb for lighting for just a quarter of USD after both discounts and government subsidies. Moreover, LED technology has attracted a huge market in the daily lives of people in the developed Yangtze Delta cities such as Shanghai, Suzhou, and Wuxi.

In the technology front, the latest LED technology delivers a highly energy-efficient solution. LED lighting uses nearly 80% less in energy consumption, has a much longer lifetime, and is better environmentally (without mercury involved in the process) than an incandescent light bulb. According to the Bright Tomorrow Lighting Prize, known as the "L Prize," hosted by the U.S. Department of Energy, there are great products such as Philips LED and Cree LED lighting. For example, the Philips' 60-watt equivalent LED bulb can provide 900 lumens but consumes less than 10 watts of energy. Its lifetime can last 17 years if used 4 hours daily. By the way, this Philips' product was the winner of the L Prize in August 2011.

In comparison, an incandescent bulb has less than 2% in energy conversion efficiency from electric energy to light energy. LED technology has better efficiency and is still maturing. There is a challenge in its thermal management for its lifetime improvement. The limitation of its temperature tolerance in current technology is being addressed through extensive efforts in research and development investments worldwide.

The market potential is huge for LED technology. Although the current market is still limited, LED technology provides home and street lighting and lights up city skylines, billboard displays, traffic lights, train and public transits signs and lighting, stage lighting, display lighting in art galleries, automotive headlights, floodlights of buildings, and growth lights for plants.

REFERENCES

Abbess, J., 2009. New reports on Monday, August 17, 2009. Wind energy variability and intermittency in the UK. Available from: < http://www.claverton-energy.com/wind-energy-variability-new-reports.html>.

Associated Press, 2009. News reports on 10/27/09, Arcadia, FL, The president has set a goal.

Clark, W., 2012. Introduction: the economics of the green industrial revolution. In: Clark, W. (Ed.), The next economics: becoming a science in energy, environment, and climate change. Springer-Verlag, New York.

DOE news 2008. Available from: http://www1.eere.energy.gov/windandhydro/pdfs/41869.pdf. The DOE wind-power news: <http://www.energy.gov/news/6253.htm>.

Energy Information Administration 2009, <http://www.eia.doe.gov/> and <http://tonto.eia.doe.gov/energy_in_brief/power_grid.cfm>.

Jin, A.J., 2010. Transformational relationship of renewable energies and the smart grid. In: Clark, W.W. (Ed.), Sustainable Communities Design Handbook. Elsevier Inc., pp. 217–231.

Jin, A., et al., 2012. DC-module-based rooftop PV system design and construction. Proceedings of the 12th China photovoltaic conference, Beijing.

U.S. Energy Information Administration, 2009. Annual energy outlook March. Available from: <http://www.eia.doe.gov/oiaf/aeo/index.html>.

CHAPTER 22

The Regenerative Community Régénérer: A Haitian Model and Process Toward a Sustainable, Self-Renewing Economy

Carl Welty
Principle Architect, Claremont Environmental Design Group, Claremont, CA, USA

Contents

Background	528
Old and New Paradigms	529
Régénérer's Business Model	529
Claremont Environmental Design Group's Team and Methods	530
Ecosystematic Analysis	531
Regenerative Design	531
Geographic Information Systems and Data Management	531
Evaluation Metrics	532
Régénérer's 25-Acre Pilot Project	532
Agriculture	532
Water	533
Energy	533
Passive Design	534
Building Systems	534
Waste	535
Summary	535
Partners in Research to Build a Regenerative Haiti	536
Haitians and Californians Benefit and Learn from Each Other	537

Régénérer Haiti was initiated by the landowner, prominent Haitian, Georges Garnier, and developed in partnership with the California-based Regenerative Development Group, LLC, whose partners have been involved in sustainable developments through the United Nations and other international organizations. In 2011, the Regenerative Development Group reached out to the Claremont Environmental Design Group (CEDG), an architecture and landscape architecture firm known for its pioneering work in sustainable, regenerative design and integrated solutions to environmental problems. CEDG is now acting as the principal planners and designers for the project.

BACKGROUND

Régénérer Haiti goals to build a regenerative, plant-based business that helps replenish the local ecosystem will emulate nature's closed-loop systems through off-the-grid renewable energy sources, wise water management and soil restoration practices, and recycling of by-products for energy and building materials—and that moreover is sustainable within local social and cultural life. The business plan that will make Régénérer Haiti economically self-sustaining is to produce premium quality, organic, dehydrated fruit and vegetable powders. This process will permit economical transport and renders energy-intensive refrigeration unnecessary, increasing produce value and community earnings.

Régénérer Haiti plans for two sites: a 25-acre pilot project, which will be an agricultural development with a mixed-use component (known as *L'Avenir*, meaning "future"), and, about 5 years later, a full 2500-acre regenerative community and agricultural development of 50,000 people. The land for both sites is owned by Georges Garnier. The smaller site offers an opportunity for intensive analysis, experimentation, design, technical development, and evaluation to create a prototype for integrated solutions designed specifically for Haiti—knowledge that will then fuel the development of the larger site, as well as potentially the entire Central Plateau region of Haiti.

Régénérer itself is a for-profit development: part of its value, indeed, must be to demonstrate that innovations in sustainable design are viable financially. However, one of the most important goals is to allow the project to be replicable elsewhere in Haiti and beyond by means of systematizing our techniques of ecosystematic analysis and developing a Web-based data management plan.

This systematic approach will require an intensive research and design phase through strong partnerships within California's sustainable industries.

Régénérer can create a road map for a 21st-century sustainable economic development. The research phase, whose purpose is to make the advances of Régénérer replicable in different environmental and social situations, will make our project of widespread interest and literally transformative value. We will show how "green design"—development in harmony with natural ecosystems—can be more cost-effective than conventional development and be more resilient to climate change and its associated severe weather, while promoting environmental restoration.

OLD AND NEW PARADIGMS

Haiti is the poorest country in the Western hemisphere. High poverty, low literacy, poor community infrastructure, and meager cultural supports are worsened by widespread ecological devastation. The January 2010 earthquake catastrophically compounded this already-existing need for systemic rehabilitation. Although over $12 billion in aid has been spent in Haiti since the earthquake, 350,000 people remain homeless and thousands more live in perpetual poverty without the resources to build a better life for themselves and their families. Haiti reveals a clear global need to build more cost-effective, ecologically sustainable, and economically empowering communities. New methods are needed to escape the cycle of dependence on foreign financial aid "drips" that keep communities on life support but fail to build local capacity. A more comprehensive perspective can root out entrenched economic relations that perpetuate poverty and exacerbate climate change.

The 20th-century's development paradigm considered nature as a commodity—a source of "natural resources" to be extracted or used for profit—and neglected the complexity and interconnectedness of all life in nature. In contrast, we believe that real, lasting solutions are regenerative: they mimic nature's closed-loop cycles of self-restoration and balance, but centuries of an extractive mentality cannot be changed by words; concrete examples of the new paradigm are essential.

Haiti's stark poverty and environmental degradation provide an opportunity to demonstrate that these overlapping problems can be solved simultaneously. For instance, Régénérer's economic potential lies not only in creating opportunities for local prosperity and return on investment, but also in increased economic stability through improved resilience to severe weather. Economic stability, ecological balance, and local culture can, and must, develop together. The innovative planning power of ecosystematic analysis and regenerative design work in tandem to show that a community integrated with the local ecosystem can maintain long-term equitable employment, efficient use of resources, and pride of place.

RÉGÉNÉRER'S BUSINESS MODEL

The partnership between Georges Garnier and Régénérer Group has set ambitious new ideals for its business model. The development is committed to a "Quadruple Bottom Line": Economic Prosperity, Environmental

Regeneration, Social Progress, and Cultural Vitality. Régénérer Haiti—on both the pilot and the larger sites—will involve both agriculture and a mixed-use development to provide facilities for other sustainable businesses. The primary agricultural and processing business of Régénérer will be premium quality, organic, dehydrated fruit and vegetable powders. With a combined global market potential of $67B in 2014, target markets for organic dehydrated food powders and prepared foods include baby food, sports nutrition, and nutrition supplements.

In addition to the agriculture and processing business owned by Régénérer, the project will also include the beginning of a town—a mixed-use residential, commercial, and industrial community—that will attract associated, similarly minded companies that satisfy Régénérer's sustainability standards. Both sites are envisioned as "sustainable enterprise zones" (SEZ), maintaining and demonstrating a commitment to a diverse, sustainable enterprise. Targeted industries include organic prepared foods, organic farming products, natural fabrics/clothing, eco-building materials, universities, hospitals, technology firms, renewable energy companies, sustainable pulp/paper companies, sustainable finance, and community banking. The region's residents will benefit from career opportunities in construction, renewable energy systems, hospitality, education, manufacturing, finance, healthcare, and retail, as well as agriculture. The mixed-use residential development will be designed to promote a rich cultural and social life, with cafés, a cultural center, and local businesses.

CLAREMONT ENVIRONMENTAL DESIGN GROUP'S TEAM AND METHODS

The CEDG has assembled a diverse technical team to address economic, ecological, and sociocultural problems. Included in the analysis and master planning will be expertise not only in ecosystematic analysis and regenerative design, but also cloud-based data sharing and management, geographic information systems (GIS), evaluation procedures, and community engagement for inclusionary problem solving. The team includes select academic institutions and nonprofit organizations, including members of the John T. Lyle Center for Regenerative Studies faculty who will contribute to analysis and design protocols; the Advanced GIS Lab at Claremont Graduate University, which will develop software to document, synthesize, and manage data to optimize design solutions; and the Los Angeles chapter of the United States Green Building Council, which will aid in outreach and connect us to the broader community of southern California's green design industries.

ECOSYSTEMATIC ANALYSIS

The CEDG considers environmental, social, and economic goals as parts of a shared, functioning ecosystem: integration of these concerns into an efficient system enables us to insert human ecosystems into the larger whole in which all benefit from the growth of each part. It is a scientific study of the local and regional ecosystem that reveals the connections between multiple species.

It requires a region-wide perspective, because on-site resources are connected to broader systems. Thus, ecosystematic analysis includes the local watershed, generating strategies to reduce erosion, increase ground water recharge, and improve soil health, especially crucial in Haiti. It also includes an analysis of community practices and patterns, and economic and cultural needs. *Ecosystematic analysis precedes regenerative design.*

REGENERATIVE DESIGN

Regenerative design sits at the intersection of nature, society and technology. It creates human ecosystems in equilibrium with nature—buildings, compounds, and communities that generate more energy than they consume. Developed by John T. Lyle and others in the 1970s, self-renewing and biologically based regenerative design is rooted in ecosystematic analysis.

This research and regional ecosystematic analysis will provide comprehensive ecological data that will foster more regenerative agricultural communities in the Central Plateau. Moreover, the project will tangibly demonstrate how important it is to think comprehensively about the ecological context.

GEOGRAPHIC INFORMATION SYSTEMS AND DATA MANAGEMENT

The GIS is a computerized information system designed specifically for managing geographical data. It merges complex statistical analysis with mapping and can be immensely useful in clarifying relationships and correlations within a geographical area. When managed well, GIS can powerfully accelerate creative collaboration and innovation. GIS use in organic agriculture has been limited until now. Our team, supported by Claremont Graduate University's Advanced GIS Lab, will develop an information

system combining cutting-edge GIS, data management, evaluation protocols, and systems monitoring technologies. The system will:
- Allow all users and stakeholders to engage in the original vision
- Create a seamless collaboration between all parties involved in the project
- Coordinate data from diverse experts
- Permit ongoing adjustment to optimize performance
- Evolve with the occupancy and operation of the development

Transforming the data plan into a powerful digital dashboard will provide decision makers' data necessary to "drive" the development by connecting initial fieldwork to final design through synthesis, master planning, and building design to evolve better and more efficient second-generation projects.

EVALUATION METRICS

Measuring success entails a rigorous application of scientific methods to assess and improve program design, implementation, and outcomes. The master plan will include metrics to provide on-site users sharpened insight into original research and design choices, facilitating operational monitoring and allowing users to fine-tune systems in order to increase efficiencies. These metrics will also help in guiding the design of the 2500-acre site. Additionally, the evaluation procedure itself will establish a shared knowledge base that will encourage the building of future regenerative agriculture communities.

RÉGÉNÉRER'S 25-ACRE PILOT PROJECT

Agriculture

Ecosystematic analysis of the site, local watershed, and region will gauge the health of the existing environment and help us develop a plan for the 25-acre site to optimize agricultural performance in harmony with nature for long-term sustainable production. The first analysis will include erosion and soil stability in the local watershed and will develop recommendations for a low-impact erosion control infrastructure project (which will also benefit the farmers already located in the watershed). This research will establish and document the importance of retaining topsoil and maximizing groundwater recharge to increase the agricultural productivity overall. Agricultural systems will include:
- Soil rebuilding through leguminous cover cropping, organic soil amendments such as biochar, and organic matter composting
- Intercropping of organic produce with native habitat
- Terraced poly-culture farming

    Agroforestry
    Aquaculture
    Native habitat restoration
    Small animals for local consumption and to produce compost materials for soil development

Demonstrating these regenerative agricultural practices will benefit other farmers across Haiti. Georges Garnier's original vision included increasing the productivity of the entire region, and it is an important ideal of Régénérer Haiti to share this research through community engagement and outreach programs.

Water

Water is a pressing regional issue. The scope of our research goes beyond the site to develop policies that benefit the entire region and local watershed. In the first research phase, CEDG will document rainfall, extent of erosion across the local watershed, and availability and quality of groundwater. Policies and design strategies will include:

    Low-impact infrastructure to control erosion and increase groundwater recharge
    Conservation
    Diverting greywater to landscape irrigation
    Cisterns for collecting and storing rainwater
    Chemical-free treatment of groundwater for drinking
    Low-flow, efficient plumbing systems
    On-site wastewater treatment

Régénérer Haiti presents an opportunity to begin analysis of the complete regional watershed of the Central Plateau, to evaluate the realistic carrying capacity of the region, and to suggest policies to prevent extracting unsustainable amounts of groundwater—perhaps the first such planning policy in the world.

Energy

This regenerative agricultural pilot project will model a path to Haiti's energy independence. CEDG's phase one research will include strategies for decentralized and renewable energy generation, both local and on-site. As Haiti lacks a solidly functioning energy grid, decentralized self-generation is needed for long-term economic independence and ecological equilibrium. On-site power generation will come from a mixed package of systems to include:

    Solar panels
    Wind turbines
    Geothermal

Refrigeration through heat transfer into adjacent river
Small-scale hydroelectric
Burning agricultural waste and biofiber in controlled biochar furnaces that sequester carbon rather than releasing it
Converting collected plastic waste into energy, combining waste management with energy production

A healthy multiplicity of decentralized energy sources increases resiliency, consistency, and safety. In case of decreased production from one source, energy remains available from the other systems.

Passive Design

Building maintenance, heating, cooling, and lighting, accounts for over 30% of worldwide energy consumption. Often overlooked, the building shell itself can provide lighting and thermal comfort without consuming energy or increasing monthly operational costs. Incorporating lessons from nature and preindustrial cities, passive design can easily reduce energy consumption by more than 50%. Further, given the severe weather that climate change is already bringing, passive design allows us to maintain thermal comfort and safety independent of electrical sourcing. These design strategies include:

Ecosystematic analysis to understand the site's unique microclimate
Optimizing of buildings' siting and orientation to the sun
Taking advantage of prevailing breezes for ventilation and cooling
Locating windows properly
Providing elements to allow winter sun in and to keep summer sun out
Incorporating fountains and water features in hot, dry climates
Ventilation towers to increase natural ventilation and natural cooling

Incorporating these simple but effective techniques, the 25-acre regenerative agricultural pilot project will leapfrog conventional energy design standards of developed countries and show how Haiti can become energy independent by 2030.

Building Systems

In wealthy countries, premanufactured panelized construction improves quality, efficiency, and cost-effectiveness of construction. Leading-edge, computer-controlled systems provide considerable efficiencies: some manufacturers report constructing buildings six times faster and 30% cheaper.

In the first phase, through a strategic partnership with Enviro Board, Inc., CEDG will begin developing a panelized building system out of natural agricultural waste materials combined with other components. Sourcing from California will allow us to build "First World–Leadership in Energy and Environmental Design Standard" buildings cost effectively in Haiti with a moderately skilled local work force. The long-term goal, beyond that first phase, is to establish in-country manufacturing facilities that use local agricultural waste. This system will prove advantageous to the community in many ways:

    Empowering local citizens to build better buildings
    Increasing resource efficiency and energy efficiency
    Providing new jobs and training
    Promoting a new manufacturing industry in Haiti

This technology could prove transformative beyond the bounds of this project. A limited number of panel types can yield a variety of building designs adaptable to different ecologies and climate zones. Additionally, the panelized systems developed for the pilot project will also be usable for building communities in response to natural or man-made disasters. Pre-manufactured building systems can be integrated with the shipping containers in which they were delivered to assemble a "village" anywhere in the world quickly and efficiently, creating regenerative communities in response to disaster relief or other pressing needs. This system can be used as well for a single building or small complex, such as a school, orphanage, or medical clinic.

Waste

Waste management is a prime opportunity to design a "closed system" that emulates nature's efficient energy loops. Régénérer Haiti will direct waste back into the local energy cycle without polluting or diminishing the health of down-cycle species. Waste management practices will include:

    Composting organic waste for soil remediation
    On-site wastewater treatment
    Processing agricultural waste into building materials for local use
    Collecting or mining plastic waste to create energy or marketable products

Summary

The power of this unique project derives from the combination of project and process. CEDG's goal is to document the analysis and design process and create a road map to simplify the development of other regenerative projects. We seek philanthropic investment for the crucial task of systematizing

and documenting our ecosystematic analysis, planning, and design strategies. Funding will support:
> Phase I: data gathering, community engagement, and analysis
>> Ecosystematic analysis: fieldwork to collect diverse ecological information, and community engagement to understand the social/cultural background and needs. Combining, analyzing, and integrating this information to bring us to a design that will solve multiple problems simultaneously.
>> GIS data mapping and information management technology that will enable us to plan efficiently and on a large scale
>> Synthesizing information to develop regenerative design strategies
>> Developing evaluation matrices and procedures
>> Begin development of panelized building systems appropriate for the climate and geography
>
> Phase II: strategies and design solutions
>> Integrated watershed-level solutions that control erosion, increase regional groundwater recharge, reforestation, and soil remediation, and directly benefit local farmers and small property owners, reducing erosion and improving agricultural productivity
>> Research and initial development of a top-quality, energy-efficient panelized building system affordable in Haiti and other developing countries
>> With support from the Haitian government, this research can be expanded to become innovative sustainable design and development guidelines, establishing Haiti as an international leader in regenerative development and a model for energy and food independence
>> The broad applicability of these studies could even prompt California to improve upon sustainable guidelines and practices, building regenerative communities for a more sustainable integrated future

Through making our work applicable to a variety of sites, we want to create an opportunity for Haiti to become a model for development worldwide by achieving energy independence, food security, and environmental restoration.

PARTNERS IN RESEARCH TO BUILD A REGENERATIVE HAITI

Now is the time to bring 21st-century ecological analysis, geographic information systems, and manufacturing technology to the forefront to increase efficiency and to reduce the cost of philanthropic development.

California is home to an enormous wealth of talent, research, and innovation in the fields of technology and sustainability. These innovations can be channeled both to empower Haiti to build an economic and ecological sustainable future and to make California an international leader in ecosystematic analysis and regenerative design.

With over 30 years of applied research experience in the United States and abroad, CEDG is uniquely qualified to lead the effort. Previous work in ecologically appropriate architecture and land-use planning includes the following.

- Japan: Rokko and Ohi Energy Villages, prototype regenerative-designed communities for the Japan Federation Housing Organization. Authorized by the Ministry of Land, Infrastructure, and Transport for improving housing quality and the environment. Educated government agencies on the power of regenerative design, contributing to new legislative priorities and procedures for land development and affecting energy and environmental regulation and development codes. Development of this project utilized GIS systems.
- Vietnam: Bien Ho Orphanage, a 100-acre regenerative agricultural community for which CEDG received a sustainability award from the Boston Society of Architecture in 1999. An ecologically and economically self-supporting orphanage powered by renewable energy and served by green infrastructure. A change in political leadership prevented the project from moving into the construction phase.
- Windom, Minnesota: Shalom Hill Farm, an education and conference center developed as a response to community, economic, family, and environmental problems created by industrial agriculture. Shalom Hill Farm's program teaches regenerative, organic farming as a means to add value to the produce from small family farms, enabling them to maintain their way of life while improving air, soil, and water quality.

HAITIANS AND CALIFORNIANS BENEFIT AND LEARN FROM EACH OTHER

This visionary project will succeed because of a committed team that started with a local Haitian community leader and has expanded to include a large and diverse team of professionals and institutions in California. Both sides of this partnership will benefit from developing and implementing the ideas of the project.

Even before construction begins on the 25-acre pilot project, the benefits to the community of Hinche, to Haiti, and to the California team will include the following.

Creating jobs to construct low-impact erosion control and water capture system over the local watershed and restore local native habitat and reforestation of pilot project site and adjacent properties. Other jobs available directly from the pilot project include construction, hospitality, agricultural industries, food processing, and renewable energy systems. By employing ecologists, hydrologists, agronomists, and other scientific and technical specialists, the project can demonstrate the value of these specialists in other community-planning efforts.

Planning and documenting the process to develop Régénérer Haiti could be the basis for ecologically sustainable development guidelines for all of or part of Haiti.

Régénérer Haiti's business model will attract other green manufacturing and provide more diverse sustainable jobs. For instance, Enviro Board, Inc., a CEDG strategic partner located in California, is already interested in creating an assembly plant to build wall panels from agricultural waste in Haiti.

Green design and building industries in California will benefit from their involvement in planning, designing, developing technologies, and implementing the pilot project. The tremendous growth in on-the-ground, concrete knowledge will enable them to take the lead in developing regenerative communities and regenerative agriculture techniques that will be applicable to sustainable communities not only in Haiti but also in California and elsewhere. California green industries will benefit further from exporting services and materials to other developing regions.

The 25-acre pilot project for Régénérer Haiti promises to be transformative: a tangible, site-specific project in the Hinche area, started by Haitians in partnership with California industry. It will bring cutting-edge ecological analysis and information technologies to develop solutions and a replicable model for building other regenerative communities in Haiti. More broadly, it will demonstrate that "green by design" can address regional economic and environmental problems simultaneously.

CHAPTER 23

Microcities

Naved Jafry[1], Garson Silvers[2]
[1]Chairman, ZEONS Group, La Jolla, CA, USA
[2]CEO, ZEONS Group, Sustainable Developer, Beverly Hills, CA, USA

Contents

Microcities: Helping Mitigate the Rise of the Underground Society	540
How the City Will be Sustained Economically	545
The Importance of Implementing Good Laws	548
Infrastructure and Environmental Planning	550
The Health Aspect of a Microcity	553
Discussion	556

Fifty percent of the world's population now lives in cities. Cities now account for 75% of the world's energy consumption, 90% of the global population growth, 80% of all CO_2 emissions, and most of the world's economic productivity. Every year nearly 70 million people migrate from rural areas to the slums and ghettos of the world's cities. Most of these new immigrants end up in these illegal townships, where they collectively become part of the world's second largest, $1 trillion informal economy. Moreover, this trend does not seem to slow down, as impoverished immigrants are actually escaping a life of crime, poverty, disease, and a backbreaking lifestyle of the rural economy.

One of the several challenges for these host cities is that that they cannot catch up fast enough to develop an adequate infrastructure and public services to regulate and maintain an acceptable standard of living for all its residents. The cities' regulated free-market economy is quickly becoming an unregulated flea market. Most local residents now feel that their traditional culture is overrun by the new immigrants, creating serious social, environmental, and economical pressures. Even large and medium size legitimate trade and businesses are now impacted negatively with the unfair competition the informal sector enjoys from the tax, regulations, and standard labor costs it bypasses. Therefore, microcities, if established, can totally revolutionize our societies, our economies, and, more importantly, our cities, which we are connected to, while giving the legitimate world an effective tool to mitigate the underground society effectively.

MICROCITIES: HELPING MITIGATE THE RISE OF THE UNDERGROUND SOCIETY

When living conditions become unfavorable in rural parts of the world, most people consider opting for a new lifestyle. Unfavorable conditions may relate to economic, social, political, or security considerations. Imagine that you are among those who are impacted by these conditions but you do not have the necessary means to be able to pack your belongings in hopes of searching for a better future elsewhere. If people had the option of leaving their situation permanently, they would take it in a heartbeat, but voters in countries with favorable rules and living conditions will not let them settle there unless the new residents can contribute something substantial intellectually and economically. According to Neuwirth (2011), every year 70 million people are moving into cities, primarily squatter or shadow cities (Figures 1 and 2).

Should individuals be forced to become squatters and just be another item on their respective country's list of burdens? Neuwirth (2011) states that one billion people already live in shanty towns and by 2050, a third of

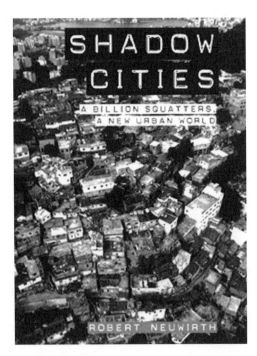

Figure 1 Shadow cities by Robert Neuwirth. Neuwirth shares his experiences visiting a vast amount of squatter areas and how the majority of urban populations are living in slums.

humanity will be living in these poor conditions. Being a squatter in itself requires an exasperated effort, as so-called residents run the risk of having their illegal homes removed (Figure 3).

These shanty towns and illegal townships are now exploited by violent criminals, corrupt corporations, and politicians to source their cheap labor and votes from. Many of these organizations operate partially or wholly outside the law by underreporting employment, avoiding taxes, ignoring product quality and safety regulations, infringing copyrights, and even failing to register some of their divisions as legal entities. Even developed

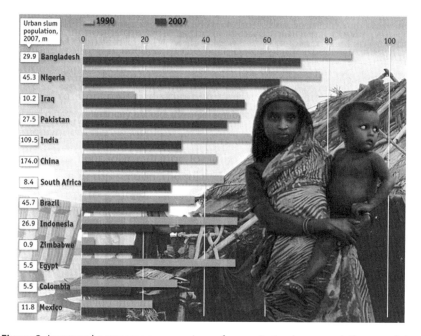

Figure 2 Increased percentage comparison of a country's urban population now living in slums during 1990 and 2007.

Figure 3 Villages like these are depopulating and moving to urban centers and settling into slums, such as the one in Mumbai India. *(Courtesy of Stewart Brand in 2006 TED talk.)*

economies such as Portugal and Italy have significant portions of their economy operating in the gray area. According to estimates from the World Bank, the informal economy comprises 40% of the gross national product (GNP) of low-income nations and 17% of the GNP of high-income ones. Economic sectors such as retail and construction have been known to be up to 80% operating in the informal or underground economy. Diana Farrell, the director at the McKinsey Global Institute, explains how the informal sector hurts economic growth by preventing larger, more productive formal companies from gaining a market share. The cost benefit of avoiding taxes and regulations often amounts to more than 10% of the final price. This advantage takes away market share from legitimate entities and discourages them from making important investments and bringing in new technology and sophisticated operating methods (Figure 4).

For self-preservation reasons, even several legal businesses choose willingly to operate in the gray area so as to prevent government and competitor's scrutiny and attacks. But sadly these very good companies also lose their ability to enhance their operations and finance their growth. As informal companies aren't legal entities, they lose their capability to raise or borrow capital from the legal economy and instead rely on illegal moneylenders that charge exorbitant rates and advance only small amounts. Informal businesses are also limited to doing business within their immediate underground circles. These underground networks of suppliers and customers are needed and are a necessity, as these illegitimate entities can't seek legal recourse during disagreements to

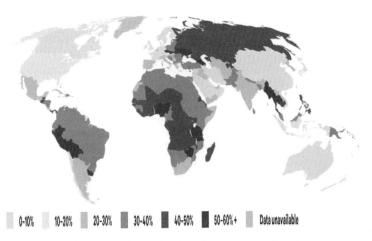

Figure 4 Colors on this map indicate the size of each country's underground economy as a percentage of its GDP. *(Source: Schneider et al. (2010).)*

enforce their contracts, protect property rights, or resolve disputes with the outside community within the legitimate legal systems (Figure 5).

The social costs of the informal society are also very high. Unlike workers of the formal sector, workers of an informal economy relinquish their rights to generous social security benefits and labor rights and earn, on average, lower wages and receive poorer health and safety protections. Moreover, consumers have less choice too. In developing countries, they can typically buy either very expensive, high-quality goods or services such as those found in rich countries or cheap, low-quality goods and services lurking with hazards and risks from shady businesses and informal enterprises.

Paul Romer, a renowned economics professor at New York University and former professor at Stanford University, formulated an idea similar to microcities called charter cities. Just like microcities, a charter city is a special reform zone, but on a larger scale with inhabitants sourced both locally and internationally. The reforms of microcities involve considering the needs of the new settlement, as well as supporting a set of rules that allow a modern market to thrive. Through the power of their operating contracts, microcities could make corporations and citizens accountable to their environment and their communities. Microcities ensure that there is better accountability all through the production, supply, and marketing chain,

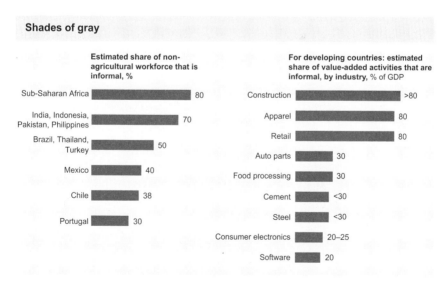

Figure 5 Percentage of nonagricultural workers in an underground economy by region and industry sectors. (*Source*: International Lobour Organaization; World Bank; McKinsey analysis).

while maintaining acceptable standards for security of life and property in the new city. Just like Romer (2010a,b) suggested, microcities could allow cross-national government partnerships to take place, as parent countries, cities, and communities transfer existing rules to new settlements as needed, hence having the residents abide by a new and more favorable set of rules, while still possessing local legitimacy (Figure 6).

One of the initial decisions to be made when building a microcity is a desired location. In his 2011 TED talk, Romer suggests that "The laws and rules of any city have to be made to attract people into the city." A microcity operates under a unique set of laws that is fitting to the lifestyle of the new settlement, but those laws should not conflict with the parent country's constitutional, state, or local laws. The new city is entitled to produce revenue in their method of choice, but a certain percentage of that money is given back to the parent country. In order for such a situation to be successful, the parent countries, cities, and communities (CCC) of the government must cooperate with the drafters of the microcity.

Key items in the creation of a microcity consist of laws and rules that must be abided by in order to maintain structure within the society. Distinct roles in participating countries, cities, and communities include a host, source, and guarantor (Charter cities, 2012). The host provides the land, while a source country has the potential residents that will move into the new city. The guarantor country ensures that the charter will be enforced for as long as the city is sustained. Other factors that play a crucial role in the creation of a city include implementation and creation of the necessary

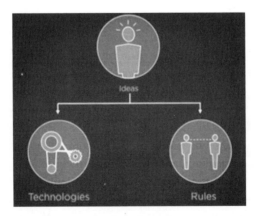

Figure 6 Components of a microcity. Rules and technologies fitting a new city's new ideas are major components of a successful microcity. *Inspired from Paul Romer's concept (Charter city 2012).*

infrastructure and the welfare of citizens through health, security, and financial well-being.

A society should be organized to serve and facilitate the growth and development of four primary groups such as those who defend us (lawyers, police, military), those who innovate (teachers, scientists, engineers, other intellectuals), those who trade (individuals involved in small business companies and firms), and those who support (blue-collar workers). Creation of a microcity implements these dimensions of society and makes certain that these groups can bring their ideas together to create a city that fulfills its needs.

HOW THE CITY WILL BE SUSTAINED ECONOMICALLY

It is imperative for microcities to attract and retain a diverse group of service, industrial, manufacturing, and knowledge creating firms and entities. By increasing its tax base through incentives and better tax enforcements, microcities can increase their competiveness further by lowering taxes, which in turns hinders the growth of the informal sector and brings the unregulated workforce into its system. It is no accident that developed nations have much more advanced and efficient tax regulatory and audit systems in place. Governments and city policymakers often forget that high taxes, complex tax systems and regulations, weak enforcement, and social norms actually breed more corruption, economic inefficiencies, and growth of an underground economy.

By merely collecting taxes from more companies, cities could enable a government to cut tax rates without reducing its tax revenue. In Turkey, for instance, McKinsey Global Institute found that the state collects just 64% of the value-added tax (VAT) revenue it is owed on retail sales. If it increased enforcement and collected 90%, the VAT rate could be lowered to 13% (from 18%) without decreasing government revenues. In many countries, the collection of retail value-added taxes is a good place to start, as it enables the government to gain information about the revenues of the companies that supply the retailers and therefore improves enforcement among suppliers as well (Figure 7).

Because cities with significant portions of informal sectors can provide the biggest opportunities in building a microcity, the cities' management must attract investors willing to build the infrastructure, such as roads, power system, airports, and buildings. Industrial, service, and manufacturing firms will need to be attracted as well, as they have the power to hire people who will move into the city in the first place. In Paul Romer's (2009) TED talk, he discusses that when employees are hired, their families will move to the city and become permanent residents, have children, get an education, and enter the workforce—this will result in an ongoing cycle (Figures 8–10).

Although investors and the private sector may provide financial means to build a microcity, they cannot shoulder all financial burdens. Aid from

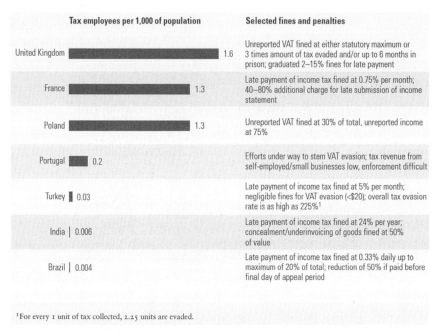

	Tax employees per 1,000 of population	Selected fines and penalties
United Kingdom	1.6	Unreported VAT fined at either statutory maximum or 3 times amount of tax evaded and/or up to 6 months in prison; graduated 2–15% fines for late payment
France	1.3	Late payment of income tax fined at 0.75% per month; 40–80% additional charge for late submission of income statement
Poland	1.3	Unreported VAT fined at 30% of total, unreported income at 75%
Portugal	0.2	Efforts under way to stem VAT evasion; tax revenue from self-employed/small businesses low, enforcement difficult
Turkey	0.03	Late payment of income tax fined at 5% per month; negligible fines for VAT evasion (<$20); overall tax evasion rate is as high as 225%[1]
India	0.006	Late payment of income tax fined at 24% per year; concealment/underinvoicing of goods fined at 50% of value
Brazil	0.004	Late payment of income tax fined at 0.33% daily up to maximum of 20% of total; reduction of 50% if paid before final day of appeal period

[1] For every 1 unit of tax collected, 2.25 units are evaded.

Figure 7 (*Source*: Economist Intelligence Unit; Organisation for Economic Co-operation and Development (OECD); tax authorities of countries shown; McKinsey analysis).

Figure 8 Shadow cities by Robert Neuwirth. Neuwirth shows that informal economies in the slums are now collectively the second largest economy in the world.

developed countries alleviates financial loads, but trade allows the new settlement to sustain itself and restore dignity to its people. It is true that international firms collecting fees for their services may possibly provide a large portion of the new settlement's infrastructure, but the city's development authority must finance the remaining public services. This includes border security, police, firefighting, courts, and other necessary services. Usually cities rely solely on income and property taxes to generate the funds needed for these operations. However, microcities employ a different approach. The

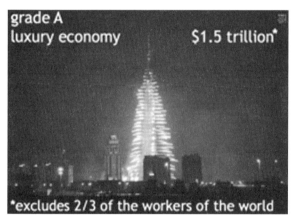

Figure 9 The shortfall of the legitimate economy, which excludes two-thirds of the workforce.

Figure 10 Major companies such as Proctor & Gamble derive 20% of all its revenues from the informal sector. It could be ideal for companies such as these to position themselves into the microcities of the world without supporting the underground economy.

city may make long-term leases to private developers, causing rent to fluctuate parallel to land value. The price of rent will increase as land value rises (Fuller & Romer 2010). Such a method would not only provide income for the city's government, but also give the government an incentive to maintain the city. Maintenance ensures that living and working conditions remain attractive since revenues rely on the city's land value (Romer 2009).

THE IMPORTANCE OF IMPLEMENTING GOOD LAWS

When a microcity is built, the foundation of its laws originates from the country, city, or community in which it is built. Creating effective laws gives residents the option to choose the lifestyle they prefer. Administrative laws should be simple enough that they can be interpreted properly by law enforcement and citizens. Siegel (1983) insists that tax forms, credit agreements, health care legislations, and other laws are incomprehensibly long and strongly suggests a sensible approach to simplifying legal paperwork. Regarding trade laws, using the corporation's purchasing power of their contracts, companies can now keep the global supply chain honest, as they have a moral obligation to society. These contracts could be instrumental in delivering human rights and safe goods and services, effectively having corporations take the place of failed governments. In general, laws should be kept straightforward in order for citizens to trust and abide by them. Laws should impact society as a whole to prevent citizens with hidden agendas from finding loopholes in litigations with teachers, doctors, and other professionals. Kee-Cheok Cheong (2010, p. 165) discusses the effectiveness of aiding parent countries by using charter cities: "Aid can be broken down in four elements: changes must be taken in a smaller scale to have a chance to succeed; rules must be transparent and have incentives; the residence has the power of choice; and the government of the developing country must have the power of choice."

Laws are the foundation of any settlement, especially a country. Without laws, the world would be in a state of anarchy. Ring et al., (2005, p. 308) make four arguments to address this issue:

> The extent to which government action can foster industry creation and economic development, the impact of corrupt governments on firm-level decision managing by management of multinational enterprises, the concept of attractiveness of political markets and the impacts they have on a firm-level strategies, and how deregulations can affect the governance mechanisms of firm.

Government is essential at both the federal and the local level. Therefore, it is imperative that a microcity establishes a strong government before it implements any laws. Although the government will never reach perfection, it must function to the best of its ability in order for the microcity to become successful.

The uniqueness of microcities lies in the interpretation and implementation of efficient laws with strict penalties and consequences for businesses facilitating corruption. The Romans believed that local laws could be implemented without affecting the Roman Empire negatively; this is where the idea of a city was conceived. The cities governed themselves but were required to remain within the legal parameters of the Roman Empire. The laws created in a municipal were called "municipal charters." This laid the foundation for American law. Keller (2002, p. 57) later states, "Local governments in the U.S. are creatures of the state government; charters are like laws made in the city; and cities can practice some home rule."

A microcity may be placed anywhere as long the parent nation's government approves. Third world countries would be the best places for these cities to thrive, such as the Philippines, India, and Thailand. Third world countries have a common denominator—their governments are not strong. India is an example. Although structure is present in India, the country has a major problem with corruption. Quah (2008) states that in 2007, nearly a quarter of the elected members of congress were charged with crimes, including rape and murder to name a few.

In order to alleviate this corruption issue, India can create several microcities. First, drafters will need to find land that is able to occupy at least 100,000 people. This size allows the city to make a profit for the state. The space cannot be too large because it will be more difficult to manage. As stated by Cheong (2010, p. 165), "Changes must be taken on a smaller scale to have a chance to succeed." Implementing a government provides structure within the city. Once a government is settled, laws may be implemented—this is the foundation of the city. The charter commission must be strong enough to implement the correct laws and rules in order for the city to thrive. Hassett (2011) discusses the process involved with this tedious endeavor and explains the position and responsibility of the commission. Laws will make or break a city. Once all items are agreed upon, the city can be built to prosper.

The government should work hand-in-hand to enforce laws. In "police, prosecutors, and judges," Kremel (1958, p. 43) states, "[t]he law, however, is ineffective without enforcement." In other words, without law enforcement, the law made by the government would have no meaning. Law enforcement is created to deter people from breaking the rules. Security

Figure 11 Robert Neuwirth shares his study on how respectable companies such as Siemens paid $1.9 billion in bribes during a period between 2001 and 2007, making a case for better accountability among governments and corporations for the benefit of all.

personnel must be noncorrupt and well trained because if they cannot do their job, then the security force is ineffective and unnecessary. They have to maintain a balance of how to enforce the law since they want to win the "support of the people." This would make their job safer and less troublesome. Once the government and law enforcement are established, the city can start accepting new residents. (Figure 11).

Cheong uses Hong Kong as an example of a successful charter city. Hong Kong was once ruled by a combination of the British and Chinese governments. Since Hong Kong was mainly under British control, they did not adopt Chinese governmental tactics. Instead, they utilized the United Kingdom's structure of government. The laws and way of life in Hong Kong were mostly derived from the British, which attracted residents from mainland China. By operating this way, Hong Kong gradually became one of the richest cities ever created. A certain percentage of their revenue was given back to mainland China. During the 1990s, the government of Hong Kong switched gears when the city was turned over to mainland China. Laws, policies, and holidays were changed when the Chinese declared possession of Hong Kong. Some residents of Hong Kong opted to leave because they initially sought refuge from the Chinese government. Turmoil was present in the beginning, but Hong Kong was able to break free again and become the city it is today.

INFRASTRUCTURE AND ENVIRONMENTAL PLANNING

Modern microcities need to be compact, so they must grow vertically (Bhaskar 2010). Since microcities need to be built in a timely manner rather than evolving over a period of time, there should be an extensive use of fast building materials such as steel, glass, and prefabricated structures. Citizens should be able to design their own communities before construction takes place while keeping cities sustainable under required building codes. Wanns and Chiras (2003) suggest that when designing a neighborhood, drafters must create a community with zero carbon emission. Instead of having private vehicles, more emphasis should be placed on public transportation. A small-scale city allows residences to be closer to work and school settings. By having a smaller city, residents will now be able to sustain their everyday lives without impacting the ecosystem dramatically.

With climate change, skyrocketing energy costs, and a weak economy on people's minds, Van Jones' (2008) book, *The Green Collar Economy*, addresses ongoing issues of social inequality. He discusses the environment and arrives at large-scale solutions that focus on improving the "greenness" of individual corporations. By examining case studies of prospective companies' green initiatives and their effects on marketing and consumers, Jones demonstrates how going green can be a win–win situation for both the bottom line and the environment, hence making the case for microcities, as it provides the perfect opportunity to implement all innovative designs, that could sustain the green collar economy.

A major challenge that a microcity faces is being able to have an efficient and advanced infrastructure that brings both industry and people into the new area (Bhaskar 2010). Typically, people flock to places with job openings. However, jobs are created by businesses only when there is good infrastructure with rules favoring trade and security. Potential residents will not move until the microcity has reliable hospitals, health centers, schools for their children, and other necessities. The success of a microcity will depend on the methods in which it is managed and promoted. This requires a strong administrator who works like a city's chief executive officer, similar to the job description of a mayor.

By implementing modern sustainable town planning, microcities can become benchmarks for other conventional cities. Microcities can reverse the flow of migration as underdeveloped CCC have been losing significant portions of their populations to mass immigration. Fan and Yakita (2011) argue that brain drain hurts growth in these communities as intellectuals move across state lines in search of better, well-planned townships across the developed world. As cities age, they become less attractive, property prices may plummet, stakeholders lose, and voters may become clouded by delusions. A microcity can reverse all of those challenges through its modern, improved, affordable, and durable infrastructure that stimulates growth and investments. These factors have significantly slowed the external flow of key populations around the world.

Figure 12 Newer technologies such as foldable electric cars and bike share programs could be used to make more compact and cleaner cities possible.

Figure 13 An artist's conception of how the same living space can be used for multiple occasions to make urban living more compact while reducing the CO_2 footprint and benefiting the economy and ecology.

Designing cities with the idea of sustainability should be implemented during policy making. Sometimes the very difference between first and third world countries is infrastructure. As infrastructure creates the space for growth and development, innovators and founders of several sustainable organizations, such as Alex Steffen, discuss how cities have the potential to save the future. Steffen (2012) sheds light on sustainably designed, neighborhood-based green projects that will expand our access to a higher standard of living, while reducing the time spent in vehicles. He emphasizes the importance and urgency of reducing humanity's ecological footprint as the Western consumer lifestyle spreads to developing countries (Figure 12).

Implementing eco-friendly designs into microcity planning would set great examples for future city developments. Danish architect Bjarke Ingels is another champion of eco-friendly designs. In his TED talk (2009) he demonstrates that his buildings not only look like nature, but also act like it. His buildings block winds, collect solar energy, and create stunning views. Architects of his caliber usually implement a hands-on, ground-up understanding of the needs of the building's occupants and surroundings, while taking in considerations of the environment (Figure 13). Advocates such as Majora Carter (2009) demonstrate how environmental and social activism

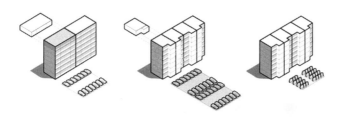

Figure 14 The effects of conventional and compact communities compared and how compact cities and cars can be made possible in the near future.

can influence policy changes on present and future city planning. As a founder and executive director of the Sustainable South Bronx, Carter devotes her life to environmental and economic justice for disenfranchised communities. She redefines the field of environmental equality by leading several local economic development movements across the United States. Carter brings several stories of people who are saving their own communities while saving the planet, calling it "hometown security." Pioneers such as Carter are key players in promoting and including ideas of environmental sustainability in city planning through local entrepreneurial and local governmental supervision.

Overall, microcities are here to stay. The demand from both external and internal migration of more than six million people a month has posed a serious challenge to city administrators all over the world. In looking at the bigger picture, Van Jones' idea of a green collar economy calls for a sustained rebuilding of infrastructure and creating alternative energy sources, which would boost the economy through increased employment and higher wages while decreasing our dependence on fossil fuels (as cited in Morris 2008, p. 73)(Figure 14).

THE HEALTH ASPECT OF A MICROCITY

It is also important to understand the essence of health care because of the large role it plays in the lives of those inhabiting the microcity. Because of the alarming growth rate, microcities need an efficient health care model.

Governmental programs, such as Medicare, are not accessible to everyone. In this day and age, one must meet a certain threshold in order to prove they qualify for health services. This notion instills a sense of unfairness, ultimate dependency, and helplessness. In a microcity, residents will not have to experience this. This is why home-based health care is a suitable alternative to traditional health care. Home-based health gives individuals the power to control their health. This feeling is also important in stabilizing one's morale in the patient's environment.

According to Searles (2011, p. 139), "sickness does not carry a passport." Neglecting underserved populations increases the risk of contamination and the spread of disease for all persons in the community. It is crucial that physicians "take the Good Samaritan approach, no matter what the circumstances are; compassion must rule legalism" (Searles 2011, p. 139). In order for health care to reach underserved populations, microcities must review developed nations' records, separate which items work in each health care system, implement the necessary services that are fitting to the new settlement, and increase access.

Alternatives mentioned in the past may not work. Universal health care sounds ideal and fair to everyone, but rising costs make this approach seem unrealistic. Reducing disease also sounds like a solution, but might be too time-consuming and impractical.

Moreover, it is important to examine locations that can serve as a model for these microcities. In Japan, it is illegal to be obese. Japanese citizens take major preventative actions to ensure the health of their population. Although Japan is a country, it is essential to fit preventative laws into a city, specifically a microcity. Examining how health works on a larger scale, such as Japan, can foster a stronger health system. Japan introduced the "metabo law," which implements the prevention of obesity through certain health precautions. Their "metabo law" is cost-efficient by reducing other health risks related to obesity. Japan also ranks highest in life expectancy (Yamataga et al. 2008). The use of safety screenings prevents disease before it becomes worse. Preventive care is just as essential as reversing the condition.

In "Improving Economic Equality and Health: The Case of Postwar Japan," Bezruchka and colleagues (2008) discuss how Japan overcame poverty after World War II and became the country with the highest life expectancy in the world. This is a noteworthy article because it identified the link between a country's mortality rate and a country's income level. They further state, "changes in a society's economic hierarchy can have profound health effects," and "Japan's example is remarkable" (Bezruchka, Namekata & Sistrom 2008, p. 593). These interesting Japanese values serve as an example of measures to increase health and longevity. However, other successful cases unique to their country should not be overlooked.

Other successful models include NxStage and Philips Healthcare (formerly Philips medical systems). NxStage has created over 17,000 health care homes, all providing cost-efficient rates and quality care (Nagpure & Prashant 2008). A microcity values the importance of cost-efficiency and adequate levels of quality. Home-based health care may not replace traditional health care, but may enhance it. Clearly, an efficient health care system must be incorporated into the city to sustain the population. This system will best support the population because health services will be more sporadic and, in turn, there will be less reliance on hospitals. Patients and those in need of these services will access health care when they need it, and it will always be of high quality.

Searles (2011, 146) states further, "illnesses are not stopped by borders." Impoverished people are dealing with numerous barriers in their search for medical care. Sadly, those barriers are man-made but can always be brought down if someone is willing to do so. If those barriers are confronted by a microcity, new development will allow medical care to be delivered in ways that are fitting to citizens' needs. Microcities will be able to demonstrate efficiency in health care to the rest of the globe (Figure 15).

Figure 15 These images document dangers to health and wellness if a proper infrastructure for sanitation and pollution control is not put in place from the very beginning.

DISCUSSION

Microcities can aid in the growth of a country, but it is crucial that appropriate laws are enacted and enforced in order to attract potential residents. Without residents there will be no consumerism, which hinders the possibility of a booming economy. More homes are needed to house the rapidly growing population. This microcity is not just a mere idea, it can saves lives and better the economy. Through the approaches listed earlier, a fresh, efficient lifestyle can ensure sustainability. When cities are governed by a good set of rules "They can be cities where people are safe from crime, safe from disease and bad sanitation, [and] where people have a chance to get a job" (Romer 2009). The concept of a microcity is gradually gaining acceptance in India. If all goes well, India should have at least 30 private cities across the country by the end of this decade. This number could increase, depending on the manner in which India's policymakers allow this concept to germinate.

The concept of microcities is a promising and symbolic movement of our time. It is a new lifestyle with a transformed vision of the future, consisting of fitting laws and regulations, a healthy environment and population, and the financial means to sustain its existence. It is where the present and future well-being of humanity, as well as the environment, is a promising one. This revolutionary idea has the potential to impact communities around the world in profound ways. If the world is committed to a behavioral change, it will not be long until we live in cities whose residents experience high levels of satisfaction (Figures 16–18).

Figure 16

Figure 17

Figure 18 If a commitment is placed to build these microcities, a transition can happen from slums to mega cities, just as Dubai was transformed from the desert to a super city through oil. The slums of the world can use human energy to transform.

REFERENCES

Bezruchka, S., Namekata, T., Sistrom, M.G., 2008. Improving economic equality and health: the care of postwar Japan. Am. J. Public Health 98 (4), 589–594.

Bhaskar, R., 2010. The new cities of India. Available from: < http://forbesindia.com/printcontent/19662 >.

Bjarke, I., 2009. Bjarke Ingels: 3 warp-speed architecture tales. [video file]. Available from: < http://www.ted.com/talks/bjarke_ingels_3_warp_speed_architecture_tales.html >.

Carter, M., 2009. Greening the ghetto. Angelican Theological Rev. 91 (4), 601–605.
Charter cities, 2012. Available from: < http://chartercities.org/concept >.
Cheong, K., 2010. Charter cities: an idea whose time has come or should have gone. Malaysian J. Econ. Stud. 47 (2), 165–168.
Components of a charter city. Available from: <http://www.urenio.org/2009/11/21/video-paul-romer-on-charter-cities/>. (30.03.12.).
Dubai. Available from: < http://ssqq.com/archive/vinlin23dubai.htm>. (30.03.12.).
Fan, X., Yakita, A., 2011. Brain drain and technological relationship between skilled and unskilled labor: brain gain or brain loss. J. Popul. Econ. 24 (4), 1359–1368.
Fuller, B., Romer, P., 2010. Cities from scratch. City J. 20 (4).
Hasset, W.L., 2011. Reviewing a city charter. National Civic Rev. 100 (2), 48–57.
Jones, V., 2008. The green collar economy: how one solution can fix our two biggest problems. Harper One, New York.
Keller, L.F., 2002. Municipal charters. National Civic Rev. 91 (1), 55–62.
Kremel, F.M., 1958. Police, prosecutors, and judges. Ann. Am. Acad. Political Social Sci. 320, 42–52.
Morris, A., 2008. Taking leadership in the green economy: a conversation with Van Jones. Harv. J. Afr. Am. Public Policy 14, 73–78.
Nagpure, P., 2008. Home based healthcare: issues and challenges. Available from: < http://dspace.mit.edu/handle/1721.1/45232 >.
Neuwirth, R., 2011. Stealth of nations: the global rise of the informal economy. Publishers Weekly 258 (34), 58–59.
Quah, J., 2008. Curbing corruption in India: an impossible dream. Asian J. Political Sci. 16 (3), 240–259.
Ring, P.S., Bigley, G.A., D'Aunno, T., Khanna, T., 2005. Perspectives on how governments matter. Acad. Manag. Rev. 30 (2), 308–320.
Romer, P., 2009. Paul Romer's radical idea: charter cities. [video file]. Available from: < http://www.ted.com/talks/paul_romer.html >.
Romer, P., 2010a. For richer, for poorer. Available from: < http://www.prospectmagazine.co.uk/2010/01/for-richer-for-poorer >.
Romer, P., 2010b. What parts of globalization matter for catch-up growth. Am. Econ. Rev. Papers Proc. 100 (2), 94–98.
Schneider, F., Buehn, A., Montenegro, C.E., 2010. New estimates for the shadow economies all over the world. Int. Econ. J. 24 (4), 443–461.
Searles, C., 2011. Beyond health care reform: immigrants and the future of medicine. Ethnic Racial Stud. 35 (1), 135–149.
Shadow cities by Robert Neuwirth. Available from: <http://www.21stcenturychallenges.org/focus/robert-neuwirth/>. (30.03.12.).
Siegal, A., 1983. Plain English. Vital Speeches Day 49 (8), 249–252.
Steffen, A., 2012. How to save the global economy: build green cities. Foreign Policy 191, 1–3.
Urban population living in slums. Available from: <http://www.economist.com/node/15766578>. (30.03.12.).
Wanns, D., Chiras, D., 2003. Superbia!: 31 ways to create sustainable neighborhoods. New Society Publishers, Gabriola Island, BC, Canada.
Yamagata, K., Iseki, K., Nitta, K., Imai, H., Iino, Y., Matsuo, S., Makino, H., Hishida, A., 2008. Chronic kidney disease perspectives in Japan and the importance of urinalysis screening. Clin. Exp. Nephrol. 12 (1), 1–8.

CHAPTER 24

Conclusion

Woodrow W. Clark II
Qualitative Economist, Managing Director, Clark Strategic Partners, Beverly Hills, CA, USA

Contents

Lessons Learned from Developed and Developing Nations	560
Economic Themes, Strategies as Opportunities for Renewable Energy	561
Social Capitalism	562
Sustainable Development	563
Agile Systems	564
Infrastructures	564
Renewable Energy	565
Finance	566
Knowledge Capital	567
Conclusions as Opportunities	568

This book covered the basic issues and needs for creating sustainable communities from a variety of examples around the world. Economic changes are critical today as a result of the global economic crisis that started in the United States in October 2008. Now with a Green Industrial Revolution (GIR) fully moving around the world, economic changes are applied to finance sustainable communities and a very new and multidisciplinary dynamic is taking place. In *The Next Economics* (Clark 2012), there were chapters on how to finance the GIR.

Facts are that climate change has gotten more and more destructive, resulting in a great intensity of weather changes that destroy land, kill people, and paralyze entire regions and nations. Costs around the world today are astonishing and growing more each day. Because there is no value that can be placed on an individual's life, the issue is *not* to adjust or even tolerate climate change. People at the local level need to take action *now*. They cannot afford to wait for national governments or even international organizations to act. It is too late. Local sustainable communities are a key element in preventing and reserving climate change.

LESSONS LEARNED FROM DEVELOPED AND DEVELOPING NATIONS

The California energy crisis produced many lessons that were learned (Clark 2001, 2002). These lessons need to be passed on to other developed and developing nations (Clark & Bradshaw 2004). Three critical lessons were noted. First is the need for reform to be clearly defined. Also, as other nations and states such as California discovered, there needs to be a continuing role that is reflected in the concept of civic markets between public and private sectors (Clark & Lund 2001)—be that role one of creating regulatory schemes, programs, or governance, it must be legally defined and in operation. Neither market forces alone nor government in actions can be tolerated in a globally competitive work for key sectors such as energy. Market manipulation of supply, as well as illegal financial actions, has stopped most deregulatory and privatization actions in developed nations (Clark & Demirag 2005).

Second, there are inherent differences and serious conflicts between the national government and almost all regional, city, and states within a nation. California is an excellent example of this. For example, it established a Renewable Energy Standard (RES) of 20% by 2017. Eighteen other states did so as well, but the national government in Washington did not and, even if it did, the percentage of renewables would be lower for the nation as politicians tend to negotiate to "lower" rather than higher regulations and percentages. Moreover, California under a new governor even expedited the time frame, moving it to 2013, which has not been achieved. Instead, the time frame has been pushed off for another decade to 2023.

The issue of areas is also regionalism, which goes beyond states versus nations. In the United States, California is part of the western states versus south, middle, and eastern regions. Each is different. Moreover cities, communities and counties are subsets of states, and provinces have established their own "green cities" alliances and even more aggressive renewable portfolio standards. In southern California, Santa Monica has declared itself "green," whereas Los Angeles has a RES of 20% by 2010 and Santa Monica will be "fossil free by 2030". The former claims to have met that goal but it is very questionable and lacks verified data, but Santa Monica appears to be headed toward its goal.

Finally, there is the issue of long-term commitment—be it in public policy, finance, or simply regulations. Government and business alike need to know if there will be policies in place along with funding so that they can plan. Without long-term financial commitments, neither public nor private sectors can plan and implement. This key reason why the "civic market" approach to policy making and planning is important is demonstrated by a variety of decisions to get California through its energy crisis (Clark & Bradshaw 2004; Clark & Morris 2002).

When Clark and Lund (2001) first articulated the concept, much theory on public–private partnerships existed. However, few practical examples were explored and rarely put into practice. The California energy crisis provided that need and application. There was no way that the state would get through the crisis without such collaborations and mechanisms. There needed to be consensus through constructive debate, confrontation, and dialogue to construct public policies and then implement them. The basic solution is for any community to have a plan and then the funding to make it happen. In agile energy systems, Clark and Bradshaw (2004) make the point that there is not just one way to generate and distribute power, for example. Central plants are needed, but so are on-site and local power on buildings and areas that could be used to produce power through renewable energy power generation.

ECONOMIC THEMES, STRATEGIES AS OPPORTUNITIES FOR RENEWABLE ENERGY

When lessons learned in developed nations and in Brazil, Russia, India, and China are compared, applied, and implemented in other nations, for example, a number of themes appear. As the chapters in the book make clear, there are different ways in which "sustainable communities" are defined. Thus one of the most critical themes is a definition of terms and numbers (Clark & Fast 2005, 2008). This is not just a matter of language translation, but goes even deeper in understanding basic ideas. For example, one of the most often used terms in renewable energy today is "market." The term means different things to different people who are even native speakers of the same language.

English speakers from the United Kingdom and the United States often have different meanings attached to the word, which can be confusing to native language speakers in China and other countries. However, even among native English speakers, the concept has different meanings. One interesting example as seen in the book are that many economists see market forces as open and free business arenas in which to compete without government interference and regulation. While this is a particular neoclassical economic argument, it does not exist in reality anywhere. And when tried, it has failed sooner or later, as the current global economic crisis demonstrates.

The United Kingdom and the United States are prime examples of promoting this definition of market forces, but practicing something very different. For Americans particularly, there are market forces or businesses that have both economic and political domination in sectors. But, more importantly, developed countries such as Germany, France, Nordic countries, and others have "government-driven markets" where goods and services are funded, procured, and demonstrated well before the "mass market." In fact, this "market" model is also true of the U.S. government's massive debt mechanism used to purchase goods and services, many of which are *not* and perhaps should *not* (for military purposes) be in the mass market, as they are clearly a different definition and meaning of the market than what most economics would expose.

A clear basic definition of renewable energy generation, technologies, policies, and programs is needed in any nation–state for providing guidelines in finances and public policies. The government market provides structure and certainty in its ability to stipulate market rules, regulations, and standards in order for the business to respond through procurement, demonstrations, and, finally, mass-market business opportunities. Moreover, as noted earlier, public policies and economic mechanisms of nation–states will differ from one another and within their own regions, cities, and communities. Nonetheless, at one level there must be clear and basic definitions.

From this book and in almost all chapters, seven themes emerge in any effort to achieve sustainable communities.

Social Capitalism

Social capitalism (Clark & Li 2003, 2010, 2012) itself is the key theme for nations, regions, and communities, but in a different manner by most developed nations (Clark & Li 2003). The exception is that the People's Republic of China (PRC), Germany, and Nordic nations share the same theme in terms of both definitions and policies. In short,

> *Social capitalism is conceptualised as a hybrid or combination of the private market mechanisms characterised by the ownership of common industries for the good of all citizens. The three forms of global industrialisation do not reflect this new economic paradigm. When the state considers what it must "control" for the good of all citizens, social capitalism becomes very apparent: water, waste, environment and energy. Some social capitalist countries might also argue that medicine and education must be guaranteed to all citizens as well.*

For example, the current debate in most countries over the telecommunications or digital infrastructure focuses on a similar economic debate: can any country have a 'digital divide' between its citizens—the haves and have-nots—dependent on whether someone can afford the technology? The energy crisis in California can be viewed in the same manner: can some citizens (or even businesses and other infrastructure sectors) afford power while others cannot? Social capitalism would agree that denying its citizen's power, for example, based on market or other forces should not exist (Clark & Li 2003, p. 8).

Definitions of social capitalism are critical for communities and nations because it is a major economic paradigm shift:

What the social capitalism paradigm argues is, that states or governments cannot be "invisible" or leave certain societal areas and sectors open to "market forces." Government must be active and even protective in certain areas that impact on all citizens, including businesses and new enterprises. Government's role is to provide guidance through some regulation, oversight and investment stimulation policies and programs (Clark & Li 2003, p. 10).

Basic definitions are one reason why the third theme in renewable energy financing is focused on "sustainable development" (SD) itself. The concept is often expressed as a key theme but for the United States it means a focus on environmental concerns, whereas it often means economic development in other nations. The two concepts and definitions can be totally opposite in practice and a cause of considerable concern. While not reviewing all the definitions of SD, there does appear to be more and more consensus globally (Clark & Lund 2007).

Sustainable Development

Second, sustainable development is very much a comprehensive theme in that like its original usage in the Brundtland report (1987), it retains the more general meaning of:

The objective of sustainable development and the integrated nature of the global environmental/development challenges pose problems for institutions, national and international, that were established on the basis of narrow preoccupations and compartmentalized concerns ... The challenges are both interdependent and integrated, requiring comprehensive approaches and popular participation.

The concept of sustainable development provides a framework for the integration of environment policies and development strategies—the term "development" being used here in its broadest sense. The word is often taken to refer to the processes of economic and social change in the Third World. But the integration of environment and development is required in all countries, rich and poor. The pursuit of sustainable development requires changes in the domestic and international policies of every nation.

As a minimum, sustainable development must not endanger the natural systems that support life on Earth: the atmosphere, the waters, the soils, and the living beings.

Agile Systems

A third theme that has arisen in the energy literature is "agile systems" (Clark 2007). The reform of the energy sector has numerous results. Sinton et al. (2005) argue for one model, while others propose others for China. If there are lessons learned from developed nations about reform as being the extreme of deregulation, privatization, or liberalization, there are very distinctive results that have already begun to form its own unique energy infrastructure model. For one thing, the China model is based on its philosophical approach to everything: social capitalism. That is the concern of every citizen's social welfare instead of leaving energy or any sector wide open to private individuals or businesses.

While this basic approach to energy or any other infrastructure does not follow developed nations, it certainly benefits from their mistakes. The goal of energy independence or security applies to regions and communities, as well as nations. The conflicts and wars today over increasingly scarce natural resources caused by "oil and gas," for example, are bound to get worse. These costs, let alone those for reversing global warming and climate change (if that is possible), will take an entirely new "paradigm" from that of developed nations thus far.

A combination of on-site power through distributed systems using renewable energy power and a central power plant are critical. This "agile" or flexible combination of both central plant power and local power allows sustainable communities to be carbon neutral and limits their impact on the environment and atmosphere. When communities have the capacity to create power from renewable energy source, then their local demands are met but with sources that are part of their local plans, systems, and economics for infrastructures and related areas such as waste, water, transportation, and Wi-Fi with smart local grids to make them efficient, cost-effective, and conserve the need for energy.

Infrastructures

Infrastructures are a fourth theme that needs to include renewable energy generation in their calculations and systems. Models and software need to be used in such calculations. DeLaquil and colleagues (2003) used the

MARKAL model specifically to look at China's energy needs. They were able to analyze China for a 55-year period of time in terms of energy demands and alternative future scenarios compared to a "business as usual strategy that relied on coal combustion technologies (no matter how advanced) [which] would not enable China to meet all of its environmental and energy security needs" (DeLaquil, Wenying & Larson 2003, p. 40).

What is interesting about this model and others (for the META NET economic model, see Barry 1996, 1999; Hong & Möller 2012; Hong, Lund & Möller 2012; Isherwood 2000; Lund 2009; Lund & Mathiesen 2009) is that they provide baseline data from which to plan. In particular, they look at the variations of needs for infrastructures that include energy along with other structures such as water, waste, and transportation. The California Commission on the 21st Century Economy did that in 2002 in order to set goals and strategies for the needs of the entire state (California's Commission on the 21st Century Economy 2002).

Then, in 2005, the state did the same thing by focusing on a hydrogen highway roadmap (California Hydrogen Highway Roadmap 2005). When looking at the financing of infrastructures, the civic market approach is a key organizing and implementation element. From that analytical framework, policymakers can begin to define ideas and strategies that can become part of the planning and implementation process (Clark & Yago 2005). However, after more than 5 years, the state did not provide financial leadership in large part due to its own financial crisis that worsened in the fall of 2008.

Renewable Energy

The fifth theme is renewable energy itself. Once defined as solar, wind, geothermal, biomass, hydropower, and other sources or combinations, renewable energy power today is created from a variety of areas. More often than not, however, each of these areas is seen as separate technologies (Clark 1997, 2000). Wind farms stand alone in large numbers located remotely from the communities that need their power. Rarely are renewable technologies combined, linked, or operated as hybrid technologies, much like the hybrid cars today: electric combined with gasoline motors. The problem is that the costs of these isolated systems become extremely high. Solar and wind, for example, are "intermittent resources" requiring storage technologies to make them dependable for base load financial or other monetary calculations (Clark & Morris 2002).

A far more economic approach, for example, would be to combine wind and solar power with storage devices or other technology in order to create a base load and hence a constant source of funds (Clark & Paolucci 2000). When generated in local communities and combined with other technologies (biomass or storage), wind becomes economically dependable and financially competitive with fossil fuels (Lund & Clark 2002). A number of studies had analyzed this approach to renewables for financial and environmental protection purposes (American Bar Association 2006). In remote communities, villages, island nations, or defined urban communities, such distributed systems are cost-effective, efficient, and environmentally friendly (Isherwood et al. 2000).

More recently, some experts and scholars have advocated "green hydrogen" as a hybrid solution that focuses on energy storage for local power needs today (Clark et al. 2005), while the hydrogen fuel cell car came to market in 2010 as several automobile manufacturers announced. Throughout California, green (and others, including natural gas reforming) hydrogen stations are coming online and operating (California Hydrogen Highway Roadmap 2006). The city of Santa Monica opened one in mid-June 2006 as part of its declaration to be a "green city" (Clark 2009).

Finance

Finance is the sixth theme. The place to start with finance is again definitions. While this chapter is not intended to review all of the literature in economics to define terms, a few critical comments need to be addressed. For one, sustainable development must be defined, including its economic component as described earlier. However, there are significant other issues that relate specially to finance. Consider the strategic planning concepts of short- and long-term goals when related to finance. More often than not, few studies define short and long terms as costs and time. This is a serious mistake for many reasons. But the primary one is that the concepts mean different things to different people, which will lead to misunderstandings and ultimately to inaction.

Furthermore, short and long term need to be defined in order to consider the costs for energy today for the long term, that is, a power plant costs today need to be repaid over a long period of time. Such finance, through either debt or equity capital, means that a region or nation will make a commitment to certain energy-generating power plants over others. Hence, an energy policy that promotes coal power generation means costs are sunk for decades in this energy source and its technological enhancements (Atwood, Fung & Clark 2002).

This short-term demand, for example, is clear in the increasing worldwide demand for natural gas. The short-term solution (defined as 3–5 years) for natural gas is liquefy [liquefied natural gas (LNG)] it and ship around the world. LNG terminals are being proposed and built in almost every area of the world. But that strategy means enormous stranded costs (in the tens of billions) for regions and nations over the long term (15–20 years) before the facilities are paid for. Moreover, the communities are now "energy dependent" on fuel supplies that are often not secure and, due to increasing demand, likely to be volatile and costly.

Other strategies are far more appropriate for both short- and long-term energy generations that do not strand or lock regions and nations into a single and hence costly energy generation infrastructure. Renewables are one solution for many reasons. As a new paradigm, instead of continuing to be dependent on fossil fuel, a conventional central plant paradigm is a significant approach to achieving national and local energy goals and objectives. The use of hybrid technologies for infrastructures (such as green hydrogen stations) as noted previously enhances, ensures, and makes that short and long term viable and cost-effective.

Knowledge Capital

The last theme is knowledge capital and concerns the future commercialization of technologies. Research and development (R&D) must continue, but become more applied as well through the commercialized and demonstration of technologies for a government-driven market. What that means is that communities have an opportunity to create new businesses through technologies that are beyond the conceptual and patent period of R&D and ready to be demonstrated in the public sector market. Once that is done, then commercial mass markets will result. This approach to innovation and economic development is exactly what the Japanese, British, French, Germans, and Americans do but in different ways and to different strategies.

For example, due to the U.S. congressional Bayh–Doyle Act in the late 1980s, the U.S. government permitted and encouraged the commercialization of knowledge capital (patents, inventions, publications) from U.S. tax-generated funds for national laboratories, academic researchers, and others into the private sector (Clark 1996). The model had various degrees of success and failure (Clark & Paolucci 1997a,b). One of the critical conclusions and lessons learned then was that the government had to be a "market force" and play a role in taking innovation to the mass market (Clark & Jensen 1997).

With the American idea of converting public funds to private sector commercialization, the problem was that the government needed to continue to play a "supportive role" (e.g., "social capitalism"). The Europeans and Japanese in particular had figured that strategy out. What they did was look at the government as the "market driver" for innovation. Hence, financial structures and mechanisms were put into place, including purchase of goods and services from companies demonstrating and commercializing new technologies.

After the Framework Convention on Climate Change was formed, Clark became director of a six-nation study on the transfer of "environmentally sound technologies" from developed to developing nations (Clark 1997, 2000). The focus was entirely on renewable energy technologies and documented the government as the market driver. The finance mechanism for the U.S. research community was likewise documented as to how intellectual property becomes commercialized (Clark & Demirag 1998).

There have been many examples since the early 2000s about how this social capitalism approach to innovation and technologies can work (Clark & Demirag 1998), but the critical strategy was to link innovation and education. Clark and Feinberg (2003) produced a "white paper" for the governor of California, entitled "California's Next Economy," that identified both the upcoming economic development areas and the needs for education and training with specific universities and programs.

CONCLUSIONS AS OPPORTUNITIES

Under the social capitalism paradigm, countries are able to develop reforms in the energy sector that impact the short- and long-term development of its infrastructures. A specific focus on renewable energy in the energy sector means a sizable financial commitment over many years through the RES. This strategy re-enforces the national and local goals of the nation so that renewable energy becomes a significant element in the energy infrastructure, as well as providing environmental mitigation and reversing global warming and climate change.

This chapter discussed the need for definitions of concepts, terms, and numbers as they mean different things to different people and identified seven themes that enable communities to implement renewable energy projects, which meet and even exceed the national energy and environment goals and objectives. Sustainable development is clearly one of the key areas and, when defined, must consider both specific sectors but also be agile and flexible in how energy systems can be financed and implemented. Clearly the fourth theme of infrastructures allows renewable energy (theme five) technologies and plans to be hybrids or combined in order to be financed, operated, and maintained more efficiently.

The last two areas of finance and knowledge capital are perhaps the "bottom line" in terms of making renewable energy a driving force. Debt and equity funding have now become viable in communities as long as the government maintains oversight and even considerable ownership in the enterprises. In order for the nation to achieve and exceed its goals, it must provide long-term leadership from a government perspective for the private sector to have both confidence and stability. This "investment" from the community must be for the long term (defined as from 10+ years) in which finance and education (knowledge capital) are the most critical themes.

REFERENCES

American Bar Association, 2006. Renewable Energy Resources Committee and the International Energy & Resources Transaction Committee, section of environment, energy, and resources, renewables in the international arena: Kyoto and beyond, Sub-committee International Teleconference.

Atwood, T., Fung, V., Clark II, W.W., 2002. Market opportunities for coal gasification in China. J.Clean Prod.

Berry, G., 1996. Hydrogen as a transportation fuel, Lawrence Livermore National Laboratory, UCRL-ID-123465.

Berry, G., Lamont, A., Watz, J., 1999. Modeling renewable energy system using hydrogen for energy storage and transportation fuels, unpublished paper, Lawrence Livermore National Laboratory, U.S. Department of Energy.

Brundtland Report United Nations World Commission on Environment and Development. 1987.

California's commission on the 21st century economy, 2002. Available from: < http://www.bth.ca.gov/invest4ca/FullReport.pdf >.

California hydrogen highway roadmap 2005 and updated 2006. Available from: <http://www.governor.ca.gov/hydrogen>.

Chadbourne, Parke, L.L.P., 2005. China moves to encourage renewables projects. Project Finance Newswire, 42–43.

Clark II, W., 2009. Sustainable communities. Springer Press.

Clark II, W., 2012. The next economics: global cases in energy, environment, and climate change. Springer Press.

Clark II, W., Bradshaw, T., 2004. Agile energy systems: global lessons from the california energy crisis. Elsevier Press.

Clark II, W., Fast, M., 2008. Qualitative economics: toward a science of economics. Coxmoor Press, Oxford.

Clark II, W., Li, X., 2010. Social capitalism in renewable energy generation: California and China comparisons. Utilities Policy J. 18 (1), 53–61.

Clark II, W., Li, X., 2012. Social capitalism: China's economic rise, Chapter 7, The Next economics: global cases in energy, environment, and climate change. Springer Press, pp.143–164.

Clark II, W.W., 1996. A technology commercialization model. J. Technol. Transfer.
Clark II, W.W., 2000a. Developing and diffusing clean technologies: experience and practical issues. OECD conference, Seoul, Korea.
Clark II, W.W., 2000b. Markets in transitional economics. J. Marketing.
Clark II, W.W., 2000c. Transfer of publicly funded R&D programs in the field of climate change for environmentally sounds technologies (ESTS): from developed to developing countries - A summary of six country studies, UN Framework Convention for Climate Change, Germany. 1997.
Clark II, W.W., 2001a. The California challenge: energy and the environmental consequences for public utilities. Utilities Policy 10 (2), 57–61.
Clark II, W.W., 2001b. California energy challenge: from crisis to opportunity. American Western Economic Confer, San Francisco, CA.
Clark II, W.W., 2002. Greening technology. Int. J. Environ. Innovation Manag.
Clark II, W.W., 2004. Hydrogen: the pathway to energy independence. Utilities Policy.
Clark II, W.W., 2007. Partnerships in creating agile sustainable development communities. J. Clean Prod. 15 (3), 294–302.
Clark II, W.W., Demirag, I., 2005. US financial regulatory change: the case of the California energy crisis. special issue , J. Banking Regul. 7 (1/2).
Clark II, W.W., Demirag, I., Bline, D., 1998. Financial markets, corporate governance and management of research and development: reflections on US managers' perspectives. In: Demirag, I. (Ed.), Corporate governance, accountability and pressures to perform: an international study. Oxford University Press, Oxford.
Clark II, W.W., Fast, M., 2005. Qualitative economics: toward a science of economics. CA. Western Economics Association Conference, San Francisco.
Clark II, W.W., Feinberg, T., 2003. California's next economy. Governor's Office of Planning and Research, Sacramento, CA. Available from: < http://www.lgc.org >.
Clark II, W., Jensen, J.D., 1997. Economic models: the role of government in business development for the reconversion of the American economy.
Clark II, W.W., Li, X., 2003. Social capitalism: transfer of technology for developing nations. Int. J. Technol. Transfer.
Clark II, W.W., Lund, H., 2001. Civic markets in the California Energy crisis. Int. J. Global Energy Issues 16 (4), 328–344.
Clark II, W.W., Lund, H., 2007. Sustainable development in practice. J. Clean Prod. 15 (3), 253–258.
Clark II, W.W., Morris, G., 2002. Policy making and implementation process: the case of intermittent resources. J. Int. Energy Policy.
Clark II, W.W., Paolucci, E., 1997a. Environmental regulation and product development: issues for a new model of innovation. J. Int. Product Dev. Manag.
Clark II, W.W., Paolucci, E., 1997b. An international model for technology commercialization. J. Technol. Transfer.
Clark II, W.W., Paolucci, E., 2000. Commercial development of environmental technologies for the automobile. Int. J. Technol. Manag.
Clark II, W.W., Rifkin, J., O'Connor, T., Swisher, J., Lipman, T., Rambach, G., 2005. Hydrogen energy stations: along the roadside to the hydrogen economy. Utilities Policy 13 (1), 41–50.
Clark II, W.W., Yago, G., 2005. Financing the hydrogen highway, public policy. Milken Institute, Santa Monica, CA.
DeLaquil, P., Wenying, C., Larson, E., 2003. Modeling China's energy future. Energy Sustainable Dev. VII (4), 40–56.
Han, W., Jiang, K., Fan, L., 2005. Reform of China's electric power industry: facing the market and competition. Int. J. Global Energy Issues 23 (2/3).

Hong, L., Lund, H., Möller, B., 2012. The importance of flexible power plant operation for Jiangsu's wind integration. Energy 41, 499–507.
Hong, L., Möller, B., 2012. Feasibility study of China's offshore wind target by 2020. Energy 48, 268–277.
Isherwood, W., Smith, J.R., Aceves, S., Berry, G., Clark II, W.W., Johnson, R., Das, D., Goering, D., Seifert, R., 2000. Economic impact on remote village energy systems of advanced technologies, University of Calif., Lawrence Livermore National Laboratory, UCRL-ID-129289, January 1998, published in Energy Policy.
Lewis, J., 2006. Renewable energy in China, American Bar Association Energy Resource Subcommittee International Teleconference.
Lund, H., 2009. Renewable energy systems – the choice and modeling of 100% renewable solutions. Academic Press.
Lund, H., Clark II, W.W., 2002. Management of fluctuations in wind power and CHP: comparing two possible Danish strategies. Energy Policy 27 (5), 471–483.
Lund, H., Mathiesen, B.V., 2009. Energy system analysis of 100% renewable energy systems: the case of Denmark in years 2030 and 2050. Energy 34 (5), 524–531.
Martinot, E., 2006. Global renewable energy markets and policies, based on Renewables 2005 Global Status Report. REN21 Renewable Energy Policy Network (www.ren21.net) American Bar Association Energy Resource Subcommittee International Teleconference.
National Commission on Energy Policy, 2004. Ending the energy stalemate: a bi-partisan strategy to meet America's energy challenges, summary of recommendations. Washington, DC.
Siegel, J., 2006. Financing renewable energy in developing countries, American Bar Association Energy Resource Subcommittee International Teleconference.
Sinton, J., Stern, R., Aden, N., Levine, M., 2005. Evaluation of China's energy strategy options, China Sustainable Energy Program.
Xinhua News, 2004. China still in dire need of electricity power in 2004. Available from: < http://news.xinhuanet.com/english/2004-02/25/content_133140.htm >. (25.02.04.).

INDEX

Note: Page numbers with "f" denote figures; "t" tables.

A

Abu Dhabi Investment Authority (ADIA), 413
Abu Dhabi Investment Council (ADIC), 413
Advocacy groups, 95
Agenda 21, 44–45, 138–139, 141–143
Agile, sustainable communities, 279
    appraisal of, 282–283
    balance, 284
    cooperation, 285–286
    decentralization, 284–285
    drake solar landing community, 280
    earthship biotecture, 280–282, 281f
    greywater recycling system, 281f
    microcosmic initiatives, 282
    resilience, 283
    sustainable design in Canada, 286–287
    sustainable housing, 280–282
Agile energy system, 470–471, 476
Agile systems, 564
Agricultural systems, 532–533
Almere Zoneiland, 45–46
Alternating current (AC), 511
Asosiasi Pemerintah Kota Seluruh Indonesia (APEKSI), 129, 172
Auction system, 350
Automated teller machine (ATM), 430
Autonomous house, 220–221
Axial coding phase, 455–456
    economic extravagance, 459–460
    family structures, 457–458
    gender, 458–459
    Islamic religion, 460–461

B

Bacterial fuel cells, 32
Balikpapan, 126, 136–138, 137f–138f.
    See also East Kalimantan
    Balikpapan Bay, 165b
    education, 157
    environmental policy, 163b–164b
    field research, 129
    local recommendations, 174–177
    planned community model in, 145f
    policy priorities in, 162
    private companies role, 168–169
    private protected forest outside, 143f
Bien Ho Orphanage (Vietnam), 537
Biofuels, 273
Biogenerator, 32
Biomass, 6–7, 30, 102, 273
    energy generation
        in Germany, 365b
        in Italy, 364t, 366t
    use in energy sector, 339–341
Biomass, 273
Branch office, 418
Brazil Russia, India, and China members (BRIC members), 3–4, 9
Building Control Act, 296
Building materials, 48, 550–551
Building systems, 534–535
Business/corporate role, 96–98

C

California, 13, 537–538, 560
California energy crisis, 470–471, 560–561
Canada
    solar potential map of, 257f
    wind energy potential in, 258f
Canadian Biomass Innovation Network (CBIN), 276
Canadian Energy Efficiency Alliance (CEEA), 265
Canadian Wind Energy Association (CanWEA), 256, 272–273
CanmetENERGY, 275–276
CanWEA. *See* Canadian Wind Energy Association
CAP. *See* Climate Action Plan
Carbon capture and storage (CCS), 107

CBA. *See* Cost–benefit analysis
CCC. *See* Countries, cities, and communities
CCS. *See* Carbon capture and storage
China
 chinese-style smart grid, merits of, 522–523
 economic system, 36–37, 487
 power industry, 524
 smart grid systems, 521
China Meteorological Administration (CMA), 54–57
Chlorofluorocarbon (CFC), 42–43
CIVETS. *See* Colombia, Indonesia, Vietnam, Egypt, Turkey, and South Africa
Claremont Environmental Design Group (CEDG), 527
 ecosystematic analysis, 531
 phase one research, 533
 team and methods, 530
Clean energy sources, 509–510
Clean tech, 475, 509–510
Climate Action Plan (CAP), 127–129, 155
Climate change, 217–218, 472, 501
 in Indonesia, 180
 issues, 155
 in Mauritius, 293–294
 planning, 154–156
 in United States, 2
CMA. *See* China Meteorological Administration
Coal, 20
Colombia, Indonesia, Vietnam, Egypt, Turkey, and South Africa (CIVETS), 134
Community-level government, 485
Compound annual growth rate (CAGR), 351
Conto Energia, 354
 fifth, 358, 358t
 first, 354
 fourth, 357–358, 357t
 second, 354, 354t, 356t
 third, 355–356
Cooperation Council for the Arab States of the Gulf. *See* Gulf Cooperation Council (GCC)

Corporate interests, 469
Corporate social responsibility (CSR), 96–97, 156–157, 162, 168
Cost–benefit analysis (CBA), 489–490
Cost–supply curves, national and provincial, 67–75
Countries, cities, and communities (CCC), 544
Cross-cultural management approach, 395–397

D

Daimler Motoren Gesellschaft (DMG), 22
Data management, 531–532
Data response system, 516–518
Democratization of energy sector, 118
Denmark, 471
 assessment methods, 228–231
 building simulation data, 226–228, 227t–228t
 information about existing building, 226
 investment estimation results, 231–236
 investment evaluation for renovation, 225–226
Department of Energy (DOE), 509
DESD. *See* UN Decade of Education for Sustainable Development
DFM. *See* Dubai Financial Market
DFSA. *See* Dubai Financial Services Authority
DHCOG. *See* Dubai Holding Commercial Operations Group
DHIG. *See* Dubai Holding Investment Group
DIC. *See* Dubai International Capital
DIFC. *See* Dubai International Financial Center
Direct current (DC), 511
Distributed renewable energy generation, 486–487
DMG. *See* Daimler Motoren Gesellschaft
DOE. *See* Department of Energy
Doing business report, 411–413, 412t
Drake solar landing community, 280
DTI. *See* U. K. Department of Trade and Industry
Dubai Financial Market (DFM), 437

Dubai Financial Services Authority (DFSA), 436
Dubai Financial Support Fund, 407
Dubai Holding Commercial Operations Group (DHCOG), 414
Dubai Holding Investment Group (DHIG), 414
Dubai International Capital (DIC), 414
Dubai International Financial Center (DIFC), 416, 434
  financial needs and requirements, 435
  financial services sectors, 436
Dubai World (DW), 408

E

EAP. *See* Environment action programs
East Kalimantan, 126, 134–136, 135t
  Arial view of refineries in, 142f
  observation and information gathering, 129
  open pit coal mining in, 147f
  provincial recommendations, 174
  Wehea Community, 136b
Economic extravagance, 459–460
Economic implications, 497–500
Economic response mechanism, 497–498
Ecosystematic analysis, 531, 536
Educational programs for sustainable societies, 387
  cross-cultural management approach, 395–397
  organizational and national culture, 397
  program activities, 393, 394t
    final conference, 395
    pilot projects development, 395
    pilot projects implementation, 395
    strategic visioning conference, 393–394
    training of trainers, 394
    workshop for professions, 394
  research objective and study design, 397–398
  in Serbia, 390–397
  sustainable development methodology, 392–393
EEMO. *See* Energy efficiency management office
EEZ. *See* Exclusive economic zone

EIA. *See* Energy Information Administration; Environmental impact assessments
Emissions Trading System (ETS), 375
Energiewende, 369
  achievements of, 376–379
  CAP and TRADE, 374–376
  challenges and impediments to, 379–385
  CO_2 emissions in, 371f
  community ownership, 110–111
  EEG, 373–374
  electricity prices in, 381f
  electricity transition, 109
  federal energy expenditures, 372f
  FIT scheme, 372–374
  fledgling movement, 106–109
  German citizens, 108f
  German energy transition, 110f
  gross production of electricity, 378f
  heat sector, 109–110
  industry payment, 382f
  investments in RES, 380f
  primary goals of, 370t
  public perception of, 383f
  renewable energy deployment, 377f
EnergiParcel project, 226
EnerGuide rating, 262–263, 263f
Energy, 99, 533–534
  conservation and efficiency, 258
    federal appraisal and provincial efforts, 267–270
    federal commitments, 259–264
    provincial commitments to, 265–267
  conventional
    coal, 100
    natural gas, 101
    oil, 100
  renewable
    biomass, 102
    hydropower, 102
    nuclear power generation, 102
    wind energy, 101–102
  technology, 508
Energy efficiency Act, 294
Energy efficiency management office (EEMO), 294
Energy Information Administration (EIA), 24, 99

ENERGY STAR rating, 264
Energy system of Baltic States, 305–306
   electrical power industry, 307
   energy grid and power industry system, 317, 318f
   energy industry, 306–307
   energy intensity of economy, 314f
   energy sector, 307, 316
   Estonian energy system, 318–321
   EU energy policy, 316–317
   EU's energy sources, 311f
   gross final electricity consumption, 312t
   gross inland consumption, 311f
   Latvian energy system, 322–327
   Lithuanian energy system, 327–333
   renewable energy production technologies, 333–344
   renewable resource contribution, 312f
   sustainable development objectives, 309–316
ENTSO-E. *See* European Network of Transmission System Operators for Electricity
Environment action programs (EAP), 388
Environmental impact assessments (EIA), 146, 152b
Environmental implications, 500–503
Environmental-friendly cultivation practices, 502
Erosion, 501
Estonian energy system, 318–321, 319f. *See also* Latvian energy system
ETS. *See* Emissions Trading System
EU. *See* European Union
Europe 2020 strategy, 309–310
Europe's carbon market, 377
European Network of Transmission System Operators for Electricity (ENTSO-E), 308–309
European Union (EU), 4, 473, 499
   policies, 478–482
Evaluation metrics, 532
Exclusive economic zone (EEZ), 55–56

F

Faculty for Management Zajecar (FMZ), 391
Family structures, 457–458
Federal energy efficiency regulations, 262
Feed-in Act, 106
Feed-in premium, 349
Feed-in tariff program (FiT program), 28, 67–69, 349, 372, 478
Finance, 566–567
Finance companies, 437
Financial project analysis, 228
Finland, 481
First Industrial Revolution (1IR), 18. *See also* Second industrial revolution (2IR)
   American communities, 20
   coal, 20
   literacy, 19
   Newcomen engine, 19
   Western civilization, 20
Fledgling movement, 106–109
FleetSmart program, 264
Food prices, 497
Fossil fuel energy systems, 489
Free trade zones (FTZ), 415
Fuel cells, 31–32

G

Gaviotas, Colombia, 47–49
GCC. *See* Gulf Cooperation Council
GDP. *See* Gross domestic product
Gender, 458–459
General Agreement on Tariffs and Trade (GATT), 406–407
Geographic information system (GIS), 55, 530–532
Geothermal power, 30–31
Germany, 479–480, 512–513
   FiT policies in, 479f, 481–482, 482f
GHG. *See* Greenhouse gas
GIR. *See* Green Industrial Revolution
GIS. *See* Geographic information system
GNP. *See* Gross national product
Government, 548–549
Green certificates, 362
Green collar economy, 551
Green Hydrogen Economy, 473–474
Green Industrial Revolution (GIR), 1–2, 471, 473–478, 559
   additive manufacturing, 34–35
   Climate Group, 28
   CO_2-consuming algae, 14
   communications tools of, 27

FiT program, 28
foreign oil producers, 17
GDP, 15
  in Japan and South Korea, 482–485
  natural resources, 16–17
  renewable energy generation, 16, 29–32
  renewable energy sources, 27
  second industrial revolution, 15
  social capitalism, 35–38
  storage and intermittent technologies, 32–34
  sustainability, 28–29
  third industrial revolution, 15
  world leaders in energy development, 487–488
Greenhouse gas (GHG), 2, 26
  emission, 155, 481, 501, 507
  reduction, 503
Grid transformation, 516–517
Gross domestic product (GDP), 15, 82–83, 129, 294–295, 408
Gross national product (GNP), 313–314, 541–542
Gulf Cooperation Council (GCC), 406–407, 451
  axial coding phase, 455–456
  coding phases, 455
  goals and strategic objectives, 454
  governments of, 452
  mutual economic and social development, 453–454
  open coding phase, 455
  selective coding phase, 456
  single European currency, 453
  UAE, 453

H
Hong Kong, 550
Human settlements, 145–146
Hydroelectricity, 272

I
IBE. *See* International Bureau of Education
ICD. *See* Investment Corporation of Dubai
IDV. *See* Individualism
IEC standards. *See* International Electrotechnical Commission standards
IISD. *See* International Institute for Sustainable Development
Illegal townships, 541–542
IMF. *See* International Monetary Fund
Immersion pumps, 343–344
India, sustainable growth/innovation issues in
  business/corporate role, 96–98
  energy, 100–102
  society and governance, 94–96
  socioeconomic issues, 93–94
  trends in energy use, 99–103
Indicators, 158t
Individualism (IDV), 396
Indonesia, NSDS in, 140
  climate change planning, 154–156
  millennium development goal report, 154
  planning mechanisms, 149
    development planning overview, 149–152
    local annual planning stages, 150f
    national medium-term, 149f
    planning chart development, 151f
    spatial planning, 152–153
  poverty reduction strategy report, 154
  process evaluation principles, 158t
  sectoral strategy, 140
    Arial view of refineries, 142f
    Borneo Orangutan survival foundation, 143f
    energy, 141–142
    forestry, 142–144
    honeycombs in trees, 143f
    human settlements, 145–146
    mining, 146–148
    planned community in Balikpapan, 145f
    planning, 141
    tourism, 148–149
Industrial revolution, third, 15, 508
Informal economy, 539
Integrated Power System (IPS), 308–309
Interdisciplinarity, 121
Internal combustion engine, 21
Internal rate of return (IRR), 229
International Bureau of Education (IBE), 461–462

International Electro-technical Commission standards (IEC standards), 63–64
International Institute for Sustainable Development (IISD), 139
International Monetary Fund (IMF), 408
Interstate power-transmission superhighway, 513
Investment Corporation of Dubai (ICD), 414
IPS. *See* Integrated Power System
IRR. *See* Internal rate of return
Islamic religion, 460–461
Italy
    biomasses, cases of, 362–366
    photovoltaics, cases of, 354–362
    renewable energy generation, 351–353, 352t–353t

J

Jaffa Slope Park, 84, 88f
    climate considerations, 87
    design codes, 89, 89f
    and neighborhoods, 85f
    outdoor activities, 87
    program and design, 85–86
    public participation process, 87
    stretch of beach, 86f
    UHI effect, 86–87
Japan International Cooperation Agency, 464
"Jumping spark" ignition system, 21

K

Kaunas hydropower plant (KHPP), 334–335, 335f
Knowledge capital, 567–568
Komisi Pemberantasan Korupsi (KPK), 169
Kruonis pumped storage power plant (KPSPP), 334–335, 335f
Kyoto accord, 259
Kyoto protocol, 6–7, 347, 514

L

L Prize, 525
L'Avenir, 528
Latvian energy system, 322–327, 322f
Law enforcement, 549–550

Laws, 548
LCA. *See* Life cycle analysis
Leadership in Environment and Energy Design (LEED), 10
Lembaga Peubedayan Masyarkat (LPM), 150
Levelized production cost (LPC), 61–63, 62f
Life cycle analysis (LCA), 489–490
Light-emitting diode (LED), 485, 525, 525
Limited liability companies (LLCs), 417
Liquid natural gas (LNG), 117, 567
Lithuania
    geothermal energy use in Lithuania, 343–344
    Lithuanian energy system, 327–333, 329f
    solar energy industry in, 341–343
LNG. *See* Liquid natural gas
Local service agents, 419
Logging Moratorium, 166b

M

Manufacturing, additive, 34–35
Masculinity index (MAS), 396
Maurice Ile Durable (MID), 115
Mauritius, 5, 293
    air conditioning, 296
    buildings, 296–299
    climate change, 294–296
    energy efficiency assessment, 299–302
    energy intensity of, 121
    energy management, 294–296
    MID vision statement, 116
    small-island developing economy, 302–303
    virtuous circle, 300t–301t
Mega cities, 557f
Metabo law, 554
Microbial fuel cells. *See* Bacterial fuel cells
Microcities
    artist's conception, 552f
    colors on map, 542f
    companies, 547f
    components of, 544f
    conventional and compact communities effects, 553f
    free-market economy, 539
    guarantor country, 544–545
    health aspect of, 553–555
    implementing good laws, 548–550

informal businesses, 542–543
infrastructure and environmental planning, 550–553
newer technologies, 552f
nonagricultural workers, 543f
reforms of, 543–544
shadow cities by Robert Neuwirth, 540f, 546f
shanty towns and illegal townships, 541–542
shortfall of legitimate economy, 547f
Siemens company, 550f
sustained economically, 545–548
urban centers and settling, 541f
urban population percentage comparison, 541f
world's population, 539
Millennium development goal (MDG), 138–139, 154
Mixed-use residential development, 530
Mortgages, 491
Multidisciplinary integrated development, 115
  energy, 116–119
  engineering, 120–121
  interdisciplinarity, 121
  limitations, 119–120
  sustainability, 116
Municipal charters, 549

N

National Development and Reform Commission (NDRC), 67–69
National Energy Administration (NEA), 53
National Renewable Energy Action Plans (NREAP), 109
National Renewable Energy Laboratory (NREL), 55
National smart grid investment funds, 523
National Strategy Sustainable Development of Serbia (NSOR), 390
National sustainable development strategy (NSDS), 126
  criteria, 138
  technical aspect of, 139
  types of, 139
Natural Resources Canada (NRC), 259–260

Natural Step Framework (TNS), 392
Near-zero energy buildings
  climate change, 217–218
  construction sector, 218
  renovation process, 218
Net present value (NPV), 229
Newcomen engine, 19
NextGEN biofuels fund, 274–275
Nongovernmental organizations (NGOs), 80, 95, 129, 388
Nuclear power generation, 102
NxStage, 555

O

Ocean current devices, 31
Ocean thermal energy conversion devices, 31
Ocean waves, 31
Offshore wind, 2–3
Offshore wind potential, resource assessment of, 53
  cost–supply curves, 67–75
  economic potential, 61–63
  GIS, 55–57, 56f
  spatial constrained potential, 59–61
  technical potential, 57–59, 59t
  tropical cyclone risks, 63–67
Open coding phase, 455–457
Open pit coal mining, 147f

P

Passive decentralization, 259–262
People's Republic of China (PRC), 562
Persian gulf markets, 441–442
Philips Healthcare, 555
Photovoltaics (PV), 16, 106, 374, 509
Planning and budget analysis charts
  in Balikpapan, 158t
  in East Kalimantan, 158t
  in Indonesia, 158t
Plus energy house, 220–221
Poverty reduction strategy report, 154
Power density, 57, 58f
Power distance (PD), 396
Power distance index (PDI), 396
Power pumping, 513
Power purchase agreement (PPA), 490
Power system management, 516

Power transmission lines, 516–518
PPA. *See* Power purchase agreement
Prince Edward Island (PEI), 266–267
Printing, three-dimensional, 34–35
Property Assessed Clean Energy program (PACE program), 491

Q

Qualitative economics, 474
Quebec, 265–266
Quota obligations, 349–350

R

R-2000 standard, 263–264
R-COS. *See* Regional codes for outdoor spaces
R&D. *See* Research and development
RAM. *See* Renewable Auction Mechanism
Real-time data, 519
Reducing Emissions from Deforestation and Development program (REDD program), 156
Regenerative community Régénérer
  business model, 529
  CEDG, 530
  ecosystematic analysis, 531
  evaluation metrics, 532
  Haitians and Californians benefit, 537–538
  partners in research to build, 536–537
  regenerative design, 531
  25-acre pilot project, 532–536
Regenerative design, 531
Régénérer Haiti, 527, 529
  green design, 528
  old and new paradigms, 529
  plant-based business, 528
Regional codes for outdoor spaces (R-COS), 89–90
REN21. *See* Renewable Energy Network
Rencana Kerja Pemerintah Daerah (RKPD), 151
Rencana Pembangunan Jangka Menengah (RPJM), 130
Renewable Auction Mechanism (RAM), 37
Renewable energy, 471
  agile systems, 564
  finance, 566–567
  government-driven markets, 562
  green hydrogen, 566
  guidelines, 562
  infrastructures, 564–565
  intermittent resources, 565
  knowledge capital, 567–568
  market forces, 561
  SD, 563–564
  social capitalism, 562–563
  sustainable communities, 561
  technology, 507
Renewable Energy Act, 4, 106
Renewable energy generation, 29
  action plan, 348
  auction system, 350
  incentives, 366–367
  incentives forms for, 347, 351f
  in Italy, 351–353, 352t–353t
  Kyoto protocol, 347
Renewable Energy Law, 372–373
Renewable Energy Network (REN21), 277–278
Renewable energy production technologies, 333
  biomass use in energy sector, 339–341, 340f
  geothermal energy use in Lithuania, 343–344
  hydropower use, 334–336
  solar energy industry in Lithuania, 341–343
  wind energy development process, 336–339
Renewable energy sources (RES), 347
Renewable Energy Standard (RES), 560
Renewable energy systems
  China, 470–472
  corporate and business influences, 469–470
  costs, finances and ROI, 489–491
  developing world leaders in, 487–488
  green industrial revolution, 473–478
  international cases, 470
  Japan and South Korea, 482–485
  for sustainable communities, 486–487
  western economic paradigm, 472–473
Renewable energy technology (RET), 269–270
  national strategy on, 275
  production in Canada, 271–273

provincial commitments to, 276–277
regulations and policies, 273–275
renewables contribution, 271f
research initiatives, 273–275
Renewable Power Production Incentive (RPPI), 278
Representative office, 419
Republic of South Sudan, 49–51
RES. *See* Renewable energy sources; Renewable Energy Standard
Research and development (R&D), 567
Residential buildings renovation, 223–225
Resilience, 283
RET. *See* Renewable energy technology
Return of investment (ROI), 229, 489–490
Rio declaration, 44–45
Rizhao (China), 46–47
RKPD. *See* Rencana Kerja Pemerintah Daerah
ROI. *See* Return of investment
RPJM. *See* Rencana Pembangunan Jangka Menengah
RPPI. *See* Renewable Power Production Incentive
Run-of-the-river systems, 31
Russia–Japan war, 16

S

Satuan Kerja Perangkat Dearah (SKPD), 150–151
SD. *See* Sustainable development
SDTC. *See* Sustainable Development Technology Canada
Second Industrial Revolution (2IR), 15, 20–21, 471, 490–491
  analog communications, 22
  carbon-based, 16–17
  commercialization of telephone, 22
  constant pressure combustion, 22
  DMG, 22
  EIA, 24
  energy requirements, 24
  fossil fuels of, 23
  internal combustion engine, 21
  petroleum, 21
  world energy consumption, 23–24
Selective coding phase, 456, 461–464
SEZ. *See* Sustainable enterprise zones

Shalom Hill Farm, 537
Shanghai Typhoon Institute (STI), 56–57
Shanty towns, 541–542
SIGMA. *See* Sustainability Integrated Guidelines for Management
Simple payback time (SPT), 229
SKPD. *See* Satuan Kerja Perangkat Dearah
Small and medium enterprise (SME), 409
Smart grid system, 517f, 518–521
  in China, 521–525
  on Chinese cases, investment and forecast, 523–524
  Chinese-style smart grid, merits of, 522–523
  LED and energy efficiency case, 525
  third-generation grid, 524–525
Smart microgrid power system, 522
SME. *See* Small and medium enterprise
Social capitalism, 35–38, 470–471, 562
  communities and nations, 563
  private market mechanisms, 562
  SD, 563
Social capitalist business model, 487
Social implications, 496–497
Society and governance, 94–96
Solar electricity systems, 509–510. *See also* Wind power
  clean tech, 509–510
  disperse energy sources, 512
  energy production, 509
  map of west energy, 510f
  solar PV cell, 511f
  solar PV electricity, 512–513
  solar technologies, 510
  solar thermal generator, 512f
  STPP, 511
  United States, 511–512
Solar energy, 271–272
Solar generation systems, 30
Solar PV systems, 511
Solar technologies, 510
Solar thermal power plant (STPP), 511
South Sudan Development Plan (SSDP), 49
Spatial constrained potential, 59
  comparison of technical and, 61t
  individual exclusion areas, 61, 61t
  for offshore wind farms, 60f

Spatial planning, 152–153, 153f
Sponsors. *See* Local service agents
SPT. *See* Simple payback time
SSDP. *See* South Sudan Development Plan
State capitalism, 472
STI. *See* Shanghai Typhoon Institute
Stockholm conference, 44
Storage devices, 32–33
STPP. *See* Solar thermal power plant
Strengths, weaknesses, opportunities, and threats analysis (SWOT analysis), 446–448
Sun Island. *See* Almere Zoneiland
Sustainability, 28–29, 495
   developing new solutions, 504–505
   economic implications, 497–500
   environmental implications, 500–503
   food security, 495–496
   meat consumption in Western countries, 502f
   natural regeneration, 495
   production of food, 495
   resources in meat and cereal production, 503f
   social implications, 496–497
   trends in production and inputs in, 498f
Sustainability Integrated Guidelines for Management (SIGMA), 392
Sustainable development (SD), 392–393, 405–406, 563–564
Sustainable development evaluation in Indonesia, 126, 132–134
   Balikpapan overview, 136–138
   baseline data, 127
   budget allocations, 157–158, 158t
   East Kalimantan overview, 134–136
   Google Earth image, 127f
   indicators, 156–157
   local analysis, 162–164, 162f
   methodology, 129–132
   national analysis, 158–160, 159f
   overarching issues, 165
      audiences, 166
      corruption and illegal activity, 169
      integration, 165–166
      legal obstacles, 167
      private sector, 167–169
      planning priorities, 157–158, 158t
      policy and budget evaluation, 126–127
      provincial analysis, 160–162, 161f
   recommendations, 127–129, 170
      cooperation between local and national governments, 177–179
      local, 174–177
      national, 170–173
      provincial, 174
Sustainable Development Technology Canada (SDTC), 274–275
Sustainable enterprise zones (SEZ), 530
SWOT analysis. *See* Strengths, weaknesses, opportunities, and threats analysis

T

Third-generation grid, 524–525
Tidal power devices, 31
Tidal waves, 31
TNS. *See* Natural Step Framework
Transmigration, 145–146
Transmission systems operators (TSO), 308–309
Tropical cyclone risks
   annual expected economic loss, 64–67, 65f
   economic costs of, 63–64
   economic risk in investment cost, 66f
   offshore wind turbines, 64
   spatial distribution of LPC, 67, 68f
Turbine components, 516
25-acre pilot project, 528, 532–536, 538
   agricultural systems, 532–533
   analysis and design process, 535–536
   building systems, 534–535
   energy, 533–534
   passive design, 534
   waste management, 535
   water, 533

U

U. K. Department of Trade and Industry (DTI), 392
UN Decade of Education for Sustainable Development (DESD), 389
UNCED. *See* United Nations Conference on Environment and Development

Uncertainty avoidance (UA), 396
Uncertainty avoidance index (UAI), 396
UNESCO. *See* United Nations Educational, Scientific and Cultural Organization
Unified Power System (UPS), 308–309
United Arab Emirates (UAE), 8, 406, 453
  banking sector, 421–434
  business cooperation with Poland, 439
    Arabic investments, 446
    entry barriers for polish entrepreneurs, 442–444
    expansion of polish enterprises, 444–445
    market chances and threats, 445–446
    Persian gulf markets, 441–442
    polish enterprises, 440, 440t
  commercial banks operation in, 428t
  Doing Business report, 411–413
  finance companies in, 437, 438t
  financial investment companies, 439t
  financial market key players, 434–438
  foreign banks in, 431t–433t
  government financial support, 409–411
  investment bodies, 413–415
  legal aspects, 415–421
  monetary and banking indicators, 423t
  national banks and branches, 424t–426t
  SME, 409–411
  sustainable development, 405–406
  SWOT analysis, 446–448
United Nations Commission on Sustainable Development (UNCSD), 132
United Nations Conference on Environment and Development (UNCED), 42, 50, 138–139
United Nations Educational, Scientific and Cultural Organization (UNESCO), 389
United Nations Environment Programme, 42–43
United Nations Environmental Program (UNEP), 154
United States Green Building Council, 530
UPS. *See* Unified Power System
Urban heat island (UHI), 82
Urban planning in developing countries, 79–80
  challenges, 80–84
  design codes, 80
  future of publicness, 91
  Jaffa slope park, 83–89
  outdoor urban spaces, 89–90
  public space, 83f
  urban park in Jaffa, 84f
Utility-scale renewable power plants, 513

V

Value-added tax (VAT), 545
Vestas, 480

W

Waste management, 535
Water, 533
Watt's steam engine, 19
Wave power conversion devices, 31
Wehea community, 136b
Western economic paradigm, 472–473
Wind, 272–273
Wind energy, 101–102, 272–273
  balance (Lithuania), 338f
  industry development, 336–339
  potential in Canada, 258f
Wind generation, 30
Wind power, 513–514
  clean energy, 515
  congestion of grid, 513
  conversion, 514
  integrated energy, 520f
  Kyoto protocol, 514
  renewable energies, 513
  transmission lines, 513
  turbine components, 516
  variables, 514–515
  wind turbine manufacturing, 514
World Bank and the International Energy Agency (WB/EIA), 299–301
World Commission on Environment and Development (WCED), 387
World Summit on Sustainable Development (WSSD), 138–139
World Trade Organization (WTO), 406–407

Z

Zero energy buildings (ZEB), 5, 219–220
  assessment of knowledge, 238f
  detached house construction, 242f
  energy savings, 222
  global economic crisis, 222–223
  indoor thermal comfort, 222
  key barriers and drivers, 241, 241f
  passive house for Lithuania, 221
  reduction of heating demands, 243f
  residential buildings renovation, 223–225
  survey on possibilities, 236–245
Zero energy house (ZEH), 220–221, 239f
  advantages and disadvantages, 236, 241f
  characteristics, 221
  design, 222
  Lithuania's climate, 244f
  parts of Europe, 244f
  rate construction costs, 242f
Zero net energy buildings, 220
Zhangbei project, 522